SPACECRAFT
SYSTEMS
ENGINEERING

SPACECRAFT SYSTEMS ENGINEERING

Third Edition

Edited by

Peter Fortescue
University of Southampton, UK

John Stark
Queen Mary, University of London, UK

Graham Swinerd
University of Southampton, UK

WILEY

Other Wiley Editorial Offices

John Wiley & Sons Inc., 111 River Street, Hoboken, NJ 07030, USA

Jossey-Bass, 989 Market Street, San Francisco, CA 94103-1741, USA

Wiley-VCH Verlag GmbH, Boschstr. 12, D-69469 Weinheim, Germany

John Wiley & Sons Australia Ltd, 33 Park Road, Milton, Queensland 4064, Australia

John Wiley & Sons (Asia) Pte Ltd, 2 Clementi Loop #02-01, Jin Xing Distripark, Singapore 129809

John Wiley & Sons Canada Ltd, 22 Worcester Road, Etobicoke, Ontario, Canada M9W 1L1

Wiley also publishes its books in a variety of electronic formats. Some content that appears
in print may not be available in electronic books.

Library of Congress Cataloging-in-Publication Data

Spacecraft systems engineering / edited by Peter Fortescue, John Stark, Graham
Swinerd.—3rd ed.
    p. cm.
  Includes bibliographical references and index.
   1. Space vehicles—Design and construction. 2. Astronautics—Systems engineering. I.
Fortescue, Peter W. II. Stark, John. III. Swinerd, Graham.

  TL875 .S68 2003
  629.47′4—dc21

                        2002028089

British Library Cataloguing in Publication Data

A catalogue record for this book is available from the British Library

ISBN: 978-0-470-85102-9 (HB)
ISBN: 978-0-471-61951-2 (PB)

Typeset in 10/12pt Times by Laserwords Private Limited, Chennai, India
Printed and bound in Great Britain by TJ International, Padstow, Cornwall

Dedicated to the memory of H Mervyn Briscoe

CONTENTS

14 GROUND STATIONS
Richard Holdaway

15 SPACECRAFT MECHANISMS
H. Mervyn Briscoe and Guglielmo S. Aglietti

16 SPACECRAFT ELECTROMAGNETIC COMPATIBILITY ENGINEERING
Ken M. Redford

17 PRODUCT ASSURANCE
Thomas A. Meaker

18 SMALL-SATELLITE ENGINEERING AND APPLICATIONS

Martin N. Sweeting and Craig I. Underwood

19 SPACECRAFT SYSTEM ENGINEERING

Adrian R. L. Tatnall, John B. Farrow and C. Richard Francis

LIST OF CONTRIBUTORS

EDITORS

Peter W. Fortescue
Aeronautics and Astronautics, School of Engineering Sciences, University of Southampton, UK (ret'd)

John P. W. Stark
Department of Engineering, Queen Mary & Westfield College, University of London, UK

Graham G. Swinerd
Aeronautics & Astronautics, School of Engineering Sciences, University of Southampton, UK

AUTHORS

Guglielmo S. Aglietti
Aeronautics & Astronautics, School of Engineering Sciences, University of Southampton, UK

H. Mervyn Briscoe (deceased)
European Space Agency

Graham E. Dorrington
Department of Engineering, Queen Mary, University of London, UK

John B. Farrow
International Space University, Strasbourg, France

Nigel P. Fillery
Astrium, Portsmouth, UK

C. Richard Francis
European Space Research and Technology Centre (ESTEC), European Space Agency

Richard Holdaway
Rutherford Appleton Laboratory, Chilton, UK

John M. Houghton
Astrium, Stevenage, UK

Thomas A. Meaker
Applied Technologies Business Unit, Verhaert, Belgium

J. Barrie Moss
School of Engineering, Cranfield University, UK

Ken M. Redford
British Aerospace, Bristol, UK (ret'd)

Chris J. Savage
European Space Research and Technology Centre (ESTEC), European Space Agency

Ray E. Sheriff
Department of Electronics and Telecommunications, University of Bradford, UK

Howard Smith
Astrium Space, Portsmouth, UK

David Stanton
Keltik Ltd, Hampton Hill, UK

John P. W. Stark
Department of Engineering, Queen Mary, University of London, UK

Martin N. Sweeting
Surrey Space Centre, University of Surrey, Guildford, Surrey, UK

Graham G. Swinerd
Aeronautics & Astronautics, School of Engineering Sciences, University of Southampton, UK

Adrian R. L. Tatnall
Aeronautics & Astronautics, School of Engineering Sciences, University of Southampton

Craig I. Underwood
Surrey Space Centre, University of Surrey, Guildford, Surrey, UK

PREFACE TO THE THIRD EDITION

Graham Swinerd, my friend and colleague, took over the running of the Space Technology short courses at Southampton University when I retired in 1989. Who would be a better choice than Graham to take over the role of principal editor for this new edition of Spacecraft Systems Engineering? I am sure that Graham will build on the reputation that the past editions have achieved, and I wish him success in his new role. Over to you, Graham...

Peter Fortescue, *Southampton, July 2002*

Since the publication of the previous edition, Dan Goldin's 'Faster, Better, Cheaper' space mission philosophy has had a major impact upon American activities. As a consequence, the last of the heavyweight interplanetary spacecraft, Cassini, was launched in October 1997 on its mission to Saturn. Programmes such as NEAR Shoemaker, which launched a relatively small but capable spacecraft in February 1996 to orbit and ultimately to land on a small body—the asteroid 433 Eros—have substituted this type of mission. These 'small missions' have significantly influenced current and proposed planetary exploration programmes.

In the same interim period, we have also seen the launch of constellations into low Earth orbit, for global mobile communications using handheld telephones—in particular, the Iridium constellation, the first satellites of which were lofted in May 1997. Although financial problems have impacted this programme, it nevertheless heralds large-scale use of constellation systems in many application areas. There are great benefits to the usage of these distributed systems, not only in communications and navigation applications but also in improving the temporal coverage of Earth observation. There is also an implicit trend here to use a number of small, but capable, spacecraft to do the job of one or two large satellites.

The principal driver for the development of small satellite technology is the reduction in cost associated with access to space. The elements contributing to this philosophy are low launch costs, a short design, build and test period, a less complex ground interface and operations, and the recognition of a means of testing new spacecraft technologies in a relatively low financial risk environment.

At the other end of the size spectrum, December 1998 saw the first elements of the International Space Station (ISS) being brought together in orbit. If development continues as originally planned, around 2005, the ISS will become the largest structure (~400 tonnes) ever to be deployed in Earth orbit, marking the beginning of permanent habitation in space.

These various developments have had a significant influence on the structure of the new edition of the book. The major changes involve the removal of the chapter on atmospheric re-entry, and the addition of a new chapter on small satellite engineering and applications. Much of the removed material has been redistributed in other chapters, however, for example, Earth atmosphere re-entry is included in Chapter 7 (Launch Vehicles), and sections on aero-manoeuvring have been included in Chapter 5 (Mission Analysis). The new Chapter 18 on Small Satellites has been contributed by Martin Sweeting and Craig Underwood of the Surrey Space Centre, based at the University of Surrey, UK. Both individuals are recognized internationally for their expertise in this field. The chapter, built on the huge expertise of the Surrey Space Centre, gives insights into small satellite systems engineering in general. Given the growing activity in this area, no textbook of this kind is complete without such a contribution.

Other chapters have been rewritten—in particular, Chapter 8 (Spacecraft Structures), Chapter 11 (Thermal Control), Chapter 16 (Electromagnetic Compatibility) and Chapter 19 (Spacecraft Systems Engineering)—and most of the others have been substantially revised, including a discussion of constellation design and small-body missions in Chapter 5 (Mission Analysis).

Some of the authors of the second edition have retired, and new names have appeared in the contributors list. The editors are grateful to all of them for their contributions. It is also sad to report that three of our previous authors have died in the interim—Howard Smith (Telecommunications), Les Woolliscroft (Spacecraft Electromagnetic Compatibility Engineering) and Mervyn Briscoe (Spacecraft Mechanisms). Each of them will be sadly missed.

The reader may have noticed the dedication at the front of the book to one of these authors, Mervyn Briscoe, who was actively involved in revising his chapter on Mechanisms when he died in 2001. Our thanks are also due to Guglielmo Aglietti who jumped into the hot seat to complete the revision of Mervyn's chapter as a co-author. Mervyn gave loyal service as a contributor to the short course activity at Southampton over many years, and we would like to acknowledge this by dedicating this edition to his memory.

Finally, it is appropriate to thank both Peter Fortescue and John Stark for their pioneering work in bringing the previous editions to fruition, and for their valued assistance with this one.

Graham Swinerd
Southampton, July 2002

PREFACE TO THE SECOND EDITION

This second edition comes in response to a phone call that we editors had been dreading. 'Had we thought of producing a second edition?' After much consideration our answer was 'Yes', and here it is.

Not only has it given our contributing authors a chance to update the material in their chapters—the technology is developing all the time and five years is a long time! It has also given us the opportunity to rectify some of the errors in the first edition (and possibly to introduce some new ones), and to respond to suggestions from readers about the content and to our inevitable 'second thoughts' on the matter. As a result there are two new chapters.

The first is on Mechanisms—important equipment on spacecraft. They are an essential part of many of the systems that are covered in the other chapters, but having their own requirements we have given them chapter status here. They are a specialist topic, involving the problem of moving one mechanical part relative to another. For an application that has a long life, no servicing, no disturbance to the structure, and ideally no single point failure as design objectives, mechanism designers are faced with a challenging task. Chapter 16 tells us how they have responded to it.

The second additional chapter addresses the subject of System Engineering. The first edition has no hyphens in its title. Those who read the title as meaning 'The Engineering of Spacecraft Systems' will probably have found that the content was much as they had expected. Indeed there have been enough satisfied readers to cause the dreaded question of a second edition to be raised. However, it could also be read as meaning 'Systems Engineering of Spacecraft', and those who interpreted it as such will no doubt have been disappointed to have found little on the discipline of System Engineering in the book.

So our response is to retain the same ambiguous title, and to retain the same thrust as in the first edition. But we have added a new chapter (No. 19), which focuses on the subject of Systems Engineering of spacecraft. It is written by authors within the spacecraft industry who have experience in that activity. We hope it will bring together the pieces of the jigsaw puzzle that are to be found among the other chapters, and will show how they can be fitted together harmoniously to form a viable whole—a spacecraft that meets its design objectives in a viable manner.

Since the first edition some of our authors have moved to new locations; some have retired. New names have appeared in our list of contributors. We editors are grateful to them all—new and old—and trust that this edition presents 'second thoughts' that are an improvement on the first.

PREFACE TO THE FIRST EDITION

This book has grown out of a set of course notes, which accompany a series of short courses given at Southampton University. These courses started in 1974 with a two-week 'space technology' course, and they are aimed at the recent science or engineering graduate who wishes to become a spacecraft engineer. The courses are still thriving, now serving much of European industry, with one-week versions for experienced engineers, sometimes senior ones, who are specialists in their own fields.

On the courses, the attendees work in competing teams on a project that involves designing a spacecraft in response to an overall objective. Over the years, mission designs have been directed at all application areas: science, astronomy, communications and Earth observations. There is now a 'museum' of models that demonstrate vehicle layouts and support the attendees' presentations covering operation, subsystem specification and launch constraints. These models demonstrate system viability rather than detailed design. The projects are designed at 'system level', and their supervision has provided a basis for deciding the level of detail that should be included in this book.

The coverage in this book is therefore aimed at giving the breadth that is needed by system engineers, with an emphasis on the bus aspect rather than on the payload. The specialist engineer is well served with textbooks, which cover many of the subsystems in detail and in depth. He is unlikely to learn very much about his own specialist topic from this book. But he may well learn something about other specialists' disciplines, and, it is hoped, enough for him to appreciate the trade-offs that affect his own subsystem in relation to others.

Chapters 2 to 5 set the general scene for spacecraft, and particularly for satellites. They must operate in an environment that is generally hostile compared to that with which we are familiar on Earth, and the main features of this are described in Chapter 2. Chapters 3 and 4 address the dynamics of objects in space, where the vehicles will respond to forces and moments that are minute, and which would be discounted as of no significance if they occurred on Earth. Indeed, most of them do occur here, but we do not often operate in a fully free state, and our Earth-bound vehicles are subject to other, much larger forces. Chapter 5 relates the motion of the spacecraft to Earth rather than to the inertially based reference system of celestial mechanics.

Chapters 6 to 15 address the main subsystems. Chapters 7 and 8 cover the subjects of getting off the ground and returning through the atmosphere. Chapters 6, 9 to 12 and 14 deal with the main subsystems on board the spacecraft, that include the on-board end of the telemetry and control link (Chapter 14) with ground control (Chapter 15). The communication link is covered in Chapter 13 in which the fundamentals of the subject are included together with their rather special application to spacecraft. This is relevant to the telemetry and control link and to a communications payload.

Chapter 16 introduces electromagnetic compatibility (EMC), one of the subjects that must be addressed by the systems engineer if the various subsystems are to work in harmony.

Product assurance is of vital concern to spacecraft engineers. Their product(s) must survive a hostile launch environment and then must last many years without the luxury of any maintenance. It does great credit to the discipline they exercise, that so many of their products do so.

We editors would like to express our thanks to the authors who have contributed chapters in the book. Most of them have lectured on the courses mentioned above. Our task has been to whittle down the material they have provided since they have been very generous. We are grateful too for their patience. The conversion of course notes into a book was expected to be a short process. How wrong we were!

We would also like to thank colleagues Graham Swinerd and Adrian Tatnall, who read some of the texts and gave advice. And finally our thanks to Sally Mulford, who has converted some much-abused text into typescript with patience and good humour.

LIST OF ACRONYMS

AATSR	Advanced Along-Track Scanning Radiometer		ATCS	Active Thermal Control System
ABM	Apogee Boost Motor		ATSR	Along-track Scanning Radar
AC	Alternating current		AU	Astronomical Unit (mean
ACS	Attitude Control System			distance from Earth to Sun)
A/D	Analogue to digital			
ADEOS	Advanced Earth Observing System		BCDT	Binary code data transfer
AGC	Automatic gain control		BCH	Bose-Chaudhuri-Hocquenchem
AIT	Assembly, integration and test		BCR	Battery Charge Regulator
AIV	Assembly, integration and verification		BDR	Battery Discharge Regulator
			BER	Bit error rate
AKM	Apogee Kick Motor		BMDO	Ballistic Missile Defence Organisation
AM	Amplitude modulation			
AMI	Active Microwave Instrument		BMU	Battery Management Unit
			BOL	Beginning of life
AMOOS	Aero-Manoeuvring Orbit-to-Orbit Shuttle		BPF	Band-pass filter
			BPSK	Bi-phase-shift keying
AO	Announcement of Opportunity		BRTS	Bilateration Ranging Transponder System
AOCS	Attitude and Orbit Control System		BSF	Back-Surface Field
			BSR	Back-Surface Reflector
AOP	Announcement of Opportunity Package			
AOS	Acquisition of signal, also Advanced Orbiting Systems		C&DH	Control and Data Handling
			CADU	Channel Access Data Unit
			CAM	Civil, Aircraft, Military
			CCB	Configuration Control Board
AOTV	Aero-assisted Orbital Transfer Vehicle		CCD	Charge coupled device
APM	Antenna Pointing Mechanism, also Attached Pressurized Module		CCIR	Comité Consultatif International de Radiocommunication
			CCITT	Comité Consultatif International de Téléphonie et de Télégraphie
ARQ	Automatic report queuing			
ASAP	Ariane Structure for Auxiliary Payloads		CCSDS	Consultative Committee for Space Data Systems
ASAR	Advanced Synthetic Aperture Radar		CCU	Central Communications Unit
			CDMA	Code-division multiple access
ASIC	Application-specific integrated circuit		CDR	Critical Design Review
			CFDP	CCSDS File Delivery Protocol
ASK	Amplitude-shift keying			
ASS	Antenna Support Structure		CFRP	Carbon fibre reinforced plastic
ASW	Address and synchronization word			

CHRIS	Compact High Resolution Imaging Spectrometer		EMP	Electromagnetic pulse
CIS	Confederation of Independent States		EOL	End of lifetime
			EOS	Earth Observing System
			ER-MIL	Established Reliability-MIL
CLA	Coupled loads analysis		ERS	Earth Resources Satellite
CLTU	Command Link Transfer Unit		ESA	European Space Agency
CMG	Control moment gyroscope		ESATAN	European Space Agency Network Analyser
CMOS	Complementary metal oxide semiconductor		ESD	Electrostatic discharge
COMSAT	Communications Satellite		ESTEC	European Space Research and Technology Centre
COTS	Commercial off the shelf			
CPL	Capillary-pumped loop		ESTL	European Space Tribology Laboratory
CPM	Coarse pointing mechanism			
CR	Corrosion resistance		EUMETSAT	European Meteorological Satellite Organisation
CSG	Centre Spatial Guyanais			
CTM	Collapsible Tube Mast		EURECA	European Retrievable Carrier
CVCM	Collected volatile condensable materials		EUTELSAT	European Telecommunications Satellite Organisation
CW	Continuous-wave			
			EVA	Extra-vehicular activity
DARPA	Defence Advanced Research Project Agency		EWSK	East-West Station Keeping
DC	Direct current			
DCP	Data Collection Platform		FBC	Faster, Better, Cheaper
DoD	Department of Defence (USA)		FDIR	Fault detection, inspection and recovery
DOF	Degree of freedom		FDMA	Frequency-division multiple access
DORIS	Doppler Orbitography and Radio positioning Integrated by Satellite		FE	Finite element
			FEA	Finite element analysis
DPL	Declared Parts List		FEEP	Field emission electric propulsion
DPSK	Differential phase-shift keying		FEM	Flight Engineering Model
DRS	Data Relay Satellite		FET	Field effect transistor
DS-CDMA	Direct Sequence Code Division Multiple Access		FFSK	Fast frequency-shift keying
			FIFO	First-In-First-Out
DSBSC	Double side-band suppressed carrier modulation		FIRST	Far Infra-Red Space Telescope
DSN	Deep Space Network		FITS	Failures per 10^9 hours
DSP	Digital signal processing, also Digital Signal Processor		FM	Frequency modulation
			FMECA	Failure Mode Effects and Criticality Analysis
e.p.	Equivalent particle		FOG	Fibre optic gyroscope
ECSS	European Cooperation for Space Standardisation		f.o.r.	Frame of reference
			FOV	Field of view
EDA	Electrically Despun Antenna		FPGA	Field Programmable Gate Array
EDAC	Error detection and correction		FPM	Fine Pointing Mechanism, also Fine-pointing mode
EGSE	Electrical Ground Support Equipment		FRR	Flight Readiness Review
EIRP	Equivalent isotropic radiated power		FRSI	Flexible reusable surface insulation
EMC	Electro-magnetic compatibility		FS	Fail safe
			FSK	Frequency-shift keying
EMI	Electro-magnetic interference		FTA	Fault Tree Analysis

G/T	Ground track	IRIG	Inter-Range Instrumentation Group
GEM	Giotto Extended Mission	IRTF	Internet Research Task Force
GEO	Geostationary Earth orbit		
GLONASS	Global Navigation Satellite System	ISO	Infrared Space Observatory
GMT	Greenwich Mean Time	ISS	International Space Station
GNSS	Global Navigation Satellite System	ITU	International Telecommunications Union
GOES	Geostationary Orbit Environmental Satellites		
GOMOS	Global ozone monitoring by the occultation of stars	JGM	Joint Gravity Model
		JHUAPL	Johns Hopkins University Applied Physics Laboratory
GPS	Global Positioning System		
GRO	Gamma Ray Observation	JPL	Jet Propulsion Laboratory
GSFC	Goddard Space Flight Center		
		KSA	K-band Steerable Antenna
GTO	Geostationary transfer orbit	KSC	Kennedy Space Center
HEO	Highly elliptical orbit	LAM	Liquid Apogee Motor
HGA	High Gain Antenna	LBR	Low bit-rate
HGAS	High Gain Antenna System	LDEF	Long Duration Exposure Facility
HOTOL	Horizontal take-off and landing		
		LED	Light emitting diode
HPA	High power amplifier	LEO	Low Earth Orbit
HRG	Hemispherical Resonator Gyroscope	LET	Linear energy transfer
		LHP	Loop heat pipe
HRSI	High-temperature reusable surface insulation	LISA	Laser Interfers meter Spaceborne Antenna
HST	High Speed Telemetry, also Hubble Space Telescope	LISN	Line impedance stabilization network
		LNA	Low noise amplifier
		LO	Local oscillator
IC	Integrated circuit	LOS	Loss of signal
ICBM	Inter-Continental Ballistic Missile	LPF	Low-pass filter
		LRR	Laser retro-reflector
ICU	Instrument Control Unit, also Intelligent Control Unit	LRSI	Low-temperature reusable surface insulation
IDHT	Instrument data-handling and transmission	LST	Low speed telemetry
		LTP	Long-haul Transport Protocol
IF	Intermediate frequency		
IFOV	Instantaneous field of view		
IFR	Inertial frame of reference	MA	Multiple access
IFRB	International Frequency Registration Board	MAC	Medium Access Control
		Mbps	Mega bits per second
IM	Intermodulation	MCC	Mission Control Centre
INMARSAT	International Maritime Satellite Organisation	MCPC	Multi-Channel Per Carrier
		MCU	Mode Control Unit
INTELSAT	International Telecommunications Satellite Organisation	MDBS	Mesh double bumper shield
		MEMS	Micro Electro Mechanical Systems
IP	Internet Protocol		
IPN	Interplanetary Internet	MEO	Medium height Earth Orbit
IPNRG	IPN Research Group	MERIS	Medium Resolution Imaging Spectrometer
IQ	In-phase and Quadrature		
IRAS	Infra-Red Astronomical Satellite	MFR	Multi-function receiver
		MHS	Micro Humidity Sensor

MIPAS	Michelson interferometer for passive atmospheric sounding	OTV	Orbital Transfer Vehicle
MIPS	Million instructions per second	P-MOS	P-type metal oxide semiconductor
MLI	Multi-layered insulation	PA	Product Assurance
MMBS	Multiple mesh bumper shield	PAEHT	Power-Augmented Hydrazine Thruster
MMH	Mono-methylhydrazine	PAM	Payload Assist Module
MMIC	Monolithic microwave integrated circuit	PAM-A	Payload Assist Module—Atlas-sized
MMS	Multi-mission Modular Spacecraft	PAM-D	Payload Assist Module—Delta-sized
MOD	Ministry of Defence	PC	Personal computer
MOP	Multiple On-line Peripheral, also Meteosat Operational Programme	PCB	Printed circuit board
		PCDU	Power Control and Distribution Unit
MOS	Metal oxide semiconductor	PCM	Pulse Code Modulation, or Phase Change Material
MOSFET	Metal oxide semiconductor field effect transistor	PDF	Probability Density Function
MPD	Magneto-plasma-dynamic	PDHT	Payload data handling and transmission
MPG	Multi Point Grounding	PDR	Preliminary Design Review
MSG	Meteosat Second Generation	PDUS	Primary Data User Station
MSH	Micro Humidity Sounder	PEB	Payload Equipment Bay
MSS	Multishock shield	PEEK	Polyether Ether Ketone
MST	Micro systems technology	PEM	Payload Electronics Module
MTBF	Mean time between failures	PF	Protoflight
MTTF	Man-Tended Free Flyer	PID	Proportional, integral and differential
MVS	Machine Vision System		
MW	Momentum wheel	PIM	Passive intermodulation product
MWR	Microwave Radiometer		
		PLC	Payload Carrier
N-MOS	N-type metal oxide semiconductor	PLM	Payload Module
NASA	National Aeronautics and Space Administration	PM	Phase modulation, also Pulse modulation
NASP	National Aero-Space Plane	PMC	Payload Module Computer
NBFM	Narrow-band frequency modulation	PMP	Parts, materials and processes
NDT	Non-destructive testing	PN	Pseudo-random noise
NEAR	Near Earth Asteroid Rendezvous	POEM	Polar Orbiting Earth-observation Mission
NERVA	Nuclear engine for rocket vehicle applications	PPF	Polar Platform
		PPL	Preferred Parts List
NSSK	North-South Station Keeping	PRARE	Precise Range and Range-rate Equipment
OBC	On-Board Computer	PRK	Phase-reversal keying
OBDH	On-board data handling	PSK	Phase-shift keying
OBS	On-board software	PT	Prototype
OCC	Operations Control Centre	PTCS	Passive Thermal Control System
OMS	Orbital Manoeuvring Subsystem	PTFE	Polytetrafluoroethylene
OSCAR	Orbiting satellite carrying amateur radio	PU	Pattern Unit
		PUS	Packet Utilisation Standard
OSR	Optical Solar Reflector	PVG	Piezo-electric Vibratory Gyroscope
OTA	Optical Telescope Assembly		

QA	Quality Assurance
QF	Quality Factor
QPL	Qualified Parts List
QPSK	Quadrature phase-shift keying
RA	Radar altimeter
RAL	Rutherford Appleton Laboratory
RAM	Random Access Memory, or Radio-frequency Anechoic Material
RAMS	Reliability, Availability, Maintainability and Safety
RARR	Range and range rate
RBI	Remote Bus Interfaces
RCC	Reinforced carbon-carbon
RCE	Reaction Control Equipment
RF	Radio Frequency
RFA	Request for approval
RFC	Radio frequency compatibility
RIG	Rate-Integrating Gyroscope
RLG	Ring Laser Gyroscope
RLV	Reusable Launch Vehicle
RMS	Remote Manipulator System
ROM	Read-only Memory
Rpm	Revolutions per minute
RS	Reed-Solomon
RTG	Radioisotope Thermoelectric Generator
RTU	Remote Terminal Unit
RW	Reaction Wheel
SA	Single access
SAD	Solar Array Drive
SADM	Solar array drive mechanism
SAO	Smithsonian Astrophysics Observatory
SAR	Synthetic Aperture Radar
SAS	Solar Array System
SAW	Surface Acoustic Wave
SBE	S-Band Exciter
SCC	Stress corrosion cracking
SCF	Satellite Control Facility
SCIAMACHY	Scanning Imaging Absorption Spectrometer for Atmospheric Cartography
SCPC	Single channel per carrier
SCPS	Space Communications Protocol Standards
SCPS-FP	SCPS File Protocol
SCPS-NP	SCPS Network Protocol
SCPS-SP	SCPS Security Protocol
SCPS-TP	SCPS Transport Protocol
SCRAMJET	Supersonic Combustion RAMJET

SDUS	Secondary Data User Station
SEE	Single event effect
SEL	Single-event latch-up
SERC	Science and Engineering Research Council
SEU	Single event upset
SGL	Space-to-ground link
SL	Safe life
SLR	Satellite laser ranging
SMM	Solar Maximum Mission
SNAP-19	System for Nuclear Auxiliary Power
SNR	Signal-to-noise ratio
SOHO	Solar and Heliospheric Observatory
SOP	Spacecraft overhead pass
SP-L/PM	Split Phase-Level/Phase Modulation
SPA	Solar Power Array
SPE	Solid Polymer Electrolyte
SPELDA	Structure Porteuse pour Lancement Double Ariane
SPG	Single-point grounding
SPOT	Satellite Pour l'Observation de la Terre
SPS	Satellite Power System
S^3R	Sequential switching shunt regulation
SRE	Spacecraft Ranging Equipment
SRP	Solar radiation pressure
SSA	S-band Steerable Antenna
SSB	Single side-band
SSLV	Standard Small Launch Vehicle
SSM	Second Surface Mirror
SSMA	Spread-spectrum multiple access
SSME	Space Shuttle Main Engine
SSPA	Solid State Power Amplifier
SSTL	Surrey Satellites Technology Limited
SSTO	Single Stage To Orbit
SSUS	Solid Spinning Upper Stage
STDN	Spaceflight Tracking and Data Network
STRV	Space Technology Research Vehicle
STS	Space Transportation System
SVM	Service Module
SW	Software
SYLDA	Système de Lancement Double Ariane
TC&R	Telemetry, Command and Ranging

TCP/IP	Transmission Control Protocol/Internet Protocol	USAF	United States Air Force
TCS	Thermal Control System	USB	Upper side-band
TDM	Time-Division Multiplexed	USNO	US Naval Observatory
TDMA	Time-division multiple access	UT	Universal Time
TDPS	Tracking and Data Processing Station	UTC	Universal Time Co-ordinated
TDRS	Tracking and Data Relay Satellite	VCDU	Virtual Channel Data Unit
		VCHP	Variable-Conductance Heat Pipe
TDRSS	Tracking and Data Relay Satellite System	VCO	Voltage-Controlled Oscillator
TIU	Time Interval Unit	VCXO	Voltage Controlled Crystal Oscillator
TLM	Telemetry	VHDL	Very High level Design description Language
TM/TC	Telemetry/Telecommand		
TML	Total Mass Loss	VHF	Very high frequency
TMM	Thermal Mathematical Model	VLSI	Very large scale integrated
TPS	Thermal Protection Subsystem		
TRR	Test Readiness Review	WARC	World Administrative Radio Conference
TT&C	Tracking, Telemetry and Control	WBFM	Wide-Band Frequency Modulation
TTL	Transistor-transistor logic	WRC	World Radio Conference
TWT	Travelling Wave Tube	WSGT	White Sands Ground Terminal
TWTA	Travelling Wave Tube Amplifier	WTR	Western Test Range
		WWW	World Wide Web
UARS	Upper Atmosphere Research Spacecraft		
UART	Universal Asynchronous Receiver and Transmitter	XMM	X-ray Multi-mirror Mission
UHF	Ultra High Frequency	XPD	Cross-Polar Discrimination
UOSAT	University of Surrey Satellite		
UQPSK	Unbalanced quadrature phase-shift keying	YSM	Yaw-steering mode

1 INTRODUCTION

John P. W. Stark[1], Graham G. Swinerd[2] and Adrian R. L. Tatnall[2]

[1]*Department of Engineering, Queen Mary, University of London*
[2]*Aeronautics and Astronautics, School of Engineering Sciences, University of Southampton*

Man has only had the ability to operate spacecraft successfully since 1957, when the Russian Sputnik I was launched into orbit. In a few decades technology has made great strides, to the extent that the Americans' manned expedition to the Moon and back is already history. In little more than four decades, unmanned explorer spacecraft have flown past all the major bodies of the solar system except for Pluto. Vehicles have landed on the Moon, Venus and Mars, and the Galileo spacecraft probe 'landed' on the gaseous 'surface' of Jupiter in 1995. A lander mission to Titan, one of Saturn's moons, is underway with the launch of the Cassini/Huygens spacecraft in October 1997. Minor bodies in the solar system have also received the attention of the mission planners. The first landing on such a body was executed by the Near Earth Asteroid Rendezvous (NEAR) Shoemaker spacecraft, when it touched down on the Eros asteroid in February 2001. Similarly, a prime objective of the ambitious Rosetta programme is to place a lander on a cometary body in around 2012. Current manned space activity sees the ongoing construction in orbit of the International Space Station (ISS), and this represents a major step for both the technology and the politics of the space industry. The United States, Europe, Russia and Japan are all involved in this ambitious, long-term programme.

Many countries have the capability of putting spacecraft into orbit; satellites have now established a firm foothold as part of the infrastructure of society. There is every expectation that they have much more to offer in the future.

Before the twentieth century, space travel was largely a flight of fantasy. Most authors during that time failed to understand the nature of a spacecraft's motion, and this resulted in the idea of 'lighter-than-air' travel for most would-be space-farers [1,2]. At the turn of the twentieth century, however, a Russian teacher, K. E. Tsiolkovsky, laid the foundation stone for rocketry by providing insight into the nature of propulsive motion. In 1903, he

Spacecraft Systems Engineering (Third Edition). Edited by P. W. Fortescue, J. P. W. Stark and G. G. Swinerd
© 2003 John Wiley & Sons Ltd

published a paper in the *Moscow Technical Review* deriving what we now term the rocket equation, or Tsiolkovsky's equation (equation 3.20). Owing to the small circulation of this journal, the results of his work were largely unknown in the West prior to the work of Hermann Oberth, which was published in 1923.

These analyses provided an understanding of propulsive requirements, but they did not provide the technology. This eventually came, following work by R. H. Goddard in America and Wernher von Braun in Germany. The Germans demonstrated their achievements with the V-2 rocket, which they used towards the end of World War II. Their rockets were the first reliable propulsive systems, and while they were not capable of placing a vehicle into orbit, they could deliver a warhead of approximately 1000 kg over a range of 300 km. It was largely the work of these same German engineers that led to the first successful flight of Sputnik 1 on 4 October 1957, closely followed by the first American satellite, Explorer 1, on 31 January 1958.

Four decades have seen major advances in space technology. It has not always been smooth, as evidenced by the major impact that the Challenger disaster had on the American space programme. Technological advances in many areas have, however, been achieved. Particularly notable are the developments in energy-conversion technologies, especially solar photovoltaics, fuel cells and batteries. Developments in heat-pipe technology have also occurred in the space arena, with ground-based application to the oil pipelines of Alaska [3] as a spin-off. Perhaps the most notable developments in this period, however, have been in electronic computers and software. Although these have not necessarily been driven by space technology, the new capabilities that they afford have been rapidly assimilated, and they have revolutionized the flexibility of spacecraft. In some cases they have even turned a potential mission failure into a grand success, as evidenced by Voyager 2.

But the spacecraft has also presented a challenge to Man's ingenuity and understanding. Even something as fundamental as the unconstrained rotational motion of a body is now better understood as a consequence of placing a spacecraft's dynamics under close scrutiny. Man has had to devise designs for spacecraft that will withstand a hostile space environment, and he has come up with many solutions, not just one.

1.1 PAYLOADS AND MISSIONS

Payloads and missions for spacecraft are many and varied. Some have reached the stage of being economically viable, such as satellites for communications, weather and navigation purposes. Others monitor Earth for its resources, the health of its crops and pollution. Determination of the extent and nature of global warming is only possible using the global perspective provided by satellites, and the monitoring of ozone holes over both poles is of great importance to mankind. Other satellites serve the scientific community of today and perhaps the layman of tomorrow by adding to Man's knowledge of the Earth's environment, the solar system and the universe.

Each of these peaceful applications is paralleled by inevitable military ones. By means of global observations, the old 'superpowers' acquired knowledge of military activities on the surface of the planet and the deployment of aircraft. Communication satellites serve the military user, as do weather satellites. The Global Positioning System (GPS) navigational satellite constellation is now able to provide an infantryman, sailor or fighter pilot with his location to an accuracy of about a metre. These 'high ground' space technologies have become an integral part of military activity in the most recent terrestrial conflicts.

Table 1.1 presents a list of payloads/missions with an attempt at placing them into categories based upon the types of trajectory they may follow. The satellites may be categorized in a number of ways such as by orbit altitude, eccentricity or inclination.

It is important to note that the specific orbit adopted for a mission will have a strong impact on the design of the vehicle, as illustrated in the following paragraphs.

Consider geostationary (GEO) missions; these are characterized by the vehicle having a fixed position relative to the features of the Earth. The propulsive requirement to achieve such an orbit is large, and thus the 'dry mass' (exclusive of propellant) is only a modest fraction of the all-up 'wet mass' of the vehicle. With the cost per kilogram-in-orbit being as high as it currently is—of the order of $50 000 per kilogram in geostationary orbit—it usually becomes necessary to optimize the design to achieve minimum weight, and this leads to a large number of vehicle designs, each suitable only for a narrow range of payloads and missions.

Considering the communication between the vehicle and the ground, it is evident that the large distance involved means that the received power is many orders of magnitude less than the transmitted value. The vehicle is continuously visible at its ground control station, and this enables its health to be monitored continuously and reduces the need for it to be autonomous or to have a complex data handling/storage system.

Low Earth orbit (LEO) missions are altogether different. Communication with such craft is more complex as a result of the intermittent nature of ground station passes. This resulted in the development, in the early 1980s, of a new type of spacecraft—the tracking and data relay satellite system (TDRSS)—operating in GEO to provide a link between craft in LEO and a ground centre. This development was particularly important because the Shuttle in LEO required a continuous link with the ground. More generally, the proximity of LEO satellites to the ground does make them an attractive solution for the provision of mobile communications. The power can be reduced and the time delay caused by the finite speed of electromagnetic radiation does not produce the latency problems encountered using a geostationary satellite.

Table 1.1   Payload/mission types

Mission	Trajectory type
Communications	Geostationary for low latitudes, Molniya and Tundra for high latitudes (mainly Russian), Constellations of polar LEO satellites for global coverage
Earth resources	Polar LEO for global coverage
Weather	Polar LEO, or geostationary
Navigation	Inclined MEO for global coverage
Astronomy	LEO, HEO, GEO and 'orbits' around Lagrange points
Space environment	Various, including sounding rockets and HEO
Military	Polar LEO for global coverage, but various
Space stations	LEO
Technology demonstration	Various

Note: GEO: Geostationary Earth orbit; HEO: Highly elliptical orbit; LEO: Low Earth Orbit; MEO: Medium height Earth Orbit.

The power subsystem is also notably different when comparing LEO and GEO satellites. A dominant feature is the relative period spent in sunlight and eclipse in these orbits. LEO is characterized by a high fraction of the orbit being spent in eclipse, and hence a need for substantial oversizing of the solar array to meet battery-charging requirements. In GEO, on the other hand, a long time (up to 70 min) spent in eclipse at certain times of the year leads to deep discharge requirements on the battery, although the eclipse itself is only a small fraction of the total orbit period. Additional differences in the power system are also partly due to the changing solar aspect angle to the orbit plane during the course of the year. This may be offset, however, in the case of the sun-synchronous orbit (see Section 5.4 of Chapter 5), which maintains a near-constant aspect angle—this is not normally done for the benefit of the spacecraft bus designer, but rather because it enables instruments viewing the ground to make measurements at the same local time each day.

It soon becomes clear that changes of mission parameters of almost any type have potentially large effects upon the specifications for the subsystems that comprise and support a spacecraft.

1.2   A SYSTEM VIEW OF SPACECRAFT

This book is concerned with spacecraft systems. The variety of types and shapes of these systems is extremely wide. When considering spacecraft, it is convenient to subdivide them into functional elements or subsystems. But it is also important to recognize that the satellite itself is only an element within a larger system. There must be a supporting ground control system (Figure 1.1) that enables commands to be sent up to the vehicle and status and payload information to be returned to the ground. There must also be a launcher system that sets the vehicle on its way to its final orbit. Each of the elements of the overall system must interact with the other elements, and it is the job of the system designer to achieve an overall optimum in which the mission objectives are realized efficiently. It is, for example, usual for the final orbit of a geostationary satellite to be achieved by a combination of a launch vehicle and the boost motor of the satellite itself.

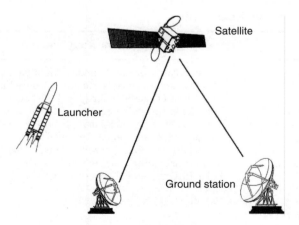

Figure 1.1 The total system—the combined space and ground segments

This starts us towards the overall process of systems engineering, which will be treated in detail in the final chapter of this book. Figure 1.1 shows the breakdown of the elements needed to form a satellite mission. Each of these may be considered to perform functions that will have functional requirements associated with them. We can thus have an overriding set of mission requirements that will arise from the objectives of the mission itself. In the process of systems engineering, we are addressing the way in which these functional requirements can best be met, in a methodical manner.

Chambers Science and Technology Dictionary provides the following very apt definition of the term 'system engineering' as used in the space field:

'A logical process of activities that transforms a set of requirements arising from a specific mission objective into a full description of a system which fulfils the objective in an optimum way. It ensures that all aspects of a project have been considered and integrated into a consistent whole.'

The 'system' in question here could comprise all the elements within both the space and the ground segments of a spacecraft project, including the interfaces between the major elements, as illustrated in Figure 1.1. Alternatively, the system approach could be applied on a more limited basis to an assembly within the space segment, such as an instrument within the payload. In the case of an instrument, the system breakdown would include antenna elements or optics and detectors as appropriate, and the instrument's mechanical and electrical subsystems.

The *mission objectives* are imposed on the system by the customer, or user of the data. They are statements of the aims of the mission, are qualitative in nature and should be general enough to remain virtually unchanged during the design process. It is these fundamental objectives that must be fulfilled as the design evolves.

For example, the mission objectives might be to provide secure and robust three-dimensional position and velocity determination to surface and airborne military users. The *Global Positioning System (GPS)* is a method adopted to meet these objectives.

An illustration of the range of methods and the subsequent requirements that can stem from mission objectives is given by the large number of different concepts that have been proposed to meet the objective of providing a worldwide mobile communication system. They range from an extension of the existing *Inmarsat* spacecraft system to schemes using highly eccentric and tundra orbits (see Chapter 5 for the definitions of these), to a variety of concepts based around a network of LEO satellites, such as *Iridium*, which was originally conceived as a constellation of 77 satellites (hence the name) and subsequently became an operational constellation of 66 satellites with additional in-orbit spares.

This example demonstrates an underlying principle of system engineering, that is, that there is *never* only one solution to meet the objectives. There will be a diverse range of solutions, some better and some worse, based on an objective discriminating parameter such as cost, or mass or some measure of system performance. The problem for the system engineer is to balance all these disparate assessments into a single solution.

The process that the system engineer first undertakes is to define, as a result of the mission objectives, the mission requirements. The subsequent requirements on the system and subsystems evolve from these initial objectives through the design process. This is illustrated in Figure 1.2, which shows how a hierarchy of requirements is established. In Chapter 19 this hierarchy is further explained and illustrated by considering a number of

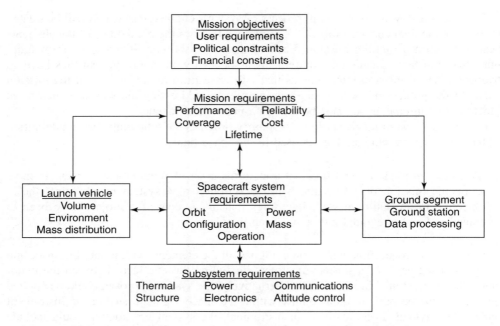

Figure 1.2 Objectives and requirements of a spacecraft mission

specific spacecraft in detail. At this point, however, it is important to note the double-headed arrows in Figure 1.2. These indicate the feedback and iterative nature of system engineering.

We turn now to the spacecraft system itself. This may be divided conveniently into two principal elements, the payload and the bus. It is of course the payload that is the motivation for the mission itself. In order that this may function it requires certain resources that will be provided by the bus. In particular, it is possible to identify the following functional requirements:

1. The payload must be pointed in the correct direction.
2. The payload must be operable.
3. The data from the payload must be communicated to the ground.
4. The desired orbit for the mission must be maintained.
5. The payload must be held together, and on to the platform on which it is mounted.
6. The payload must operate and be reliable over some specified period.
7. An energy source must be provided to enable the above functions to be performed.

These requirements lead on to the breakdown into subsystems, which is shown in Figure 1.3. Inset in each of these is a number that relates it to the functions above.

The structure of this book recognizes this overall functional breakdown, shown in Figure 1.3. The individual subsystems are covered separately in the chapters. Thus, in Chapter 8 the structural subsystem is considered, and in Chapter 15, mechanism design is outlined. The power subsystem, including the various ways in which power can be raised on a spacecraft, is described in Chapter 10. The main elements of an attitude control subsystem are indicated principally in Chapter 9, although the underlying attitude motion of

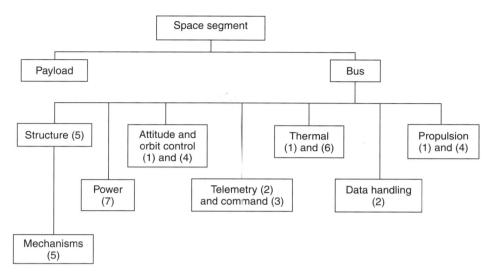

Figure 1.3 Spacecraft subsystems

a free body such as a satellite is covered in Chapter 3. Telemetry and command subsystems may be conveniently considered alongside on-board data handling (OBDH); these topics are covered in Chapter 13, with the underlying principles and practice of spacecraft communications in the previous chapter. The thermal control subsystem appears in Chapter 11. Propulsion, as it relates to on-board systems, is described in Chapter 6, while its application to launch systems is described in Chapter 7.

One facet of these subsystems is that the design of any one has impacts and resource implications on the others. A most important feature of spacecraft system design is to identify what aspects of the mission and what elements of the design provide the major influences on the type of satellite that may meet the specific mission requirements. This process is the identification of the *'design drivers'*. In some cases the drivers will affect major features of the spacecraft hardware. The varied mission requirements, coupled with the need to minimize mass and hence power, has thus led to a wide variety of individual design solutions being realized. However, the spacecraft industry is now evolving towards greater standardization—in the shape of the specific buses that may be used to provide the resources for a variety of missions (e.g. the *Multi-mission Platform*, the *Eurostar* bus and the *SPOT* bus—see Chapter 19).

It is not simply the nature of its payload that determines the design that is selected for a given mission, although this will have a considerable influence. Commercial and political influences are strongly felt in spacecraft engineering. Individual companies have specialist expertise; system engineering is dependent on the individual experience within this expertise. This was perhaps most notably demonstrated by the Hughes Company, which advanced the art of the spin-stabilized satellite through a series of *Intelsat* spacecraft. Spacecraft systems engineering is not all science—there is indeed an art to the discipline.

This leads to another major feature of spacecraft system design, namely, the impact of reliability. The majority of terrestrial systems may be maintained, and their reliability, while being important, is not generally critical to their survival. If a major component fails, the maintenance team can be called in. In space, this luxury is not afforded and while the Shuttle can provide in-orbit servicing for a limited number of satellites, this is an extremely expensive option. This requires that the system must be fault-tolerant, and when this tolerance is exceeded the system is no longer operable and the mission has ended.

There are two principal methods used to obtain high reliability. The first is to use a design that is well proven. This is true for both system and component selection.

The requirement to validate the environmental compatibility of components (Chapter 2) leads to relatively old types being used in mature technology, especially in electronic components. This tends to lead to a greater demand for power than the terrestrial 'state-of-the-art' technology. At system level a 'tried and tested' solution will minimize development risk, reducing system cost while also achieving high reliability.

The second method of achieving high reliability is *via* de-rating (Chapter 17). By reducing the power of the many electronic components, for example, a greater life expectancy can be obtained. This leads to an overall increase in mass.

The net effect of designing for high reliability is that spacecraft design is conservative—'if it has been done before then so much the better'. Much of satellite design is thus not state-of-the-art technology. Design teams evolve a particular design solution to meet varied missions—because it is a design they understand—and hence system design is an art as well as a science.

In making the selection of subsystems for the spacecraft, the designer must have a good grasp of the way in which the subsystems work and the complex interactions between them, and he/she must recognize how the craft fits into the larger system. Further, the designer must be able to trade off advantages in one area with the disadvantages in another and achieve a balance in which the end result will work as a harmonious whole. While each subsystem will have its own performance criterion, its performance must nevertheless be subordinated to that of the system as a whole.

1.3   THE FUTURE

As we approach the centenary of heavier-than-air manned aviation, we are also approaching a new frontier in space. Up until now we have been able to demonstrate no more than a rudimentary access to the space environment—our utilization of it is still limited. Beyond the frontier, however, we shall have an established space infrastructure, including the prime elements of transportation and communications, with a permanent presence of man in space—initially on space stations, but perhaps also on the Moon and Mars. Over the last thirty years space exploration has had to adapt to changes in world politics. The funding available for the Apollo missions was motivated by the 'Space Race' and current levels of spending are well below that available during the Apollo era.

Clearly, to set up the infrastructure there must be a transportation system. The first phase—getting off the ground—already sees the vertical-launch reusable vehicle (Shuttle) competing with expendable launchers, and it in turn is likely to be overtaken by next-generation launchers, which are truly reusable, having aircraft-like operational characteristics. Already retrieval of spacecraft from LEO has been demonstrated and further developments in two-way traffic between ground and LEO may be expected, with reductions in the cost of this operation.

The turn of the twenty-first century has seen the initial stages in the building of the ISS, with the projected completion around 2005. The use of the ISS as a staging post, where 'a new team of horses' can be obtained, is perhaps not too far away. Surely such staging posts will eventually become assembly and servicing posts too, so that spacecraft do not all have to be designed to withstand the full rigours of launch (Section 2.2 of Chapter 2) when their subsequent stages of travel can be relatively stress-free. Perhaps manufacturing in space will be a way of proceeding, not only for exotic materials such as are currently being considered but also for lightweight structural materials extruded in zero gravity, for use in zero gravity. A communications infrastructure is already in being, with systems such as TDRSS. There will need to be accommodation units fulfilling both scientific and space tourism needs. This is made possible by the store of knowledge of Man's ability to live in space being gathered each year. There needs to be a power generating and supply system . . . and so on.

Such a space infrastructure would surely open up the solar system to manned exploration. This new frontier will be achieved only if it is the will of the people of this planet, and the politicians commit the resources apolitically for durations far greater than the term of office of any one administration.

There is a whole new exciting arena waiting to be explored, occupied and used for the benefit of terrestrial mankind.

REFERENCES

[1] de Bergerac, Cyrano (1649) *Voyage dans la lune*.
[2] Winter, F. H. (1983) *Prelude to the Space Age*, Smithsonian.
[3] Briscoe, M. and Toussaint, J. (1989) *European Space Technology*, ESA BR-55.

THE SPACECRAFT
2 ENVIRONMENT AND ITS
EFFECT ON DESIGN

John P. W. Stark

Department of Engineering, Queen Mary, University of London

2.1 INTRODUCTION

Spacecraft operation is characterized by its remoteness from the Earth and thus the loss of the Earth's protective shield, namely, the atmosphere. This atmosphere evidently provides a suitable stable environment in which the human species has been able to evolve. Coupled with the gravitational force of the Earth, 'the one-*g* environment', it provides familiarity in design, and its removal has significant and sometimes unexpected implications. The aim of this chapter is to introduce the reader to the nature of the spacecraft's environment and the implications that it has on spacecraft design by considering both the way materials behave and the way in which systems as a whole are influenced. The final section looks at the implications for manned space flight.

Before considering the environment in detail it should be noted that the different phases in the life of a space vehicle, namely, manufacture, pre-launch, launch and finally space operation, all have their own distinctive features. Although a space vehicle spends the majority of its life in space, it is evident that it must survive the other environments for complete success. Whilst the manufacturing phase is not specifically identified in the following section, it has an effect upon the reliability and the ability to meet design goals. Cleanliness, humidity and codes of practice are critical in the success of spacecraft missions, and these issues are addressed in Chapter 17.

2.2 PRE-OPERATIONAL SPACECRAFT ENVIRONMENTS

2.2.1 Pre-launch environment

The design, manufacture and assembly of a large spacecraft, and its final integration into a launch vehicle is a lengthy process, lasting typically 5 to 10 years. Components and

Spacecraft Systems Engineering (Third Edition). Edited by P. W. Fortescue, J. P. W. Stark and G. G. Swinerd
© 2003 John Wiley & Sons Ltd

subsystems may be stored for months or even years prior to launch (e.g. the Galileo mission). Careful environmental control during such periods is essential if degradation of the spacecraft system as a whole is to be avoided.

2.2.2 The launch phase

From an observer's viewpoint, the launch of a space vehicle is evidently associated with gross noise levels. This impinges on the structure of a satellite contained within the launcher's shroud. The launch sequence entails high levels of vibration, associated both with the noise field and structural vibration, modest-to-high levels of acceleration during ascent, mechanical shock due to pyrotechnique device operation, a thermal environment that differs from both laboratory and space environments, and for most launch vehicles, a rapidly declining ambient pressure. These features are described separately below.

The *severe acoustic/vibration environment* during launch is due to both the operation of the launch vehicle's main engines, and also the aerodynamic buffeting as the vehicle rises through the lower region of the Earth's atmosphere. Two peak levels occur.

The first peak occurs at the moment of lift-off. The overall build-up of the rocket motor firing and the exhaust products reflected from the ground reaches a peak at launcher release. During ascent, the contribution from ground reflection decreases, but a variety of mechanical components, such as liquid fuel turbopump operation, continue to excite the vehicle's structure. The vibration is not only directly transmitted to the spacecraft through structural components, but it also excites the launch shroud to generate a secondary acoustic field. For light, flexible components such as the solar array, the acoustic environment may be more severe than the mechanically induced vibration [1].

The second peak in the acoustic field occurs during transonic flight. The launch shroud is again excited, this time by the unsteady flow field around the vehicle.

Measurement of the field is generally made in dB with reference to a pressure of 2×10^{-5} Pa. The sound pressure level I is then measured in decibels, given by

$$I = 20 \log_{10} \left(\frac{F}{2 \times 10^{-5}} \right) \text{ dB}$$

where F is the acoustic field intensity. The frequency spectrum of the noise field will be dependent on the launch vehicle and will change during the launch sequence. Data is provided for design purposes, thus enabling the designer to quantify the vibrational inputs that individual components and the whole space vehicle will experience. The design noise spectrum for Ariane 4 and 5 is indicated in Figure 2.1 [2,3]. The random vibration is indicated in Figure 2.2 for Ariane 4. Data for the Ariane 5 launch vehicle shows a similar trend, but with the peak amplitude shifted to a lower frequency.

The *steady component of launch acceleration* must achieve a speed increase of about 9.5 km/s. Its time history is dependent on the launch vehicle used. Low-mass payload vehicles such as Scout, sounding rockets, and air-launched systems such as Pegasus generally exhibit high peak acceleration, whereas those of large payload/manned vehicles are smaller. For multi-stage vehicles, the acceleration increases during the burn of each stage and peaks at burn-out/stage-separation. This is demonstrated for Ariane 4 in Figure 2.3. The air-launched Pegasus, a three-stage solid rocket vehicle, shows peak static acceleration

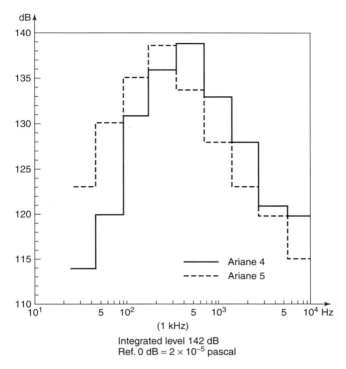

Integrated level 142 dB
Ref. 0 dB = 2×10^{-5} pascal

Figure 2.1   Ariane noise spectrum under SPELDA fairing, Ariane 4 [2] and SPELTRA, Ariane 5 [3]

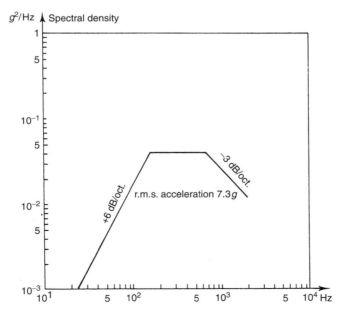

Figure 2.2   Ariane 4 random vibration spectrum (Reproduced by permission of Arianespace [2])

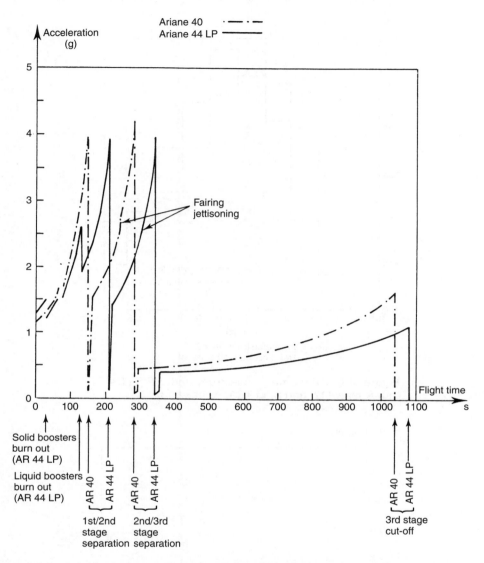

Figure 2.3 Ariane static acceleration profile (Reproduced by permission of Arianespace [2])

at stage burnout, with the actual value being heavily dependent upon the payload mass. Thus for a 50 kg payload at stage 3 burn-out, the 3σ acceleration is $13g_0$ [4]. However, for the near maximum payload of 600 kg this has fallen to nearly $4.5g_0$ (g_0 is the surface gravitational acceleration, 9.8 m/s^2). These figures may be compared to Shuttle, which has a peak acceleration of $3\,g_0$. Design data currently available for Ariane 5 indicates peak acceleration values during various mission events. These are provided in Table 2.1.

For manned flight, it is necessary to place the astronaut in a suitable position to withstand peak acceleration levels; Figure 2.4 shows the typical maximum levels that may

Table 2.1 Acceleration values for Ariane 5 launch vehicle [3]

Flight Event	Acceleration (g_0)		
	Longitudinal		Lateral
	Static	Dynamic	Static dynamic
Lift-off	−1.7	+/−1.5	+/−1.5
Maximum dynamic pressure	−2.7	+/−0.5	+/−2
P230 oscillations	−4.25	+/−1.75	+/−1
H155 thrust tail-off	−0.2	+/−1.4	+/−0.25

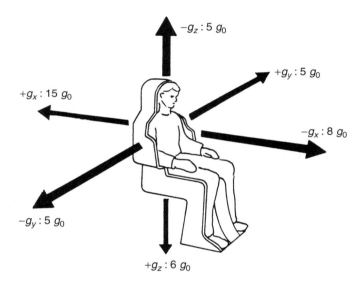

Figure 2.4 Typical tolerance to sustained linear acceleration as a function of direction of acceleration

be withstood. Greatest protection is provided when the astronaut is on his or her back during launch.

Mechanical shock is experienced when devices such as latches or explosive bolts are used, or at ignition of rocket motor stages and their subsequent separation, launch vehicle/payload separation, or when docking or landing.

These instantaneous events can provide extremely high-acceleration levels lasting only a few milliseconds locally, or in some cases, extending to the complete system. Their frequency spectrum is characterized by high-frequency components. In the case of Ariane 4, during payload separation the peak excitation that the satellite must survive is some $2000g_0$ at frequencies above 1.5 kHz. For Ariane 5, the peak shock occurs above 3 kHz, with a design requirement to survive $10^4 g_0$. For Pegasus, the highest shock loading is experienced above a frequency of 1.5 kHz where the design load is $3500g_0$.

The *thermal environment* experienced during launch is determined generally by the temperature reached by the launch shroud. Its high temperature arises from the aerodynamic frictional forces of the vehicle moving at high velocity through the atmosphere. The temperature reached is determined by the specific heat of the shroud material and a balance between friction heating and radiative and convective heat losses. The subsequent temperature rise of the payload within the shroud is dominated by radiative and heat conduction paths between shroud and payload. Once the shroud has been jettisoned, payload heating arises directly as a result of frictional forces, but the low density of the atmosphere then results in only modest heat fluxes.

For Ariane 4, the peak heat flux experienced by the payload within the shroud is less than 500 W/m² at any point. Following shroud jettison this rises to a maximum of 1135 W/m² [2]. For comparison, Figure 2.5 shows the heating experienced during the Ariane 5 launch ascent into geostationary transfer orbit (GTO) [3]. The same design of aerothermal flux limit as that for Ariane 4 is used to control the altitude at which the fairing is jettisoned.

The *ambient atmospheric pressure* declines during launch. The rate at which depressurization occurs depends on the venting of the shroud volume. Generally, this is fixed by the inclusion of venting ports; for example, on Ariane the static pressure declines at a rate of 10 mbar/s, but for Shuttle the venting of the cargo bay can be controlled. For Shuttle, which has both pressurized and unpressurized elements, venting control is particularly important because of possible adverse static loads being placed on structural members. A detailed description of its venting system is given in Reference [5]. It should be noted that venting of electronic boxes is also generally required within the vehicle.

Great care is required during payload integration to ensure that *electromagnetic interference* (EMI) does not present a hazard. Hazards may be in a variety of forms but the most severe are cases in which EMI may result in the activation of part of the payload, which could lead to death of attendant personnel, perhaps *via* the ignition of an on-board

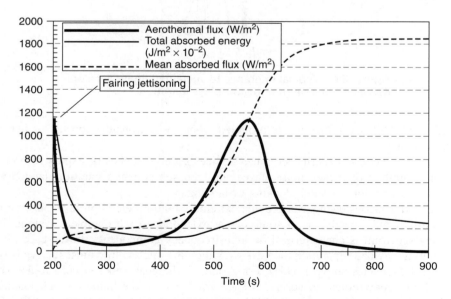

Figure 2.5 Aerothermal fluxes on a standard Ariane 5 trajectory. ⎯⎯⎯ Aerothermal flux (W/m²). ⎯⎯⎯ Total absorbed energy (J/m² × 10⁻²). - - - - Mean absorbed flux (W/m²). Fairing jettison and second flux peak constrained at 1135 W/m²

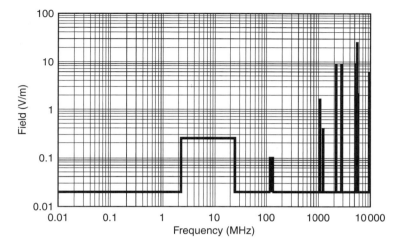

Figure 2.6 Electromagnetic interference for Pegasus vehicle at Western Test Range [4] Source: Orbital Sciences Corporation www.orbital.com

propulsion system. Figure 2.6 shows the EMI environment anticipated for the Pegasus launch vehicle whilst undergoing integration at the Western Test Range.

2.3 OPERATIONAL SPACECRAFT ENVIRONMENTS

2.3.1 Solar radiation

The Sun provides all the heat input to the solar system (excluding planetary radioactive decay processes), and its mass is 99.9% of the total. It is not surprising that it dominates the space environment of the whole solar system, and it is therefore appropriate to outline its significant features as they affect the near-Earth environment.

The Sun itself is not a particularly significant body in the universe. It is a star of mass $\sim 2 \times 10^{30}$ kg, modest by stellar standards, and is one of $\sim 10^{11}$ stars that form our galaxy. It is classified as a G2V star, having a yellowish appearance because its radiated light peaks at ~ 460 nm, and it is termed as *a yellow-dwarf star*. Its radius is 7×10^8 m.

After the Sun, the nearest star is 3.5 light years away (1 light year = 9.46×10^{12} km) and between the stars the gas density is low, with hydrogen as the dominant species. The density amounts to only 3 atoms/cm^3, in comparison to the nominal number density of our own atmosphere at sea level of $\sim 3 \times 10^{19}$ molecules/cm^3.

The Sun is fundamentally a giant thermonuclear fusion reactor whose surface temperature is ~ 5800 K. The photosphere is optically thick, and its spectrum approximates to that of a black body.

The *solar spectrum* is shown in Figure 2.7, with a 5900 K black body for comparison. It is evident that it departs from the black-body spectrum at some wavelengths, these discrepancies arising in the solar atmosphere. There are two primary regions of this. The lower, or chromosphere, extends to a few thousand kilometres above the photosphere and is a region of increasing temperature, peaking at $\sim 10\,000$ K; it is responsible for

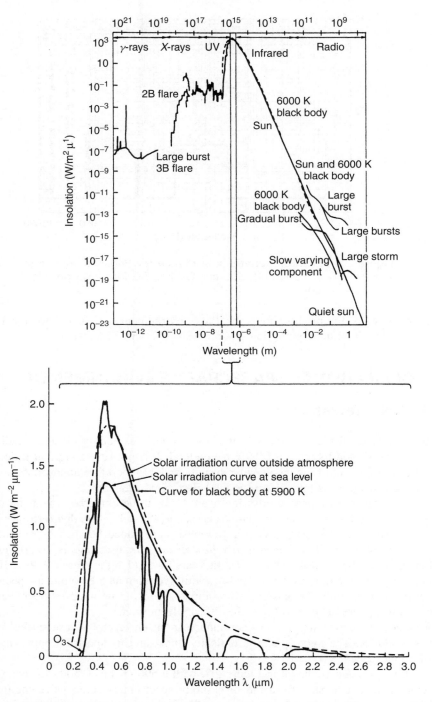

Figure 2.7 Solar spectrum (Reproduced by permission from Hynek, J. A. (1951) *Astrophysics*, McGraw-Hill)

enhanced ultraviolet (UV) emission. The upper atmosphere, called the corona, becomes more tenuous and extends to several solar radii. Its nominal temperature is around 2×10^6 K, and it emits substantial amounts of X-rays. The nominal release of energy from the Sun is at a rate of 3.85×10^{26} W. In order to gain an insight to this power level, if the Earth's fossil fuel resources were to be burnt at this rate, then they would be exhausted in a mere 50 milliseconds.

The *solar wind* is another outward flux from the Sun. It is a flow of plasma expelled at high velocity. In reality it forms the outermost layer of the solar atmosphere, being continuously driven outward as a result of the Sun's radiation pressure. At Earth, the speed of the wind is $\sim$450 km/s, its density is $\sim$9 protons/cm^3 and its kinetic temperature is $\sim$100 000 K.

Sunspots are an indication that there are significant disturbances taking place on the Sun's surface and through its atmosphere. These, first observed by Galileo, are regions of its disc, which are cooler than the surrounding surface. They emit less radiation and thus appear as dark spots. Periods of high solar activity occur when there are a large number of sunspots and then enhanced emission of radiation occurs, most notably at radio wavelengths and at X-ray and γ-ray energies. This enhanced emission is generally associated with solar flares, which occur at sites near sunspots. They may last from a few minutes to several hours and occur as frequently as one every two hours during high solar activity.

The Zurich sunspot number R_z is used to quantify the overall number of sunspots on the Sun at any time. It is defined as $R_z = K(10g + f)$, where f is the number of sunspots that exhibit umbrae, and g is the number of groups into which these spots fall. K is a factor that relates to the observing instrument and is used as a normalization factor.

The detailed prediction of individual flares is not yet possible, but the general level of activity has a well-defined 11-year cycle as shown in Figure 2.8. However, owing to magnetic pole reversal of the Sun at peak solar activity, the real period is 22 years. It may be seen that the magnitude of the peak varies from cycle to cycle. Thus the cycle commencing in June 1954 shows a significantly higher peak than all other cycles that were completed in the 20th century. Regular updated sunspot activity may be readily accessed at a number of web sites including *http://www.dxlc.com/solar*, which provides daily data on a variety of forms of measuring solar activity, including sunspot number, the geomagnetic index and maps of solar active regions. The site *http://sidc.oma.be/index.php3* provides a long-term historical record of solar activity from 1700 when the first regular recording of sunspot activity took place.

Table 2.2 shows the typical intensity variability of the Sun at particular wavelengths where it has been established. There is good correlation between sunspot number and radiated power at some specific wavelengths and this is used to provide an additional measure of the overall level of solar activity. The most frequently used monitor is the solar flux at 10.7 cm. One source for such data is from NOAA, in the US.

An enhanced flux of radiation and high-energy particles is also associated with major solar flares. Two components may be distinguished when it arrives at the Earth. The first occurs approximately 20 min following the flare, when the electromagnetic emissions first increase and then a more prolonged component arrives about a day after it. These latter particles appear to be an enhanced component of the solar wind, travelling at velocities of $\sim$10^3 km/s. Figure 2.9 shows the variability of particle fluxes in the interplanetary medium caused by solar activity. Detailed descriptions of ionizing radiations are given in the next section.

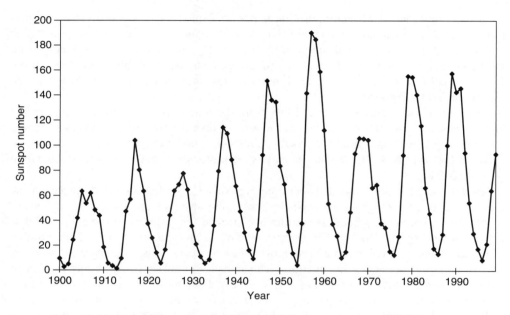

Figure 2.8 Zurich sunspot number archive for last 100 years

Table 2.2 Solar variability [6] (Reproduced by permission of NASA)

Spectral region	Wavelength	Flux (J/(m² s μm))	Variability
Radio	$\lambda > 1\,\mathrm{mm}$	$10^{-11}-10^{17}$	$\times 100$
Far infrared	$1\,\mathrm{mm} \geq \lambda > 10\,\mu\mathrm{m}$	10^{-5}	Uncertain
Infrared	$10\,\mu\mathrm{m} \geq \lambda > 0.75\,\mu\mathrm{m}$	$10^{-3}-10^{2}$	Uncertain
Visible	$0.75\,\mu\mathrm{m} \geq \lambda > 0.3\,\mu\mathrm{m}$	10^{3}	$<1\%$
Ultraviolet	$0.3\,\mu\mathrm{m} \geq \lambda > 0.12\,\mu\mathrm{m}$	$10^{-1}-10^{2}$	$1-200\%$
Extreme ultraviolet	$0.12\,\mu\mathrm{m} \geq \lambda > 0.01\,\mu\mathrm{m}$	10^{-1}	$\times 10$
Soft X-ray	$0.01\,\mu\mathrm{m} \geq \lambda > 1\,\text{Å}$	$10^{-1}-10^{-7}$	$\times 100$
Hard X-ray	$1\,\text{Å} \geq \lambda$	$10^{-7}-10^{-8}$	$\times 10- \times 100$

2.3.2 Earth orbit environment

The Earth orbits the Sun at a mean distance of one astronomical unit (AU), equal to 1.496×10^{8} km. It is a nearly spherical body having a mass of only 3×10^{-6} times that of the Sun, but having a gravitationally bound atmosphere and a significant magnetic field. Each of these features is important in determining the near-Earth environment, in which the majority of space vehicles operate.

The *atmosphere* at sea level is predominantly molecular nitrogen (78%) and molecular oxygen (21%), with a variety of trace elements, the most significant being argon. Standard sea-level pressure is accepted to be 1.013×10^{5} Pa [7]. The lower atmosphere, up to ∼86 km, is sufficiently turbulent to result in a homogeneous gas mixture, albeit with a pressure decreasing exponentially. Above this height, photochemical processes disturb the

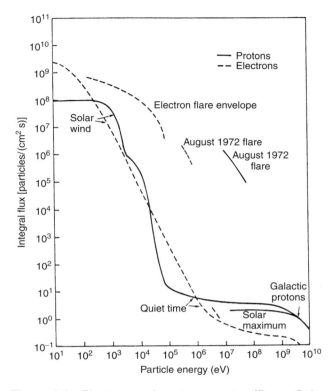

Figure 2.9 Electron and proton spectra (From *Solar Cell Array Design Handbook* by H. S. Rauschenback, Copyright © 1980 by Van Nostrand Reinhold. All rights reserved)

homogeneity. Of particular importance is the absorption of UV radiation from the Sun, leading to the dissociation of oxygen in the upper atmosphere. The details of the processes are beyond the scope of this chapter and the interested reader is referred to Reference [7]; however, the resultant atmosphere above ~120 km is one in which each atmospheric constituent is decoupled from all the others. For each species, it is then possible to write down an equation of the diffusive equilibrium of the form

$$n_i v_i + D_i \left(\frac{\mathrm{d}n_i}{\mathrm{d}Z} + n_i \frac{(1 + \alpha_i)}{T} \frac{\mathrm{d}T}{\mathrm{d}Z} + \frac{g n_i}{R^*} \frac{M_i}{T} \right) = 0 \tag{2.1}$$

where n_i is the number density of species i, having a molecular weight M_i at altitude Z, v_i is the vertical transport velocity of the species, and D_i and α_i are its molecular and thermal diffusion coefficients, T is the atmospheric temperature, R^* is the universal gas constant and g is the height-dependent acceleration due to gravity. If negligible vertical transport takes place, and for species in which thermal diffusion is negligible, equation (2.1) reduces to a hydrostatic equilibrium equation wherein the number density profile is driven by the atmospheric temperature. Figure 2.10 shows the variation of number density with height for different species.

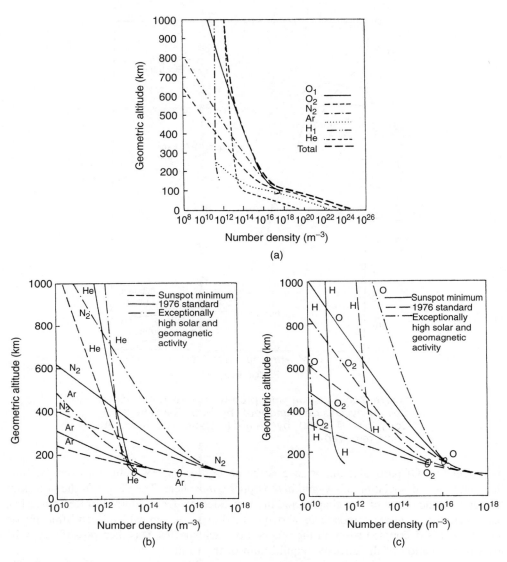

Figure 2.10 Species concentration (a) in US standard atmosphere and (b, c) under extreme conditions

For detailed calculation of the atmosphere, a variety of models are available (e.g. References [8,9]). Each of these requires the specification of a temperature at some height, and then by using a pre-defined temperature profile, the density may be determined through equation (2.1). Figure 2.11 shows the general form of such profiles. It can be seen from this that at extreme altitudes the temperature tends to a limiting value, the so-called exospheric temperature, T_∞. Within the exosphere the atmosphere is effectively isothermal. T_∞ rises through increased solar activity; most models relate it to the flux of solar radiation at 10.7 cm ($F_{10.7}$) through an algorithm, and also include the effects of geomagnetic

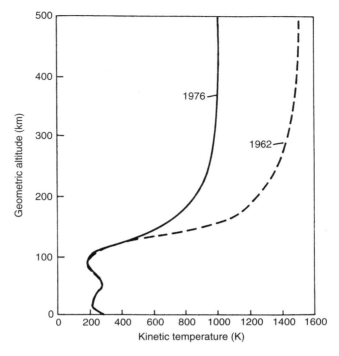

Figure 2.11 Temperature profile of atmosphere

activity, stimulated by the interaction of the solar wind and the Earth's magnetic environment or magnetosphere (see below). Since under conditions of hydrostatic equilibrium, the density falls from that at sea level, ρ_{SL}, at a rate

$$\rho = \rho_{SL} \exp\left(\frac{-gM_i}{R^*T} z\right) \tag{2.2}$$

then for a given altitude, the atmospheric density will increase with solar activity and this will result in reduced lifetimes for low orbiting vehicles.

Equation (2.2) provides the simplest formulation for air density as a function of altitude and other more sophisticated analytic descriptions have been developed to yield expressions for the rate of orbit decay in low Earth orbiting (LEO) vehicles. The pioneering work of King-Hele in this field [10] provides the most comprehensive of these analytic models, which have the potential to describe the variation of scale height in the atmosphere with altitude, and the effects of oblateness and atmospheric rotation. Whilst these expressions can become complex, for the case wherein only scale height variations around a single orbit are concerned, the equivalent density may be written:

$$\rho = \rho_p \left[1 + \frac{\mu_0}{H_p}(r - r_p)\right]^{-1/\mu_0} \tag{2.3}$$

where suffix p refers to conditions at perigee of the orbit, H is the scale height and μ_0, the altitude gradient of H, is a constant of order of the eccentricity of the orbit (see Chapter 4).

For circular orbits in which e is zero, clearly equation (2.3) provides difficulty. However, in such cases the altitude of the satellite remains constant, and it is then inappropriate to consider the variation in scale height.

The US standard atmosphere 1976 [7] is shown in Figure 2.10. At altitudes typical of LEO vehicles (300–900 km) the density is low, but not insignificant. At geostationary altitude, $\sim$36 000 km, the density of the neutral atmosphere is approximately the same as in the interplanetary medium, namely, 10^{-20} kg/m^3, and the pressure is $\sim$10^{-15} Pa [11]. The dominant species of the atmosphere in LEO is generally either atomic oxygen or helium. The effects of atomic oxygen on surfaces are considered in Section 2.4.1.

One of the most frequently used models of the thermosphere is the MSIS model [8]. This was originally produced in 1986 but has been revised in the MSISE-90 model more recently to include both shuttle flight data and incoherent radar scattering data. It is now possible to run the model for specific historical dates through the NASA web site at *http://nssdc.gsfc.nasa.gov/cgi-bin/models/msis_m*. This can be used to produce density profiles of the major atomic and molecular species together with temperature and total density information on a longitude, latitude specific basis.

Few *atomic/molecular collisions* take place between components of the atmosphere and spacecraft in either LEO or GEO. Table 2.3 shows that above 200 km altitude the mean free path is significantly greater than the dimensions of most space vehicles. This has two consequences.

Firstly, the ability to exchange heat energy with the environment is solely as a result of radiation. Thus the dominant radiative heat input is due to solar radiation, which in the near-Earth environment is 1371 ± 5 W/m^2. Secondary input occurs as a result of Earth albedo (the reflection of solar radiation from the top of the atmosphere) and Earth shine (the black-body radiation of the Earth), and has a magnitude of $\sim$200 W/m^2. The neutral atmosphere at $\sim$10^3 K and the solar wind at $\sim$2 $\times$ 10^5 K provide negligible heating. The temperature that a space vehicle reaches is thus dependent upon a balance of radiative heat input and output (see Chapter 11).

Secondly, the aerodynamics of spacecraft at orbiting altitudes must be based upon free molecular flow as briefly described in Chapter 4. Since the density is low, the frictional heating forces are negligible, even though the relative velocity approaches 8 km/s.

The *ionosphere*, above $\sim$86 km, is a region of increasing plasma density caused by photoionization, due to incident UV photons. The plasma has a significantly lower density than the neutral density below an altitude of $\sim$1000 km, even though its peak value occurs at 300 to 400 km. It has significant influences upon the propagation of radio waves; for an electron density n_e electrons cm^{-3}, frequencies below the plasma frequency, given

Table 2.3 Mean free path λ_0 as a function of altitude

Altitude (km)	λ_0 (m)	Altitude (km)	λ_0 (m)
100	0.142	300	2.6×10^3
120	3.31	400	16×10^3
140	18	500	77×10^3
160	53	600	280×10^3
180	120	700	730×10^3
200	240	800	1400×10^3

by $f_p \sim 9000\sqrt{n_e}$ Hz, cannot propagate, and thus radio waves of lower frequency incident upon such a plasma will be reflected. If there is additionally a magnetic field present, as is the case in the near-Earth environment, then the polarization of any electromagnetic radiation propagating through the plasma will be rotated due to Faraday rotation; inefficiencies can then occur in communication systems if linearly polarized radio waves are used (see Chapter 12).

The Earth's magnetic field has two primary sources. The dominant one at its surface is due to currents circulating within its core, whilst at higher altitudes the currents caused by the differential motion of electrons and ions in the magnetosphere play a significant role. The solar wind plasma, carrying its own magnetic field, distorts the Earth's simple dipole field into the shape shown in Figure 2.12, with both open and closed magnetic field lines.

From Figure 2.12 it is apparent that at high altitude the magnetic field structure is complex; however, at lower altitude it is possible to make certain observations. Firstly, the overall strength of the magnetic field is not constant, but is decreasing at ~0.05% per year. This field is weakest on the equator, and Figure 2.13 shows its dependence on both latitude and altitude.

Ionizing radiations

In the near-Earth environment, there are several types of ionizing radiations that impact upon spacecraft systems.

The Van Allen radiation belts contain energetic protons and electrons that are trapped in the Earth's magnetic field and generally follow the magnetic field lines. There may also be significant fluxes of heavy ions such as of helium (He), nitrogen (N) and oxygen (O) whose atmospheric densities depend on solar and geomagnetic activity. The

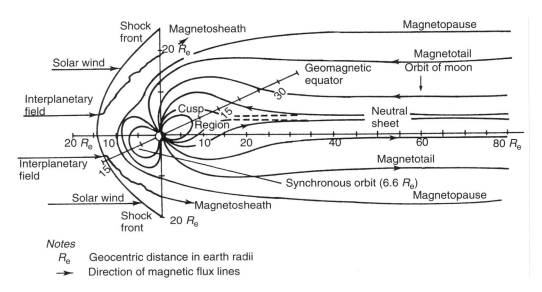

Notes
R_e Geocentric distance in earth radii
→   Direction of magnetic flux lines

Figure 2.12 The Earth's magnetosphere (Reproduced by permission of Kluwer Academic Publishers)

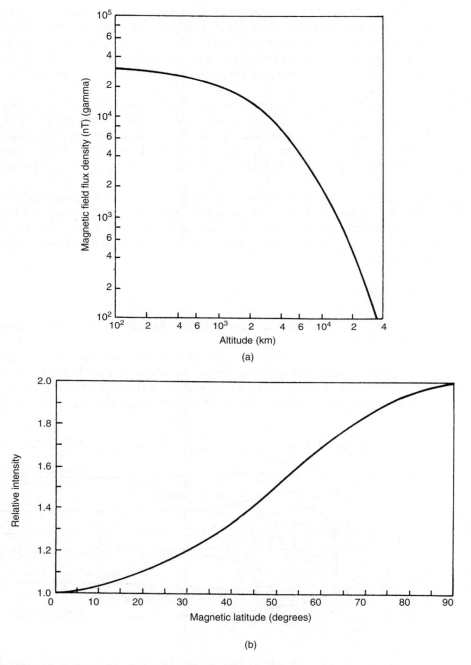

(a)

(b)

Figure 2.13 (a) Earth's magnetic field intensity at the magnetic equator as a function of altitude (Adapted from Reference [12]) and (b) relative intensity of the Earth's magnetic field as a function of magnetic latitude (Reproduced by permission of Kluwer Academic Publishers)

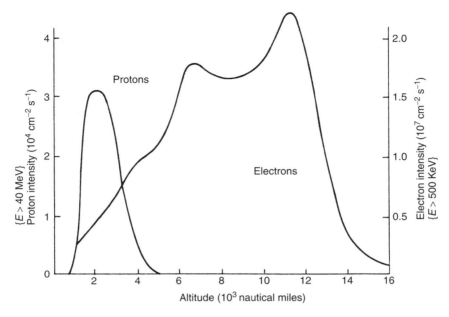

Figure 2.14   Van Allen radiation belts (idealized)

overall structure of the Van Allen belts can be seen from the highly simplified diagram in Figure 2.14.

Proton energies range from 0.01 to 400 MeV with fluxes in the range of 10^8 to 600/cm^2 s, respectively. Electron energies are in the range from 0.4 to 4.5 MeV with fluxes from 4×10^8 to 100/cm^2 s, respectively. One of the most notable features of the radiation belt particles is the large variation with both altitude and latitude. Figure 2.15 shows typical contour plots of the electron and proton fluxes at various locations [13].

To predict electron and proton fluxes for LEO missions, the NASA models AE8 [14] and AP8 [15] are used. Since these particle fluxes vary with the solar cycle, there are two variants for each model, the maximum and minimum, corresponding to the maximum and minimum of the solar cycle, respectively. However, it should be noted that the solar cycle variations of the fluxes are not well understood. Furthermore, the data on which these models are based are old (circa 1960s and 1970s), and the model should only really be used to predict integrated fluxes (fluences) for periods greater than about six months.

At low altitudes and low inclinations, the dominant feature of the radiation environment is the region known as the South Atlantic Anomaly. Because of the offset and tilt of the geomagnetic axis relative to Earth's rotation axis, this is a region of enhanced radiation in which parts of the radiation belt are brought to lower altitudes (Figure 2.16).

The effects of the trapped particles are degradation of electronic parts due to accumulated dose, degradation of solar array performance due to displacement damage, single-event upsets (SEUs) and dielectric charging. Whilst the detailed processes associated with these damage types is somewhat complex, the key phenomena that take place result from the impact of the high-energy particle upon both the energy structure and lattice structure of the semiconductor material. The damage influence of an individual massive particle, such as a proton, is significantly greater than that from an electron due to the higher

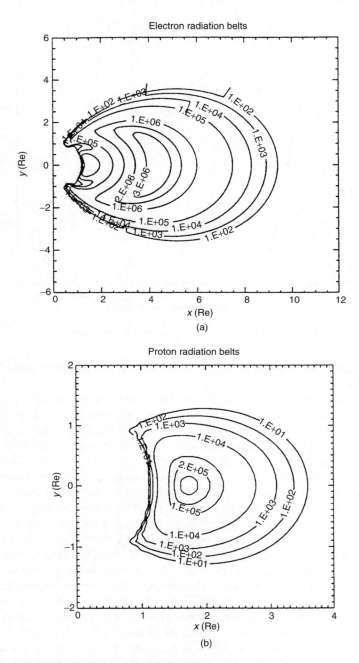

Figure 2.15 Radiation belt model particle fluxes. Contour plots in 'dipole space' in a plane containing the geo-dipole axis (y): (a) electron fluxes at greater than 1 MeV and (b) proton fluxes at greater than 10 MeV. (Axes are calibrated in Earth radii, Re) [13]. (Reproduced by permission of the International Astronautical Federation)

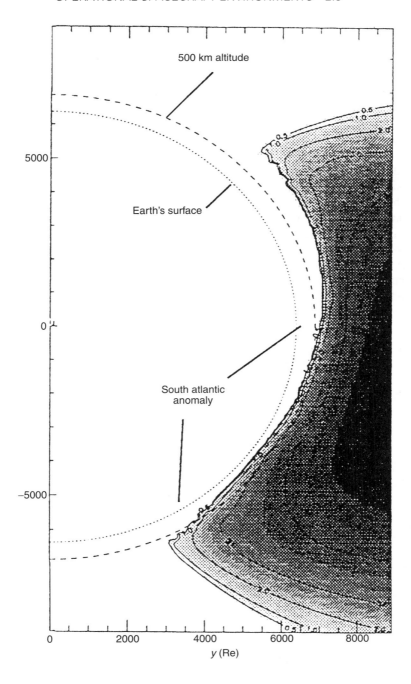

Figure 2.16 Proton radiation belt at low altitude, derived from AP8 model. Flux contours at longitude 325°. Tilt and offset brings contours below the 500 km altitude line. This is the South Atlantic anomaly [13]. (Reproduced by permission of the International Astronautical Federation)

momentum carried by the former. In particular, if, for example, a proton is stopped in the material, because of the momentum exchange a significant *displacement* of a lattice atom will occur. This displacement can result in local ionization and the disruption, again locally, in the energy structure, possibly with the creation of additional energy states between the conduction and valence bands. The ionization process results in an impulsive release of charge, leading to an SEU. Changes to the energy structure result in reduction in the efficiency of solar cells converting sunlight to electricity (see Chapter 10). Both electrons and protons contribute to the total dose and the major contribution will depend on the particular orbit; in general, low altitude orbits (less than about 800 km) will be dominated by protons, whereas high altitude orbits including geosynchronous orbits, will be dominated by electrons (Figure 2.17). Both types of particles will cause displacement damage, but protons are more effective. Because of the relative magnitudes of electron and proton fluxes, it is not easy to generalize about their contributions to the displacement damage (usually expressed as a 1 MeV equivalent fluence), but as with total dose, protons usually dominate at low altitude and electrons at high altitude.

Galactic cosmic radiation is composed of high-energy nuclei, believed to propagate throughout all space unoccupied by dense matter. Its origin is still a matter of scientific debate and may have both galactic and extragalactic sources.

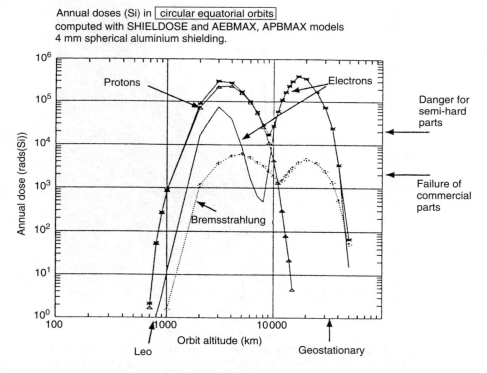

Figure 2.17 Variation of computed annual dose on circular orbits as a function of altitude. Computed with AE8 and AP8 models and SHIELDOSE [13]. (Reproduced by permission of the International Astronautical Federation)

Primary cosmic radiation is by definition that which is incident on the Earth's atmosphere. Cosmic rays propagating through the atmosphere undergo nuclear collisions and generate secondary rays consisting of all known nuclear and sub-nuclear species. The flux of galactic cosmic radiation is believed to be essentially isotropic outside the heliosphere; inside it, propagation effects result in an isotropy of approximately 1%.

High-energy cosmic ray particles have a large amount of kinetic energy and this can have a permanent effect upon the material through which they pass.

Primary cosmic radiation observed at the Earth's orbit consists of approximately 83% protons, 13% alpha particles, 1% nuclei of atomic number greater than two and 3% electrons. The composition extends over an energy range from a few hundred MeV to greater than 10^{20} MeV. Figure 2.18 [16] shows typical quiet-time spectra for several elements.

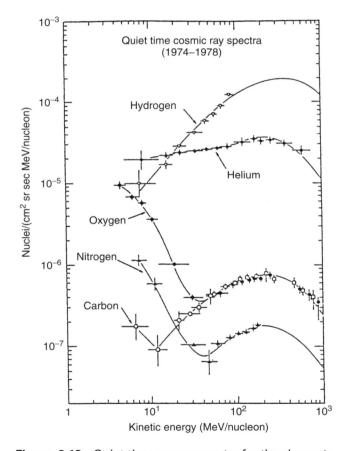

Figure 2.18 Quiet-time energy spectra for the elements H, He, C, N and O. Measured at 1 AU over the solar minimum period from 1974 to 1978 [16]. Note the 'anomalous' enhancements in the low-energy spectra of He, N and O. The data are from Caltech and Chicago experiments on IMP-7 and IMP-8 [16]. (Reproduced by permission of the Jet Propulsion Laboratory, California Institute of Technology, Pasadena, California)

The intensity of cosmic rays observed is dependent on the solar cycle, decreasing as the sunspot number cycle increases.

During their travel from their source regions to the vicinity of the Earth, cosmic rays interact with the interstellar medium. This causes fragmentation to occur, resulting in a depletion of the heavy charge primary rays and an increase in the lighter nuclei. The low-energy portion of the cosmic ray spectrum is quite variable, reflecting its dependence on solar modulation; from solar cycle minimum to maximum, the energy density of near-Earth primary galactic rays decreases by about 40%. The ray intensity observed at Earth does not change smoothly from maximum to minimum values, but decreases in a series of sharp drops followed by a partial recovery until the minimum intensity is reached near to the maximum in the solar activity.

Solar energetic particle events

Part of the energy in solar flares is in the form of nuclei accelerated to high energies and released into space. These are commonly referred to by descriptive names such as solar cosmic events, solar proton events, solar electron events, solar cap absorption events and ground level events. The most commonly measured components of these are the proton and electron events. Major solar particle events occur at random, with a frequency that varies from approximately one every two months to one every two years. Events initiated by solar flares may also contain a small apparently variable flux of heavier elements.

Solar proton events are statistical in nature, showing a wide variation in characteristic parameters such as integrated flux (fluence), peak flux and energy spectrum. The most recent and accurate method for predicting proton fluence for space missions is that of Feynman *et al.* [17] and is based on a combination of observations made from the Earth's surface, from above the atmosphere between 1956 and 1963 and from spacecraft in the vicinity of Earth between 1963 and 1985. Analysis of these data shows that there are seven hazardous years during each solar cycle, beginning two years before the solar maximum year and extending to four years after it; the remaining four years of the cycle contribute negligibly to the expected mission fluence.

Figure 2.19 shows the probability of exceeding a given fluence level over the life of a mission, assuming a constant heliocentric distance of 1 AU. If a mission is in space for more than one solar cycle, the best method for finding the total expected fluence is to estimate the fluence per cycle from the seven-year curve and add the appropriate number of additional yearly fluences to this line. The use of the one-year curve to estimate the additional fluence expected on a long mission is not recommended because it will lead to an overestimate. Correction for the dependence on the distance from the Sun (r) may be made by assuming that it varies as r^{-3} for r less than 1 AU, and as r^{-2} for r greater than 1 AU.

Solar protons are accelerated in solar active regions during solar flares, and X-ray, radio and optical emissions are indicators that particle acceleration is occurring. The decaying portion of a flare is normally exponential in character, the flux decay typically having a time constant of two to three days, but there is a slight dependence on the energy, and there can be a large variability between events.

In addition to energetic protons, solar energetic particles produce heavy ions. Their fluxes and fluences can be found by extrapolation from those of protons, using elemental abundance ratios normalized to hydrogen. Typical values are shown in Table 2.4 [18].

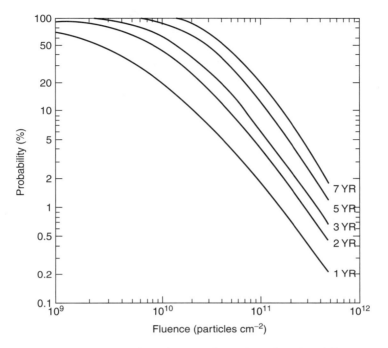

Figure 2.19 Probability of exceeding given levels of fluence energy >10 MeV *versus* active years of solar cycle. (Reproduced by permission of the Jet Propulsion Laboratory, California Institute of Technology, Pasadena, California)

Table 2.4 Normalized abundances of solar energetic particle events [18] (Reproduced by permission of the Jet Propulsion Laboratory, California Institute of Technology, Pasadena, California)

		1 MeV	1–20 MeV	10 MeV	6.7–15 MeV
1	H	1.0	1.0	1.0	1.0
2	He	2.2 E-2	1.5 E-2		1.5 E-2
3	O	3.2 E-4	2.2 E-4	2.2 E-4	2.8 E-4
4	Al	3.5 E-6	3.5 E-6	3.1 E-6	3.3 E-6
5	Ca	2.3 E-6	2.6 E-6	3.1 E-6	3.2 E-6
6	Fe	4.1 E-5	3.3 E-5	3.4 E-5	

Electrostatic charging of a spacecraft travelling through the near-Earth space environment will occur, whether it is in or out of the radiation belts. As a consequence, currents will occur between the space vehicle and the plasma, imbalance of which will cause spacecraft to develop a charge that may be returned to balance through arcing. The two major sources of currents are [12] the ambient plasma itself and photoelectron emission due to solar sunlight, and in particular, the short wavelength component of this radiation. The latter phenomenon is particularly important when the spacecraft enters and leaves eclipses [19]. Severe problems arise if differential charging of the spacecraft surface occurs. The simplest method of preventing this is to use conductive surfaces wherever possible. One primary area in which this is not possible is on the solar array; an alternative solution is then to apply a near transparent coating of indium oxide to the cell cover glass material, which typically reduces the resistivity of the glass surface to less than $5000\,\Omega/cm^2$ [20].

Meteoroids and micrometeoroids occur with a frequency that varies considerably with the type of space mission. These are solid objects whose mass and size vary over many orders of magnitude. Their mass spectrum is shown in Figure 2.20. Near large gravitational masses such as the Earth their fluxes tend to be enhanced. The asteroid belt is also a region of enhanced meteoroid density. Impact of micrometeorites generally causes a degradation of surface thermal properties, although the possibility of component failure clearly exists. The most dramatic evidence of particle impacts on a spacecraft is from the Giotto spacecraft and its passage near to Halley's comet during 1986. Particle impacts led to the failure of some experiments and a change in the attitude of the vehicle at closest encounter.

Man-made space debris, consisting of aluminium oxide dust particles (from solid rocket exhausts), instrument covers, nuts and bolts, rocket upper stages, and so on is in addition

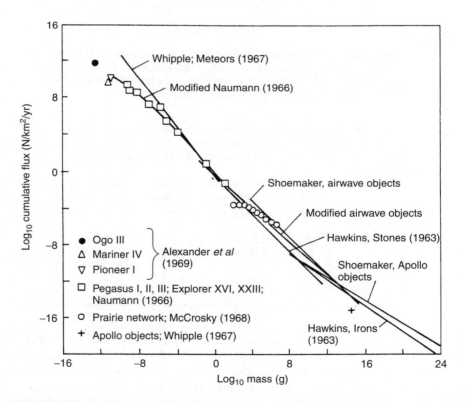

Figure 2.20 Terrestrial mass-influx rates of meteoroids. *N* is the flux of particles with mass greater than *m* [21] (Reproduced by permission of NASA; see Reference [6] for details)

to the naturally occurring micrometeoroid environment. The size varies from 0.001 mm to 10 m in diameter. Figure 2.21 compares the predicted flux of debris with that of meteoroids for a 500 km, 30° orbit. Space debris varies with the orbit's altitude, eccentricity and inclination, and exhibits a strong directional dependence. The average velocity (relative to the orbiting spacecraft) is about 11 km/s, but it is a function of altitude, ranging from zero to twice the orbital velocity. Since the majority of the particles originate from satellite and launcher components, the average density is assumed to be that of aluminium alloys—about 2.8 g/cm^3. The threat to space stations such as the ISS due to the larger sizes of particles, greater than 1 mm, has caused a great deal of interest in this environment, both in modelling and in measuring it.

The smaller particles, in the range from 10^{-3} to 10^{-9} g, have a flux that is high enough to erode surfaces and have enough energy to penetrate protective coatings. Of particular concern is their effect on large solar arrays, sensitive optical surfaces and detectors. The possibility of penetrating protective coatings necessitates the consideration of synergistic effects between the particulate and the atomic oxygen environments. See Section 2.4.1.

System requirements for meteoroid and debris protection amount generally to ensuring the safety of the crew for manned spacecraft and the operational availability for unmanned craft. Usually, these requirements are expressed as the probability of no damage over a given time. For example, typical probability values for the Columbus Attached Laboratory range from 0.995 for endangering the crew or module's survivability for a 10-year life, to 0.999 for penetration or bursting over a one-year period. Having set the system requirement probabilities, the critical debris size can be calculated using the environment models and the spacecraft geometry. For Columbus, the critical debris size ranges from about 0.5 to 1.0 cm in 1995 [23].

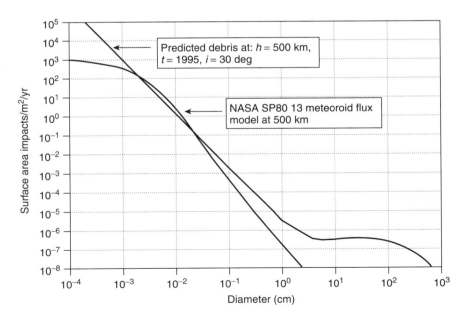

Figure 2.21 Comparison of the fluxes of predicted orbital debris and meteoroids [22] (Reproduced by permission of NASA)

Effective shielding can be achieved by using a double-walled bumper shield, in which the first wall fragments the impacting particle into a cloud of smaller, slower moving debris, which are then stopped by the second wall (See Chapter 8).

2.3.3 Solar system environments

The *solar system* may be considered to be the region of space that is dominated gravitationally by the Sun. In Chapter 5, this will be defined in greater detail. However, at this point, we can consider the solar system to be that region containing the nine major planets and their natural satellites, together with all the other minor bodies including asteroids, comets and dust, and the residue of the solar wind. Following the detailed environment of the near-Earth region of space, in which the majority of space vehicles operate, it is appropriate to consider briefly the major features of those other solar system bodies, which have been the focus of space missions.

Table 2.5 lists the mass properties of the major solar system bodies together with the orbital characteristics of these about the Sun. The definition of the orbital parameters may be found in Chapter 4, together with additional data on these objects. The key features discernible from the data presented in Table 2.5 are firstly, with the exception of Pluto, and to some extent Mercury, all the planets nearly lie in a plane. This can be seen from the angle of inclination i, being no more than a few degrees away from the ecliptic plane. Secondly, again with the exception of these two planets, and evidenced by the eccentricity e of the planetary orbits, the planetary orbits are nearly circular. Whilst these two features provide evidence of the history of the solar system, from the consideration of space vehicle design, they lead to significant design implications. Thus, the near planar nature of the solar system results in only relatively minor plane changes being required to send a probe to another solar system body. This has significant influence on the fuel load for the vehicle's propulsion system; details of this are provided in Chapter 5. The circularity of most planetary orbits means that following insertion about a planet, the spacecraft will not be subject to any additional major changes in the thermal environment. Clearly the overall thermal environment to a first approximation will scale as $(r_E/r_P)^2$, where r_E is the radial distance from the Earth to the Sun, and r_P is the distance of the planet from the Sun.

The majority of time, during most planetary space missions, is spent in interplanetary space. This environment is dominated by the solar wind, for which a description has already been provided. Both solar insolation and the density of the solar wind decrease with distance from the Sun, again with an approximate $1/r^2$ relationship. For missions beyond Mars, spacecraft need to traverse the asteroid belt, for which there is enhanced collision risk with sizeable objects.

The planetary environments themselves are very diverse. Table 2.6 provides data on those planets having an atmosphere, together with what is known about such atmospheres and any associated ionosphere. Only Mercury of the four inner planets does not have an atmosphere. The atmospheres of Venus, Earth and Mars are radically different, both in their composition and thermodynamic properties of pressure and temperature. It is particularly noteworthy that lander missions to Venus are required to operate both in the vacuum of space, and then on arrival at the surface of that planet survive pressures over 90 times that at the surface of the Earth, whilst the temperature is of order 750 K.

Titan is also included in Table 2.6, being the sole natural satellite of a planet that possesses a significant atmosphere. Our current knowledge of the atmospheric constituents

Table 2.5 Solar system data: mass characteristics and orbit

Object	Mass $(10^{24}$ kg)	$R_{equator}$(m) Equatorial radius	Density $(10^3$ kg/m^3)	Angular momentum $(10^{39}$ kg m^2/s)	Average speed (km/s)	a(AU) Semi-major axis of orbit[2]	e Orbital eccentricity	i (°) Orbital inclination[3]	Sidereal period (years)
Sun	1.99×10^6	0.696×10^9	1.409	170[1]	—	—	—	—	—
Mercury	0.33	2.44×10^6	5.46	0.906	47.9	0.3871	0.206	7.00	0.241
Venus	4.87	6.05×10^6	5.23	18.5	35.1	0.7233	0.007	3.39	0.615
Earth	5.97	6.38×10^6	5.52	26.7	29.8	1.0000	0.017	0.00	1.000
Mars	0.642	3.40×10^6	3.92	3.52	24.2	1.5237	0.093	1.85	1.881
Jupiter	1899	7.15×10^7	1.31	19 400	13.1	5.2028	0.048	1.31	11.862
Saturn	568	6.03×10^7	0.7	7840	9.64	9.5388	0.056	2.49	29.46
Uranus	87.2	2.56×10^7	1.3	1700	6.81	19.1914	0.046	0.77	84.01
Neptune	102	2.48×10^7	4.66	2500	5.44	30.0611	0.010	1.77	164.79
Pluto	0.66	1.20×10^6	4.9	17.9	4.75	39.5294	0.248	17.15	248.43

[1] Spin angular momentum of the Sun.
[2] 1 AU = 1.496×10^{11} m.
[3] Inclination of orbit plane relative to the ecliptic.
See also Tables 2.6, 2.7 and 4.1.

Table 2.6 Solar system data: approximate atmospheric parameters for solar system bodies

Planet/ Moon	Composition %	Surface pressure (Bar)	Surface temperature (K)	Temperature @ 200 km (K)	Ionosphere (Electrons/cm^3)
Mercury	None	—	—	—	
Venus	CO_2 (96); N_2 (3.5)	92	750	100–280	$\sim 10^6$
Earth	N_2 (77); O_2 (21); H_2 (1)	1	285	800–1100	$\sim 10^6$
Mars	CO_2 (95); Ar (1.6); N_2 (2.7)	0.006	220	310	$\sim 10^5$
Jupiter	H_2 (89); CH_4 (0.2); He (11)	Gaseous planet	165[1]		$\sim 10^5$
Saturn	H_2 (93); CH_4 (0.2); He (7)	Gaseous planet	130[1]		
Titan	N_2 (65–98); CO_2 (2–10)	1.5	95	150	$\sim 10^3$
Uranus	H_2 (85); $CH_4 (< 1)$; He (15)	Gaseous planet	80[1]		
Neptune	H_2 (90); $CH_4 (< 1)$; He (10)	Gaseous planet	70[1]		
Pluto	N_2 CH_4/CO (traces only)	—	40	—	—

[1]Temperature quoted where pressure is the same as Earth sea level (P = 1 Bar).
See also Tables 2.5, 2.7 and 4.1.

of Titan's atmosphere is still somewhat uncertain, although spectroscopic evidence shows that it is very different from that found on the Earth. The atmospheric surface pressure is similar to that of the Earth. The Cassini-Huygens probe, at the time of writing already beyond the orbit of Jupiter, will in 2004 provide a substantially enhanced understanding and description of this atmosphere.

The four gas giants are listed here as having atmospheres. The surface properties noted are however for an assumed surface, on which the pressure of the atmosphere is the same as that on the Earth at sea level.

A necessary condition for a planet to have an *ionosphere*—a region of space that contains free electrons and ions—is that it also has an atmosphere. The major energy source for the ionization of the upper regions of an atmosphere is solar radiation through the process of photo-ionization. Typical peak electron densities in each of those bodies having an ionosphere is provided in Table 2.6. As might be expected, Venus and Earth, having the most significant atmospheres of the terrestrial planets, and also being relatively close to the Sun, have the highest ionospheric electron number density.

If a planet also possesses a magnetic field, then this will influence the properties of the ionosphere. As noted for the Earth, a magnetic field provides an opportunity to both trap and add energy to charged particles, which in general may originate from the planetary atmosphere/ionosphere or from the solar wind. The magnetic properties of the major solar system bodies are listed in Table 2.7. As we have already noted for the Earth, its magnetic field leads to the formation of the Van Allen radiation belts. The most significant magnetosphere in the solar system is however that of Jupiter, which acts as an energy source for the generation of relativistic particles that radiate by synchrotron emission. Whilst Jupiter's magnetosphere is far more complex than that of the Earth, its overall structure is similar to that shown in Figure 2.12. The magnetosphere for Jupiter is vast—the magneto-tail extends to the orbit of Saturn. This magnetic field co-rotates with Jupiter, which has a period of just less than 10 hours. This provides an additional source of energy to accelerate particles. Further, the satellite Io passes through the plasma sheet formed within the magnetosphere. Io contributes ions, principally from the dissociation

Table 2.7   Solar system data

Planet/ Moon	Number of moons	Presence of rings	Magnetic field-Dipole moment (Relative to Earth)	Magnetopause height (units of $R_{equator}$)	Major missions
Mercury	—	—	0.0007	1.5	Mariner 10
Venus	—	—	<0.0004	—	Mariner 2/5/10 Venera 3–16 Pioneer Venus 1/2
Earth	1	—	1	10	—
Mars	2	—	<0.0002	—	Mariner 4/6/7/9 Mars 2/3 Viking 1/2 Mars Pathfinder
Jupiter	16	Yes	20 000	80	Pioneer 10/11 Voyager 1/2 Galileo
Saturn	18	Yes	600	20	Pioneer 11 Voyager 1/2 Cassini
Titan	N/A	—	None detected	—	Cassini/Huygens probe
Uranus	17	Yes	50	20	Voyager 2
Neptune	8	Yes	25	25	Voyager 2
Pluto	1	—	Unknown	Unknown	None

Notes: See also Tables 2.5, 2.6 and 4.1.

products of SO_2, to this plasma sheet from intense volcanic activity on its surface. At this stage, it is not clear however whether Io has its own magnetic field. The overall interaction yields resultant particle fluxes that are several orders of magnitude greater than those at the peak flux of the Earth's own radiation belts. These particles in the Jovian system, coupled with the motion of Io's motion through the magnetosphere also provide an intense electromagnetic radiation source, extending from VLF ~10 MHz through to 3 GHz. Not surprisingly therefore, all missions that transit the Jovian magnetosphere have shown signs of radiation damage in electronic circuits and the darkening of exposed optical systems. The radiation dose absorbed on the two grand-tour Pioneer spacecraft was a thousand times beyond the level of human tolerance [24]. As a result the Jovian system, having the additional feature common with the other gas giants of a ring system, proves to be one of the most hostile environments in the solar system.

2.4   ENVIRONMENTAL EFFECTS ON DESIGN

2.4.1   Effects on materials

Outgassing or *sublimation* refers to the vaporization of surface atoms of a material when it is subjected to an ambient pressure that is comparable with its own vapour pressure. Such a pressure, 10^{-11}–10^{-15} Pa, occurs at spacecraft altitudes. This process occurs at an increasing rate as temperature rises. A summary of rates for different metals at a variety of temperatures is shown in Table 2.8 but they do not apply to alloy materials or those having a covering surface layer of a different material. Since the precise surface conditions of orbiting vehicles are difficult to determine, particularly for vehicles in LEO, exact figures for mass loss are not at present available. Whilst structural problems arising because of outgassing are unlikely, the subsequent deposition of the material is hazardous to both optical and electrically sensitive surfaces. Thin plastic layers and oxide coatings are particularly sensitive to mass loss, especially if the material is used for its thermal properties; if mass loss is associated with specific constituents then modified emissive properties may result.

For plastic materials, high-vapour-pressure components evaporate rapidly, although initially mass loss is usually associated with the loss of adsorbed gases and water vapour.

Traditional lubricants used on the ground are clearly not appropriate to spacecraft operation. Generally, they have high vapour pressure and would thus outgas rapidly, but in many the lubricative action arises because of the presence of adsorbed gases and water vapour. Whilst low-volatility oils are used, solid lubricant coatings such as MoS_2 are frequently adopted. Reference [11] provides a detailed listing of NASA recommended lubricants (but see also Chapter 15).

Atomic oxygen erosion

The atmospheric composition shown in Figure 2.10 indicates that atomic oxygen forms the major atmospheric species in LEO. Following the severe erosion noted for Kapton on STS3 [25], it has become apparent that atomic oxygen provides an aggressive environment for materials used on space vehicles in LEO. This arises not only from its chemical activity, but most significantly from the fact that its atoms are travelling at ~8 km/s relative to the vehicle (due to the vehicle's orbital velocity). Silver is one of the few metals attacked by this environment, so due to its extensive use on solar arrays, it is important to avoid bare silver exposure.

The interactions between the oxygen atoms and the spacecraft surfaces are erosion, formation of stable oxide, scattering or reflection and chemiluminescent glow. When erosion takes place volatile products are formed, causing surface recession. Erosion rates are usually quoted in terms of yield in units of 10^{-24} cm^3/atom. Erosion yields range from 0.01 to 0.09 for materials like aluminium-coated kapton, FEP teflon and silicones at the low end, to between 2 and 4 for polyethylene, Kapton-H and Z-302 (glossy black) [26]. For a comprehensive compilation of yields, see Reference [27].

Table 2.8 Temperature for given sublimation rate ($^{\circ}$C)

Element	0.1 μm/yr	10 μm/yr	1 mm/yr
Cd	38	77	122
Zn	71	127	177
Mg	110	171	233
Au	660	800	950
Ti	920	1070	1250
Mo	1380	1630	1900
W	1870	2150	2480

The net effect of this erosion interaction is to degrade the material properties (optical, thermal, mechanical and electrical) irreversibly, and so this must be taken into account when choosing materials for optical and thermal control coatings, structural material, thermal blankets, solar panels and optical components. Analysis of the NASA Long Duration Exposure Facility (LDEF), which spent 5.8 years in LEO and was retrieved in January 1990, has increased our understanding of the mechanism of atomic oxygen degradation.

Stable oxide formation can lead to significant property changes. For example, molybdenum disulphide, which is a lubricant, can oxidize to create an abrasive oxide [28]. Dimensional changes can also occur, leading to cracking in the formation of a surface oxide layer on silicone (due to contraction and expansion), and spalling in the case of oxidation of silver.

Indirect impingement caused by ram atoms scattered from adjacent surfaces can lead to erosion of surfaces 'shadowed' from direct attack [27]. The chemiluminescent glow at visible wavelengths, observed on the Space Shuttle and around small spacecraft [29], is currently thought to be caused by surface mediated $O + NO$ recombination, forming excited NO_2 molecules that then radiate [30].

Some work has been done on the use of protective coatings that are resistive to the attack of atomic oxygen [26], but the problems of either manufacture-induced pinholes or those created by debris/micrometeoroid impact must be overcome.

Simulation of the atomic oxygen environment has become particularly important and there is much active research due in part to the material requirements for the international space station (ISS). Further information on this critical feature of the environment may be found in References [31] and [32].

Material strength and fatigue life are also affected by a high-vacuum environment. Generally, fatigue life is improved. For many materials, it may be extended by more than an order of magnitude [33], although in some cases, for example, pure nickel and Inconel 550, the improvement is uncertain [33,34]. The physical reasons for such changes in mechanical properties, and also changes in strength, are believed to arise because of one of two principal reasons. One of these is that gases absorbed into surface cracks either aid or hinder crack propagation. The second possible physical cause is that oxidation and gas diffusion absorbed into the material bulk influences material properties. A well-documented example of a material whose strength is improved in vacuum is that of glass, wherein a threefold improvement in strength has been noted at a pressure that is one thousandth that of atmospheric pressure.

Embrittlement is a form of material damage that is caused by exposure to UV radiation. Many polymers are particularly sensitive to such photons, whose energy is high enough to modify their chemical bonding structure.

Ultraviolet exposure also causes *electrical changes* in the form of resistivity modification, and *optical changes* affecting both thermal characteristics and opacity. A particularly UV-sensitive element is the solar array. More specifically the solar cell coverglass and its attendant adhesive are subject to darkening. This results in reduced cell illumination and an enhanced operating temperature, both being deleterious to cell operation (see Chapter 10).

Radiation damage affects all materials to some extent, and man. High-energy corpuscular radiation is experienced most severely in the Van Allen radiation belts, but is also at a significant level in any space operation. Under even modest radiation doses, some metals such as cadmium and zinc may form metallic whiskers.

Semiconductor materials and biological tissue are particularly sensitive to damage caused by high-energy charged and neutral particles. This generally arises because of both the displacement of atoms from crystal lattice sites and the attendant local ionization, together with an ionization track caused by the penetrating radiation. Protection is a complex task, since screening material will itself generate secondary radiations due to the passage of a primary high-energy particle through it. It should be noted that 'heavy' particles such as protons and neutrons cause significantly more damage than light ones such as electrons. They cause a dense track of ionization through a material until they are stopped by an atomic collision. When this occurs, a significant displacement of the atom arises, leading to further ionization, which does not occur for electrons.

Radiation shielding analysis

To ensure that electronic components such as transistors, diodes and so on are capable of surviving the radiation environment, the total dose inside the spacecraft, in rads has to be calculated. This is done for the specific orbit in question by first calculating the fluence spectrum of the external (unshielded) particles—the number of particles/cm^2 at specific energies. The shielded environment is then determined by transporting the external fluence through the spacecraft mass. Usually, the first step is to calculate a one-dimensional dose—depth curve based on a spherical shell geometry of aluminium. The dose behind a thickness typical of the spacecraft (usually about 2.5 to 3.8 mm) is read from this curve. This dose is then used with some design margin, typically between 1.3 and 2, to set the rad hardness requirement for electronic parts. If parts to be used fall below this requirement, several options are available; a more detailed three-dimensional analysis can be carried out to determine the dose at the actual location of the 'soft' parts. This usually results in the lowering of the dose with the inclusion of more individual spacecraft components further reducing the dose values. If the dose is still too high, then spot shielding can be implemented (i.e. the placement of a shield of tantalum or tungsten at the location of the actual part), or a rad-hard version of the part can be sought.

Other radiation effects including single-event phenomena, noise in charge-coupled devices CCDs (and optical detectors), and radiobiological interactions also need to include the transport of the primary particles through the spacecraft structure. Whilst, in general, shielding is beneficial and reduces the hazard caused by the radiation, in some cases large thicknesses of shielding can worsen the effects. For example, in considering single-event effects (SEEs), a heavy ion passing through a certain thickness of material will be slowed down to such an extent that its linear energy transfer (LET), and therefore its ability to produce ionization, is increased. Consideration must also be given to secondary particles produced as a result of the interaction of the primaries with the atoms of the shielding material. For example, secondary neutrons produced by proton fluxes can pose a problem for manned missions in orbits where the proton flux is high (trapped proton fluxes at altitudes greater than 500 km, and solar protons in high inclination orbits). Protons can also be responsible for SEEs in the South Atlantic Anomaly and in polar regions during solar flares, by producing secondary recoiling nuclei and light fragments [35].

The production of radioactive materials, which is called activation, can also be a problem. This may lead to dose problems on long, manned missions such as the Space Station, and can also induce limiting background noise levels in low-level measurements made by radiation detectors.

In semiconductor materials, two damaging effects have been noted. Primarily, radiation damage reduces the effectiveness of semiconductor operation. Specifically in solar cells, it results in a reduction in the efficiency of conversion from sunlight to electrical energy. In order to quantify the effect of this, the spacecraft designer usually resorts to tables of fluence for particular orbits. The upper, exposed surface of the cell may be protected to some extent by the use of a coverglass. The lower surface is generally protected by the substrate on which it is mounted. Manufacturers' calibration data is generally available to enable the expected degradation in solar cell performance to be assessed during the mission lifetime. This topic will be returned to in greater detail in Chapter 10.

Single event effects (SEEs)

In addition to total dose effects, which are caused by the deposition of energy by many particles, there is a set of phenomena that are caused by single particles. A SEU occurs when a heavy ion is incident on the sensitive area of an integrated circuit, producing sufficient charge in the form of electron—hole pairs to cause a change in the logic state of the device. This type of error is known as a *'soft' error* since it is reversible and causes no permanent damage. However, if it occurs in critical circuitry such as a control system or decision-making logic, then it can have serious consequences on the spacecraft operation—generating false commands such as thruster firings.

A single-event latch-up (SEL) occurs when the passage of a single charged particle leads to a latched low impedance state in parasitic PNPN devices in bulk complementary metal oxide semiconductor (CMOS) material, and it can result in burn-out. A more serious effect is single-event burn-out, which occurs when an incident ion produces a conducting path—in a metal oxide semiconductor field effect transistor (MOSFET), for example, that causes the device to latch-up; if this condition continues for a sufficiently long time then the device could be completely burnt out and destroyed.

Finally, single-event-induced dark current is caused by the passage of a particle which causes displacement damage in a single pixel.

Particles that cause SEUs include heavy ions, protons and neutrons. Both galactic cosmic rays and solar flares contain these. High-energy trapped protons can also cause SEUs, not by direct ionization but by the recoiling heavy reaction products. The parameter that characterizes a particle's ability to cause a SEE is its energy deposition rate, known as dE/dx, stopping power, or LET, and is measured in units of $MeV/g\,cm^2$ or $MeV/\mu m$.

SEU sensitivity is largely determined by the critical charge for a sensitive node, which is the quantity of charge needed to alter the device's memory status. The critical charge is a function of the feature size and roughly scales with the square of its dimension. SEU rates can be determined by combining the LET spectrum of the environment (i.e. the number of particles with a given energy loss) and the device dimensions.

In general, systems approaches to SEU-hardening can be divided into three categories: error toleration, error correction and error prevention. The first is costly and involves specifying tolerance levels for various parts of the system, determining maximum permissible error rates, and designing and constructing each part of the system within these constraints. Tests must be conducted at the subsystem and at the total system levels to ensure that the tolerance levels are met and that catastrophic failure is impossible. Error-correction techniques applied at the systems level include: redundant units, self-checking circuits, error-detecting and error-correcting codes, and serial calculation with error correction (hardware solution), concurrent programme execution and breakpoint reasonableness testing, checkpoint storage and roll-back for recovery (software solution), and repetitive execution and watchdog timers (time-related solutions) [36]. The last approach involves the choice of components that will not upset; however, there are only a limited number of SEU-hard devices for very large scale integrated (VLSI) levels of complexity and speed.

The potential for software errors is becoming more apparent as the processing power of on-board computers is increasing. This is a challenging area of present spacecraft design for which solutions are being sought.

2.4.2   Effects on man

The removal of Man from his natural environment or habitat necessarily introduces deleterious effects. A suitable life-support system will guard against the more obvious ones such as lack of atmosphere, his thermal environment and possible high-acceleration loads. Details applicable to the design of a suitable life-support environment may be found in References [37] and [38]. Here we consider the environmental problems for which a life-support system offers no solution. It should be noted that the response of Man to extended exposure to the space environment is neither well documented nor well understood, principally owing to the modest experience of manned flight to date. A review may be found in Reference [39].

In general terms, the response of Man to spaceflight may be classified under one of four categories: transient effects lasting for short periods on initial exposure to, or removal from, the space environment; flight duration adaptation; cumulative effects dependent upon the length of the flight, which are reversible on return to the ground; cumulative effects that are irreversible.

The dominant effect of the space environment, which cannot be removed through life-support systems, is zero or microgravity. This causes a major disturbance to the human system, with effects that fall into each of the first three categories above. The most notable of these are the following:

1. *Blood volume redistribution*. On Earth the blood pressure of a person whilst standing decreases with height above the feet. Typically the pressure in the brain is only one third that in the feet. The immediate effect of a 'zero g' environment is to cause a redistribution of the blood volume, resulting in a 'puffy' face. Whilst on Earth, the body system is familiar with orientation changes (lying down, etc.), and a complex system of hormonal secretion results in a control system that adjusts heart rate to these conditions. In zero gravity, however, this hormonal action appears to result in a loss of sodium with consequent fluid loss. Typically stabilization occurs after ~4 days when a loss of some 2–4 kg has occurred. Readaptation on return to Earth occurs over a short period of time, during which giddiness may be expected.

2. *Muscular atrophy*. Atrophy of all muscles occurs through long periods of inactivity. In zero g the heart itself is required to do less work since it is not pumping blood against gravity. Reduction in both heart muscle mass and heart rate occurs. It is believed that anaemia (a reduction in red blood cell count) is an additional side effect that has been noted in astronauts. To combat general atrophy of muscles astronauts spend much of their time in space undergoing physical exercise.

Table 2.9 Radiation dose in space-laboratory-type orbit (free space)

Source	Dose rate
Galactic radiation	0.01–0.05 rad/day ($\sim$0.3 rem/day)
Radiation belts	protons: 1–10 rad/h [behind 1 cm Al] electrons: 10^2–10^3 rad/h [at surface of S/C] ($\sim$1–30 rem/h)
Solar flares	12–350 rad/event ($\sim$10 500 rem)

Note: Mean dose on earth 250–300 millirem/year.

3. *Vestibular problems.* The human vestibular system is dependent upon both visual and inner ear sensors. In the inner ear the sensor has two orthogonal components, horizontal and vertical, which rely on gravity for their operation. Removal of gravity results in enhanced sensitivity to acceleration (including rotation) because of fluid motion in these sensors. Conflict between visual and inner ear sensors therefore arises during motion of the astronaut. The most notable effect is motion sickness, but astronauts also experience enhanced clumsiness. Adaptation generally takes three to four days.

4. *Locomotor system.* The major effect of zero gravity on the locomotor system is cumulative bone decalcification. This directly results in bone fragility and indirectly leads to problems of recalcification, external to the bones in, for example, the kidneys, forming kidney stones. High calcium diets do not appear to stop this increased calcium mobility and on long duration flights such as Salyut 6, a 2 to 8% loss of calcium was noted. Restabilization occurs on return to Earth.

The corpuscular radiation environment noted in preceding sections evidently provides a hazardous environment for Man. Shielding materials must be chosen with care, as poor design can lead to a secondary radiation that is more hazardous than the primary radiation [37]. Typical unscreened radiation dose rates are indicated in Table 2.9. From this it is evident that solar flares provide a particularly hazardous environment for Man, and thus permanently manned space stations will require some type of flare shield for personnel protection. For a geostationary orbit base, it has been estimated [40] that an equivalent shield protection of 21 g/cm^2 of aluminium would be required. This would result in a mass of $\sim$3.6 tons for a six-man habitat. For comparison purposes, it should be noted that for the Apollo command module's lunar flights, the shield, made of aluminium, stainless steel and phenolic epoxy, provided a shield thickness equivalent to 7.5 g/cm^2 of aluminium. The maximum radiation dose experienced was $\sim$1.0 rad. Radiation dose in Man is a cumulative hazard that cannot be reversed on return to the ground.

ACKNOWLEDGEMENT

Thanks is due to Dr. Stephen Gabriel for compiling material for this chapter.

REFERENCES

[1]   Rauschenback, H. S. (1980) *Solar Cell Array Design Handbook*, Van Nostrand, New York.
[2]   Ariane 4 Users Manual Issue 1 (1983) Arianespace.
[3]   Ariane 5 Users Manual (1998) Arianespace.
[4]   Pegasus Users Guide, available at www.orbital.com/launchvehicles/pegasus/peg-user-guide.pdf.
[5]   Lufti, M. S. and Neider, R. L. (1983) NASA CP 2283, 231.
[6]   NASA TM 82478 (1982) Volume 1.
[7]   US Standard Atmosphere (1976) NOAA, Washington, DC.
[8]   Hedin, A. E. (1991) Extension of the MSIS thermospheric model into the middle and lower atmosphere, *J. Geophys. Res.*, **96**, 1159.
[9]   Jacchia, L. G. (1977) SAO Special Report 375.
[10] King-Hele, D. G. (1987) *Satellite Orbits in an Atmosphere*, Blackie, London.
[11] NASA SP-8021 (1973).
[12] Garrett, H. B. (1979) *Rev. Geophys. Space Phys.*, **17**, 397.
[13] Daly, E. J., Adams, L., Zehnder, A. and Ljungfelt, S. L. (1992) ESA's radiation monitor and its theological role, IAG-92-0799, *43rd Congress of the International Astronautical Federation*, 28 Aug to 5 Sept, Washington, DC.
[14] Vette, J. I. (1991) The AE-8 Trapped Electron Model Environment, NSSDC-91-24, NASA-GSFC.
[15] Sawyer, D. M. and Vette, J. I. (1976) AP-8 Trapped Proton Environment for Solar Maximum and Solar Minimum, NSSDC-76-06, NASA-GSFC.
[16] Mewaldt, R. A., Spalding, J. D. and Stone, E. C. (1984) The isotopic composition of the anomalous low energy cosmic rays, *Astrophys. J.*, **283**, 450.
[17] Feynman, J. L., Armstrong, T. P., Dao-Gibner, L. and Silverman, S. M. (1990) Solar proton events during solar cycles 19, 20 and 21, *Solar Phys.*, **126**, 385.
[18] Feynman, J. and Gabriel, S. (eds) (1988) *Interplanetary Particle Environment: Proceedings of Conference*, JPL Publication 88-28, Jet Propulsion Laboratory, Pasadena, USA.
[19] Garrett, H. B. and Gaunt, D. M. (1980) *Prog. Astro. Aero.*, **71**, 227.
[20] Pilkington Space Technology (1984) Coverglass Specification PS 292.
[21] Gault, D. E. (1970) *Radio Sci.*, **5**, 273.
[22] Kessler, D. J., Reynolds, R. C. and Anz-Meador, P. D. (1989) Orbital Debris Environment for Spacecraft Designed to Operate in Low Earth Orbit, NASA TM 100471.
[23] Lambert, M. (1993) Shielding against natural and man-made space debris: a growing challenge, *ESA J.*, **17**, 31–42.
[24] Van Allen, J. A. and Bagenal, F. (1999) Planetary magnetospheres and the interplanetary medium, in *The New Solar System* (4th edn), ed. Beatty, J. K., Peterson, C. C. and Chaikin, A. L., Cambridge University Press, New York.
[25] Leger, L. J. (1983) AIAA Paper No. AIAA-83-0073.
[26] Tennyson, R. C. (1993) Atomic oxygen and its effect on materials, in *The Behaviour of Systems in the Space Environment*, ed. DeWitt, R. N., Duston, D. and Hyder, S. K., Kluwer Academic Publishers, Dordrecht, p. 233.
[27] Banks, B. (1990) Atomic oxygen, *Proc. LDEF Materials Data Analysis Workshop*, NASA CP 10046.
[28] Cross, J. B., Martin, J. A., Pope, L. E. and Koontz, S. L. (1989) Oxidation of MoS_2 by thermal to hyperthermal atomic oxygen, *3rd Int. SAMPE Electron. Conf.*, **3**, 638.
[29] Garrett, H. B., Chutjian, A. and Gabriel, S. (1988) Space vehicle glow and its impact on space systems, *J. Spacecraft Rockets*, **25**, 321–340.
[30] Greer, W. A. D., Pratt, N. H. and Stark, J. P. W. (1993) Spacecraft glows and laboratory luminescence evidence for a common reaction mechanism, *Geophys. Res. Lett.*, **20**, 731–734.
[31] Fourth European Symposium on Spacecraft Materials in the Space Environment (1988), CERT.
[32] Brinza, D. E. (ed.) (1987) Proc. NASA Workshop on Atomic Oxygen Effects, JPL Publication 87-14, Jet Propulsion Laboratory, Pasadena, USA.
[33] NASA TN-D 2563 (1965).
[34] NASA TN-D 2898 (1965).

[35] Harboe-Sorensen, R., Daly, E. J., Underwood, C., Ward, J. and Adams, L. (1990) The behaviour of measured SEU at low altitude during periods of high solar activity, *IEEE Trans. Nucl. Sci.*, NS-37, 1938–1946.

[36] Nagle, H. T., Santago, P., Miller, T. K., McAllister, D. F. and Mehrotra, R. (1989) Fault tolerance for single events, *J. Rad. Effects Res. Eng.*, **6**, 2.

[37] NASA SP-3006 (1973).

[38] Sharpe, M. R. (1969) *Living in Space*, Aldus Books 1, London.

[39] ESA BR-17 (1984).

[40] NASA SP-413 (1977).

3 DYNAMICS OF SPACECRAFT

Peter W. Fortescue

*Aeronautics and Astronautics, School of Engineering Sciences,
University of Southampton*

3.1 INTRODUCTION

This chapter serves as a general introduction to the subject of the dynamics of bodies and sets a framework for the subjects of celestial mechanics and attitude control (Chapters 4 and 9). For both of these, Newtonian dynamics will provide a sufficient means of forecasting and of understanding a spacecraft's behaviour. The summary presented here is chosen with a view to its relevance to spacecraft.

The approach adopted is to develop an understanding of dynamics in two stages. The first is to express the dynamics of both translation and rotation in terms of the appropriate form of momentum—linear or angular. Momentum becomes an important concept, in terms of which it is relatively easy to determine the consequences of forces or moments.

The second stage is to interpret the momenta in terms of the physical movement—the velocities, linear and angular. This is straightforward for linear momentum since momentum and velocity are in the same direction. More difficult is the relationship between rotational movement and angular momentum.

3.1.1 Translation/rotation separation

One feature that is peculiar to spacecraft is that their translational (trajectory) motion is virtually independent of their rotational motion. This is due to the fact that the moments or torques that cause their rotation are not dependent upon their direction of travel, and the gravitational forces that determine their trajectory are not dependent on their attitude. Whilst this is not entirely true, it is approximately so, and spacecraft designers will normally aim to preserve this independence.

At a fundamental level it is convenient to think of a spacecraft as being a collection of those particles and bodies that lie within a closed surface S (Figure 3.1). This concept allows the surface to be chosen at one's convenience, to embrace the complete craft or

Spacecraft Systems Engineering (Third Edition). Edited by P. W. Fortescue, J. P. W. Stark and G. G. Swinerd
© 2003 John Wiley & Sons Ltd

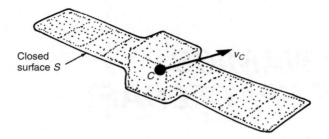

Figure 3.1 The closed surface S

just part of it or, when considering a docking manoeuvre or a tethered configuration, it may be chosen to embrace more than just the one spacecraft. But for the following, assume that it envelops just one spacecraft, unless stated otherwise.

The dynamics of the craft (or whatever is within the surface S) may now be described in terms of its momenta—its linear momentum **L** and its angular momentum **H**—the former leading to equations that describe the trajectory and the latter leading to equations that describe the attitude motion.

In particular, it is the centre-of-mass C whose trajectory will be described. It may seem to be pedantic to pick on one specific point for this purpose, but the centre-of-mass has special properties that lead to it being chosen for the development of Newtonian dynamics, properties that make it sensible to separate the motion into

1. the motion of the centre-of-mass, C and
2. the motion relative to the centre-of-mass.

For highly specialized purposes, such as when using the orbit to improve man's knowledge of the geoid, the accuracy with which the orbit is determined may be less than the dimensions of the spacecraft, and then it is no longer pedantic to consider to which point of the vehicle its orbit refers.

3.1.2   The centre-of-mass, C

The centre-of-mass of the particles in S, relative to an arbitrary point O, is the point whose position vector $\mathbf{r}_{OC}$ obeys

$$M\mathbf{r}_{OC} = \sum (m\mathbf{r}_{OP}) \tag{3.1}$$

where $\mathbf{r}_{OP}$ is the position vector of a general particle P (see Figure 3.2),
             m is the mass of the general particle
and M is the total mass within S.

When the location of C is being determined for a spacecraft, then it is useful to note that equation (3.1) can be applied to objects such as items of equipment rather than to particles. In that case, $\mathbf{r}_{OP}$ refers to the centre-of-mass of the item and m is its mass.

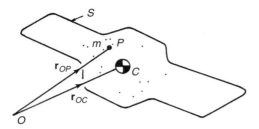

Figure 3.2   The centre-of-mass, C

When an object has a continuous mass distribution, the integral equivalent of equation (3.1) should be used, this being

$$M\mathbf{r}_{OC} = \int \mathbf{r}_{OP}\, dm \tag{3.2}$$

In the development of the theory of dynamics *the centre-of-mass C is useful as an origin or reference point* for the particles in S. Putting the origin there leads to $\mathbf{r}_{OC} = \mathbf{0}$, and so equations (3.1) and (3.2) become

$$\sum (m\mathbf{r}_{CP}) = \mathbf{0} \quad \text{and} \quad \int \mathbf{r}_{CP}\, dm = \mathbf{0} \tag{3.3}$$

Since this is true at all times, its derivatives are also zero, leading to

$$\sum (m\mathbf{v}_{CP}) = \mathbf{0} \quad \text{and} \quad \int \mathbf{v}_{CP}\, dm = \mathbf{0} \tag{3.4}$$

where $\mathbf{v}_{CP}$ is the velocity of a particle at P relative to the centre-of-mass C.

3.2   TRAJECTORY DYNAMICS

For trajectory purposes, it is convenient to treat a spacecraft as a particle, as is done in Chapters 4 and 5.

This *equivalent particle (e.p.)* has mass M equal to that of the spacecraft and is situated at its centre-of-mass C. Therefore, the e.p. moves with the velocity of C and its *momentum* $\mathbf{L}$ is the aggregate of all the particles and bodies that comprise the spacecraft (Figure 3.3).

$$\text{Momentum} \quad \mathbf{L} = M\mathbf{v}_C \tag{3.5}$$

This interpretation admits the possibility that the spacecraft may have moving parts or appendages. The surface S (Figure 3.1) may in fact be chosen to embrace any objects whose joint orbit(s) are of interest, although it will here refer to a single spacecraft or celestial body unless otherwise stated.

According to Newtonian mechanics—of a single particle—the only way of changing its momentum $\mathbf{L}$ is to apply a force. For a spacecraft—whose e.p. represents many particles—there are two ways:

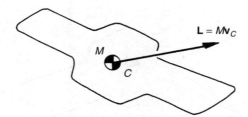

Figure 3.3 The equivalent particle at C

1. *Application of external forces.* These include gravitational attraction towards the heavenly bodies, solar radiation pressure, aerodynamic forces and so on.

Internal forces, acting between any two particles or bodies *within S*, will not produce any change in the total momentum. It follows that objects moving within a spacecraft, such as fuel, astronauts, mechanisms or flexing structures and so on, will not cause the total linear momentum to change. Likewise, the total momentum of two docking spacecraft will not be changed by any forces between them.

This mechanism applies to a spacecraft that has no propulsion; its characteristic is that the mass M remains constant.

2. *Ejection of some of the particles from within S.* This occurs during rocket propulsion. The ejected particles take away both their mass and their momentum, leading to a change in the residual mass and momentum of the spacecraft.

In the above text it is important to recognize that the momentum must in all cases involve the velocity relative to inertial space rather than to the spacecraft.

It is convenient to deal separately with the two cases above, that is, with and without propulsion.

3.2.1 Translational motion with no propulsion

The effect of external forces $\mathbf{F}_{\text{ext}}$ on the momentum can be forecast from a diagram such as Figure 3.4. Their effect will depend upon not only the magnitude and direction of the forces but also upon the magnitude of the momentum $\mathbf{L}$ when the forces act. The additional momentum $\delta\mathbf{L}$ produced during an interval of time δt is the additional momentum

$$\delta\mathbf{L} = \mathbf{F}_{\text{ext}}\delta t \tag{3.6}$$

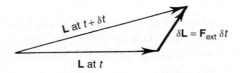

Figure 3.4 The incremental increase in momentum caused by an external force

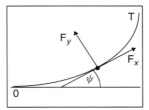

Figure 3.5 The force components along and normal to the trajectory

This must be added vectorially as shown in Figure 3.4 in order to see the effect upon the momentum **L** during the time interval.

Equation (3.6) may be written as the Newtonian equation:

$$\mathrm{d}\mathbf{L}/\mathrm{d}t = \mathrm{d}(M\mathbf{v}_C)/\mathrm{d}t = \mathbf{F}_{\text{ext}} \tag{3.7}$$

To establish the effect that an external force has upon the motion, it is best to resolve the force into two components—F_x along the trajectory and F_y at right angles to it—as shown in Figure 3.5. Then

1. F_x along the trajectory will change only the *magnitude* of the momentum, and hence only the speed—from v_0 to v_T say—where

$$M(v_T - v_0) = \int_0^T F_x \, \mathrm{d}t \tag{3.8}$$

2. F_y normal to the trajectory will change only the *direction* of the momentum. The rate at which the direction changes is $\mathrm{d}\psi/\mathrm{d}t = F_y/Mv$, and so a change of direction is best effected when the speed is least, such as at an apofocus. If the speed remains constant ($F_x = 0$), then during a time interval from $t = 0$ to T, the trajectory will turn as in Figure 3.5, through an angle ($\psi_T - \psi_0$), where

$$Mv(\psi_T - \psi_0) = \int_0^T F_y \, \mathrm{d}t \tag{3.9}$$

If F_y is constant, then the trajectory is an arc of a circle.

In Section 3.3.2, it will be seen that angular momentum **H** responds to a torque **T** in exactly the same way as linear momentum **L** responds to a force **F**; this also applies to the moment-of-momentum $M\mathbf{h}_I$ in response to a force whose moment is $\mathbf{M}_I$ (Section 3.2.3).

3.2.2 Moment-of-momentum *m*h and angular momentum H

Moment-of-momentum is a useful concept in celestial mechanics (Section 4.2 of Chapter 4). It is also a useful stepping stone towards angular momentum and the study of attitude motion.

The *moment-of-momentum* $m\mathbf{h}_O$ referred to a point O may be defined as

$$m\mathbf{h}_O = \mathbf{r} \times m\mathbf{v} \tag{3.10}$$

$\mathbf{r}$ is the position vector from O to any point on the line of action of the momentum vector $m\mathbf{v}$ (Figure 3.6) and $\mathbf{v}$ is the velocity relative to the reference point O. This definition is equivalent to the product of the momentum with the perpendicular distance from O to the momentum vector.

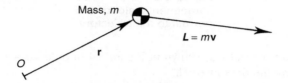

Figure 3.6 The situation leading to equation (3.10)

For the orbit of (the e.p. of) a spacecraft whose mass is M, it is useful to take the centre of gravitational attraction—the centre of the Earth for Earth orbits, for example—as the reference point, O say. Then the moment-of-momentum vector $M\mathbf{h}_O$ will be normal to the orbit plane (Figure 3.7).

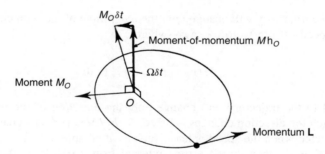

Figure 3.7 Precession of an orbit's plane due to a moment $\mathbf{M}_O$ lying in the plane. Its normal turns towards the direction of the moment vector

In the study of attitude dynamics, the aggregate of all the moments of the momenta of the particles comprising the spacecraft is a useful concept and it is defined as the *angular momentum*, $\mathbf{H}$. In this case the most useful reference point is the centre-of-mass, C. *This subject is covered in Section 3.3.1.*

Note that 'moment-of-momentum' $M\mathbf{h}_O$ is often loosely referred to as 'angular momentum' $\mathbf{H}_O$. This can lead to some confusion. Only in special cases are the two identical, such as when there is only one particle. In the more general case of a spacecraft its angular momentum referred to O is the sum of the moment-of-momentum of its e.p. referred to O and its angular momentum referred to C:

$$\mathbf{H}_O = M\mathbf{h}_O + \mathbf{H}_C \tag{3.11}$$

In practice, $\mathbf{H}_C$ is very much less than $M\mathbf{h}_O$, and it is neglected in orbit studies.

3.2.3   The rate of change of moment-of-momentum Mh_O

The moment-of-momentum of a particle $M\mathbf{h}_O$ *referred to an inertially fixed point O is only changed if the forces on it have a moment* $\mathbf{M}_O$ *about O.* The relationship may be expressed in the form of the Newtonian equation

$$\mathrm{d}(M\mathbf{h}_O)/\mathrm{d}t = \mathbf{M}_O \qquad (3.12)$$

This is also true for the e.p. of a spacecraft.

The forces on a spacecraft are conventionally separated into

(1.) the dominant central gravitational force and
(2.) the additional perturbative forces.

The former acts towards a point O that may be taken to be inertially fixed, such as the centre of the Earth for Earth orbiters, and so it contributes nothing towards the moment $\mathbf{M}_O$.

The similarity between equations (3.12) and (3.7) indicates that the moment-of-momentum responds to the moment $\mathbf{M}_O$ in the same way as linear momentum $\mathbf{L}$ responds to a force $\mathbf{F}_{\mathrm{ext}}$. When this is applied to an orbiting spacecraft, the consequences are

1. The component of $\mathbf{M}_O$ that is *normal* to the orbit's plane—in the same direction as $M\mathbf{h}_O$—will change the magnitude of the moment-of-momentum but not its direction, and so the orbit will remain in the same plane, but will change its shape. (See Section 4.4 of Chapter 4: Apsidal precession.)
2. The component of $\mathbf{M}_O$ *in* the orbit plane—at right angles to the moment-of-momentum $M\mathbf{h}_O$—will affect the direction of the normal to the plane, turning it towards $\mathbf{M}_O$ as shown in Figure 3.7. This represents a rotation of the orbital plane. The rate of turn Ω is given by

$$\Omega = M_O/Mh_O \qquad (3.13)$$

(See Section 4.4 in Chapter 4: Regression of the line of nodes, for example.)

3.2.4   The impulses

If a force $\mathbf{F}$ acts for only a limited duration—from $t = 0$ to τ, say—then its impulse $\mathbf{I}$ is defined as

$$\mathbf{I} = \int_0^\tau \mathbf{F}\,\mathrm{d}t \qquad (3.14)$$

From equation (3.6) it may be seen that $\mathbf{I} = \mathbf{L}_\tau - \mathbf{L}_0$, so that *an impulse is equal to the change in momentum that it causes.*

The impulse is used in situations in which there is an insignificant amount of movement during the time that the force acts. A collision is a common application. The firing of a rocket when in Earth orbit may also occur over a very small orbital arc, and if so then it may be treated as an impulse. A swing-by manoeuvre in heliocentric space is another application since the passage through the sphere of influence of a planet involves very little movement in the heliocentric orbit (see Chapter 5).

A torque impulse $\mathbf{I}_T$ is defined in terms of torque $\mathbf{T}$ in a manner similar to equation (3.14).

$$\mathbf{I}_T = \int_0^\tau \mathbf{T}\,dt \tag{3.15}$$

Since the relationship between torque $\mathbf{T}$ and angular momentum $\mathbf{H}$ (equation 3.28) is identical to that between force $\mathbf{F}$ and linear momentum $\mathbf{L}$ (equation 3.7), it follows that $\mathbf{I}_T = \mathbf{H}_\tau - \mathbf{H}_0$, so that *a torque impulse is equal to the change in angular momentum that it causes.*

The restriction that there is little movement—angular in this case—during the torque impulse means that collisions and landing impacts are the main situations in which the torque may be deemed to be sudden.

In the case of a body in motion—linear and/or rotational—the impulses are a measure of what would cause the motion from an initially stationary state. It follows that the negative of the impulses will bring the body to rest.

3.2.5   Translational motion under propulsion

When a rocket motor is fired, there is an expulsion of particles out of the spacecraft—out of the surface S referred to above. Each particle of the exhaust gases takes away its contributions to two of the properties of the e.p., its mass M and its momentum $M\mathbf{v}_C$—the total momentum within S.

If the rocket's mass flow is σ, then this is the rate at which the mass of the e.p. decreases, that is,

$$dM/dt = -\sigma \tag{3.16}$$

Also, the absolute velocity of the exhaust is $(\mathbf{v}_C + \mathbf{v}_{ex})$, and so the rate at which the momentum of the e.p. changes is

$$d(M\mathbf{v}_C)/dt = -\sigma(\mathbf{v}_C + \mathbf{v}_{ex}) \tag{3.17}$$

Here $\mathbf{v}_{ex}$ is the exhaust velocity relative to the centre-of-mass, taken to be positive in the direction of travel.

In the general case there will be an external force $\mathbf{F}_{ext}$ acting in addition, part of this being due to the back-pressure if the rocket operates in an atmosphere. Then the total rate of increase of momentum obeys the Newtonian equation

$$d(M\mathbf{v}_C)/dt = \mathbf{F}_{ext} - \sigma(\mathbf{v}_C + \mathbf{v}_{ex}) \tag{3.18}$$

From equations (3.16) and (3.18) it follows that the absolute acceleration $\mathbf{a}_C$ of the e.p. obeys

$$M\mathbf{a}_C = \mathbf{F}_{ext} - \sigma\mathbf{v}_{ex} \tag{3.19}$$

Clearly when the rocket is restrained, as on a test bed, there will be no acceleration, and the restraining force $\mathbf{F}_{ext}$ will be equal to the thrust, $\sigma\mathbf{v}_{ex}$.

The rocket equation is the integral of equation (3.19) for the case of no external forces, and the thruster points in a constant direction. The velocity increment $\Delta\mathbf{v}$ due to the burn is then

$$\Delta\mathbf{v} = -\mathbf{v}_{\text{ex}} \ln(M_0/M_1) \tag{3.20}$$

M_0/M_1 is the *mass ratio*, the ratio between the masses M_0 before and M_1 after the burn.

Equation (3.20) is frequently used as an approximation when the burn is short enough that it may be considered to be impulsive. Then, by pointing the rocket in different directions, it may be used to achieve an increase or decrease in speed, or a change of direction or any combination that is consistent with the vector diagram shown in Figure 3.8.

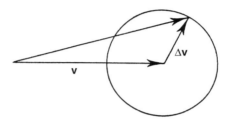

Figure 3.8 Possible changes of velocity due to a rocket's $\Delta\mathbf{v}$

A *course change* $\Delta\psi$, for example, without any change of speed, will occur if the initial and final velocity vectors in Figure 3.8 form an isosceles triangle with $\Delta\mathbf{v}$ as base. The angle turned through will be

$$\Delta\psi = 2 \arcsin\left(\tfrac{1}{2}|\Delta\mathbf{v}|/v\right) \tag{3.21}$$

and the thrust needs to be directed at an angle $(90 + \psi/2)$ degrees to the initial direction of travel. An assumption above is that the thrust vector passes through the centre-of-mass. If this is not so, then its moment about this point will produce a possible course change (see Section 3.3.2).

3.2.6 Translational kinetic energy

An energy equation may be obtained as the first integral of equation (3.7), and this relationship can be very useful in the subject of celestial mechanics.

The translational kinetic energy may be expressed as

$$\text{KE} = \tfrac{1}{2}Mv_C^2 \quad \text{or} \quad \tfrac{1}{2}M\mathbf{v}_C\cdot\mathbf{v}_C \quad \text{or} \quad \tfrac{1}{2}\mathbf{L}\cdot\mathbf{v}_C \tag{3.22}$$

The change in this is equal to the work done by the external forces $\mathbf{F}_{\text{ext}}$, and it may be expressed as

$$\Delta\text{KE} = \Delta\left(\tfrac{1}{2}M\mathbf{v}_C\cdot\mathbf{v}_C\right) = \int \mathbf{F}_{\text{ext}} \cdot d\mathbf{s} \tag{3.23}$$

where $d\mathbf{s}$ is the incremental change in position (see Figure 3.9).

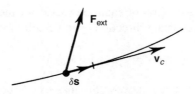

Figure 3.9

The conservative forces may be separated from the non-conservative ones since their contribution to the integral may be expressed in terms of potential energy (PE). The resulting energy equation becomes

$$\Delta KE + \Delta PE = \int \mathbf{F}_{nc} \cdot d\mathbf{s} \tag{3.24}$$

where $\mathbf{F}_{nc}$ is the non-conservative contribution to the total external force $\mathbf{F}_{ext}$.

When the only force on the spacecraft is that due to gravitational attraction, a conservative force whose potential energy is $-\mu M/r$, then the orbital energy equation becomes

$$\tfrac{1}{2} M \mathbf{v}_C \cdot \mathbf{v}_C - \mu M/r = \text{constant} \tag{3.25}$$

3.3 GENERAL ATTITUDE DYNAMICS

Trajectory dynamics supplies rules governing the motion of the centre-of-mass C relative to some inertially fixed frame of reference. Attitude dynamics on the other hand uses the centre-of-mass C as a reference point.

In terms of momentum, attitude dynamics is mathematically identical to trajectory dynamics. That is to say angular momentum $\mathbf{H}$ responds to a torque $\mathbf{T}$ in exactly the same way as linear momentum $\mathbf{L}$ does to a force $\mathbf{F}$. But the physical motions associated with the two types of momentum are quite different.

In order to establish the fundamental principles in terms of momentum, it is convenient to refer once more to the closed surface S (Figure 3.1), the boundary that separates particles that are of interest from those that are not. These results are then quite general, covering spacecraft with fluids and moving parts as well as the important cases of a rigid body and multiple bodies; the rules for the physical motions of the latter are dealt with later.

3.3.1 Angular momentum H

The angular momentum $\mathbf{H}_O$ referred to a point O is defined as the aggregate of the moments of the momenta of all the particles within S. In mathematical terms, angular momentum referred to O is defined as

$$\mathbf{H}_O = \sum (\mathbf{r} \times m\mathbf{v}) \tag{3.26}$$

where both $\mathbf{r}$ and $\mathbf{v}$ are relative to O. It is a product of rotation in much the same way as linear momentum is a product of translation.

In physical terms the angular momentum of a rigid body is a measure of the torque impulse that is needed to create its rotational motion. Thus the rotation of a body will be brought to rest by the application of a torque impulse that is equal and opposite to its angular momentum $\mathbf{H}$.

The reference point that is most useful for attitude dynamics is the centre-of-mass C. For example, the rotational motion of a spacecraft is normally derived from equation (3.28), using its momentum $\mathbf{H}_C$ referred to its centre-of-mass.

The rule governing the transfer of reference point from C to some other point O is

$$\mathbf{H}_O = \mathbf{H}_C + M\mathbf{h}_O = \mathbf{H}_C + (\mathbf{r} \times M\dot{\mathbf{r}}) \tag{3.27}$$

Figure 3.10 illustrates the terms used. The transfer involves adding on the moment-of-momentum of the e.p., a process similar to the transfer of inertia from C to O (Appendix, equation 3.A9). Equation (3.27) could be used, for example, for expressing the contribution to a spacecraft's total angular momentum referred to *its* centre-of-mass, which arises from a momentum wheel (see also equation 3.35).

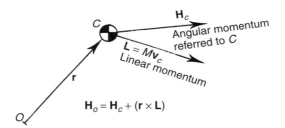

Figure 3.10   Transfer of reference point from C to O

3.3.2   Rate of change of angular momentum, H

The following explanation applies only when the reference point is the centre-of-mass C or an inertially fixed point I. It does not apply for other points.

The angular momentum may be changed in two ways, the first being

- by applying an external couple or a force that has a moment about the reference point. The term 'external torque' ($\mathbf{T}$) will be used as a general term to cover both of these. Its consequence is a rate of change of $\mathbf{H}$, referred to C or I, described by the Newtonian equation

$$d(\mathbf{H})/dt = \mathbf{T} \tag{3.28}$$

Internal torques, acting between particles or bodies, will *not* change the total momentum. Thus mechanisms, fuel movement and so on will not change the total angular momentum of a spacecraft. Similarly, the forces between two docking spacecraft will not affect their combined angular momentum.

There is an important consequence for spacecraft designers since there will always be naturally occurring external disturbance torques (see Section 9.4 of Chapter 9). Their mean level will therefore cause a progressive build-up of the angular momentum over the lifetime of the craft. The rotational motion associated with this would be quite unacceptable. It follows that spacecraft *must* be fitted with means of controlling this build-up, and only external torquers are capable of doing so.

The second means of changing **H** is

● by the ejection from *S* of some particles whose momenta have moments about the reference point. This will occur during the firing of rockets, when their thrust vector does not pass precisely through the centre-of-mass. An analysis of this process will not be covered in this book. (A procedure similar to that used above in the development of equations (3.16), (3.17) and (3.20) can be followed if required.)

Equation (3.28) takes the same form as equation (3.7), and so the effect that a torque **T** has upon angular momentum **H** is the same as the effect that a force **F** has upon linear momentum **L**, stated as follows:

1. The component of a torque **T** that is in the same direction as the angular momentum **H** will change only the magnitude of the momentum.
2. The component T_N of the torque **T** that is at right angles to the angular momentum **H** will cause it to change direction towards the direction of the torque **T**.
   In this case, the torque impulse during a small interval of time δt is $T_N \, \delta t$ and is the change in the angular momentum during this time (see equation 3.15). This is shown in Figure 3.11.

The best-known physical illustration of this is the precession of a gyroscope. From Figure 3.11 it may be seen that the rate of precession is

$$\mathrm{d}\psi/\mathrm{d}t = T_N/H \tag{3.29}$$

As the angular momentum **H** becomes very large, the effect of a torque impulse becomes less, leading to the property known as *gyroscopic rigidity*.

This characteristic is made use of by spacecraft designers when they give their craft *momentum bias*, as a means of making the bias direction insensitive to disturbance torques. Making **H** large causes the precession rate to become small in response to a given torque.

The rate of change of the components of **H** depends upon the rotation of the axis system chosen, in addition to the change described by the Newtonian equation (3.28). When the

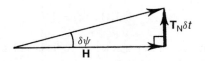

Figure 3.11 Precession due to an external torque

axis system has an angular velocity $\mathbf{\Omega}$ relative to inertial space, then equation (3.28) must be interpreted using the Coriolis theorem, as

$$\mathrm{d}(\mathbf{H}_C)\mathrm{d}t = \mathrm{d}/\mathrm{d}t_{\mathrm{compts}}(\mathbf{H}_C) + (\mathbf{\Omega} \times \mathbf{H}_C) = \mathbf{T} \tag{3.30}$$

where $\mathrm{d}/\mathrm{d}t_{\mathrm{compts}}(\ )$ means the rate of change of the components of $(\ )$.

This equation is used in the remainder of this chapter in order to develop the scalar dynamic equations for the rotational motion of spacecraft.

3.3.3 Angular momentum of rigid bodies

The angular momentum of a rigid body may be expressed in terms of its angular velocity $\boldsymbol{\omega}$. The angular momentum equations given above can then be used in order to develop equations in $\boldsymbol{\omega}$ that describe the attitude and rotational motion of such a body, and of systems made up of rigid bodies.

The angular momentum $\mathbf{H}_C$ of a single rigid body referred to its centre-of-mass C may be expressed as

$$\mathbf{H}_C = [I_C]\,\boldsymbol{\omega} \tag{3.31}$$

where $\boldsymbol{\omega}$ is its angular velocity relative to an inertial (non-rotating) frame of reference and $[I_C]$ is the inertia matrix based upon the centre-of-mass C.

In general, $[I_C]$ may be expressed as

$$[I_C] = \begin{bmatrix} I_{xx} & -I_{xy} & -I_{zx} \\ -I_{xy} & I_{yy} & -I_{yz} \\ -I_{zx} & -I_{yz} & I_{zz} \end{bmatrix} \tag{3.32}$$

where I_{xx}, I_{yy} and I_{zz} are the moments of inertia.

I_{xy}, I_{yz} and I_{zx} are the products of inertia, broadly representing a measure of the lack of mass symmetry, leading to cross-coupled behaviour, as will be seen later.

The Appendix to this chapter contains a full treatment of the subject of inertia, including the dependency of the elements upon the choice of axes. It is sufficient to record here that every body has a set of orthogonal axes at each point, for which the products of inertia are zero. These are called *principal axes*, and there may well be more than one set of them, depending upon the mass symmetry of the body. Principal axes are eigenvectors of the inertia matrix.

It is evident from equation (3.31) that the components of the angular momentum may in general be expressed as

$$\mathbf{H}_C = \begin{bmatrix} (I_{xx}\omega_x - I_{xx}\omega_y - I_{zx}\omega_z) \\ (I_{yy}\omega_y - I_{yz}\omega_z - I_{xy}\omega_x) \\ (I_{zz}\omega_z - I_{zx}\omega_x - I_{yz}\omega_y) \end{bmatrix} \tag{3.33}$$

When principal axes are used, then

$$\mathbf{H}_C = \{I_{xx}\omega_x, I_{yy}\omega_y, I_{zz}\omega_z\}^{\mathrm{T}} \tag{3.34}$$

It is usual to develop the rotational equations for a body by choosing axes in which the inertias are constant. Axes fixed in the body will always achieve this, but there are other options if there is a mass symmetry. For example, a spinning object that has cylindrical mass symmetry about its spin axis, such as a typical wheel, will have constant inertias in non-spinning coordinate axes.

The angular momentum of a rigid body with spinning wheels, such as a spacecraft fitted with momentum or reaction wheels, can be expressed as the sum of the angular momentum of the rigid body containing the wheels in their non-spinning state (equations 3.33 or 3.34), together with the extra momentum due to the angular velocities of the wheels *relative to the body*.

For example, suppose a spacecraft is fitted with a wheel that is spinning with an angular velocity ω_{wh} relative to the craft. The vector ω_{wh} will be along the wheel's axis of symmetry, which will be in a fixed direction in the spacecraft's axes. If its moment of inertia about its own axis is I_{wh}, then an additional angular momentum equal to $I_{wh}\omega_{wh}$ must be added to that of the spacecraft's body. If there are several wheels and as a whole they contribute the additional momentum components $\{H_x, H_y, H_z\}^T$, say, then the total angular momentum of the body plus wheels becomes

$$\mathbf{H}_C = \begin{bmatrix} I_{xx}\omega_x - I_{xy}\omega_y - I_{zx}\omega_z + H_x \\ I_{yy}\omega_y - I_{yz}\omega_z - I_{xy}\omega_x + H_y \\ I_{zz}\omega_z - I_{zx}\omega_x - I_{yz}\omega_y + H_z \end{bmatrix} \tag{3.35}$$

If the body's principal axes are used, then this becomes

$$\mathbf{H}_C = \{(I_{xx}\omega_x + H_x), (I_{yy}\omega_y + H_y), (I_{zz}\omega_z + H_z)\}^T \tag{3.36}$$

Multiple rigid bodies of a more general nature need to be addressed in order to deal with the docking manoeuvre, ejecting payloads from a launcher's cargo bay, astronauts repairing spacecraft, tethered spacecraft and so on and indeed to the Earth/Moon combination. For these multi-body situations, each separate body will obey the dynamic equations that are stated elsewhere in this chapter. But it is worth noting that their combined momenta will not be changed by any forces or moments of interaction between them, and so will not be affected by a collision, separation or tethering. This applies to their absolute linear momentum, and also to the angular momentum about their combined centre-of-mass, C.

For example, consider two bodies with masses M_1, M_2, as in Figure 3.12, and total mass M. Assume that their centres-of-mass at C_1, C_2 have absolute velocities $\mathbf{v}_1$, $\mathbf{v}_2$ and that referring to these, they have angular momenta $\mathbf{H}_1$, $\mathbf{H}_2$.

Then, during a collision or separation or due to a tether

- their total absolute linear momentum remains constant, so

$$M\mathbf{v}_G = M_1\mathbf{v}_1 + M_2\mathbf{v}_2 \text{ remains constant} \tag{3.37}$$

- their angular momenta referred to C remains constant, so

$$\mathbf{H}_G = (M_1M_2/M)(\mathbf{r}_{12} \times \mathbf{v}_{12}) + \mathbf{H}_1 + \mathbf{H}_2 \text{ remains constant} \tag{3.38}$$

where $\mathbf{r}_{12}$, $\mathbf{v}_{12}$ are the position and velocity vectors of C_2 relative to C_1.

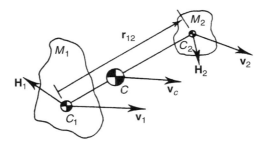

Figure 3.12 The combined momenta of two rigid bodies

These conservation of momentum laws also apply to the erection of solar arrays and so on, and when using pointing mechanisms.

3.3.4 Rotational kinetic energy

The rotational energy of a rigid body may be expressed as

$$E = \tfrac{1}{2}(\mathbf{H}_C \cdot \boldsymbol{\omega}) \text{ or } \tfrac{1}{2}([I_C]\boldsymbol{\omega} \cdot \boldsymbol{\omega}) \tag{3.39}$$

This is additional to the translational energy of the e.p. quoted in equation (3.22).

Work is done by a torque $\mathbf{T}$ at a rate $\mathbf{T} \cdot \boldsymbol{\omega}$, and the rate at which the rotational energy is increased by a torque $\mathbf{T}_C$ about C is $\mathbf{T}_C \cdot \boldsymbol{\omega}$. It follows that in the absence of any external torque $\mathbf{T}_C$ the rotational energy will remain constant.

When the body is not rigid, however, or when there are moving parts, the energy level can change in the absence of any external torque. In a spacecraft there will be internal dissipative mechanisms such as flexure, passive nutation dampers and so on that lead to the loss of kinetic energy, but at the same time the angular momentum $\mathbf{H}_C$ remains constant. A notable consequence of this is that the long-term tendency of a spinning body is towards spinning about its axis of maximum inertia, this being its minimum-energy state (see Section 3.4.2).

3.4 ATTITUDE MOTION OF SPECIFIC TYPES OF SPACECRAFT

The general theory in Section 3.3 may now be applied to specific types of spacecraft. For this purpose it will be assumed that all craft have rigid bodies, and have rigid moving parts unless otherwise stated. Whilst the response of their angular momentum to a torque is well ordered and straightforward, as explained above, the associated rotational motion is by no means necessarily so well behaved.

Spacecraft may be classified for convenience, as shown in Figure 3.13. The main subdivision depends upon whether the spacecraft has momentum bias or not. Bias means that the craft has a significant amount of angular momentum due to the spin of part or all

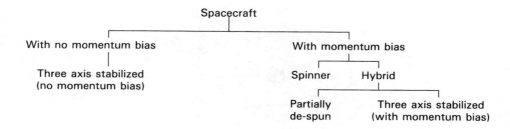

Figure 3.13 Categories of spacecraft

of it. It then behaves like a gyroscope, with the associated characteristics of gyroscopic rigidity and a precessional type of response to a torque, as described in equation (3.29).

Some spacecraft may be spun-up for only a short time. It is common practice to do so prior to the firing of a high thrust rocket, for example. This is primarily to prevent any thrust offset from causing the craft to veer off course, the spin causing the mean path to be straight. Subsequently, the craft will be spun-down to the level of bias that is required for normal operation.

*Cross-coupling** is an essential feature of precessional response. The precession axis is at right angles to the axis of the torque that caused it. It is nevertheless a systematic and orderly effect. Normally, large cross-couplings are to be avoided so that control about each axis can proceed without interfering with the other axes, and vice versa. The rigid body dynamic equations developed in this section enable the sources of cross-couplings to be identified.

3.4.1 Three-axes-stabilized spacecraft with no momentum bias

Spacecraft in this class are usually large, with extensive solar arrays. They have a variety of different shapes such as the notional one shown in Figure 3.14 or that of the Hubble telescope. Their angular velocity is normally small, perhaps one revolution per orbit in order to maintain one face pointing towards Earth. The solar arrays may have even less angular movement, being required to point towards the Sun, and they may be mounted on a bearing to permit this.

The craft will be treated as a single rigid body. This is a reasonable approximation for the main structure, but the solar arrays are very flexible (see Section 3.5.2). Nevertheless, the presentation brings out some important points that are relevant to the 'rigid body' responses of these craft.

Using principal axes that are fixed in the body, their angular velocity $\mathbf{\Omega}$ will be that of the body, $\boldsymbol{\omega}$. Inserting $\mathbf{H}_C$ from equation (3.34) into equation (3.30) leads to

$$\left.\begin{array}{l} I_{xx}\dot{\omega}_x - (I_{yy} - I_{zz})\omega_y\omega_z = T_x \\ I_{yy}\dot{\omega}_y - (I_{zz} - I_{xx})\omega_z\omega_x = T_y \\ I_{zz}\dot{\omega}_z - (I_{xx} - I_{yy})\omega_x\omega_y = T_z \end{array}\right\} \tag{3.40}$$

---

* 'Cross-coupling' here means that a cause about one axis, such as a torque about the pitch axis, produces a response about another axis—the yaw axis, for example—and vice versa. (In electrical circuits it implies an undesired effect.)

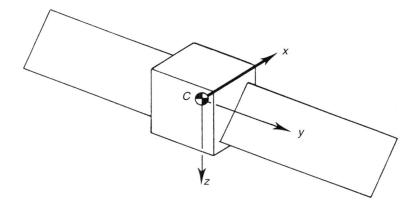

Figure 3.14 The three-axis-stabilized spacecraft

From these equations it may be seen that *cross-coupling* will potentially be present in attitude control systems (ACS) for a variety of reasons. The term is used when the acceleration in response to a torque does not take place solely about the torque axis.

For an initially stationary spacecraft a torque about a principal axis will produce a response about this axis, without any cross-coupling into the other axes. The response is an angular acceleration given by

$$\dot{\omega}_x = T_x / I_{xx} \tag{3.41}$$

Since torquers produce couples, their axes do not need to pass through the centre-of-mass, C.

In the more general case, when there are products of inertia, the initial acceleration from rest will be

$$\dot{\boldsymbol{\omega}} = [I_C]^{-1}\mathbf{T} \tag{3.42}$$

Cross-coupling will occur, and if a pure response $\dot{\omega}_x$ is required, then potentially all three torquers must be used.

Cross-coupled motion will also result if the x-component of torque is used when there is already an angular rate about the y- or z-axes. It must be assessed when a repointing manoeuvre is needed. It may be that the manoeuvre will best be carried out as a sequence of separate rotations about the principal axes, but there are a number of possible routes that may be taken in changing from one attitude to another.

In the absence of any external torque, a stationary spacecraft with no bias will remain in that condition indefinitely. Once it is rotating, its motion may become divergent. The stability of rotating motion is dealt with in Section 3.4.2.

3.4.2 Spinning spacecraft

When the whole of the spacecraft structure spins, it has angular momentum. At very low values this leads to minor cross-couplings such as are mentioned above. But when the momentum becomes large and deliberate it is called *momentum bias*, and then the cross-couplings become so great that the behaviour needs to be looked at afresh. The behaviour of the spacecraft becomes like that of a gyroscope rotor, or usually its designer would like it to be so. This will only be achieved if he observes constraints upon the choice of spin axis and upon the mass distribution, as indicated below.

The *choice of spin axis* is important. From the mission designer's point of view this axis will be the direction of the momentum bias, gyroscopically rigid as described in Section 3.3.2, and so it is ideally orientated in a direction that is not required to change during the mission—normal to the orbit plane, for example. But the dynamics of the spacecraft requires that constraints on the mass distribution are observed.

If a body is spinning with angular velocity ω, then the spin axis will not in general remain in a fixed direction in space. It is the angular momentum $\mathbf{H}_C$ that has the fixed direction. If it is required that ω is also fixed, then it follows that it must be aligned with $\mathbf{H}_C$, that is, $[I_C]\omega$. This can only happen if the spin axis is along an eigenvector of the inertia matrix, which is to say that the spin must take place about a principal axis.

If rotation is initiated about some axis other than a principal axis, then the physical behaviour depends upon a number of factors. In general, the angular velocity vector varies in direction and magnitude. The motion may be anything from an untidy tumbling type of motion to an oscillatory coning motion, depending upon the magnitudes of the initial rotation, the departure of its axis from a principal axis and the disparity between the inertias. When faced with this type of behaviour, and also the instabilities referred to in the following paragraphs, it is difficult to believe that the angular momentum vector $\mathbf{H}_C$ is fixed in magnitude and direction.

The *stability of a spinning body* may be explored by using equation (3.40), with no torque components present. It will be assumed here that the spin is about the z-axis and has a rate $\omega_z = S$, consistent with Figure 3.15. Two of equations (3.40) then become

$$I_{xx}\dot{\omega}_x - (I_{yy} - I_{zz})S\omega_y = 0$$

and   (3.43)

$$I_{yy}\dot{\omega}_y - (I_{zz} - I_{xx})S\omega_x = 0$$

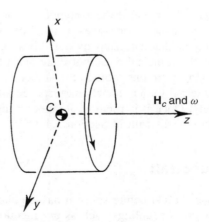

Figure 3.15 The spinning space-craft

These are linear equations whose characteristic equation in terms of the Laplace operator, s, is

$$s^2 + (1 - I_{zz}/I_{xx})(1 - I_{zz}/I_{yy})S^2 = 0 \qquad (3.44)$$

It may be seen that for the spinning motion to be stable, I_{zz} must be the maximum or the least of the moments of inertia.

Bodies that are designed to spin will normally have a mass distribution that is axisymmetric, meaning that $I_{xx} = I_{yy}$. The moments of inertia about all radii through C then have this same value.

There are many examples of such axially symmetric spinning bodies. Gyroscope rotors, vehicle wheels, propellers and so on spin about their axes of maximum inertia. On the other hand rifle bullets, expendable launchers, guided missiles and so on spin about their axes of least inertia.

Long-term stability is important for spacecraft that must remain spinning for their lifetime. If they spin about their axis of least inertia, they will be stable in the short term only. Their long-term behaviour will be unstable if there is a loss of rotational energy brought about by internal dissipation rather than by an external torque. They will eventually adopt a cartwheeling type of motion—a spin motion about the axis of maximum inertia. During this process the angular momentum $\mathbf{H}_C$ remains constant, whilst the energy $1/2\mathbf{H}_C \cdot \boldsymbol{\omega}$ decreases, and it follows that the motion must move towards the minimum-energy state in which ω is least. With $[I_C]\boldsymbol{\omega}$ remaining constant, this means that it moves towards rotation about an axis of maximum inertia.

Thus it follows that spacecraft that are total spinners will spin about their axis of maximum moment of inertia.

The effect of a torque $\mathbf{T}$ is to change the angular momentum $\mathbf{H}_C$, as explained in Section 3.3. For the physical response its effect upon the angular velocity $\boldsymbol{\omega}$ is required, and the appropriate equations may be obtained by inserting $\mathbf{H}_C$ in the form $[I_C]\boldsymbol{\omega}$ into equation (3.30).

In the following analysis it will be assumed that the spacecraft is initially spinning about its z-axis, a principal axis of maximum or least inertia. When this is the only angular motion, the craft is in a state of equilibrium, stable at least in the short term, as shown above.

Repointing of the spin axis will normally call for a rotation about an axis that is at right angles to the spin axis and that is fixed in space rather than fixed in the rotating structure. Coordinate axes that are convenient for the analysis are therefore non-spinning ones with the z-axis aligned with the spin axis. These axes have their angular velocity z-component equal to zero at all times, so that during a manoeuvre the angular velocity will be $\boldsymbol{\Omega} = \{\Omega_x, \Omega_y, 0\}^T$.

The spacecraft spins at a rate $\{0, 0, S\}^T$ relative to the coordinate axes, and so its angular velocity $\boldsymbol{\omega}$ is $\{\Omega_x, \Omega_y, S\}^T$.

The inertia matrix $[I_C]$ referred to the coordinate axes has elements that are in general changing. If the moments of inertia referred to the spacecraft's principal axes are $\{I_{xx}, I_{yy}, I_{zz}\}$ and these axes are at an angle ψ to the coordinate axes, where $\dot{\psi} = S$, then the inertia matrix $[I_C]$ is obtainable from Appendix equation (3.A10):

$$[I_C] = \begin{bmatrix} (I_+ - I_- c) & I_- s & 0 \\ I_- s & (I_+ + I_- c) & 0 \\ 0 & 0 & I_{zz} \end{bmatrix} \qquad (3.45)$$

where $I_+ = \frac{1}{2}(I_{yy} + I_{xx})$, the mean inertia orthogonal to the spin axis,

$\quad I_- = \frac{1}{2}(I_{yy} - I_{xx})$, a measure of lack of axial symmetry,

and $\quad c, s = \cos 2\psi$ and $\sin 2\psi$, respectively.

The angular momentum $\mathbf{H}_C$ is then

$$\mathbf{H}_C = [I_C]\,\boldsymbol{\omega} = \begin{bmatrix} \Omega_x(I_+ - I_-c) + \Omega_y I_- s \\ \Omega_x I_- s + \Omega_y(I_+ + I_-c) \\ S I_{zz} \end{bmatrix} \tag{3.46}$$

Noting that $dc/dt = -2Ss$ and $ds/dt = 2Sc$, then equation (3.30) leads to the following dynamic equations referred to non-spinning axes:

$$\left. \begin{aligned} I_+\dot{\Omega}_x + I_{zz}S\Omega_y + I_-\{-(c\Omega_x - s\Omega_y) + 2S(s\Omega_x + c\Omega_y)\} &= T_x \\ I_+\dot{\Omega}_y - I_{zz}S\Omega_x + I_-\{(s\Omega_x + c\Omega_y) + 2S(c\Omega_x - s\Omega_y)\} &= T_y \\ I_{zz}\dot{S} + I_-\{s\Omega_x^2 + 2c\Omega_x\Omega_y - s\Omega_y^2\} &= T_z \end{aligned} \right\} \tag{3.47}$$

A number of conclusions can be drawn:

1. *A torque T_z about the z-axis* causes a simple acceleration $\dot{S}$ about that axis when the spacecraft is in its equilibrium state. Note that in this state Ω_x and Ω_y are zero, so

$$\dot{S} = T_z/I_{zz} \tag{3.48}$$

2. *Repointing the spin axis* by means of a constant precession rate Ω_y about the y-axis, say, will require a torque T_x about the x-axis. Only if the moments of inertia I_{xx} and I_{yy} are equal, that is, $(I_- = 0)$, will the torque be constant. Under this condition,

$$\Omega_y = T_x/I_{zz}S \tag{3.49}$$

For example, a spacecraft with a moment of inertia of $400\,\mathrm{kg\,m^2}$, spinning at 10 rpm, will precess at about 0.5 arcsec per second for each newton millimetre of torque.

This represents complete cross-coupling between x- and y-axes, which is a characteristic of a gyroscope's precessional behaviour. But in addition it should be noted that a torque produces an angular *velocity* rather than the acceleration produced by T_z. Figure 3.16(a) shows how the angular momentum $\mathbf{H}_C$ is rotated towards the torque vector $\mathbf{T}$ and Figure 3.16(b) shows the total angular velocity, with contributions from ω and Ω. If I_{xx} and I_{yy} differ, then additional oscillating torques will be needed, that is,

$$T_x = 2cSI_-\Omega_y \text{ and } T_y = 2sSI_-\Omega_y \tag{3.50}$$

Furthermore, if a constant rate of spin S is to be maintained, then a torque will be needed, which is equal to

$$T_z = -sI_-\Omega_y^2 \tag{3.51}$$

Failure to provide these oscillating torques will lead to an unsteady precession in response to a constant torque T_x. Most objects that are designed to spin will be given axial mass symmetry, that is, $I_- = 0$, or $I_{xx} = I_{yy}$, and this applies to spacecraft too.

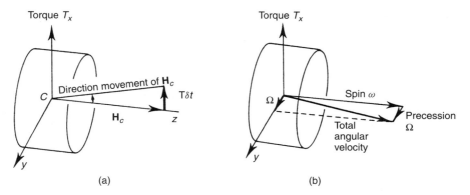

Figure 3.16 Precession of a spinner due to a torque

It should be noted that the components T_x and T_y are in non-spinning axes and that torquers fixed in the spacecraft will need to apply resolved versions of T_x, T_y, that is,

$$
\begin{array}{c}\text{Torque components} \\ \text{in spacecraft axes}\end{array} = \begin{bmatrix} \cos\psi & \sin\psi & 0 \\ -\sin\psi & \cos\psi & 0 \\ 0 & 0 & 1 \end{bmatrix} \begin{bmatrix} T_x \\ T_y \\ T_z \end{bmatrix} \tag{3.52}
$$

3. *There is an oscillatory 'nutation' mode.* This appears as a coning type of motion when the spacecraft is disturbed from its equilibrium state. Whenever precessional torques are applied, the mode will be activated; additional activation will occur when the torque ceases. If the duration of the precession is equal to an exact number of nutation periods, then the torque cessation will cancel the nutation, but engineered damping may be necessary (see Section 3.5.1).

It may already be seen from the characteristic equation (3.43) that an oscillatory mode exists, whose frequency is

$$
\omega_{\mathrm{nut}} = S\sqrt{(1 - I_{zz}/I_{xx})(1 - I_{zz}/I_{yy})} \tag{3.53}
$$

This is the frequency that may be observed on traces from rate gyroscopes that are mounted on the structure, that is, when it is observed in the spinning axes.

In the axisymmetric case, I_{zz}/I_{xx} lies in the range $1 \leq I_{zz}/I_{xx} < 2$. The *observed* nutation frequency will depend upon the rate at which the observer is spinning.

When viewed from the (spinning) spacecraft frame of reference, as measured by instruments that are fixed to the spacecraft's structure, for example,

$$
\omega_{\mathrm{nut}} = S((I_{zz}/I_{xx}) - 1), \text{ and so } 0 \leq \omega_{\mathrm{nut}} < S \tag{3.54}
$$

When viewed from a non-spinning (inertial) frame of reference,

$$
\omega_{\mathrm{nut}} = S(I_{zz}/I_{xx}), \text{ and so } S \leq \omega_{\mathrm{nut}} < 2S \tag{3.55}
$$

3.4.3 Hybrid spacecraft

Two classes of spacecraft bring together the advantages of having momentum bias and yet provide a non-spinning base for their payload and torquers. These are the partially de-spun or dual-spin craft and the 'three-axes-stabilized craft with momentum bias'. In each case momentum bias is provided by mounting a rotating body—or more than one—on the non-spinning part; it is part of the structure in the case of the dual-spin craft, but consists of high-speed purpose-built wheels—momentum wheels—in the three-axis-stabilized case.

The spinning parts of these hybrids have dynamic properties that are similar to those of the 'spinner' covered above. They will have axially symmetric mass properties for the reasons given. So their centres-of-mass lie on the spin axis, and about axes normal to this their moments of inertia will be equal. Their contribution to the total inertia $[I_C]$ of the spacecraft will then be independent of their rotation. Their rate of spin will remain constant unless there is a torque about their axis.

The non-spinning structure will behave like the three-axes-stabilized craft described in Section 3.4.1. Some of the torques on it will be applied via the bearings of the rotating parts in which there will be built-in torque motors.

The angular momentum of the hybrid spacecraft is then given by the expressions in equation (3.35) or (3.36). These may be substituted into equation (3.30) in order to obtain equations obeyed by the components of the angular velocity of the non-spinning structure, in any coordinate axis.

If the momentum bias direction is along the z-axis, a principal axis, then

$$\mathbf{H}_C = \{I_{xx}\omega_x, I_{yy}\omega_y, (I_{zz}\omega_z + H_z)\}^{\mathrm{T}} \tag{3.56}$$

The dynamic equations for the components of angular velocity along principal axes are

$$\left.\begin{aligned}
I_{xx}\dot{\omega}_x + \omega_y\omega_z(I_{zz} - I_{yy}) + \omega_y H_z &= T_x \\
I_{yy}\dot{\omega}_y + \omega_z\omega_x(I_{xx} - I_{zz}) - \omega_x H_z &= T_y \\
I_{zz}\dot{\omega}_z + \dot{H}_z + \omega_x\omega_y(I_{yy} - I_{xx}) &= T_z
\end{aligned}\right\} \tag{3.57}$$

A number of observations may be made:

1. *Stability*. The momentum bias axis may be the axis of intermediate inertia without causing instability. The constraint imposed on the total spinner is not necessary here.

The long-term stability cannot be forecast from the above equations. The hybrid will, however, be stable provided that energy dissipation in the non-spinning part exceeds that in the spinning part. The bias direction may then be along the axis of least inertia. Passive nutation dampers, for example, will be placed in the non-spinning part of the spacecraft.

2. There is a *nutation mode* whose frequency, when observed in non-spinning axes, is

$$\omega_{\mathrm{nut}} = H_z/\sqrt{(I_{xx}I_{yy})} \tag{3.58}$$

For example, a spacecraft having both moments of inertia I_{xx} and I_{yy} equal to $400\,\mathrm{kg\,m}^2$, and carrying a momentum wheel giving a bias of $40\,\mathrm{N\,m\,s}$, will have a nutation mode with a period of about one minute.

3. *The response to a torque.* The inclusion of momentum bias makes the whole structure behave as a gyrostat, with a response that is very similar to that of a spinner. A constant torque about the x-axis will cause a constant rate of precession about the y-axis, for example, whose magnitude is

$$\omega_y = T_x / H_z \tag{3.59}$$

Thus a torque of $1\,\text{N}\,\text{mm}$ will produce a rate of precession of about $5\,\text{s}$ of arc per second when the bias is $40\,\text{N}\,\text{m}\,\text{s}$. On the other hand a torque about the bias direction will produce an angular acceleration about that direction if it is parallel to a principal axis. In other cases there will be cross-coupling.

3.5 OSCILLATORY MODES

All vehicles have a tendency to oscillate and vibrate, and spacecraft are no exception. In their case the damping of the oscillatory modes is less than that experienced by Earthbound vehicles, and as a consequence any oscillation, once started, may last for a very long time—a characteristic of modes that are very close to being unstable.

It follows that it is important that these modes are identified by the designer and that artificial damping is introduced by the Attitude and/or Orbit Control system if possible; certainly the modes must not be destabilized. Care must also be taken to avoid exciting those modes whose ultimate damping remains very low.

The main oscillatory tendencies occur as either 'rigid body modes' or 'flexure modes' and are addressed below. Fuel movement in its tanks can also have an oscillatory tendency but this is normally controlled by means of baffles.

3.5.1 Rigid body modes

There are potentially two rigid body modes, libration and nutation.

The nutation mode is present in spacecraft that have momentum bias, whether this is achieved by spinning the complete structure or part of it or by using a momentum wheel. It is easiest to describe it as follows, when there is just one spinning part.

In the absence of any torque, the equilibrium state occurs when the axis of the spinning part is aligned with the angular momentum vector and they point at the same fixed direction in space. A small torque impulse will cause them to become misaligned, the momentum vector being turned to a new fixed direction. The ensuing nutation oscillation then has a pattern of motion in which the spin axis adopts a 'coning' type of behaviour around this fixed angular momentum direction.

At the start of a constant torque causing precession of the momentum bias vector, the step change in the torque will excite the nutation mode and the oscillation will be superimposed on the precession as shown in Figure 9.6. Further excitation will occur when the torque ceases; this may be timed so as to cancel the nutation.

A hand-held bicycle wheel may be used to demonstrate nutation. The wheel must be carefully balanced and fitted with a handle that is an extension of its axle. When the wheel is spinning rapidly with its axle horizontal, it may be supported by just one finger under the handle, whereupon it will rotate about a vertical axis at a constant rate. This is precession, obeying equation (3.29). If during this motion a sharp impulse is delivered to the handle, then an oscillatory motion will take place (in addition to the precession); this is nutation, the observed frequency obeying equation (3.55).

The libration mode is caused by the gravity gradient—the reduction in the gravitational field strength as the distance from Earth's centre increases.

A rigid body is in stable equilibrium in this field when its axis of minimum moment of inertia lies along the local vertical. The Moon is an example of a body that is roughly in this state. When the body in equilibrium is disturbed by a small amount, it will then oscillate like a conical pendulum if it is free to do so. The frequency of oscillation is given by equation (9.21) of Chapter 9; at the Earth's surface its period is about 48 min, about half the period of a satellite in Low Earth Orbit (LEO).

A few spacecraft use this phenomenon as a means of achieving an Earth-pointing face, and they must then incorporate damping for this mode in their ACS algorithms. (The Earth's tidal system provides damping for the Moon.) In other spacecraft the torques of this mode appear as disturbance torques (see Section 9.4.3 of Chapter 9).

3.5.2   Flexure modes

All structures are elastic and will vibrate at their modal frequencies when they are appropriately excited. The appendages of spacecraft in particular can be very flimsy structures compared with any Earthbound equivalents, and their fundamental frequency—their lowest modal frequency—can be very low indeed. The solar arrays of the Hubble telescope have a mode with frequency 0.11 Hz for example. Theory indicates that there are an infinity of modes with frequencies higher than the fundamental, each with its corresponding modal shape.

The solar arrays shown in Figure 9.3 of Chapter 9, the form in which they appear on many spacecraft, are cantilevered outwards from the central body and each will have bending and torsional modes. The frequencies of the bending modes of a single, rigidly attached uniform cantilever are given by

$$f = (K^2\pi/8L^2)\sqrt{(EI/\rho A)}\,\text{Hz} \qquad (3.60)$$

where $K \approx 1.2, 3, 5, 7, \ldots$ for the different modes,
  L = array length (m)
  E = Young's modulus (N/m^2)
  I = second moment of area (m^4) (see Figure 8.14 of Chapter 8)
  ρ = density (kg/m^3)
  A = cross-sectional area (m^2)

However, on a spacecraft it is unlikely that the attachment will be rigid, and the actual fundamental frequency may be only about 50% of the value given by this equation. A large number of the overtones' frequencies fall within the passband of the ACS.

Any oscillation of the solar arrays will be transmitted to the main structure of the spacecraft by the bending moments and shear forces at their roots. In consequence, the payload will tend to oscillate in sympathy with the flexure modes. This occurred on the Hubble telescope, the oscillation being initiated by the thermal shock that took place when it moved from being in eclipse to being in sunlight.

This simple example of an interaction is compounded by interactions between different flexure modes, affecting their modal frequencies. Forecasting the modal properties of the spacecraft as a whole is a complicated process for which specialist software packages are used (see also Sections 8.2.1 and 8.5 of Chapter 8).

It should be noted that flexural oscillations do not directly involve external forces and moments on the spacecraft and so they do not directly affect the momenta, linear or angular.

3.6 IN CONCLUSION

The material in this chapter has been aimed primarily at the dynamics of the main categories of spacecraft. The benefits of adopting certain mass distributions and torque axes have been shown, together with the consequence of doing otherwise.

There are likely to be increasingly many occasions in which the designer has to deal with non-optimum mass distributions, as, for example, when reusable launch vehicles part with their payload and when large space structures are assembled or constructed in orbit. The fundamental laws still apply but the control systems will have to adjust to the changes in the mass distribution.

The methods contained in the chapter are not confined to the conventional configurations. Any configurations that may be treated as being made up of rigid bodies may be dealt with in the manner shown. Their linear momenta, moments-of-momentum and angular momenta will obey the Newtonian equations (3.7) or (3.18), (3.12) and (3.28), respectively. The components of their angular momentum $\mathbf{H}_C$ may be found by using equations (3.31) or (3.35), having first obtained their inertia matrix using the Appendix. All the Newtonian equations should be developed by using the Coriolis theorem as illustrated in equation (3.30), enabling the motion equations to be expressed in terms of the components of their velocities, linear and angular, along any desired axes.

APPENDIX: INERTIA

A1 Introduction

The inertia matrix $[I_O]$ referred to a point O is a property of the mass distribution about O. Its elements depend upon the directions of a right-handed orthogonal set of axes x, y, z through O, and in particular it contains the moments and products of inertia associated with these axes. The matrix at the centre-of-mass C plays an important part in the rotational behaviour of a spacecraft, and it must be evaluated and controlled during its design.

This appendix defines terms that are associated with mass distributions and presents formulae that are useful for the evaluation of the inertia matrix.

A2 Definitions

- *Moments of inertia* $\mathbf{I}_{xx}, \mathbf{I}_{yy}, \mathbf{I}_{zz}$
  A moment of inertia is the second moment of mass about an axis. The contribution of an increment of mass δm at a distance d from the axis is $d^2 \delta m$ (see Figure 3.A1). The moment of inertia about the x-axis is, for example,

$$I_{xx} = \int (y^2 + z^2)\, dm \qquad (3.A1)$$

where the integral extends over the whole mass distribution.

● *Products of inertia* $\mathbf{I}_{xy}, \mathbf{I}_{yz}, \mathbf{I}_{zx}$
The product of inertia associated with the x-axis is

$$I_{yz} = \int yz\, dm \qquad (3.A2)$$

Products of inertia are measures of the lack of symmetry in a mass distribution.

If there is a plane of symmetry, then the product of inertia associated with all axes in that plane will be zero. For example, an aircraft whose xz-plane is a plane of symmetry will have I_{xy} and I_{yz} equal to zero.

If two of the coordinate planes are planes of symmetry, then all three of the products of inertia will be zero. This applies to axially symmetric bodies such as many expendable launchers.

● *Principal axes* are sets of orthogonal axes for which all three products of inertia are zero. There is always one such set at each point.
● *The inertia matrix* $[I_O]$ referred to $O(x, y, z)$ is defined as

$$[I_O] = \begin{bmatrix} I_{xx} & -I_{xy} & -I_{zx} \\ -I_{xy} & I_{yy} & -I_{yz} \\ -I_{zx} & -I_{yz} & I_{zz} \end{bmatrix} \qquad (3.A3)$$

For a single particle with mass m at (x,y,z) this becomes $[I_{Om}]$, say, where

$$[I_{Om}] = \begin{bmatrix} m(y^2 + z^2) & -mxy & -mzx \\ -mxy & m(z^2 + x^2) & -myz \\ -mzx & -myz & m(x^2 + y^2) \end{bmatrix} \qquad (3.A4)$$

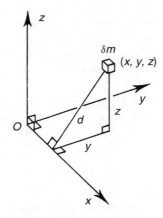

Figure 3.A1 An   incremental mass δm

● *Rotation matrix* $[R]$

If the components of a vector **V** in one set of axes are expressed as the terms in a (3×1) column matrix $\mathbf{V}_1$, say, and $\mathbf{V}_2$ consists of its components in a second set that is rotated relative to the first, then $\mathbf{V}_2$ may be expressed as $\mathbf{V}_2 = [R]\mathbf{V}_1$. Then, $[R]$ is known as a rotation matrix.

If Euler angles are used to describe the rotation as a sequence of separate rotations about the coordinate axes, then it is convenient to use a notation $[X(\phi)]$, $[Y(\theta)]$ and $[Z(\psi)]$ for the separate rotations. $[X(\phi)]$ is the rotation matrix for a clockwise rotation through an angle ϕ about the x-axis, and similarly for $[Y(\theta)]$ and $[Z(\psi)]$. These matrices are

$$[X(\phi)] = \begin{bmatrix} 1 & 0 & 0 \\ 0 & \cos\phi & -\sin\phi \\ 0 & \sin\phi & \cos\phi \end{bmatrix} \quad [Y(\theta)] = \begin{bmatrix} \cos\theta & 0 & \sin\theta \\ 0 & 1 & 0 \\ -\sin\theta & 0 & \cos\theta \end{bmatrix}$$

(3.A5)

$$[Z(\psi)] = \begin{bmatrix} \cos\psi & -\sin\psi & 0 \\ \sin\psi & \cos\psi & 0 \\ 0 & 0 & 1 \end{bmatrix}$$

For an aircraft whose attitude is defined in the conventional manner as the sequence of rotations yaw ψ, followed by pitch θ, followed by roll ϕ, as in Figure 3.A2, the matrix $[R]$ for converting the components of its velocity **V**, say, from its own axes $\mathbf{V}_{a/c}$, back to the datum axes $\mathbf{V}_{\text{datum}}$, is

$$\mathbf{V}_{\text{datum}} = [R]\mathbf{V}_{a/c}, \quad \text{where} \quad [R] = [Z(\psi)][Y(\theta)][X(\phi)] \tag{3.A6}$$

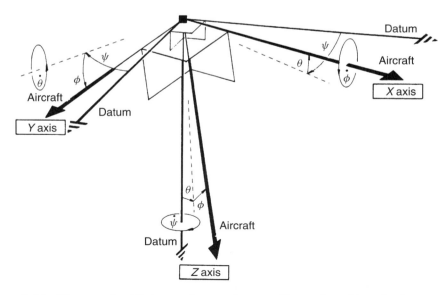

Figure 3.A2 The standard Euler angles used to specify an aircraft's attitude

A3   Useful Formulae

● *Inertia invariant*

$$I_{xx} + I_{yy} + I_{zz} = 2 \int r^2 \, dm = 2I_O, \text{ say} \qquad (3.A7)$$

where I_O is the second moment of mass about the origin O and is independent of the direction of axes.

● *Perpendicular axis theorem*—for laminas only.
For a lamina lying entirely in the yz-plane,

$$I_{yy} + I_{zz} = I_{xx} \qquad (3.A8)$$

● *Transfer of reference point* (parallel axis theorem) (see Figure 3.A3).
If an object whose centre-of-mass G is at (X, Y, Z) has an inertia matrix $[I_G]$ referred to G, then add on the inertia matrix of its e.p. referred to O, in order to obtain the inertia matrix $[I_O]$ referred to parallel axes at O, that is,

$$[I_O] = [I_{OM}] + [I_G] \qquad (3.A9)$$

where equation (3.A4) may be used for $[I_{OM}]$.

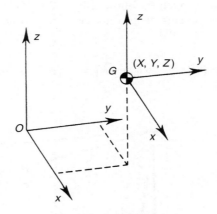

Figure 3.A3

● *Rotated axes theorem*
If the rotation matrix for transforming a vector's components from axes number 1 to axes number 2 is $[R]$ (see rotation matrices above), then the inertia matrix $[I]$ may be transformed between the same sets of axes by using

$$[I_2] = [R][I_1][R]^{-1} \qquad (3.A10)$$

A4   Contribution of a Piece of Equipment to $[I_C]$

Suppose that a piece of equipment with mass M is installed with its centre-of-mass G at (x, y, z) relative to the spacecraft's axes at C. Suppose that $[I_{eq}]$ is its inertia matrix referred to its own natural axes and that these axes are rotated through an angle ϕ about the spacecraft's x-axis when it is installed.

Then its contribution to the inertia matrix $[I_C]$ of the spacecraft will be $[X(\phi)]\,[I_{eq}]$ $[X(-\phi)]$ plus the inertia matrix of its e.p. referred to C (equation 3.A4). In general,

$$[I_C] = [R][I_{eq}][R]^{-1} + [I_{CM}] \tag{3.A11}$$

4 CELESTIAL MECHANICS

John P. W. Stark[1], Graham G. Swinerd[2] and Peter W. Fortescue[2]

[1]*Department of Engineering, Queen Mary, University of London*
[2]*Aeronautics and Astronautics, School of Engineering Sciences, University of Southampton*

4.1 INTRODUCTION

The theory of celestial mechanics underlies all the dynamical aspects of the orbital motion of spacecraft. The central feature is the mutual gravitational force of attraction that acts between any two bodies. This was first described by Newton, and together with his laws of motion (see Chapter 3), it provides us with the theoretical framework for celestial mechanics. The orbits that it forecasts will be relative to an inertial frame of reference (IFR) that is fixed with respect to the stars; the consequential motion relative to the ground will be covered in Chapter 5.

The simplified case in which the gravitational force acts between two pointlike objects gives a good approximation to orbital motion for most spacecraft situations. It may easily be shown that if a body has a uniform mass distribution within a spherical surface, then outside it the gravitational force from the body does indeed appear to emanate from a pointlike source. This so-called two-body problem has a solution—a Keplerian orbit.

Kepler, whose major works were published during the first 20 years of the seventeenth century, consolidated the observations of planetary motion into three simple laws, which are illustrated in Figure 4.1. These are

1. the orbit of each planet is an ellipse with the Sun occupying one focus;
2. the line joining the Sun to a planet sweeps out equal areas in equal intervals of time;
3. a planet's orbital period is proportional to the mean distance between Sun and the planet, raised to the power 3/2.

Newton's theory of gravity predicting the above results was not developed until more than 65 years later.

Spacecraft Systems Engineering (Third Edition). Edited by P. W. Fortescue, J. P. W. Stark and G. G. Swinerd
© 2003 John Wiley & Sons Ltd

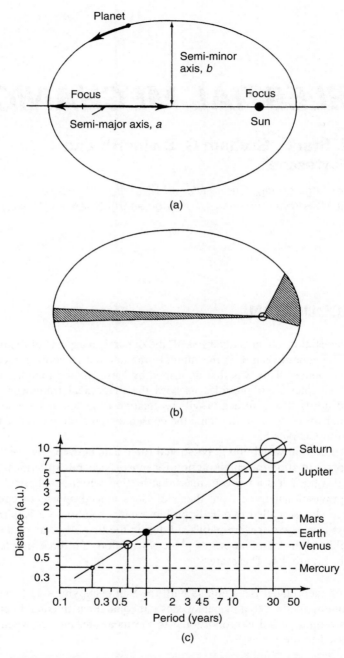

Figure 4.1 (a) Kepler's first law: orbit shape; (b) Kepler's second law: equal areas in equal time and (c) Kepler's third law: period $\tau \propto a^{1.5}$

Various perturbing forces must be included in order to produce a better approximation to a spacecraft's orbit. As the altitude of a spacecraft's orbit decreases, the departure from spherical symmetry of the Earth's mass and shape becomes more important, together with aerodynamic forces. At high altitudes such as in geostationary orbits (∼36 000 km) the gravitational pull of the other celestial bodies such as the Moon and Sun become significant, together with radiation pressure from the Sun.

The theory of celestial mechanics is required in order that the motion of a spacecraft may be predicted. In most situations the accuracy required for the vehicle's position is relatively modest—typically a few kilometres. However, there are significant deviations from this, one important case being when the space vehicle must perform a manoeuvre to change its orbit. This topic is left to the next chapter. The problems of spacecraft tracking are left until Chapter 5. It should be noted that for some vehicles, particularly those that employ active remote sensing instrumentation, precise orbit determination is required. An accuracy of the order of 1 m was required for Seasat, and values of 10 cm or less are a requirement for the European Envisat spacecraft. Such precision in orbit determination is difficult because our knowledge of the Earth's gravity field remains limited, and more importantly because of the uncertainties associated with surface forces, in particular, aerodynamic drag. The former problem will be alleviated when a dedicated spacecraft mission to determine the Earth's gravity field is flown. The solution to the latter problem of quantifying drag is much more elusive, since difficulties are associated with adequately modelling the atmospheric density, atmospheric winds, the response of the atmosphere to solar activity and the magnitude of lift and drag coefficients relevant to spacecraft. Consequently, the precise orbit can only be determined retrospectively. Fortunately, the accuracy level required for orbit prediction, derived from tracking and operational activities, is much less stringent.

4.2 THE TWO-BODY PROBLEM — PARTICLE DYNAMICS

The simplest problem in celestial mechanics concerns the predicted motion of two co-rotating point masses. It is assumed in this analysis that the sole force acting between them is their mutual gravitational attraction. Newton demonstrated that the gravitational field from any pointlike body is solely dependent upon its mass. The field's potential U at a distance R from a particle of mass M is simply

$$U = -\frac{GM}{R} \tag{4.1}$$

where G is the universal constant of gravitation, having a numerical value in SI units of $6.670 \times 10^{-11} \text{N m}^2/\text{kg}^2$.

If two masses m_1 and m_2 are a distance r apart, as shown in Figure 4.2, then the equation governing their motion due to their mutual gravitational attraction is easily obtained by combining Newton's Laws of gravity and motion. Newton's Law of gravity gives the forces $\mathbf{F}_1$ and $\mathbf{F}_2$ acting on the masses as

$$\mathbf{F}_1 = \frac{Gm_1m_2}{r^2}\left(\frac{\mathbf{r}}{r}\right), \mathbf{F}_2 = \frac{Gm_1m_2}{r^2}\left(\frac{-\mathbf{r}}{r}\right) \tag{4.2}$$

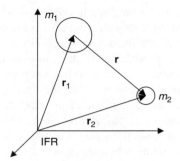

Figure 4.2 The two-body problem geometry in an inertial
frame of reference (IFR)

The motion of each mass with respect to the IFR is expressed by Newton's second law
of motion,

$$m_1\ddot{\mathbf{r}}_1 = \mathbf{F}_1, \quad m_2\ddot{\mathbf{r}}_2 = \mathbf{F}_2 \tag{4.3}$$

By noting that $\mathbf{r} = \mathbf{r}_2 - \mathbf{r}_1$, we may combine equations (4.2) and (4.3) to give

$$\ddot{\mathbf{r}} + \frac{G(m_1 + m_2)}{r^2}\left(\frac{\mathbf{r}}{r}\right) = \mathbf{0}$$

which is the equation of unperturbed motion due to gravity of mass m_2 with respect to
mass m_1.

The restricted two-body problem assumes that the mass of one body far exceeds that of
the other, $m_1 \gg m_2$ say. This applies to a spacecraft of mass m, say about 10^3 to 10^4 kg,
compared with a heavenly body of mass M; Earth's mass, for example, is 6×10^{24} kg.
In this case the equation above reduces to:

$$\ddot{\mathbf{r}} + \frac{(GM)}{r^2}\frac{\mathbf{r}}{r} = \mathbf{0} \tag{4.4}$$

Equation (4.4) is also a good approximation for the planets and the Sun, since even the
largest one, Jupiter ($\sim 2 \times 10^{27}$ kg), is dominated by the Sun ($\sim 2 \times 10^{30}$ kg).

The parameter GM, here referred to as μ, is the gravitational parameter for the body
about which the motion is taking place. Table 4.1 lists its values for a selection of solar
system bodies.

The orbit constants

Equation (4.4) may be solved and will lead to the orbit equation if the initial conditions
are known. The constants associated with a particular orbit may be thought of as being
constants of integration of equation (4.4) or they may be approached in physical terms
via conservation laws, as in Chapter 3.

Moment-of-momentum conservation follows from the fact that the only force acting has
no moment about the centre of the primary body. The moment of the momentum vector
$m\,\mathbf{h}$ is therefore constant, in both magnitude and direction.

Table 4.1 Some physical properties of the major bodies in the solar system, including Earth's Moon. See also Tables 2.5, 2.6 and 2.7 of Chapter 2. The sphere of influence (see Section 5.8.1 of Chapter 5) is expressed with respect to the Sun as the disturbing body

Parameter\ Body	μ (m^3/s^2) Gravitational parameter	Equatorial surface gravity (m/s^2)	Surface escape velocity (m/s)	Oblateness J_2	Sphere of influence $(10^6\ km)$	Axial rotation period (sidereal)
Sun	1.327×10^{20}	273.98	6.18×10^5	—	—	~ 27 days
Mercury	2.203×10^{13}	3.70	4250	—	0.09–0.14	58.646 days
Venus	3.249×10^{14}	8.87	10 360	2.7×10^{-5}	0.61–0.62	243.019 days
Earth	3.986×10^{14}	9.81	11 180	0.001083	0.91–0.94	$23^h 56^m 22.7^s$
Mars	4.283×10^{13}	3.71	5020	0.001964	0.52–0.63	$24^h 37^m 22.6^s$
Jupiter	1.267×10^{17}	23.12	59 530	0.01475	45.9–50.5	$\sim 9^h 50^m$
Saturn	3.794×10^{16}	9.05	35 560	0.01645	51.6–57.5	$\sim 10^h 15^m$
Uranus	5.780×10^{15}	7.77	21 250	0.012	49.4–54.1	$\sim 17^h 50^m$
Neptune	6.871×10^{15}	11.00	23 540	0.004	85.7–87.6	$\sim 19^h 10^m$
Pluto	1.021×10^{12}	0.40	1300	—	11.4–18.8	6.387 days
Moon	4.903×10^{12}	1.62	2380	0.0002027	0.157–0.162	27.322 days

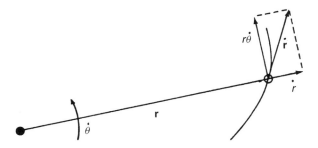

Figure 4.3 The particle's position and velocity vectors

By definition $\mathbf{h} = \mathbf{r} \times \dot{\mathbf{r}}$, and so its direction, perpendicular to both the position vector and the velocity vector, is normal to the orbit plane and is constant. The fact that the magnitude of $\mathbf{h}$ is constant, is consistent with Kepler's second law. The value of this is $r^2\dot{\theta}$ (see Figure 4.3), which is twice the areal velocity $\dot{A}$.

Energy conservation follows from the fact that the only external force on the spacecraft is due to the gravitational field, a conservative force whose potential energy per unit mass is $-\mu/r$.

The total orbital energy per unit mass ε (the sum of the kinetic and gravitational energies) remains constant, and so

$$\frac{1}{2}V^2 - \frac{\mu}{r} = \varepsilon \tag{4.5}$$

This is known as the energy equation or *vis-viva* integral.

Solution to the two-body problem

The vector solution to equation (4.4) may be obtained by first taking the cross-product with the constant $\mathbf{h}$ and integrating once with respect to time. This yields

$$\dot{\mathbf{r}} \times \mathbf{h} = \mu \left(\frac{\mathbf{r}}{r} + \mathbf{e} \right) \tag{4.6}$$

where $\mathbf{e}$ is the vector constant of integration called the eccentricity vector, which lies in the plane of the orbit as shown in Figure 4.4

The final solution to equation (4.4), obtained by taking the dot product of equation (4.6) with $\mathbf{r}$, is

$$r = \frac{h^2/\mu}{1 + e \cos \theta} \tag{4.7}$$

where θ is the angle between $\mathbf{r}$ and $\mathbf{e}$. This is the equation of a conic section and demonstrates the first of Kepler's laws.

The *eccentricity e* determines the type of conic. It is a circle when $e = 0$, an ellipse when $0 < e < 1$, a parabola when $e = 1$ and a hyperbola when $e > 1$. These shapes are shown in Figure 4.4. The dominant mass is at one of the two foci.

Elliptic and circular orbits ($0 \leq e < 1$)

The *ellipse* is the general form of a closed orbit and is shown in Figure 4.5. The major axis is called the *line of apsides*, recalling that half an ellipse is shaped like the apse of a church. The angle θ is termed *the true anomaly*. The dimensions are related and may be expressed in terms of the physical constants h and ε. For example,

● The semi-latus rectum p and semi-major axis a obey:

$$p = a(1 - e^2) = h^2/\mu \tag{4.8}$$

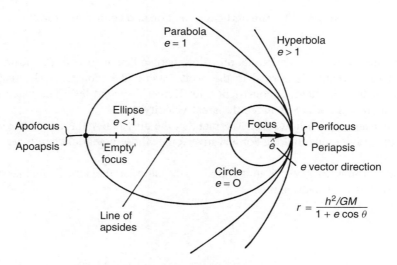

Figure 4.4   Conic sections

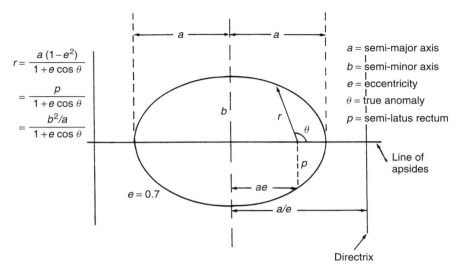

Figure 4.5 Ellipses

● The point of closest approach to the main body is termed the periapsis or perifocus and its distance is

$$r_p = p/(1 + e) = a(1 - e) \tag{4.9}$$

● The most distant point, the apoapsis or apofocus, is at a distance

$$r_a = p/(1 - e) = a(1 + e) \tag{4.10}$$

Clearly $r_p + r_a = 2a$.

The apses normally carry the name of the main body; so periapsis is called the perigee if the orbit is round the Earth, perihelion if round the Sun and so on. Equating the energy and the moments of momenta at the apses leads to

$$\varepsilon = -\mu/2a \tag{4.11}$$

Kepler's third law is simply obtained, since the orbit period is the time taken for the complete area of the ellipse to be swept out by the radius vector. Since $\dot{A} = h/2$ and the area of an ellipse is $\pi a^2 \sqrt{(1 - e^2)}$, then by making use of the relationships above it follows that the orbit period τ is given by

$$\tau = 2\pi \sqrt{(a^3/\mu)} \tag{4.12}$$

The *position versus time relationship* will be required for ground station passes, manoeuvres and other mission activities.

For *elliptical motion* $(0 < e < 1)$, a drawing of the trajectory may be achieved by constructing a circumscribing circle of radius a about the ellipse, as shown in Figure 4.6.

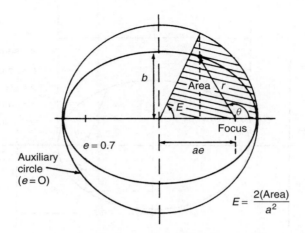

Figure 4.6 Eccentric anomaly definition

This is called the *auxiliary circle*, and it is used to transform from the variable angle θ to a new variable E, called the *eccentric anomaly*.

Some useful expressions can be obtained from the geometry:

● The position on the ellipse may be written in terms of E as

$$r = a(1 - e \cos E) \tag{4.13}$$

● The relationship between θ and E is

$$\tan \theta/2 = \tan E/2 \sqrt{\left[\frac{1+e}{1-e}\right]} \tag{4.14}$$

To find a position *versus* time relationship, differentiation of equation (4.13) with respect to time gives

$$\frac{dr}{dt} = ae \sin E \frac{dE}{dt}$$

If we differentiate equation (4.7) to find dr/dt and note from the geometry in Figure 4.6 that $r \sin \theta = a\sqrt{1 - e^2} \sin E$, we obtain

$$\frac{dE}{dt} = \frac{1}{r}\sqrt{(\mu/a)}$$

Separating the variables this gives

$$a(1 - e \cos E)\, dE = dt\sqrt{(\mu/a)} \tag{4.15}$$

which on integration yields

$$E - e \sin E = \sqrt{(\mu/a^3)}(t - t_p) \tag{4.16}$$

where t_p is the time of perifocal passage. Also, note that the *mean motion n* is defined by

$$n = 2\pi/\tau = \sqrt{(\mu/a^3)} \qquad (4.17)$$

Therefore, if we suppose that the spacecraft passes through perifocus at time zero (i.e. $t_p = 0$), then the time of flight t from perifocus to the current position is given by

$$E - e \sin E = M \qquad (4.18)$$

where the *mean anomaly* is defined by

$$M = nt \qquad (4.19)$$

Equation (4.18) is known as *Kepler's equation*. It provides a simple relationship between time and position, which can be used in conjunction with equations (4.13) and (4.14) to provide a complete specification of position with time. The problem of the determination of position given the time of flight from perifocus can be represented symbolically by the following chain of operations,

Time t—equation (4.19) $\rightarrow$ M—equation (4.18) $\rightarrow$ E—equation (4.14) $\rightarrow$ Position θ.

Conversely, to determine the time to reach a given position the above chain is reversed,

Position θ—equation (4.14) $\rightarrow$ E—equation (4.18) $\rightarrow$ M—equation (4.19) $\rightarrow$ Time t.

It is important to note that the latter chain of operations can be performed analytically. However, the $M \rightarrow E$ step of the former chain requires a computational solution of Kepler's equation (4.18). This problem has attracted a great deal of attention since Kepler first wrote down his famous equation, and over a 100 ways of solving it have been presented in the literature. Newton's method of successive approximations is well suited to the task. Given the values of M and e, successive estimates of E can be obtained iteratively using

$$E_{i+1} = E_i - \left(\frac{E_i - e \sin E_i - M}{1 - e \cos E_i} \right), \qquad E_0 = M \qquad (4.20)$$

Example: Estimate the time for a spacecraft to move through the $60°$ of true anomaly centred around the apogee point, for an orbit with $a = 24\,000$ km and $e = 0.7$.

Solution: Let perigee be point P_0, the point at true anomaly $150°$ be point P_1 and the point at true anomaly $210°$ be point P_2. We need to find the time of flight from point P_1 to point P_2. Using the chain of operations above for $\theta \rightarrow t$, we find the time from perigee to P_1. At point P_1 we have $\theta_1 = 2.6180$ rads, $E_1 = 2.0060$ rads and $M_1 = 1.3713$ rads. This gives $t(P_0 \rightarrow P_1) = 8075.61$ s, which by symmetry is identical to $t(P_2 \rightarrow P_0)$. Hence the time to traverse the apogee region from P_1 to P_2 is

$$t(P_1 \longrightarrow P_2) = \tau - t(P_2 \longrightarrow P_1) = 37\,002.25\,\text{s} - (2 \times 8075.61\,\text{s})$$

$$= 20851.03\,\text{s} \approx 5.8\,\text{h}.$$

A similar calculation shows that the spacecraft traverses a similar $60°$ interval of true anomaly centred on perigee in $807.57\,\text{s} \approx 0.22\,\text{h}$, which illustrates the significantly different velocity of the vehicle at perigee compared to apogee, for this highly eccentric orbit.

The *velocity relationships* for elliptic orbits may be obtained primarily from equation (4.7) and the *vis-viva* integral (equations 4.5 and 4.11), namely,

$$\frac{1}{2}V^2 - \frac{\mu}{r} = \varepsilon = -\frac{\mu}{2a} \qquad (4.21)$$

This equation is also the basis for velocity calculations for all conic section trajectories.

For a *circular orbit* $(e = 0)$, the circular velocity is given by

$$V_{\text{circ}} = \sqrt{(\mu/r)} \qquad (4.22)$$

Using this, one can note that for Earth-orbiting space vehicles having a 24-h period, the orbit radius will be 4.2×10^4 km, so the velocity is 3.1 km/s.

For a spacecraft to orbit the Earth at its surface (equatorial radius 6378 km), the velocity needed is 7.91 km/s. This is substantially larger than the Earth's equatorial velocity of ~ 0.46 km/s. Thus this increment in velocity, 7.45 km/s, will be required just to orbit the Earth at zero altitude. This feature of orbit attainment will be developed further in Chapter 7, which deals with launch vehicles.

For *elliptical and circular orbits* $(0 \le e < 1)$, the magnitude of the potential energy per unit mass is greater than the kinetic energy. The total energy ε per unit mass is negative, a consequence of the potential energy zero datum being at infinity. This is a necessary condition for a closed orbit, and may be used to determine planetary capture requirements for interplanetary vehicles.

Parabolic trajectories ($e = 1$)

From equation (4.8) it is apparent that as e tends to unity, the semi-major axis becomes infinitely large; the parabolic trajectory is open to infinity (see Figure 4.7). The distance

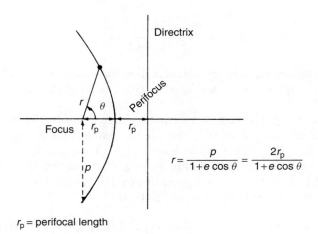

$$r = \frac{p}{1 + e\cos\theta} = \frac{2r_p}{1 + e\cos\theta}$$

r_p = perifocal length

Figure 4.7   Parabola ($e = 1$)

of its perifocus from the centre of the primary body is obtained from equation (4.7),

$$r_p = p/2 \tag{4.23}$$

A *position versus time relationship* may be found by noting that $h = r^2\dot{\theta}$ which, together with equation (4.7) yields

$$\dot{\theta} = \sqrt{(\mu/p^3)}(1 + e\cos\theta)^2$$

Whilst this equation is generally valid for all values of eccentricity, direct integration is only straightforward in the case $e = 1$, leading to Barker's equation

$$2\sqrt{(\mu/p^3)}(t - t_p) = 2M = \tan\frac{\theta}{2} + \frac{1}{3}\tan^3\frac{\theta}{2} \tag{4.24}$$

where the time of perifocal passage t_p is usually taken as zero.

The *energy of a parabolic trajectory* is zero, and as such it represents the boundary case between a captive, orbiting satellite and an escaping one. It is easy to see from equation (4.5) that the speed approaches zero at very large distances from the primary body. Consequently the parabola can be regarded as the minimum energy escape trajectory. Indeed, with $\varepsilon = 0$ in equation (4.5), the minimum speed required for escape is shown to be

$$V_{esc} = \sqrt{\left(\frac{2\mu}{r}\right)} \tag{4.25}$$

Hyperbolic trajectories ($e > 1$)

It can be seen for equation (4.8) that the hyperbolic trajectory has a negative semi-major axis, a curious attribute that influences consideration of the hyperbolic motion and geometry. The hyperbola is an open 'orbit' corresponding physically to planetary fly-by trajectories or to escape trajectories when the spacecraft retains some residual speed at a great distance from the primary body.

To obtain a *position versus time relationship* a method analogous to that applied in the elliptic motion case may be used. A hyperbolic eccentric anomaly F is introduced, the geometrical realization of which is given in terms of the shaded area shown in Figure 4.8. The position on the hyperbola is given in terms of focal distance

$$r = a(1 - e\cosh F), \tag{4.26}$$

and true anomaly θ, which is related to F by

$$\tan\frac{\theta}{2} = \tanh\left(\frac{F}{2}\right)\sqrt{\left(\frac{e+1}{e-1}\right)} \tag{4.27}$$

The required relationship between position and time is then found to be analogous to Kepler's equation,

$$M = e\sinh F - F \tag{4.28}$$

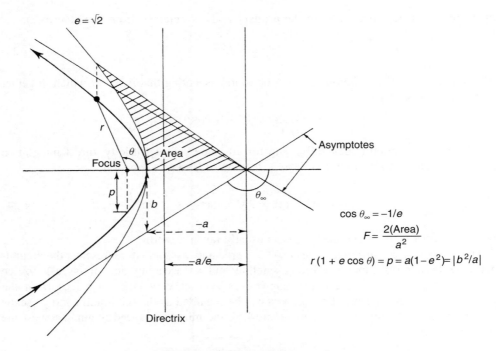

Figure 4.8 Hyperbola ($e > 1$)

where in this case

$$M = n(t - t_p)$$

Here the time of perifocal passage is usually taken as zero (i.e. $t_p = 0$), and the hyperbolic mean motion is defined by

$$n = \sqrt{\mu/(-a)^3}$$

Example: As a result of a swing-by manoeuvre (see Section 5.8.2 of Chapter 5) of Jupiter, a spacecraft is injected into a hyperbolic trajectory (relative to the Sun) to Pluto, with parameters $a = -3.5 \times 10^8$ km and $e = 3.0$. If the heliocentric distances of Jupiter and Pluto are 7.8×10^8 km and 59.0×10^8 km, respectively, produce a first estimate of the transfer time from Jupiter to Pluto, neglecting the effects of the gravitational field of each planet.

Solution: From equations (4.7) and (4.8), the true anomaly values on the hyperbola at Jupiter and Pluto are $\theta_J = 0.5291$ rads and $\theta_P = 1.7468$ rads, respectively. From equation (4.27), the corresponding values of the hyperbolic eccentric anomaly are $F_J = 0.3879$ and $F_P = 2.4698$. Using the hyperbolic analogy of Kepler's equation (4.28), the transit time can be estimated from

$$T_{\text{transit}} = t_P - t_J = \frac{(e \sinh F_P - F_P) - (e \sinh F_J - F_J)}{\sqrt{\mu/(-a)^3}} = \frac{(15.1335) - (0.8053)}{5.5633 \times 10^{-8}}$$

$$= 2.5755 \times 10^8 \, \text{s} \approx 8.2 \, \text{years.}$$

For hyperbolic trajectories, the energy ε is positive, and so the motion is dominated by kinetic energy. As r becomes very large $(r \to \infty)$, equation (4.21) shows that the speed becomes

$$V_\infty = \sqrt{\left(\frac{-\mu}{a}\right)} \tag{4.29}$$

V_∞ is called the *hyperbolic excess velocity*.

The asymptotic direction along which the spacecraft escapes is obtained from equation (4.7) and Figure 4.8 as

$$\theta_\infty = \cos^{-1}\left(-\frac{1}{e}\right) \tag{4.30}$$

and the angle δ through which the trajectory is deflected by the encounter is given by

$$\delta = 2\sin^{-1}\left(\frac{1}{e}\right) \tag{4.31}$$

To determine these geometrical aspects, the eccentricity is required. This may be related to the physical motion [1] by combining equations (4.9) and (4.29). This gives

$$e = 1 + \frac{r_p V_\infty^2}{\mu} \tag{4.32}$$

4.3   SPECIFYING THE ORBIT

The orbit equation (4.4) is equivalent to a sixth-order scalar differential equation that requires six initial conditions in order to determine the six constants of integration. At orbit injection these will be the three components of position and of velocity. A standard way of specifying an orbit is to use *orbital elements*, which refer the orbit to a frame of reference that is fixed relative to the stars. This is used both for astronomy and for satellites.

The *frame of reference* that is commonly used can be defined in terms of X-, Y-, and Z-axes (see Figure 4.9). The X- and Y-axes lie in Earth's equatorial plane and the Earth spins about the Z-axis.

The X-axis is in a direction from Earth to the Sun at the vernal equinox (21 March). The direction thus indicated is termed the *first point of Aries*. Some 2000 years ago this direction did point towards the constellation of Aries, but at present it points to Aquarius, and is moving along the equator at about $0.8'$ per year.

The Z-axis is along the Earth's spin axis, in the northerly direction. It is at an angle of $23°27'8''$ to the normal of the ecliptic plane, changing slowly with respect to the stars, with a period of about 25 725 years. The overall precessional rate is $\sim 0.12''$ per year and may be neglected for most spacecraft applications.

The Y-axis makes up a 'right-handed' orthogonal set with the X-, Z-axes.

The *celestial sphere* is the name given to a sphere with infinite radius centred on the Earth. The Z-axis meets it at a point known as the North celestial pole. The motion of

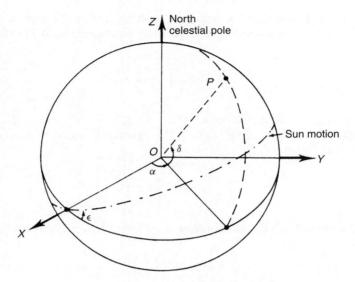

Figure 4.9 Celestial sphere

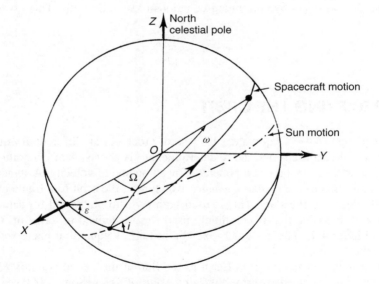

Figure 4.10 Spacecraft orbital elements

the Sun on the celestial sphere is indicated in Figure 4.9. It moves in the ecliptic plane, tilted about the X-axis through $23°27'8''$, known as the obliquity of the ecliptic, ε.

The location of a point P on the celestial sphere, strictly a direction OP, can be expressed in terms of the two angles α and δ. The great circle through P and the North celestial pole is called the hour circle of P. The angle α between OX and the equatorial radius to the hour circle is called the right ascension of P. Its declination δ is the angle between OP and the equatorial plane (see Figure 4.9).

The *six orbital elements* of a spacecraft's orbit round the Earth are chosen to represent different features. The orbital plane in Figure 4.10 intersects the equatorial plane in a line called the line of nodes; the ascending node is the point on the equator at which the spacecraft moves from the southern into the northern hemisphere. The right ascension of this node together with the inclination angle i define the plane of the orbit. The orientation within the plane is defined by the angle ω, known as the argument of periapsis, which is the angle measured in the direction of motion between the line of nodes and the vector **e**, which points to the periapsis. The shape of the orbit is defined by its eccentricity e, and its size by its semi-major axis a. The sixth element defines the position of the spacecraft in its orbit. The time (or epoch) of last passage through the periapsis may be used, or the mean anomaly M (equation 4.19).

4.4 ORBIT PERTURBATIONS

At the distances of orbiting spacecraft from the Earth its asphericity and non-uniform mass distribution result in its gravitational potential departing from the simple $1/r$ function, which was assumed in Section 4.2. Equation (4.1) is not valid in this situation and the equation of motion (4.4) must be modified to take account of the Earth's gravitational field.

There are additional forces that act on space vehicles, which were not included in the Keplerian formulation. Some are from additional masses that provide secondary gravitational fields; for Earth orbit, the Moon and the Sun provide such forces. Also, at low altitudes (typically at less than 1000 km altitude) the Earth's atmosphere imposes a drag force. Table 4.2 lists the major perturbing forces, and their relative importance, for space vehicle orbital analysis.

The equation of motion for a space vehicle about a body taking into account perturbative influences may be written in the form

$$\ddot{\mathbf{r}} = -\nabla U + \mathbf{b} \tag{4.33}$$

Table 4.2 Magnitude of disturbing accelerations acting on a space vehicle whose area-to-mass ratio is *A/M*. Note that A is the projected area perpendicular to the direction of motion for air drag, and perpendicular to the Sun for radiation pressure

Source	Acceleration (m/s^2)	
	500 km	Geostationary orbit
Air drag*	$6 \times 10^{-5} A/M$	$1.8 \times 10^{-13} A/M$
Radiation pressure	$4.7 \times 10^{-6} A/M$	$4.7 \times 10^{-6} A/M$
Sun (mean)	5.6×10^{-7}	3.5×10^{-6}
Moon (mean)	1.2×10^{-6}	7.3×10^{-6}
Jupiter (max.)	8.5×10^{-12}	5.2×10^{-11}

*Dependent on the level of solar activity

where U is the gravitational potential field and **b** is the force vector per unit mass due to other sources of perturbation, to which the vehicle is subject. A general closed solution is not possible, but there are a variety of solution methods that are appropriate for spacecraft dynamics. The 'variation of orbital elements' method is described here. Other methods such as those first proposed by Cowell and Crommelin [2] and Encke [3] are summarized by Cornelisse *et al.* [4].

The method of the variation of orbital elements may be considered in the following way. The elements referred to in the preceding section are constants for a Keplerian orbit, as derived in Section 4.2. When perturbative forces exist, they are no longer constant but for small forces they will change slowly.

A simple, physical model will serve to demonstrate this. Consider a spacecraft in circular orbit about a spherically symmetrical planet possessing an atmosphere of density ρ kg/m^3. If it is assumed that the perturbative drag force is small, then it is to be expected that the orbit will remain near circular. Now the velocity in a circular Keplerian orbit is given by $\sqrt{(\mu/r)}$. If the spacecraft's projected area in the direction of flight is S, then the work performed by the atmosphere as the vehicle moves round the orbit is given by $\sim -\pi r \rho S C_D \mu / r$, where C_D is an appropriate drag coefficient for the vehicle. Since this reduces the energy of the system, it is apparent that the energy constant ε in equation (4.21) will decrease and the orbital element a must do so too.

The variation of orbital elements method assumes that the actual orbit of a body, at any given instant, may be considered to have instantaneous values of Keplerian orbital elements. These are defined so that if the perturbing forces are removed at that instant, thus leaving only a central gravitational field whose potential is proportional to $1/r$, then the orbit will follow the Keplerian orbit that has the instantaneous orbital elements. These are called the *osculating elements*. It must be emphasized that the method is only appropriate for perturbing forces having a magnitude significantly smaller than μ/r^2.

The normal method for so describing an orbit is with recourse to Lagrange's planetary equations (see e.g. Reference [5]). One Gaussian form of these is the following [6]:

$$\left.\begin{aligned}
\frac{da}{d\theta} &= \frac{2pr^2}{\mu(1-e^2)^2}\left\{e\sin\theta\, S + \frac{p}{r}T\right\} \\
\frac{de}{d\theta} &= \frac{r^2}{\mu}\left\{\sin\theta\, S + \left(1+\frac{r}{p}\right)\cos\theta\, T + e\frac{r}{p}T\right\} \\
\frac{di}{d\theta} &= \frac{r^3}{\mu p}\cos(\theta+\omega)W \\
\frac{d\Omega}{d\theta} &= \frac{r^3\sin(\theta+\omega)}{\mu p\sin i}W \\
\frac{d\omega}{d\theta} &= \frac{r^2}{\mu e}\left\{-\cos\theta\, S + \left(1+\frac{r}{p}\right)\sin\theta\, T\right\} - \cos i\frac{d\Omega}{d\theta} \\
\frac{dt}{d\theta} &= \frac{r^2}{\sqrt{(\mu p)}}\left\{1 - \frac{r^2}{\mu e}[\cos\theta\, S - \left(1+\frac{r}{p}\right)\sin\theta\, T]\right\}
\end{aligned}\right\} \qquad (4.34)$$

where S, T, W form a triad of forces in a spacecraft-centred coordinate reference frame, S acting radially, T transverse to S in the orbital plane and directed positively in the sense of the spacecraft motion, and W normal to the orbit plane giving a right-handed system of forces. Strictly speaking, these components are those of the disturbing acceleration, though they are often referred to as components of the disturbing force.

It should be noted that the particular form of Lagrange's equations adopted depends on the type of orbit under investigation. For example, the set defined in equations (4.34) fails for circular ($e = 0$) and equatorial ($i = 0$) orbits. These may be dealt with by introducing parameter transformations of the type given, for example, by Roy [7].

Solutions of equations (4.34) must in general be performed using numerical techniques. This requires the formulation of functions for U and $\mathbf{b}$, and these are given below for certain cases of particular interest.

4.4.1 Gravitational potential of the Earth

The most convenient method for describing Earth's gravitational field outside its surface is to use a spherical harmonic expansion [8], given by

$$
U(r, \Phi, \Lambda) = \frac{\mu}{r} \left\{ -1 + \sum_{n=2}^{\infty} \left[\left(\frac{R_E}{r} \right)^n J_n P_{n0}(\cos \Phi) \right. \right.
$$

$$
\left. \left. + \sum_{m=1}^{n} \left(\frac{R_E}{r} \right)^n (C_{nm} \cos m\Lambda + S_{nm} \sin m\Lambda) P_{nm}(\cos \Phi) \right] \right\} \quad (4.35)
$$

where $U(r, \Phi, \Lambda)$ is the gravitational potential at a distance r from the centre of the Earth and Φ, Λ are the latitude and longitude. P_{nm} are Legendre polynomials. J_n, C_{nm} and S_{nm} are dependent on the mass distribution of the body, in this case the Earth. Terms of the form J_n are called *zonal harmonic coefficients*; they reflect the mass distribution of the Earth independently of longitude. C_{nm} and S_{nm} are the Earth's *tesseral harmonic coefficients* for $n \neq m$ and the *sectoral harmonic coefficients* for $n = m$.

These coefficients have mainly been determined from the motion of Earth-orbiting spacecraft. Whilst the lower-order terms were determined during the early 1960s, determination of the Earth's gravitational field continues to be an area of active research. Consequently there is a plethora of 'standard' global gravity field models, for example, the Joint Gravity Model (JGM) series [9], of which the JGM-3 model is an example. This gives the harmonic coefficient values to degree and order 70.

One of the major problems in determination of the higher-order terms is due to their rapid decrease with altitude; from equation (4.35) terms decrease with $(R_E/r)^n$. However, at low altitudes there are also difficulties, since the gravity effects are difficult to separate from other perturbations, in particular, those due to variable air drag. This situation would be greatly improved by the launch of dedicated spacecraft missions to determine the higher-order harmonic coefficients in the Earth's gravity field. Promising proposals, such as the US GRAVSAT and the European ARISTOTELES missions, were stalled in the 1990s because of fiscal problems. However, with the launch of the CHAMP mission in July 2000, and that of other similar missions (e.g. GRACE) soon afterwards, the prospect of significantly improved gravitational field models has been enhanced.

Table 4.3 gives the magnitude of some of the lower-order coefficients taken from a current geoid model [9]. From this it is apparent that the term J_2 is some three orders of

Table 4.3 Magnitude of low-order J, C and S values for Earth

J_2	1082.6×10^{-6}	C_{21}	0	S_{21}	0
J_3	-2.53×10^{-6}	C_{22}	1.57×10^{-6}	S_{22}	-0.90×10^{-6}
J_4	-1.62×10^{-6}	C_{31}	2.19×10^{-6}	S_{31}	0.27×10^{-6}
J_5	-0.23×10^{-6}	C_{32}	0.31×10^{-6}	S_{32}	-0.21×10^{-6}
J_6	0.54×10^{-6}	C_{33}	0.10×10^{-6}	S_{33}	0.20×10^{-6}

magnitude larger than the others, and to a first approximation it dominates the gravitational perturbative influences of the Earth. This term represents the polar flattening of the Earth (or equatorial bulge), and its magnitude can be summarized by noting that the Earth's poles are approximately 21 km closer to the Earth's centre than the equator. This 'excess' gravitational mass in the equatorial regions influences the motion of an orbiting spacecraft principally in two ways—*the regression of the line of nodes and the precession of the line of apsides*. Before discussing these, it is worth commenting briefly on the effects of the smaller tesseral harmonic coefficients.

As can be seen from equation (4.35), the C_{nm} and S_{nm} coefficients are associated with the longitudinal variation in the geopotential surface. These variations are generally smaller than those due to the dominant zonal harmonic terms, producing variations of the order of a few tens of metres in the height of the geopotential surface as measured above a reference spheroid. These small longitudinal variations tend to be 'averaged-out' by the effect of the Earth's rotation beneath the orbital plane of a low Earth-orbiting satellite. The resulting perturbing effect on the satellite's motion is small and periodic, generally producing a negligible net effect on each orbit. Their short-periodic effect must, however, be taken into account for spacecraft missions requiring very precise orbit determination (e.g. geodetic missions).

Tesseral harmonics can, however, produce moderate changes in the satellite's orbit if the orbit is in a *resonant state* with respect to the Earth's gravitational field. This occurs when the orbit period is such that the ground track of the satellite repeats precisely after a number of orbits and days. The orbit perturbations in this case are small but secular, causing a build-up of orbit changes while the satellite remains in the resonant state. It should be emphasized, though, that the resonant state is the exception rather than the rule, and generally the effect of the tesseral harmonic component is small for low Earth-orbiting satellites. A spacecraft in a Geostationary Earth orbit (GEO) encounters a form of resonance with respect to the lowest-order sectoral harmonic coefficients ($n = m = 2$), and this is often referred to as *triaxiality*. This is also discussed below.

Regression of the line of nodes

The equatorial bulge produces a torque that rotates the angular momentum vector. For prograde orbits ($i < 90°$), the orbit rotates in a westerly direction, leading to a regression of the line of nodes as shown pictorially in Figure 4.11. Neglecting all harmonic coefficients other than J_2, the rate of nodal regression may be written [7] to the first order in J_2 as

$$\overline{\Omega} = \Omega_0 - \frac{3}{2} \frac{J_2 R_E^2}{p^2} \overline{n} t \cos i + O[J_2^2] \tag{4.36}$$

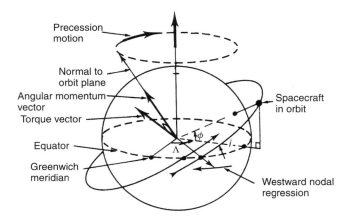

Figure 4.11 Nodal regression

where $\bar{n}$ is the mean angular velocity, $\sqrt{(\mu/a^3)}$. Thus, the secular rate of nodal regression per orbit is

$$\Delta\Omega = -\frac{3\pi J_2 R_E^2}{p^2} \cos i \text{ rad/rev} \tag{4.37}$$

Precession of the line of apsides

The second influence of the equatorial bulge may be considered in the following way. Since the mass 'seen' by the spacecraft crossing the equator is greater than the mean mass, the orbit tends to curve more rapidly. Since the gravitational field of the Earth is conservative, however, this leads to an overall rotation of the orbit, *within* the orbit plane, as demonstrated in Figure 4.12. This motion implies rotation of the semi-major axis and is termed precession of the line of apsides. The secular effect is given by

$$\bar{\omega} = \omega_0 + \frac{3}{2}\frac{J_2 R_E^2}{p^2}\bar{n}\left(2 - \frac{5}{2}\sin^2 i\right)t + O[J_2^2] \tag{4.38}$$

or per orbit

$$\Delta\omega = 3\pi \frac{J_2 R_E^2}{p^2}\left(2 - \frac{5}{2}\sin^2 i\right) \text{ rad/rev} \tag{4.39}$$

At an inclination of $\sim 63.4°$ the precession is zero. In the 1960s, the Soviet Union devised a highly eccentric orbit with this inclination, called a *Molniya* orbit. The resulting 'frozen apogee' condition was used to good effect for high-latitude communications, as discussed in Chapter 5.

Differential evolutionary rates

Each of the rates of orbital evolution noted in equations (4.36) to (4.39) indicate the absolute changes in the orbital elements relative to an inertial frame. As such, it is clear from the numerical values given in Table 4.3 why it is that the J_2 term dominates the evolution

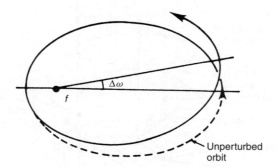

Figure 4.12   Apsidal precession

of an orbit (see also Figure 4.15). There are, however, several important situations when it is the *differential rate of evolution* of two or more orbits that should be considered. Cases to note are, for example, a satellite constellation requiring orbiting vehicles to maintain a fixed relative position, or where satellites are required to co-orbit in formation. A third example is the differential evolution of a space debris cloud [10].

Expressions for the evolution of an orbit due to higher harmonic terms can be found in Reference [11]. These may be partially differentiated to find the relative evolutionary rate of two orbits as a function of changes in each of the orbit elements, the semi-major axis, eccentricity and so on. If the semi-major axis is varied, then the differential rates of evolution in right ascension of the ascending node and the argument of perifocus may be written in the form

$$\frac{\partial \Omega_n}{\partial a} = \Delta\Omega_n \text{ and } \frac{\partial \omega_n}{\partial a} = \Delta\omega_n$$

where for the second, third and fourth harmonic terms we have

$$\frac{\partial \Delta\Omega_2}{\partial t} = -5.25a^{-9/2}(1-e^2)^{-2}\mu^{1/2}J_2R_E^2\cos i$$

$$\frac{\partial \Delta\Omega_3}{\partial t} = -1.6875a^{-11/2}e(1-e^2)^{-3}\mu^{1/2}J_3R_E^3\cot i(15\sin^2 i - 4)\sin\omega$$

$$\frac{\partial \Delta\Omega_4}{\partial t} = 1.1786a^{-13/2}(1-e^2)^{-4}\mu^{1/2}J_4R_E^4$$
$$\times \cos i[(1+1.5e^2)(7\cos^2 i - 3) - (7\cos^2 i - 4)e^2\cos 2\omega]$$

$$\frac{\partial \Delta\omega_2}{\partial t} = -5.25a^{-9/2}(1-e^2)^{-2}\mu^{1/2}J_2R_E^2(2-2.5\sin^2 i)$$

$$\frac{\partial \Delta\omega_3}{\partial t} = 1.6875a^{-11/2}e^{-1}(1-e^2)^{-3}\mu^{1/2}J_3R_E^3$$
$$\times \sin i[5\cos^2 i - 1 + e^2(35\cos^2 i - 4\operatorname{cosec}^2 i)]\sin\omega$$

$$\frac{\partial \Delta\omega_4}{\partial t} = -2.5781a^{-13/2}(1-e^2)^{-4}\mu^{1/2}J_4R_E^4$$
$$\times [\{(49\cos^4 i - 36\cos^2 i + 3) + 0.75e^2(63\cos^4 i - 42\cos^2 i + 3)\}$$
$$+ \{\sin^2 i(7\cos^2 i - 1) - 0.5e^2(63\cos^4 i - 56\cos^2 i + 5)\}\cos 2\omega]$$

Using these relationships it has been found that under some circumstances the differential evolution of the argument of perifocus due to J_3 can be more significant than that due to J_2.

Triaxiality perturbation

The terms representing the longitudinal variation of the Earth's gravitational field have their most significant influence on geostationary satellites, since non-synchronous orbits will average them out. The term J_{22} equal to $(C_{22}^2 + S_{22}^2)^{1/2}$ has the greatest influence due to the $(R_E/r)^n$ weighting of coefficients. This term represents the slight ellipticity of the Earth's equatorial cross-section, having its major axis aligned approximately along $15\,°W-165\,°E$.

The term 'triaxiality' is derived from the geometrical form of the Earth resulting from the combination of the oblateness and J_{22} characteristics. To lowest order, the Earth may be modelled as an oblate spheroid with an elliptical equatorial cross-section. This form is generated by three mutually orthogonal axes (tri-axes) centred on the Earth's centre, each corresponding to a different Earth radius. A satellite will be stable if it is at a minimum of the Earth's potential field, that is, on the minor axis of the equator. Departure from these two longitude positions provides an increasing perturbation. Agrawal [12] gives the following expression for the longitudinal acceleration:

$$\ddot{\Lambda} = k^2 \sin 2(\Lambda - \Lambda_0) \tag{4.40}$$

where $k^2 = -18J_{22}\omega_e^2(R_E^2/a_s^2) \approx -1.7 \times 10^{-3}$ deg/day^2

and $\Lambda_0 = 75°$ E or $255°$ E (stable longitudes). Here ω_e is the sidereal rotation rate of the Earth and a_s is the semi-major axis of the synchronous orbit.

Longitude drift over time is shown in Figure 4.13.

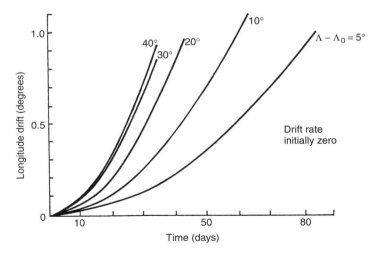

Figure 4.13 Longitudinal drift acquired over a period in geostationary orbit, as a function of the difference $(\Lambda - \Lambda_0)$ in initial operating longitude and minor axis (stable) longitude

Triaxiality can be seen to provide an East/West station-keeping problem for a geostationary satellite. This is referred to in Chapter 5.

4.4.2   Atmospheric drag

For low Earth-orbiting spacecraft, the perturbation due to the atmosphere cannot be neglected. Atmospheric effects lead to a force that may be expressed as two orthogonal components—drag along the direction of travel and lift at right angles to it. The simplest form in which to write the drag force is

$$\mathbf{F}_D = \frac{1}{2}\rho S C_D V_r^2 \left(\frac{-\mathbf{V}_r}{|V_r|}\right) \tag{4.41}$$

where $\mathbf{V}_r$ is the velocity vector of the spacecraft relative to the atmosphere, ρ is the atmospheric density, S a reference area for the vehicle and C_D the vehicle's coefficient of drag referred to the reference area.

The lifting force is obtained by replacing C_D with the coefficient of lift C_L, in equation (4.41), and by noting that the lift force acts perpendicular to the velocity vector $\mathbf{V}_r$. For most spacecraft the effects of lift are negligible compared with drag, but when precise orbit determination is required ($\sim$ few metres accuracy) its influence must be included [13].

The drag coefficient for space vehicles is somewhat complex due to the type of flow regime within which the motion takes place. Because of the large mean free path, the flow cannot be treated in a continuum manner. Indeed, the most suitable description is that of free molecular flow wherein the molecular description is such that molecules reflected from the spacecraft surfaces do not interact further with the flow field; no shock wave is therefore formed about a body moving through the atmosphere at orbit altitudes. The drag force experienced is dependent on the gas—surface interaction, for which there is only sparse experimental data at typical incident velocities. Stalder and Zurick [14], Schaff and Chambre [15] and Schamberg [16] provide the basic theory for rarefied flow, and Cook [17] provides height-related C_D data adopting the Schamberg model. Typically values of $C_D \sim 2.5$ are predicted by these theories, when the reference area is taken as the spacecraft's cross-sectional area projected normal to the velocity vector.

With reference to equation (4.41), drag is most significant at perigee where both the velocity and density are greatest. King-Hele [5] provides detailed analysis of the secular changes caused by drag. To first order, these approximate to an impulsive, negative velocity increment occurring at perigee. This will lead to a reduction in the semi-major axis of an elliptical orbit. For a circular orbit, drag will occur continuously around the orbit. Assuming that the change in the radius of such an orbit is small, the $\Delta\tau$ in orbit period for a circular orbit of radius r will be given by

$$\frac{\Delta\tau}{\tau} \approx -3\pi\rho r (S C_D / M) \tag{4.42}$$

where ρ is the density at r (measured from the centre of the Earth) and M/SC_D is called the vehicle ballistic parameter; M is the vehicle mass, and S the projected area.

The *dominant influences of drag* are thus to cause orbit contraction and circularization, with eventual re-entry. Since the Earth's atmosphere rotates approximately synchronously with the Earth, in general the drag force has a component perpendicular to the orbit plane. This results in a change in inclination of the orbit. King-Hele [5] gives expressions for the radial, transverse and normal force components for a vehicle moving through an atmosphere rotating at an angular rate α rad/s. These are

$$S = -\frac{1}{2}\rho \upsilon \delta \left(\frac{\mu}{pF}\right)^{1/2} e \sin \theta$$

$$T = -\frac{1}{2}\rho \upsilon \delta \left(\frac{\mu}{pF}\right)^{1/2} \left(1 + e \cos \theta - r\alpha \left(\frac{p}{\mu}\right)^{1/2} \cos i\right) \quad (4.43)$$

$$W = -\frac{1}{2}\rho \upsilon \delta \frac{r\alpha}{\sqrt{F}} \sin i \cos(\theta + \omega)$$

where

$$F = \left(1 - \frac{r_p \alpha}{\upsilon_p} \cos i\right)^2$$

and

$$p = a(1 - e^2)$$

Here, r_p, υ_p represent conditions at perigee, υ is the absolute velocity of the vehicle and δ is a modified ballistic parameter given by $\delta = FSC_D/M$.

4.4.3   Additional gravitational fields

Luni-solar perturbations

Other bodies in the solar system impose additional gravitational forces on spacecraft orbiting the Earth. The proximity and mass of the Moon provides the most significant influence. The Sun is a great deal further away but its larger mass generates an influence that is of a similar order of magnitude. These perturbations are collectively termed *luni-solar perturbations*. Since in general these bodies will not lie in the same plane as the vehicle orbit, their most significant influence will be to change the inclination of the orbit with respect to the equator. The formulation of this three-body interaction does not admit a general closed-form solution, and numerical techniques must be employed.

The disturbing acceleration a_d of a satellite due to a disturbing body having a mass M_d and gravitational parameter μ_d is given by

$$a_d = \mu_d \sqrt{(\mathbf{R} \cdot \mathbf{R})} \quad (4.44)$$

where

$$\mathbf{R} = \frac{\mathbf{r}_{sd}}{r_{sd}^3} - \frac{\mathbf{r}_d}{r_d^3}$$

and $\mathbf{r}_{sd}$ and $\mathbf{r}_d$ are defined in Figure 4.14.

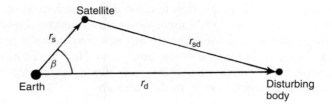

Figure 4.14 Disturbing body and satellite positions

It can then be shown [4] that the maximum value of the ratio of the disturbing acceleration a_d to the central acceleration a_c is given by

$$\frac{a_d}{a_c} = \frac{M_d}{M_c}\left(\frac{r_s}{r_d}\right)^3 \sqrt{(1 + 3\cos^2\beta)} \tag{4.45}$$

or

$$\frac{a_d}{a_c} \le 2\frac{M_d}{M_c}\left(\frac{r_s}{r_d}\right)^3$$

The angle β is shown in Figure 4.14; M_c is the mass of the central body about which the vehicle is in orbit.

At geostationary orbit this ratio provides values of 3.3×10^{-5} and 1.6×10^{-5} for the Moon and Sun, respectively. The effect of Jupiter is some five orders of magnitude lower than these.

Expressions that provide approximate average rates of change of orbital elements for a single disturbing body, given by Cook [18], are

$$\left.\begin{aligned}
\frac{da}{dt} &\approx 0 \\[2mm]
\frac{de}{dt} &\approx \frac{-15}{2}\frac{K}{n}e(1 - e^2)^{1/2}\left[AB\cos 2\omega - \frac{1}{2}(A^2 - B^2)\sin 2\omega\right] \\[2mm]
\frac{d\Omega}{dt} &\approx \frac{3KC}{4n(1 - e^2)^{1/2}\sin i}[5Ae^2\sin 2\omega + B(2 + 3e^2 - 5e^2\cos 2\omega)] \\[2mm]
\frac{d\omega}{dt} + \frac{d\Omega}{dt}\cos i &= \frac{3}{2}\frac{K(1 - e^2)^{1/2}}{n}\left[5\left\{AB\sin 2\omega + \frac{1}{2}(A^2 - B^2)\cos 2\omega\right\}\right. \\[2mm]
&\quad\left. -1 + \frac{3}{2}(A^2 + B^2) + \frac{5a}{2er_d}\left\{1 - \frac{5}{4}(A^2 + B^2)\right\}(A\cos\omega + B\sin\omega)\right] \\[2mm]
\frac{di}{dt} &= \frac{3KC}{4n(1 - e^2)^{1/2}}[A(2 + 3e^2 + 5e^2\cos 2\omega) + 5Be^2\sin 2\omega]
\end{aligned}\right\} \tag{4.46}$$

where

$$K = \frac{GM_d}{r_d^3}$$

and

$$A = \cos(\Omega - \Omega_d)\cos u_d + \cos i_d \sin u_d \sin(\Omega - \Omega_d)$$
$$B = \cos i[-\sin(\Omega - \Omega_d)\cos u_d + \cos i_d \sin u_d \cos(\Omega - \Omega_d)] + \sin i \sin i_d \sin u_d$$
$$C = \sin i[\cos u_d \sin(\Omega - \Omega_d) - \cos i_d \sin u_d \cos(\Omega - \Omega_d)] + \cos i \sin i_d \sin u_d$$

Angles $\Omega_d, u_d = (\theta_d + \omega_d)$ and i_d are the orbital elements of the disturbing body referred to the equatorial-based system of Section 4.3.

4.4.4 Solar radiation pressure (SRP) perturbations

A spacecraft moving within the solar system will experience, to some degree, a perturbation to its trajectory owing to the incidence of solar radiation upon its illuminated surfaces.

Electromagnetic radiation carries momentum, and the reflection of incident radiation at a surface represents an exchange of momentum. Consequently the radiation exerts a small but measurable pressure on the spacecraft of mean magnitude

$$P = F_\odot/c$$

where $F_\odot$ is the solar energy flux at the spacecraft, and c is the speed of light. For a spacecraft in near-Earth orbit, where $F_\odot = 1400 \, \text{W/m}^2$, the resulting mean SRP is approximately $P_E = 4.7 \times 10^{-6} \text{N/m}^2$. The effect of this pressure as a force acting on the vehicle is dependent upon the reflective characteristics of the spacecraft surfaces.

An Earth-orbiting spacecraft is also immersed in other electromagnetic radiation fields, such as those due to Earth albedo and infra-red emission, which similarly produce a perturbing force on the vehicle. However, these are subordinate in magnitude to the dominate solar radiation effect.

The perturbing effect of SRP on a spacecraft's trajectory is directly dependent upon the vehicle's area-to-mass ratio A/m, and inversely proportional to the square of its distance from the Sun. Hence, the magnitude of the SRP force per unit spacecraft mass (the disturbing acceleration) along the Sun—spacecraft line may be expressed as

$$f_{\text{SRP}} = s\frac{A}{m}P_E\left(\frac{a_\odot}{r_\odot}\right)^2 \tag{4.47}$$

where $r_\odot$ and $a_\odot$ are the spacecraft's distance and the Earth's mean distance from the Sun, respectively; s is a constant whose value, between 0 and 2, depends upon the reflective properties of the spacecraft's surface.

To estimate how the orbit elements are perturbed by SRP, Lagrange's planetary equations (4.34) may be used, with the triad of disturbing accelerations given by [19]:

$$S = f_{\text{SRP}}S_0(\theta), \; T = f_{\text{SRP}}T_0(\theta), \; W = f_{\text{SRP}}W_0 \tag{4.48}$$

where $S_0(\theta)$, $T_0(\theta)$ and W_0 are direction cosines,

$$
\begin{Bmatrix} S_0(\theta) \\ T_0(\theta) \end{Bmatrix} = -\cos^2\frac{i}{2}\cos^2\frac{\varepsilon}{2}\begin{Bmatrix} \cos \\ \sin \end{Bmatrix}(\lambda_\odot - u - \Omega)
$$

$$
-\sin^2\frac{i}{2}\sin^2\frac{\varepsilon}{2}\begin{Bmatrix} \cos \\ \sin \end{Bmatrix}(\lambda_\odot - u + \Omega)
$$

$$
-\frac{1}{2}\sin i \sin\varepsilon\left[\begin{Bmatrix} \cos \\ \sin \end{Bmatrix}(\lambda_\odot - u) - \begin{Bmatrix} \cos \\ \sin \end{Bmatrix}(-\lambda_\odot - u)\right]
$$

$$
-\sin^2\frac{i}{2}\cos^2\frac{\varepsilon}{2}\begin{Bmatrix} \cos \\ \sin \end{Bmatrix}(-\lambda_\odot - u + \Omega)
$$

$$
-\cos^2\frac{i}{2}\sin^2\frac{\varepsilon}{2}\begin{Bmatrix} \cos \\ \sin \end{Bmatrix}(-\lambda_\odot - u - \Omega), \tag{4.49}
$$

$$
W_0 = \sin i \cos^2\frac{\varepsilon}{2}\sin(\lambda_\odot - \Omega) - \sin i \sin^2\frac{\varepsilon}{2}\sin(\lambda_\odot + \Omega)
$$

$$
-\cos i \sin\varepsilon \sin\lambda_\odot. \tag{4.50}
$$

Here $u = \theta + \omega$, ε is the obliquity of the ecliptic, and $\lambda_\odot$ is the ecliptic longitude of the Sun.

A method of integrating these equations was developed by Harwood and Swinerd [20], who applied the theory to the satellites Explorer 19 and Lageos [21].

Communication spacecraft in GEO, many of which have large solar array surfaces, experience SRP perturbations to their orbits. Generally, this leads to an increase in the eccentricity of the orbit, which has implications for station-keeping activities (see Chapter 5). For Low Earth Orbit (LEO) spacecraft, below around 600 km in altitude, the effects of air drag, however, dominate those of radiation pressure.

4.4.5 Concluding remarks

Figure 4.15 illustrates the relative magnitude of the main sources of perturbation acting upon an Earth-orbiting spacecraft. For each effect, the logarithm of the disturbing acceleration, normalized to 1 g, is shown as a function of altitude.

From the diagram it is apparent that the primary inverse square law gravity field of the Earth is very much the dominant influence. The only effect to compete with this is aerodynamic drag, the drag curve reaching the 1 g level at low altitude around 80 km. In this case the spacecraft encounters an atmospheric re-entry situation, when the magnitude of inertial accelerations due to aerodynamic effects can equal and exceed the gravitational acceleration (see Chapter 5).

After primary gravity, the zonal harmonic potential terms begin to play a role. The J_2, or Earth oblateness, effects are around three orders of magnitude smaller than primary gravity, but nevertheless lead to significant perturbations in the orbit perigee and nodal positions as we have seen in Section 4.4.1. These effects must be modelled accurately when planning LEO operations, for example, rendezvous strategies. The higher-order zonal harmonics are of lesser importance, but still lead to significant long-period perturbations in

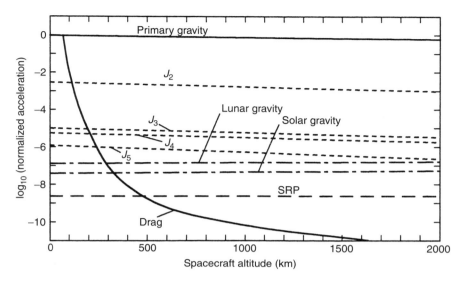

Figure 4.15 Comparisons of the disturbing accelerations for the main sources of perturbation

the orbit elements. These higher-order terms decrease rapidly in magnitude with altitude, and are comparable to the third-body forces of lunar and solar gravity.

The surface forces, drag and SRP, are dependent upon the spacecraft area-to-mass ratio, and a representative value of $A/m = 0.005\,\mathrm{m}^2/\mathrm{kg}$ was used for the purpose of compiling Figure 4.15. Furthermore, the drag curve is dependent upon the level of solar activity. Here a moderate level was chosen, although it is worth noting that at, say, 500 km altitude the drag acceleration can be an order of magnitude higher at solar maximum than at solar minimum.

The SRP curve is effectively independent of altitude, since it is a function of distance from the Sun, rather than from the Earth. The heliocentric distance of an Earth-orbiting spacecraft obviously does vary, but the effect on the disturbing acceleration is essentially second order. It is interesting to note the altitude at which the SRP and drag curves intersect. Figure 4.15 suggests that above 600 to 700 km altitude SRP dominates. To try to estimate more precisely the altitude at which this transition occurs is difficult since it is dependent upon the level of solar activity.

4.5 RESTRICTED THREE-BODY PROBLEM

The *three-body problem* refers to the problem of predicting the motion of three bodies, each influenced by the gravitational fields of the others. There is no general solution to the equations governing this. Of particular relevance to spacecraft is the *circular restricted* form of this problem, in which two of the bodies have masses that far outweigh the mass of the third, and are in circular orbits about each other.

An important example of this is when the spacecraft moves within the region of space that is dominated by both Earth and Moon; their masses are, respectively, about 6×10^{24} and 7×10^{22} kg. Another example is a rendezvous with a comet that has a mass distribution like a dumbbell. In both of these cases, the two dominant masses move in approximately circular orbits about their joint centre-of-mass—their barycentre—both lying on the same diameter, which rotates at a constant angular rate Ω.

Using a frame of reference (f.o.r.) that is centred on the barycentre and such that the two dominant masses are at fixed positions in it, then this f.o.r. is not an inertial one; it is rotating at a fixed rate Ω. A small mass m, at position $\mathbf{r}$ relative to the barycentre, and at $\mathbf{r}_1$ and $\mathbf{r}_2$ relative to M_1 and M_2 as shown in Figure 4.16, will experience gravitational forces $\mathbf{F}_1$, $\mathbf{F}_2$, and possibly a disturbance or thrusting force $\mathbf{F}_d$, where

$$\mathbf{F}_1 = -\frac{mGM_1}{r_1^3}\mathbf{r}_1 \quad \text{and} \quad \mathbf{F}_2 = -\frac{mGM_2}{r_2^3}\mathbf{r}_2$$

The equation governing the motion of the mass m within this f.o.r. is:

$$\ddot{\mathbf{r}} + 2(\Omega \times \dot{\mathbf{r}}) + (\Omega \cdot \mathbf{r})\Omega - \Omega^2\mathbf{r} = -G\left(\frac{M_1}{r_1^3}\mathbf{r}_1 + \frac{M_2}{r_2^3}\mathbf{r}_2\right) + \frac{1}{m}\mathbf{F}_d \qquad (4.51)$$

It should be noted that $\dot{\mathbf{r}}$ is not the *absolute* velocity $\mathbf{v}$; it has components that are the rates of change of the components of $\mathbf{r}$, along axes that are fixed in the (rotating) f.o.r. It is thus the velocity as it would be seen by an observer at the barycentre in Figure 4.16 and who is rotating with the same angular velocity as the f.o.r. described above. $\ddot{\mathbf{r}}$, defined in similar fashion in terms of the rate of change of the components of $\dot{\mathbf{r}}$, is the acceleration of the mass m as seen by the same observer.

When the disturbance or thrust force $\mathbf{F}_d$ is zero, equation (4.51) may be integrated once, to give the *Jacobi Integral*, the equivalent of the *vis-viva* equation (4.5) for this

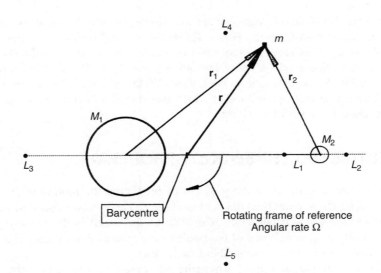

Figure 4.16 The restricted three-body situation, showing the Libration points L_1 to L_5

three-body situation:

$$\frac{1}{2}\dot{\mathbf{r}}^2 - \frac{1}{2}(\mathbf{r} \times \mathbf{\Omega})^2 - G\left(\frac{M_1}{r_1} + \frac{M_2}{r_2}\right) = J \tag{4.52}$$

J is a constant of integration having the dimensions of energy per unit mass, whose value depends on the initial conditions.

Equation (4.52) may be used in ways such as the following:

- A trajectory in the three-body region cannot be obtained analytically. J remains constant along a trajectory, however, and so equation (4.52) may be used as a 'check' when a numerical integration of equation (4.51) is used to predict a trajectory.
- *Escape criterion*: The requirement for a body to escape from the two dominant bodies is for J to be at least zero. The reasoning is similar to that used in Section 4.2, 'Parabolic trajectories'.

For each value of J less than zero, a surface surrounding the two dominant bodies may be determined when $\dot{\mathbf{r}} = \mathbf{0}$. This represents a boundary beyond which a spacecraft with this amount of J cannot go. The intersection of this zero-velocity surface with the $x - y$ plane of the rotating frame of reference is shown in Figure 4.17(a) to (f) for different values of J. The example shown is the Earth–Moon system. Forbidden regions are shaded. In (a) a spacecraft with a low energy level ($J < -1.84345$ (km/s)^2/kg) will be restricted to within the region around the Earth that is shown as approximately circular in this plane. If it is setting out from the Moon, it is restricted to a smaller circular region there. On the other hand, if an object is outside the outer circle then it cannot reach the vicinity of either the Earth or the Moon.

Increasing energy progressively reduces the size of the restrictions, as shown in (a) to (f), with transfer between Earth and Moon becoming possible from (b) onwards ($J > -1.6793$ (km/s)^2/kg), and journeys beyond the Moon to outer space from (d) onwards ($J > -1.67077$ (km/s)^2/kg). 'Gateways' open at the points L_1, L_2 and L_3 when the appropriate energy levels are reached, and all restricted areas shrink to zero when the energy level $J = -1.5738$ (km/s)^2/kg of L_4 and L_5 is exceeded. These special points are the *Lagrangian Points*. It is worth noting that in the Earth–Moon system the energy levels differ very little between these points, with only 0.5% difference between L_1 and L_2, for example.

- *Lagrangian, or Libration Points:* These are five points fixed within the rotating frame of reference at which a stationary body will be in equilibrium. Denoted by L_1 to L_5, they all lie in the plane defined by the orbits of the primary bodies around each other (see Figures 4.16 and 4.17). Three of these, L_1, L_2 and L_3, lie on the line joining the primary bodies, and correspond to unstable equilibrium positions. These have been used by spacecraft, such as the European solar observatory SOHO, which is 'orbiting' the L_1 point in the Earth–Sun system. It is a good position for observing the Sun. Because of the unstable nature of the equilibrium about the L_1 point, station-keeping activity is required to stabilize its 'orbit'. Conversely, it can be shown that the off-axis points L_4 and L_5 are stable with respect to small displacements. An impressive example of this stability is the presence of a number

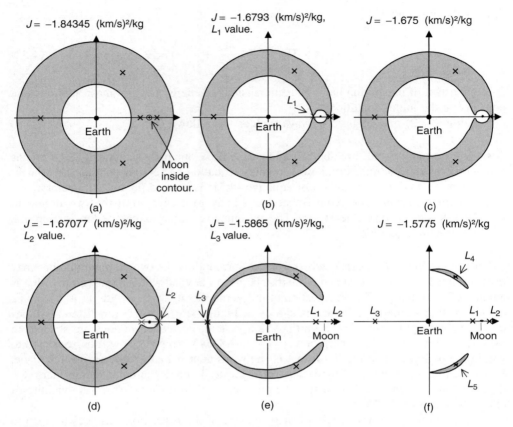

Figure 4.17 The intersection of the zero-velocity surface with the Earth–Moon $x - y$ plane as the energy level J changes

of large asteroids, called the Trojans, oscillating about the L_4 and L_5 points with respect to the Jupiter–Sun system.
- *Effect of* $\mathbf{F}_d$: The inclusion of the extra force $\mathbf{F}_d$ into the Jacobi equation (4.52) leads to the right-hand side becoming

$$J + \frac{1}{m} \int \mathbf{F}_d \cdot \dot{\mathbf{r}} \, dt \qquad (4.53)$$

From this it may be seen that the best direction for a thrust force aimed at changing the value of J is along the trajectory in this (non-inertial) frame of reference.

REFERENCES

[1]   Brown, C. D. (1998) *Spacecraft Mission Design* (2nd edn), AIAA, Reston.
[2]   Cowell, P. H. and Crommelin, A. C. D. (1910) *Investigation of the Motion of Halley's Comet from* 1759–1910, Appendix to Greenwich Observatory.
[3]   Encke, J. F. (1852) *Astron. Nachr.*, **33**, 377–398.

[4]   Cornelisse, J. W., Schoyer, H. F. R. and Wakker, K. F. (1979) *Rocket Propulsion and Space-flight Dynamics*, Pitman, London.

[5]   King-Hele, D. G. (1987) *Satellite Orbits in an Atmosphere: Theory and Applications*, Blackie, London.

[6]   Brouwer, D. and Clemence, G. M. (1961) *Methods of Celestial Mechanics*, Academic Press, London.

[7]   Roy, A. E. (1988) *Orbital Motion* (3rd edn - Student Text), Adam Hilger, Bristol.

[8]   Wertz, J. R. (1978) *Spacecraft Attitude Determination and Control*, Reidel, Dordrecht.

[9]   Nerem, R. S. *et al.* (1994) *J. Geophys. Res.*, **99**, 24421–24447.

[10] Stark, J. P. W. (2001) The evolution of debris clouds to microscopically chaotic motion, *J. Spacecraft Rockets*, **38**, 554–562.

[11] Groves, G. V. (1959) Motion of a satellite in the Earth's gravitational field, *Proc. Royal Soc. London* **254**, 48.

[12] Agrawal, B. N. (1987) *Design of Geosynchronous Spacecraft*, Prentice-Hall, Englewood Cliffs.

[13] Stark, J. P. W. (1986) ESA SP 255, 239–246.

[14] Stalder, J. R. and Zurick, V. J. (1951) NACA Tech. Note 2423.

[15] Schaff, S. A. and Chambre, P. L. (1958) in *Fundamentals of Gas Dynamics*, ed. Emmons, H., Oxford University Press, Oxford.

[16] Schamberg, R. (1959) Rand Corp. R.M. 2313.

[17] Cook, G. E. (1965) *Planet. Space Sci.*, **13**, 929.

[18] Cook, G. E. (1962) *Geophys. J.*, **6**, 271.

[19] Aknes, K. (1976) *Celestial Mech. Dynam. Astron.*, **13**, 89.

[20] Harwood, N. M. and Swinerd, G. G. (1995) *Celestial Mech. Dynam. Astron.*, **62**, 71.

[21] Harwood, N. M. and Swinerd, G. G. (1995) *Celestial Mech. Dynam. Astron.*, **62**, 81.

5 MISSION ANALYSIS

John P. W. Stark[1] and Graham G. Swinerd[2]

[1] Department of Engineering, Queen Mary, University of London
[2] Aeronautics and Astronautics, School of Engineering Sciences, University of Southampton

5.1 INTRODUCTION

Whereas celestial mechanics deals with a spacecraft's dynamics and orbit relative to the stars, one aspect of mission analysis is to relate both orbit and attitude to the ground, the Sun and so on. This ensures that operational aspects such as spacecraft–ground communications, power raising and spacecraft tracking can be achieved, which is of course crucial to mission success. However, beyond these operational issues, the mission design process is also a systems level activity, performed early in the feasibility study stage, which has a significant influence upon the spacecraft design. Generally, a careful analysis of the mission objectives, and consequently the payload operation and orbital aspects, is required. This, in turn places *design requirements* upon the spacecraft subsystem elements supporting the payload. The mission design then becomes an iterative process, wherein the impacts on the spacecraft system design may compromise an optimal solution for the mission design, and *vice versa*. These system aspects of mission analysis are touched upon briefly in this chapter, but a more detailed discussion is given in Chapter 19.

For study purposes, it is convenient to place missions into categories based upon their orbits. The following cover most applications:

- Low Earth orbit (LEO)
- Medium height Earth orbit (MEO)
- Geostationary Earth orbit (GEO)
- Highly elliptical orbits (HEO)
- Non-geocentric orbits (lunar and interplanetary)

There is also a potentially increasing category embracing space stations, on-orbit servicing and rescue. However, this chapter will focus on the five categories mentioned above since they illustrate most of the aspects that link orbits to missions.

Spacecraft Systems Engineering (Third Edition). Edited by P. W. Fortescue, J. P. W. Stark and G. G. Swinerd
© 2003 John Wiley & Sons Ltd

A history of each mission may be viewed as a sequence of events starting at pre-launch and extending to end-of-life. The *pre-launch phase* includes all those operations that are required in order to effect launch vehicle ignition and separation from the umbilical cable between space system (spacecraft plus launcher) and the ground launch facilities.

The *launch phase* involves a sequence of events, many of which are pre-programmed and automatic. Careful planning is required for placing the spacecraft into an appropriate intermediate orbit from which the operational one may be attained. One evident constraint during the launch phase is the requirement for continuous communications and tracking. These facilities may be required for a ground-centre override to abort the mission, but also there is generally a requirement to monitor the performance of a launch vehicle. Figure 5.1 shows the ground-centre network used for a typical Ariane 5 launch from the launch site in French Guiana.

Orbit transfer is the next major event. This involves transferring the spacecraft from the orbit into which the launch vehicle *actually* places it, to the one from which the operational phase of the mission can commence. There is *a priori* uncertainty in the orbit at launcher burn-out as shown by the performance envelope of the launch vehicle (see Chapter 7); but there is *a posteriori* uncertainty because of the orbit determination process that inevitably includes some errors in the range and range-rate measurements of the spacecraft trajectory (see Chapter 4). Transfer between these orbits requires propellant, and it is the task of the mission planners to determine how much is required in order to attain the desired orbit from any given initial one, with a given level of confidence. This is particularly important for geostationary spacecraft, since the final operations needed to obtain their orbit call for the use of a secondary propulsion system that may subsequently be used for attitude and orbit control. Excessive use or under-budgeting of fuel will therefore affect the available payload mass and reduce the operational life of the space system as a whole.

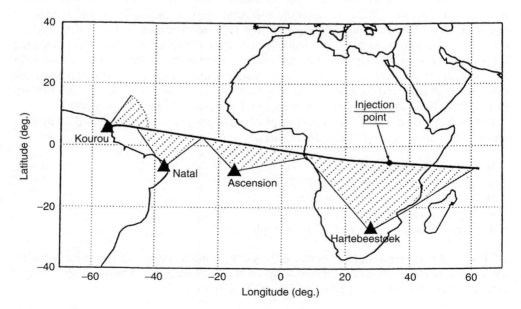

Figure 5.1 Typical Ariane 5 ground-track, showing launch monitoring ground stations (Reproduced by permission of Arianespace)

On-station operations include various mission events. Fuel budgeting for station-keeping and attitude control will be necessary.

For sky survey observatory missions, the scientific goals may require continuous changing of the orientation of the spacecraft in such a way that a wide variety of directions or even complete coverage of the celestial sphere may be achieved. Careful planning is required in order to optimize the observing programme; both the rate at which the spacecraft is reorientated and the angular distance through which its attitude is changed between observations will influence fuel consumption. Repointing operations are normally performed using reaction wheels, with thrusters (propellant) being used to periodically dump angular momentum from the wheels (see Chapter 9). Whilst contingency planning is made for unscheduled repointing, many scientific missions inevitably cease only when the fuel has been exhausted. Other constraints that will be of concern in deriving mission profiles for scientific observatories include minimum sun/telescope angle, earth limb/telescope angle, passage through the South Atlantic Anomaly in the Earth's magnetosphere and thermal balance of the spacecraft and its sensors. Analysis and planning activities associated with a number of specific types of mission during their operational phases are detailed in the following sections.

Decommissioning the spacecraft is the final event associated with a space mission. In GEO, the demand for longitude slots is increasing, and thus an uncontrolled satellite in this orbit is wasteful and also constitutes a collision hazard [1]. It has therefore become standard practice to remove an obsolete spacecraft from GEO into a higher orbit by the use of residual propellant in the secondary propulsion system at the end of its life. This manoeuvre is frequently and appropriately called a 'graveyard burn'. It is also becoming the practice in LEO missions to provide a controlled re-entry through the Earth's atmosphere. The reason for this is that uncontrolled re-entry can lead to the vehicle breaking up, providing a hazard on the ground and adding to the problem of space debris.

As mentioned earlier, a key aspect of mission analysis is to identify critical features of the mission, which have an impact upon system and subsystem design. Every space vehicle is required to meet specific mission objectives and these, thus provide design requirements.

The analysis of these, in the context of a particular spacecraft mission, usually leads to a subset of design requirements that are influential in shaping the whole nature of the vehicle. These requirements are often referred to as *design drivers*. For example, the thermal control system design may be routinely achieved, relatively speaking, in the case of a GEO communications satellite. However, for an infrared astronomical observatory spacecraft, it may become very influential in terms of governing the spacecraft configuration and operation (e.g. ISO).

To illustrate briefly how mission design influences spacecraft system design, consider the operation of a LEO spacecraft compared to one in GEO. At top level, the most obvious difference is the requirement for the GEO spacecraft to have primary propulsion (with the consequent impact upon the vehicle's mass budget) to achieve transfer from the initial orbit to the operational orbit, whereas the LEO vehicle often has no need for this. The LEO spacecraft will experience an eclipse each orbit period, whereas the GEO spacecraft remains in sunlight for the majority of its orbital lifetime. The rapid cycling of solar input experienced by the LEO spacecraft leads to differing design requirements, principally in the power and thermal control subsystem areas, compared to the GEO vehicles. The requirements in the communications subsystem areas (payload data, telemetry and command) are dissimilar for the two categories of mission in terms of both ground station coverage and communications link budget. These simple examples illustrate well the impact of the mission specification on the overall design of space vehicles.

A further example, this time of an interplanetary mission, is the ESA/NASA spacecraft Ulysses [2]. The mission objective of this vehicle, launched by the Space Shuttle in October 1990, is to explore the interplanetary medium away from the plane of the ecliptic. The spacecraft still continues to pursue this objective at the time of writing, a decade after its launch. Despite the longevity of the mission, it remains an excellent example in demonstrating the interaction between the spacecraft mission and system design. To achieve the mission objective, the probe needed to be inserted into a heliocentric orbit, the plane of which was near-normal to that of the Earth's orbit about the Sun. To acquire such an orbit a velocity increment (ΔV) of the order of 30 km/s was required after Shuttle deployment. This mission requirement became a principal design driver for the spacecraft. To achieve such a large ΔV, two mission design options were available. The velocity would be acquired either by using an electrical propulsion system, or by using the gravitational field of a major planet to swing the plane of the orbit out of the ecliptic (see Section 5.8.2 for details of swing-by trajectories). The latter option was chosen and the vehicle was injected into a trajectory, which it took to Jupiter. A swing-by manoeuvre was then performed (in February 1992) over the Jovian pole to insert it into the required mission orbit. As a consequence of this choice of mission, the design of the spacecraft's subsystems in nearly every respect (communications, thermal control, power raising, radiation hardening, etc.) was profoundly influenced by the necessity for the visit to the relatively cold and distant Jovian environment.

It is important to appreciate that the attitude history of a space vehicle is inextricably linked with the mission analysis. For example, an astronomical observatory spacecraft requires to point the payload to specified regions of the celestial sphere, the communications antenna to the ground, the solar array to the Sun, and possibly thermal radiators to 'deep space' (e.g. Hubble space telescope). Clearly for some spacecraft, pointing requirements can be very influential in shaping the vehicle configuration. The control of attitude behaviour will be discussed in Chapter 9.

5.2 KEPLERIAN ORBIT TRANSFERS

The situation frequently arises where a space vehicle must be transferred from one orbit to another. The detailed optimization of these manoeuvres is beyond the scope of this text. (For detailed analysis, see References [3,4].) However, it is appropriate to consider the nature of simple impulsive manoeuvres here, applied to Keplerian orbits.

If a single impulsive manoeuvre is performed, then the initial and final orbits intersect at the location of the manoeuvre; thus, a single manoeuvre may only transfer a vehicle between intersecting orbits. *At least* two manoeuvres will be required to transfer a vehicle between two non-intersecting orbits.

A transfer from orbit A to an intersecting orbit B will occur if at some point, say r_1, the velocity vector is instantaneously changed from its value on orbit A to that which it would be on orbit B. The simplest case is that for a coplanar transfer from a circular orbit to an elliptical orbit. If, on the circular orbit the velocity is increased, then the semi-major axis will be expanded.

A *Hohmann transfer* makes use of two such manoeuvres in order to transfer from one circular orbit to a larger coplanar one, as shown in Figure 5.2. It is the minimum energy, two-manoeuvre transfer, and is optimal (minimum velocity increment ΔV) if $r_2/r_1 < 11.8$.

The velocity increments needed for the two manoeuvres may be assessed by using the *vis-viva* integral equations (4.5) and (4.11). The first one requires an increase from the circular orbit speed $\sqrt{(\mu/r_1)}$ to the speed V_p at the periapsis of the transfer ellipse, where

$$V_p^2 = 2\mu \left\{ \frac{1}{r_1} - \frac{1}{r_1 + r_2} \right\}$$

This represents a ΔV given by

$$\Delta V = \sqrt{\left(\frac{\mu}{r_1}\right)} \left\{ \sqrt{\left(\frac{2r_2}{r_1 + r_2}\right)} - 1 \right\} \tag{5.1}$$

A second, similar manoeuvre will transfer the vehicle into a circular orbit at apoapsis, with a radius of r_2. The ΔV required is

$$\Delta V = \sqrt{\left(\frac{\mu}{r_2}\right)} \left\{ 1 - \sqrt{\frac{2r_1}{r_1 + r_2}} \right\} \tag{5.2}$$

Table 5.1 provides details of Hohmann transfers within the solar system. The transfer time is taken as half of the transfer orbit period. The velocity increments are for Earth departure only, from an assumed 300 km altitude circular Earth orbit. The Δv for the arrival phase is strongly dependent upon the insertion orbit parameters at the destination planet.

Plane rotation manoeuvres and rotation of the line of apsides may be treated similarly to the above analysis. They involve a change of direction for which the ΔV requirement is

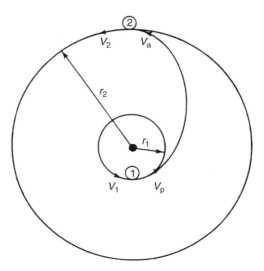

Figure 5.2 Transfer velocity definitions

Table 5.1 Hohmann transfer data from the Earth to the planets. Δv requirements are for the Earth departure phase only, from an assumed 300 km altitude circular Earth orbit

Planet	a Transfer orbit semi-major axis (AU)	$T_{transfer}$ transfer time (years)	Δv for Earth departure (km/s)
Mercury	0.694	0.289	5.55
Venus	0.862	0.400	3.48
Mars	1.262	0.709	3.59
Jupiter	3.101	2.731	6.30
Saturn	5.269	6.049	7.28
Uranus	10.096	16.040	7.98
Neptune	15.531	30.605	8.25
Pluto	20.265	45.616	8.37

given in Chapter 3, equation (3.21). This can involve appreciable use of fuel. If no change of speed is entailed, then the velocity increment required is proportional to the velocity at the time of the manoeuvre, and these transfers should, if possible, be performed when the velocity is at a minimum. For example, to execute large plane changes of an Earth orbit it is generally more efficient, from a propellant usage point of view, to perform the manoeuvre at a large geocentric distance where the spacecraft is moving slowly. This point will be developed later in this chapter.

5.3 MISSION ANALYSIS

As mentioned earlier, *mission analysis* is a 'front-end' systems level activity that takes aspects such as payload operational requirements and spacecraft system constraints as inputs, and generates as an output the mission specification. Clearly, a central feature of this process is the selection of the orbital parameters of the final *mission orbit* (as well as intermediate orbits during the early orbit-acquisition phase) that satisfy these requirements. It is important to note, however, that the mission specification also leads to design requirements on the spacecraft systems and subsystems. For example, the attitude control subsystem engineer can take the orbital characteristics and compute the on-orbit disturbance torques produced for a given spacecraft configuration, which in turn determines the angular momentum profile (see Chapter 9). It is these aspects that aid the engineer in producing design specifications for the attitude actuators and sensors.

As can be seen, mission analysis is not just concerned with the celestial mechanics discussed in Chapter 4, but also involves aspects governing operational and design issues, such as launch vehicle selection (see Chapter 7), ground station visibility, spacecraft ground coverage (see Section 5.5), eclipse duration (which impacts particularly on the thermal control and power subsystem design) and orbital lifetime. Summarizing this discussion, we can conclude that the principal spacecraft design inputs come from the payload interface with the spacecraft and its operational requirements, the spacecraft interface with the launch vehicle, and the orbit. Clearly, the mission analyst has a major role to play in this process. In this section, we will address some of these operational and system aspects related to the orbit.

5.3.1   Ground station visibility

A satellite is visible at all points on the Earth's surface within a circle that is centred on the sub-satellite point, and whose diameter increases with satellite altitude (see Table 5.2). However, signals from satellites at the horizon limit are considerably attenuated by the atmosphere, and so for practical purposes the surface coverage is restricted to the region in which the *satellite elevation* above the horizon is greater than ~5°. Figure 5.3 shows the geometry associated with a satellite S in a circular orbit of height h, passing immediately over a ground station at G. If it is visible down to elevations equal to ε, typically 5° to 10°, then the geocentric semi-angle ϕ over which it is visible is given by

$$\phi = -\varepsilon + \cos^{-1}\left(\frac{R_E}{R_E + h}\cos\varepsilon\right) \tag{5.3}$$

The range of the satellite from the ground station, referred to as the *slant range s* is given by

$$s = (R_E + h)\sin\phi/\cos\varepsilon \tag{5.4}$$

Table 5.2   Relationship between satellite altitude, geocentric semi-angle ϕ and slant range s

Altitude h (km)	Geocentric semi-angle ϕ(deg)			Slant range s (km)		
	$\varepsilon = 0°$	$\varepsilon = 5°$	$\varepsilon = 10°$	$\varepsilon = 0°$	$\varepsilon = 5°$	$\varepsilon = 10°$
100	10.08	6.24	4.16	1134	707	477
500	21.98	17.52	14.05	2574	2078	1696
1000	30.18	25.55	21.64	3709	3194	2763
5000	55.91	51.05	46.49	9423	8882	8379
35 786	81.30	76.33	71.43	41 679	41 126	40 585

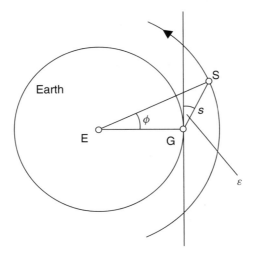

Figure 5.3   Geometry of a satellite passage directly overhead a ground station

The duration of an overhead pass with semi-angle ϕ is

$$\tau = 2\phi/\omega_{ES} \tag{5.5}$$

where ω_{ES} is the orbital angular velocity of the satellite relative to the Earth. ω_{ES} may be obtained from the Earth's angular rate ω_E (7.3×10^{-5} rad/s) and the satellite's orbital rate ω, using

$$\omega_{ES}^2 = \omega_E^2 + \omega^2 - 2\omega_E\omega\cos i \tag{5.6}$$

For LEO, the error produced by using ω in place of ω_{ES} in equation (5.5) (around 5 to 10%) may be acceptable. For a circular orbit, ω in rad/s is given by $631(R_E + h)^{-3/2}$, where R_E and h are in km. For example, a satellite in a 500 km altitude orbit will have an elevation $>5°$ for 9.2 min and $>10°$ for 7.4 min during an overhead pass. Most passes will be shorter than this, and a useful pass should be of more than four minutes duration.

For non-overhead passes, equation (5.5) may be used with ψ in place of ϕ, where

$$\psi = \sin^{-1}(\sin^2\phi - \tan^2\alpha\cos^2\phi)^{1/2}$$

ϕ is now the geocentric semi-angle of the cone within which the spacecraft is visible from the ground site, defined by the minimum elevation ε—as given by equation (5.3). Here the non-overhead pass orbit plane makes an angle α—measured at the Earth's centre—to the directly overhead one. Often non-overhead passes are characterized in terms of the maximum elevation ε_{max} of the satellite. In this case, α and ε_{max} are related by

$$\alpha = \cos^{-1}\left[\left(\frac{R_E}{R_E + h}\right)\cos\varepsilon_{max}\right] - \varepsilon_{max}, \qquad \varepsilon_{max} = \tan^{-1}\left[\cot\alpha - \left(\frac{R_E}{R_E + h}\right)\csc\alpha\right]$$

5.3.2 Eclipse duration

A spacecraft in an Earth orbit will generally encounter an *eclipse period*. This is particularly the case for orbits at low altitudes. The frequency and duration of eclipse periods is strongly dependent on orbital inclination and altitude. For example, in a low-altitude, equatorial orbit, the satellite resides in the Earth's shadow for approximately 40% of every orbit. Conversely, for dawn–dusk sun-synchronous orbits (see Section 5.4), even at low altitude, several months of wholly sunlit operation may be obtained. Spacecraft in highly elliptic orbits (see Section 5.7) with relatively low perigee altitudes will generally encounter eclipse periods when near the Earth. These are relatively brief however, compared to the orbit period, because of the spacecraft's high speed in the perigee region. For example, the eclipse period in a Molniya orbit varies from zero to a maximum of around 50 min (7% of orbit period), depending upon perigee height and season. In Geostationary Earth orbit (GEO) (see Section 5.6), a satellite spends most of its time in sunlight. The eclipse periods in this case are brief and depend on the season. In the period around the solstices, the operation is eclipse-free, whereas near the Spring and Autumn equinox the satellite-Sun vector is near the orbit plane and eclipses are encountered with durations up to a maximum of 72 minutes. This is still, however, a small fraction of the total orbit period of 24 h.

The calculation of the duration of eclipse is important from the point of view of the spacecraft design. For example, if the spacecraft's primary power source is solar arrays, backed up by a battery storage system, then the sizing of the power subsystem is strongly influenced by the length of the eclipse period (see Chapter 10). Similarly, the thermal input to the spacecraft from the Sun is governed by the eclipse period and so influences the design of the thermal control subsystem (see Chapter 11). For some orbits, such as circular orbits, which contain the Earth–Sun vector, the eclipse duration calculation is simple. However, in more general cases the process is more complicated and a computational technique is required.

The following outlines such a technique, and allows the computation of the eclipse period for a given orbit (defined generally by its Keplerian orbit elements) on any day of the year. This implied seasonal dependence is illustrated in Figure 4.9, which shows the motion of the Earth–Sun vector throughout the year. The unit dimensionless vector defining the direction from the Earth to the Sun is given by

$$\mathbf{r}_{sun} = \cos L_\odot \mathbf{i} + \sin L_\odot \cos \varepsilon \mathbf{j} + \sin L_\odot \sin \varepsilon \mathbf{k} \tag{5.7}$$

where $L_\odot$ is the Sun's ecliptic longitude, measured East along the ecliptic from the vernal equinox ($\mathbf{i}$ axis), ε is the obliquity of the ecliptic (23.45°), and ($\mathbf{i}$, $\mathbf{j}$, $\mathbf{k}$) defines the geocentric inertial frame. The unit vector defining the direction to the satellite from the Earth's centre, in the same frame of reference, is given by

$$\mathbf{r}_{sat} = [\cos \Omega \cos(\theta + \omega) - \sin \Omega \cos i \sin(\theta + \omega)]\mathbf{i} \tag{5.8}$$

$$+ [\sin \Omega \cos(\theta + \omega) + \cos \Omega \cos i \sin(\theta + \omega)]\mathbf{j} + \sin i \sin(\theta + \omega)\mathbf{k}$$

where the Keplerian elements have their usual notation. The criterion to establish whether the satellite is in eclipse is illustrated in Figure 5.4. The angle β that defines the entry and exit positions is given by

$$\beta = \sin^{-1}(R_E/r) \tag{5.9}$$

where R_E is the Earth radius, and r is the geocentric distance of the spacecraft given by equations (4.7) and (4.8). Note that we have made simplifying assumptions (e.g. the Earth's shadow is cylindrical and the Earth is spherical). The satellite is in eclipse if the angle between $\mathbf{r}_{sat}$ and the vector defining the anti-Sun direction $\boldsymbol{\rho} = -\mathbf{r}_{sun}$ is less than or equal to β. This is expressed as

$$\cos^{-1}(\boldsymbol{\rho}.\mathbf{r}_{sat}) \leq \beta \tag{5.10}$$

The computational method to determine the eclipse period on a particular day of the year is based on the following steps using equations (5.7) to (5.10):

1. Evaluate $\boldsymbol{\rho} = -\mathbf{r}_{sun}$ from equation (5.7). An approximate value of $L_\odot$ can be calculated in degrees from

$$L_\odot \approx \{(D - D_0)/365\}360° \tag{5.11}$$

where D is the day number, assuming $D = 0$ is 0.0 h on 1 January (e.g. midday on 2 January gives $D = 1.5$), and $D_0 = 79.0$ (this assumes that the Northern Hemisphere Spring Equinox occurs at 0.0 h on 21 March). Clearly, equation (5.11) is

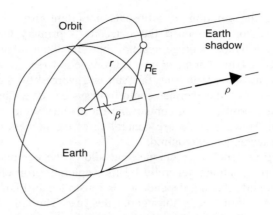

Figure 5.4 Geometry illustrating orbit's intersection with the Earth's shadow

approximate, and ignores complications such as leap years. However, it is adequate
for most design calculations. More precise expressions can be obtained, if required,
from published solar ephemeris data.

2. Cycle through values of true anomaly θ from $0°$ to $360°$ in steps of, say, $1°$ to
   calculate β and $\mathbf{r}_{sat}$ at each step, and use the condition (5.10) to determine the values
   of θ at eclipse entry and exit. To save on computational effort, an initial search with
   a larger θ step ($10°$ or $20°$ say) may be performed to find the eclipse region (bearing
   in mind that the eclipse duration may be small or zero). If the entry and exit values
   are required to a greater accuracy than $1°$, the values of θ can be further subdivided
   in the computation in the entry and exit regions. This is recommended for spacecraft
   entering eclipse when at large geocentric distance, since it will take a relatively long
   time period to traverse a $1°$ step in θ.
3. Use equations (4.14), (4.18) and (4.19) to convert true anomaly values at entry and
   exit to eclipse duration.

The eclipse period of a typical Space Shuttle orbit (200 km altitude circular orbit,
inclined at $28°$) when the orbit plane contains the Earth–Sun vector is 37.3 min, and the
above method gives 36.9 min for a $1°$ step in θ.

5.3.3   Launch windows

The time of launch of a spacecraft is often constrained by dynamical aspects related to
reaching the mission orbit, or by system requirements. Since the orbital plane is essentially
inertially fixed, the satellite can be launched only as the launch site rotates through the
orbital plane. The limited interval of time during which launch can be effected is referred
to as the *launch window*.

The dynamical constraints are emphasized for launch into a LEO sun-synchronous orbit
(see Section 5.4), for example, when the orbit's line of nodes must be set at a specified
angle to the Earth-Sun vector. In this case, the launch window amounts to only a few
minutes twice each day. Further, if there is a constraint on **launch azimuth** at the site,

only one of these opportunities may be possible. The execution of a launch through such a small window is a non-trivial exercise. For missions launched to **rendezvous** with an orbiting object [e.g. Hubble servicing or International Space Station (ISS) transit missions], the dynamical issues are again paramount, since there are not just geometrical constraints to consider, but also temporal issues to do with phasing the launch to reach the target with minimum propellant.

The time of launch to acquire a GEO, on the other hand, is additionally constrained by systems issues which usually lead to two launch opportunities each day around the launch site midday and midnight. The launch window is constructed by combining constraints related to power raising, thermal control, sun-blinding of sensors, and so on, in the transfer orbit. More detail is given in Section 5.6.1.

For interplanetary missions, the *planetary alignment* at launch becomes an issue in defining the launch window. To illustrate this, consider a mission to Mars for which the heliocentric trajectory is a Hohmann transfer (see Section 5.2). Figure 5.5 shows the outbound transfer orbit. If the mission leaves Earth at the point E_{depart}, then clearly Mars must be at point $M_{arrival}$ when the spacecraft has traversed the transfer orbit. For this to be the case, the position of Mars at the Earth departure time M_{depart} is such that Mars moves from M_{depart} to $M_{arrival}$ in the same time as the spacecraft takes to execute half a transfer orbit period (~259 days). This defines the illustrated planetary geometry at Earth departure time, with a heliocentric angle of ~45° between Earth and Mars. The opportunity to launch into a Hohmann trajectory is therefore limited by this planetary alignment, which recurs every 780 days (the Martian synodic period ~2.14 years). The planetary alignment defines a launch window of approximately two weeks duration centred around the ideal departure time, and Earth rotation during this period will impose a further restriction allowing launch for a relatively short period (~ half hour to one hour) each day. The position of Earth $E_{arrival}$ at Mars arrival is determined in the same way, and

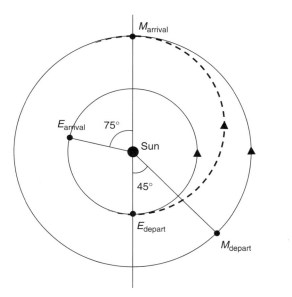

Figure 5.5 Planetary geometry at Earth departure and Mars arrival for a Hohmann transfer mission

allows an estimate of the communications distance. If the mission were to return on a Hohmann transfer, the wait for the return launch window to open would mean that a stay at Mars of approximately 16 months would be required.

5.3.4   Orbit lifetime

For LEO vehicles, aerodynamic drag will eventually result in re-entry as described in Chapter 4, Section 4.4.2. To the accuracy of a few per cent the lifetime of an uncontrolled space vehicle will be

$$\tau \sim \frac{e_0^2}{2B} \left(1 - \frac{11}{6}e_0 + \frac{29}{16}e_0^2 + \frac{7}{8}\frac{H}{a_0} \right) \tag{5.12}$$

where e_0 and a_0 are the initial values of eccentricity and semi-major axis once control has ceased, H is the scale height of the atmosphere near perigee and B is given by

$$B \sim \sqrt{\left(\frac{\mu}{a_0^3}\right) \frac{AC_D}{M} \rho_{p0} a_0 e_0 I_1 \left(\frac{a_0 e_0}{H}\right) \exp\left(-e_0\left(1 + \frac{a_0}{H}\right)\right)} \tag{5.13}$$

This expression neglects the effects of atmospheric rotation. I_1 is the Bessel function of the first kind and order 1 and ρ_{p0} is the atmospheric density at the initial perigee. A more elaborate treatment may be found in Reference [5]. For large space vehicles and for military vehicles, re-entry may cause problems as a result of both the geographical position of the re-entry corridor and the size of individual spacecraft elements on ground impact. Most notably, the effect on the public of Skylab re-entry over Australia, and Cosmos 943B over Canada, provide historical evidence of some of the problems that may be encountered in re-entry.

5.4   POLAR LEO/REMOTE-SENSING SATELLITES

5.4.1   Mission orbit design

The choice of orbit for a LEO remote sensing spacecraft is governed by the mission objectives and payload operational requirements. To achieve near-global coverage, the plane of the orbit must be inclined at about 90° to the Earth's equator. In this near-polar orbit, the spacecraft ground track will span a latitude coverage of $\pm i$ for prograde orbits, where i is the orbital inclination.

In terms of orbit altitude, this is principally established by a trade-off between instrument resolution and the fuel required to maintain the orbit in the presence of aerodynamic drag. For civil remote sensing systems, which are required to operate for an extended period, an altitude typically of the order of 600 to 900 km results. It is worth noting, however, that the mission requirements for military surveillance spacecraft may be different, leading to more complex operational scenarios.

Other payload-derived requirements often demand that the mission orbit be *Earth-synchronous*, or *Sun-synchronous*, and sometimes both (e.g. SPOT). These further constrain the choice of orbit inclination and height in a manner that is discussed in more detail below.

Consequently, for viewing all parts of the Earth's surface at close quarters it is necessary to adopt a low-altitude polar orbit. With the orbit fixed in space and the Earth rotating underneath it, the result is that the ground tracks of successive orbits cross the equator at points that move westward as shown in Figure 5.6.

An Earth-synchronous orbit results when the sub-satellite point follows a ground track identical to some previous orbit after a certain period of time. This need occurs as a result of a payload-derived requirement to revisit ground target sites or regions. The repetition occurs on a regular basis, and it can be achieved in a variety of ways. (A geosynchronous orbit is of course another example of an Earth-synchronous orbit.) Between successive orbits, the sub-satellite point on the equator will change in longitude by $\Delta\phi$ radians, this angle being determined by two effects. The first of these is as a result of the rotation of the Earth beneath the orbit, and the second is caused by nodal regression. It will be assumed that a positive $\Delta\phi$ means a move towards the east.

The Earth rotates through one revolution in its *sidereal* period of τ_E, where $\tau_E = 86\,164.1$ s, neglecting small secular variations. This sidereal period of 23 h 56 min is with respect to the stars. The rotation period with respect to the Sun is of course the mean solar day of 24 h. If the satellite's nodal period is τ, then the contributions to $\Delta\phi$, which is caused by the Earth's rotation, will be given by

$$\Delta\phi_1 = -2\pi \frac{\tau}{\tau_E} \text{ rad/orbit} \tag{5.14}$$

The regression of the line of nodes (equation 4.37) contributes

$$\Delta\phi_2 = -\frac{3\pi J_2 R_E^2 \cos i}{a^2(1 - e^2)^2} \text{ rad/orbit} \tag{5.15}$$

The total increase in longitude at the equator is

$$\Delta\phi = \Delta\phi_1 + \Delta\phi_2 \text{ rad/orbit} \tag{5.16}$$

Clearly, if we wish to have an Earth-synchronous orbit, then we will require that some integral number of orbits later the accumulated value of $\Delta\phi$ will equal 2π. In general, we may, therefore, write

$$n|\Delta\phi| = m2\pi \tag{5.17}$$

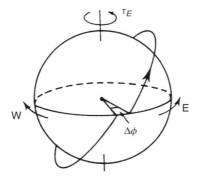

Figure 5.6   Orbit plane motion

where n is the total number of orbits performed and m is the number of Earth revolutions (equivalent to days) before an identical ground track will occur.

Sun-synchronism occurs when the orbit plane rotates in space at the same rate as the Earth moves round the Sun—at one revolution per year, or roughly one degree per day eastwards. Figure 5.7 illustrates this over a period of about three months, during which the orbit clearly needs to rotate through $90°$ in order to be synchronous.

The required rotation rate is

$$\Delta\phi_2 = 2\pi \frac{\tau_E}{\tau_{ES}} \frac{\tau}{\tau_E} \text{ rad/orbit} \tag{5.18}$$

where $\tau_{ES} = 3.155\,815 \times 10^7$ s is the orbital period of Earth round the Sun.

From this discussion, it can be seen that when an orbit is Sun-synchronous, the relative geometry of the Earth–Sun vector and the orbit plane remains approximately constant throughout the year. This implies that spacecraft in such orbits pass over particular ground sites at the same local solar times each day (one pass in daylight and one at night). The need for Sun-synchronism for some remote sensing missions is a payload-derived requirement to use this attribute. The operator of such a vehicle will often focus interest on a particular geographical region. By an appropriate choice of orbit-node position (equator crossing) with respect to the Earth–Sun vector, the solar aspect on the ground can be chosen optimally to fulfil the remote sensing mission objectives. This advantage of Sun-synchronous orbits is illustrated in Figure 5.8.

In general, configuration (a) is unattractive for Earth viewing since the Sun is always low on the horizon at the sub-satellite point, resulting in long shadows and low illumination. Indeed, during winter in the northern hemisphere the ground-track is in darkness while the spacecraft is north of the equator. However, the orbit is attractive for the power subsystem as it provides lengthy eclipse-free periods, making energy storage minimal, and the array may be fixed relative to the body of the vehicle. Further, since the array is aligned along the direction of flight, a minimum projected area to the velocity vector of the satellite is obtained, thus reducing the influence of drag and increasing satellite lifetime. For the noon–midnight orbit (c), remotely sensed images have low definition due to short shadows on the ground, and the payload sensors may experience specular reflection from sea surfaces. The choice of orbit plane is usually a compromise, as shown in Figure 5.8(b).

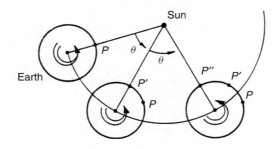

Figure 5.7 Sidereal and solar motion

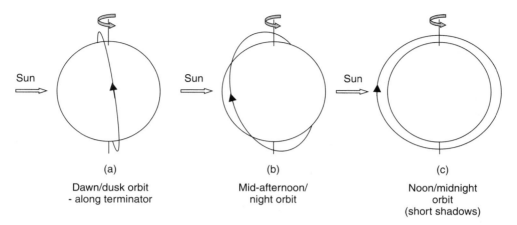

(a)	(b)	(c)
Dawn/dusk orbit - along terminator	Mid-afternoon/ night orbit	Noon/midnight orbit (short shadows)

Figure 5.8   Implications of changing local time coverage

The other feature to note about Sun-synchronous orbits is that since the orbit plane is fixed relative to the solar vector it is possible to cant the array relative to the spacecraft body, providing optimal illumination conditions for the solar array.

To achieve Sun-synchronism, the Earth-oblateness perturbation governing the regression of the line of nodes may be used without the use of fuel. Equation (5.15) indicates that an inclination in excess of 90° will be needed.

For Sun and Earth synchronism equations (5.16) to (5.18) apply, and these lead to the condition

$$n\tau \left(1 - \frac{\tau_E}{\tau_{ES}}\right) = m\tau_E \tag{5.19}$$

The angular displacement between successive orbits in a westward direction is given by

$$\Delta\phi = 2\pi\tau \left(\frac{1}{\tau_{ES}} - \frac{1}{\tau_E}\right) = -7.27 \times 10^{-5}\tau\,\text{rad} \tag{5.20}$$

From equation (4.12) we have

$$\tau = 9.952 \times 10^{-3}a^{3/2} \text{ seconds} \tag{5.21}$$

(where a is in kilometres). Since for remote sensing satellites in LEO the altitude is approximately 600 to 900 km, then $\tau \sim 96$ to 103 min, and so $|\Delta\phi| \sim 4.3 \times 10^{-1}$ rad. At the equator, this translates into a distance of $\sim$2800 km between ground tracks, although this decreases at higher latitudes. It is clear from this that wide instrument swath-widths will be required if complete Earth coverage is to be obtained. By extending the period between repetitions for a given set of ground tracks, a more densely packed set may be achieved. For example, if the requirement is for a daily repeat ($m = 1$) of a set of ($n =$) 14, 15 or 16 tracks, then from equations (5.19) and (5.21), it may be deduced that the corresponding orbit altitudes are 894, 567 and 275 km, respectively. More precisely, if account is taken of the higher order effects of Earth oblateness on the spacecraft's nodal period, then these altitudes become 888, 561 and 268 km, respectively. This type of orbit is known as a *zero drift* orbit, and has no infilling between ground tracks on subsequent days. A greater density of tracks may be achieved by infilling the tracks of Day 1 on subsequent days. The cycle repeats itself over a number of days in excess of one ($m > 1$), with a number (n) of tracks, which is not a multiple of m.

If the condition $(n \pm 1)/m = k$ is satisfied, where k is an integer, then a *minimum drift* orbit is obtained. In this situation, two successive tracks of a given day are infilled sequentially on subsequent days. Non-minimum drift orbits, $(n \pm 1)/m \neq k$, infill in a non-sequential manner.

Since τ is a function of altitude (equation 5.21), it is possible to plot the repeat period of Earth synchronization *versus* altitude. This, for minimum drift orbits, is shown in Figure 5.9. As an example, SPOT4 was launched into a 26 day repeat period, non-minimum drift orbit on 24 March 1998. The SPOT (Satellite Pour l'Observation de la Terre) spacecraft series are the space segment of a French National remote sensing programme. The characteristics of the orbit are

- Sun- and Earth-synchronous
- Ground track repeat cycle $(m, n) = (26, 369)$
- Orbit parameters.
  semi-major axis $a = 7200$ km (height $\sim$822 km, period $\tau = 101.3$ min)
  eccentricity e ≈ 0, inclination $i = 98.7°$
  Node time (descending) $= 10.30$ local solar time (LST)

At the equator the separation between successive orbits is 2838 km. However, over the 26-day repeat period, this separation is filled-in (non-sequentially) so that the distance between adjacent ground tracks is reduced to $2\pi R_E/n \sim 108.6$ km. This ground

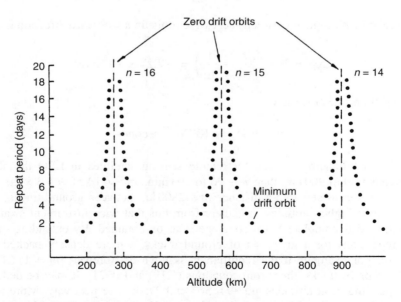

Figure 5.9 Locus of points for minimum drift orbits as a function of repeat period in days (m) and altitude

repetition geometry was chosen to ensure that the SPOT imaging system, with a nominal swath-width of 117 km, can achieve global coverage.

5.5 SATELLITE CONSTELLATIONS

5.5.1 Introduction

The use of *constellations of satellites* to achieve a mission objective is by no means a new phenomenon. For example, the United States Department of Defence (DoD) deployed the *TRANSIT system* in the 1960s, which comprised a constellation of spacecraft in polar orbits. This was used principally by US Naval ships to determine position. Because of the way the fix was determined (tracking the range rate of a single satellite), the system was unsuitable for navigation of highly dynamic vehicles, such as aircraft or missiles. This constraint was removed, however, with the implementation of another US military navigation system, the *Navstar Global Positioning System (GPS)* constellation. This has had a major impact on both military and civilian sectors. To satisfy global coverage requirements, the nominal GPS constellation comprises 24 satellites deployed in 6 orbit planes, inclined at 55° to the equator. Each orbit plane contains 4 spacecraft, equally spaced in true anomaly, and the node of each plane is spaced equally in right ascension around the equator. Each orbit is near-circular, and has an orbit period of half a sidereal day (height ∼20 200 km). A similar system—GLObal NAvigation Satellite System (*GLONASS*)—was deployed concurrently by the former Soviet Union, primarily for military use. This also comprises 24 satellites, but this time with 8 satellites in each of 3 planes with an inclination of 65°.

The prime stimulus of interest in LEO constellations, particularly during the last decade of the twentieth century, has been their use for global data and communication services. Satellite communications were formally based principally on GEO systems (see Section 5.6), relaying data between relatively large, fixed ground stations. However, the perceived market for *personal* communications led to the consideration of satellites deployed in LEO to reduce the slant range, and therefore the required radiated power of the user terminal (phone) to physiologically acceptable levels. This simple driver gave rise to a large number of proposals for *personal global communication systems*, on the basis of LEO constellation configurations. Some of these proposals have become a reality, for example *ORBCOMM*, *IRIDIUM* and *GLOBALSTAR*, and others are to be built imminently (at the time of writing). For example, the IRIDIUM constellation comprised 66 satellites (plus one spare per plane) deployed at 780 km altitude in 6 equally spaced orbit planes, inclined at 86.4° to the equator.

This activity has been dampened somewhat by the unfortunate financial demise of the IRIDIUM system, but the trend is still one of growth, and the mood one of optimism. Although communication systems have led the trend, other applications for LEO constellation systems are being proposed and implemented. These are civilian and military remote sensing systems. As sensor technology progresses resulting in greater spatial resolution, the next step is the improvement of temporal resolution through the use of constellations. One aspect that inhibits this development is, of course, the cost involved in launching multiple satellites. This, however, is offset to some degree by the development in recent years of capable, small satellites, and their use in constellations. This theme is developed further in Chapter 18.

At the mission design phase of a project, there is the need to determine whether the objectives can be achieved with a single satellite or a constellation. The main advantage of a single spacecraft is clear—the cost benefit of a single build and launch activity. On the other hand, a collection of satellites in a constellation will offer considerable improvements in *coverage*, in *reliability*—the system will degrade more gracefully in the event of a failure, and *survivability* in the case of a military system. In some instances, the requirements make the choice of a constellation mandatory—for example, the Navstar GPS user must be able to see four spacecraft with a diverse line-of-sight geometry to acquire a navigational fix.

Once the need for a constellation is clear, the design of its geometry is a non-trivial problem, since there are a large number of possibilities (see Section 5.5.3). Since coverage is the prime benefit derived from a constellation, the choice normally involves a trade-off using coverage (as a measure of performance) against the number of spacecraft (as a measure of cost). However, even this process is not straightforward. For example, a constellation with $t + 1$ members may exhibit a step function in improved performance over one with t members, but the larger constellation may be deployed at a lower height or inclination, so reducing the cost of launch.

5.5.2   Ground coverage

Figure 5.10 shows a typical ground coverage geometry, with the satellite at height h. The 'viewing angle' of the payload is 2α. This may be the field of view of an imaging system or the beamwidth of a communications antenna. This defines a *circle of coverage* on the ground of geocentric semi-angle ϕ and diameter D km. The satellite elevation and slant range at the edge of coverage (useful concepts for communications missions) are ε and ρ, respectively.

From Figure 5.10, we have

$$\varepsilon = \pi/2 - \phi - \alpha \tag{5.22}$$

and (as seen in Section 5.3.1),

$$\phi = -\varepsilon + \cos^{-1}\left(\frac{R_E}{R_E + h}\cos\varepsilon\right) \tag{5.23}$$

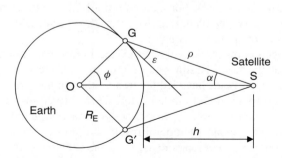

Distance over the ground GG′ = D = coverage diameter.

Figure 5.10   Typical ground coverage geometry

The coverage diameter is then

$$D = 2\phi R_{\mathrm{E}} \tag{5.24}$$

with ϕ in radians, and the slant range is given by

$$\rho = R_{\mathrm{E}} \left(\frac{\sin \phi}{\sin(\pi/2 - \varepsilon - \phi)} \right) \tag{5.25}$$

From these geometrical considerations, it is possible to relate coverage aspects to orbit height and payload characteristics.

A further issue is the degree of coverage required. For example, for communications missions, the user may require to 'see' just one satellite at any time to enable the service. However, as mentioned above, the Navstar GPS user requires *multiple coverage*—in this case four-fold coverage. Line of sight to at least four spacecraft is required at any time, to acquire the information needed to solve four unknowns—the three components of the user's position, plus the offset between the GPS satellite time and the user's clock. Figure 5.11 illustrates the concept of multiple coverage. The central horizontal line represents one orbit plane in the constellation and the dots denote the s satellites in that plane. The integers give the instantaneous levels of multiple coverage within the intersecting circles of coverage. The figure also illustrates the concept of *streets of coverage*, introduced in Reference [6] and developed by Adams and Rider [7], which is useful in analysing constellation systems requiring multiple levels of coverage. The street width giving j-fold coverage is given by

$$d_j = 2 \cos^{-1}(\cos \phi / \cos(j\pi/s)), \quad j = 1, 2, 3, \ldots, \quad s \geq 3 \tag{5.26}$$

where s is the number of satellites in each orbit plane. It should be noted that multiple coverage can also be provided by the overlapping of streets of coverage from neighbouring orbit planes, particularly at high latitudes when orbit planes converge.

Coverage is often the crucial issue when deciding upon a constellation geometry. Coverage statistics can be accumulated by numerical simulation of the candidate constellation designs and an appropriate *coverage figure of Merit* devised to distinguish between them.

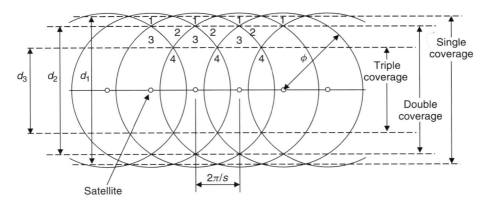

Figure 5.11 Constellation orbit plane geometry, illustrating multiple coverage and 'streets of coverage'

5.5.3   Constellation geometries

As seen below, there are a large number of possible constellation configurations that may satisfy a particular mission requirement. One factor that is common to the majority of these designs, however, is the invariance of the constellation geometry when subjected to orbital perturbations. In other words, it is usually the case that the orbit of each member satellite is similar in terms of height, eccentricity and inclination, so that each orbit is perturbed by the effects described in Chapter 4, Section 4.4 in the same manner (to first order). In this way, the constellation geometry can be preserved without recourse to an excessive station-keeping requirement. Generally, the only acceptable exception to this is the combination of inclined and equatorial orbit planes in a constellation design.

A further factor that should be considered in the design is the use of appropriate phasing of the member satellites within each orbit plane to ensure that collisions are avoided at orbit plane intersections. This is dealt with in some detail in Reference [7]. The most commonly considered constellation designs are described below.

Walker Delta pattern constellation. This geometry, first proposed by Walker [8,9], can be uniquely specified by a small number of parameters; namely, the total number of satellites t, the number of orbit planes p, an integer f that determines the relative spacing between satellites in adjacent planes, and finally the orbit inclination i. Therefore, the Delta pattern denoted by

$$i : t/p/f, \qquad 0 \le f \le (p-1)$$

comprises a design with p orbit planes, the ascending nodes of which are equally spaced in right ascension. All orbits are nominally circular, and of the same height and inclination. Each orbit plane contains $s(= t/p)$ satellites, equally spaced in true anomaly. To define the phasing of satellites within the pattern, Walker introduced a 'pattern unit' $PU = 360°/t$. The relative phasing is defined by the condition that, if a satellite is at its ascending node, then the satellite in the next most-Easterly plane will be $f \times PU$ past its node. An example, specified by $55° : 25/5/1$, is illustrated in Figure 5.12, in which the $PU = 360° / 25 = 14.4°$. In the figure, satellite one is positioned at its ascending node, with satellite six, therefore, one PU beyond its nodal position in the next most-Easterly plane.

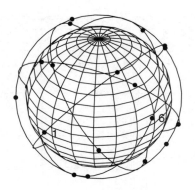

Figure 5.12   A Walker Delta pattern constellation, defined by 55°: 25 / 5 / 1

n-plane polar constellations. As the name suggests, this design comprises *n* orbit planes, each of which is of polar (or near-polar) inclination. Distributed within each of the *n* planes is an equal number of satellites. Walker [8] originally referred to this type of geometry as a (Walker) 'Star pattern' constellation. Figure 5.13 shows, as an example, the geometry of a six-plane polar constellation, as viewed when looking down on the North Pole of the Earth. As can be seen, most adjacent orbits are 'synchronized' in that the satellites are co-rotating. However, by necessity two adjacent planes must be 'non-synchronized', in which the satellites are counter-rotating. The manner in which the coverage is achieved in these two situations is dissimilar, and generally leads to different nodal spacing of the orbit planes.

Consequently, the *n*-plane polar constellation is not an example of a Walker Delta pattern. Also, given the general convergence of the orbit planes at the poles, issues of in-plane phasing become critical to avoid collisions, which in turn can further affect coverage. A good example of the development of an *n*-plane polar constellation for a personal communication satellite system with global coverage is given by Benedicto *et al.* [10]. This demonstrates well, the interactions between coverage and satellite/orbit-node separation. Figure 5.14 illustrates their solutions, specifying the total number of satellites required as orbit height varies, for minimum 'edge of coverage' elevations of 10°, 30° and 50°. This clearly demonstrates the strong dependence of the total number of satellites *t* on orbital altitude *h*, which is a general feature of constellations.

Geometries for regional coverage. The geometries discussed so far are useful candidates if global coverage is a requirement. However, if regional coverage is needed, a simpler constellation geometry should be sought, to gain the cost benefit of deploying fewer spacecraft. It is easy to see, for example, that coverage required within say 20° or 30° latitude of the equator, can be fulfilled by a number of satellites phased in true anomaly in an equatorial orbit. Similarly, regional coverage of the polar caps may be achieved by satellites optimally distributed in one (or two) near-polar orbits. To optimize coverage at mid-latitudes (30° to 60°), a region containing Europe, North America, the CIS, S. Australia and Japan, recourse to a number of orbit planes inclined at ~55° to the equator may suffice. This utilizes the useful tendency for neighbouring orbits to converge at their Northern and Southern apexes. Consideration can also be given to the use of eccentric orbits when seeking solutions to regional coverage problems, bearing in mind the need to preserve the geometry when subjected to natural perturbations.

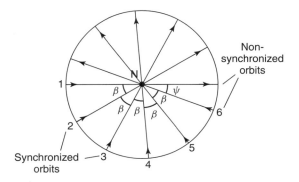

Figure 5.13 Illustration of a six-plane polar constellation

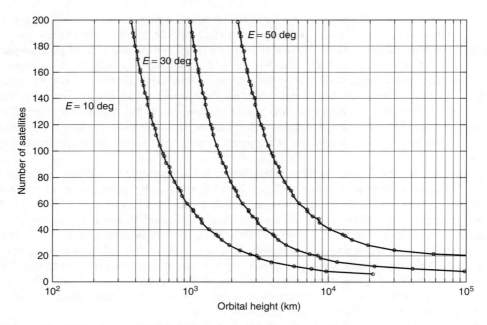

Figure 5.14 Number of satellites as a function of orbit height and minimum elevation Reproduced by permission of ESA (authors: J. Benedicto *et al.*) (from Reference [10])

5.5.4 Design issues

In terms of constellation design, there are unfortunately no general rules to aid the designer in solving what is often a complex problem. The only semblance of a rule is perhaps the above-mentioned invariance of the geometry with respect to perturbations. However, even this was not considered a constraint by the designers of the ORBCOMM constellation, which remarkably exhibits a variety of orbit heights and inclination.

An observation that may assist in the design process is that there is often an advantage in deploying more satellites in fewer orbit planes. Underlying this statement is the relative ease with which spacecraft can be moved within an orbit plane, compared to the difficulty of moving them from one plane to another. The former involves a small, in-plane phasing manoeuvre, whereas the latter requires a prohibitively expensive plane change manoeuvre. In a constellation with p orbit planes, *performance plateaux* occur with the launch of the first satellite, then with one satellite in each plane, and so on. The major plateaux therefore occur with the launch of 1, p, $2p$, $3p$, ... satellites. Clearly, these plateaux occur more frequently, and with fewer spacecraft when the number of planes p is small, giving a cost benefit. Similarly, when considering satellite failures within a constellation, there is a greater degree of *graceful degradation* in performance in systems with fewer orbit planes. For example, in a system with t satellites, equally distributed in p planes, there are $s = t/p$ satellites in each plane. If a satellite was to fail, then the whole constellation could be re-phased with little cost, to a configuration with $s - 1$ satellites in each plane, corresponding to the $p(s - 1)$ performance plateau. Lower values of p again give benefits when considering degradation. However, despite these comments, it is often the coverage requirements that dictate the number of orbit planes, so that low values of p may not be possible.

5.6 GEOSTATIONARY EARTH ORBITS (GEO)

The utility of the geostationary orbit for providing global communications was first noted by Arthur C. Clarke [11]. Its primary attribute is that the sub-satellite point is fixed at a selected longitude, with 0° latitude. It does not have dynamic tracking problems. GEO spacecraft may therefore provide fixed-point to fixed-point communications to any site within the beam of their antennas. Figure 5.15(a) shows the horizon as viewed from GEO, and the region over which the satellite appears with an elevation in excess of 10°; Figure 5.15(b) demonstrates that only three satellites are required to provide almost a global communications network.

The 24-hour geostationary orbit clearly offers unique advantages, providing almost complete global coverage (except for the immediate polar regions) from merely three satellites, and with no need for the ground antenna to switch between satellites. A potential disadvantage of GEO communication systems is the time delay for a transmission to reach its destination. This is clearly less of a problem for data services (e.g. internet, e-mail, etc.) than for live speech.

For a service of high quality and reliability, it is necessary to consider the control and replacement policy for a GEO spacecraft. The failure of a space segment would cause substantial financial penalties to the system operator, and since the ability to replace it is dependent on launcher availability, with substantial time being needed, the philosophy of having an in-orbit spare is frequently adopted. This spare is at a slightly different longitude, and offers the advantage of extra capacity should it be required.

Maintaining a spacecraft's orbit is an essential requirement for maintaining a communication link. The capability and method of achieving GEO and maintaining a specific location will now be discussed and fuel requirements for station-keeping will be indicated.

5.6.1 Geostationary orbit acquisition

The final stage of the launch vehicle will place the satellite in a nominal orbit. During the 1980s, the US Space Shuttle was used to launch commercial geostationary communication

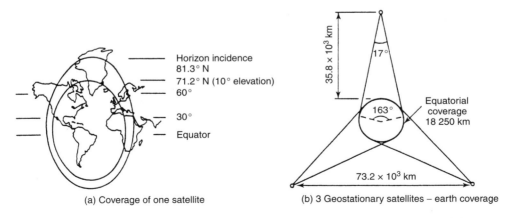

(a) Coverage of one satellite   (b) 3 Geostationary satellites – earth coverage

Figure 5.15 Geostationary views of the Earth (a) coverage of one satellite and (b) 'global' coverage using three satellites

spacecraft. In order to make up the launcher's payload complement, typically two or three such spacecraft were launched at a time. Currently, the Shuttle is used to launch those large science, communications and military payloads that warrant its large lift capability. Some of these are destined for GEO, and the launch scenario begins with injection into a near-circular orbit of a few hundred kilometres altitude with an inclination of ~28°, since the height range of Shuttle is limited. An orbit transfer propulsion unit then expands this into an elliptical orbit, the apogee of which is near geostationary altitude (~35 786 km), whilst its perigee remains at Shuttle altitude. Precision in the timing of the firing of this motor is essential in order to obtain the correct *geostationary transfer orbit* (GTO); apogee must occur at an orbit node (equator crossing) so that subsequently a single firing of an *apogee boost motor* (ABM) may both circularize and change the plane to become equatorial.

An Ariane 5 launch scenario, for example, is similar in philosophy although the details differ. Its final stage places the satellite directly into a GTO with a perigee of between 200 and 650 km altitude, and an apogee near geostationary altitude. Separation from the launch vehicle occurs after a powered flight of ~27 minutes. The transfer orbit is inclined by only 7°, significantly less than for Shuttle.

Several transfer orbit revolutions occur before injection of the satellite into a near-circular, near-GEO orbit. This period is essential for attitude manoeuvres and determination, and for tracking the satellite and determining its orbit before the ABM is fired. This motor increases the velocity of the satellite from ~1.6 km/s to ~3 km/s at apogee. Most satellites will be spin-stabilized during the entire transfer and orbit-acquisition phase, although vehicles using liquid apogee engines (low thrust) may be three-axis-stabilized. The GTO attitude and orbit control activities, and consequently the launch window (see Section 5.3.3), are very much dependent on the particular spacecraft and its system constraints. When specifying the time of launch, however, the constraints considered generally include

— the duration of eclipse periods, and the exposure of sufficient solar array surface to the Sun (particularly for spin-stabilized spacecraft) to ensure adequate power
— the Sun-relative attitude, and eclipse durations, to ensure thermal control can be maintained
— the relative Sun-Earth-spacecraft geometry, to ensure that the vehicle's sensors can provide information of adequate accuracy for attitude determination
— Sun blinding of sensors and so on.

For spinners, the orientation of the spin axis, and its control during motor firings, is particularly crucial not only for reasons of orbit attainment but also for power raising and thermal control. These latter constraints typically require a local midday or local midnight apogee firing to take place since the spin axis will be oriented along the thrust vector (see Figure 5.16).

For vehicles that utilize a *liquid apogee motor* (LAM), a single firing at apogee is insufficient to transfer the vehicle into the desired near-GEO orbit. The low thrust of such motors (~10^2 N rather than ~10^4 N for a solid rocket) would result in the need to rotate the thrust axis during firing in order to deliver the required impulse. Instead, typically three firings of the LAM are utilized during successive apogee transits. As mentioned above, these apogee manoeuvres not only circularize the transfer orbit, but also rotate it into the equatorial plane. The cost of this plane change is directly proportional to the

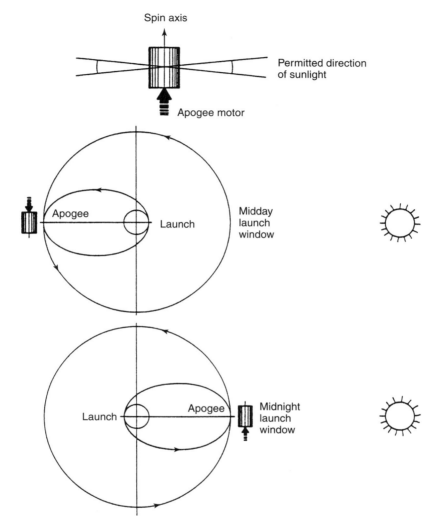

Spin axis

Permitted direction of sunlight

Apogee motor

Apogee

Launch

Midday launch window

Launch

Apogee

Midnight launch window

Figure 5.16 Vehicle orientation for injection

GTO inclination, so that the low inclination of the Ariane transfer orbit, for example, is preferable in this respect. To further reduce the Δv of this composite apogee manoeuvre, a *supersynchronous transfer orbit* is often used, which has an apogee in excess of the GEO altitude. Since the spacecraft moves more slowly at this higher altitude, the cost of the plane change is reduced. An additional in-plane burn is then required to lower the apogee, but the overall propellant mass can be reduced using this strategy.

   If the one-impulse solid propellant ABM is used, there is a requirement to place the satellite into a slightly elliptical orbit, termed a drift orbit, whose apogee is at GEO altitude. Its period is less than the GEO period, the precise value being chosen to minimize the ΔV requirements, whilst maintaining a reasonable GEO attainment strategy. As a consequence, the satellite progressively drifts eastward. Through a sequence of manoeuvres in both latitude and longitude, it is eventually brought to rest at the required GEO location. These manoeuvres are similar to those described below for station-keeping. Final *station acquisition* may be up to a month after satellite launch. Table 5.3 summarizes some of the attributes of the LAM [12].

5.6.2 GEO station-keeping requirements

The primary disturbances on a satellite orbit were described in Chapter 4. The dominant effects for GEO are luni-solar perturbations, Earth triaxiality and solar radiation pressure.

Solar radiation perturbation is complex to model and depends critically upon both the ratio of projected area to mass of the vehicle and the surface characteristics. Its predominant influence is upon the orbit eccentricity vector **e**. Generally, an effective area-to-mass ratio, which includes the reflectivity coefficient to solar illumination, is used. When this is less than $\sim 0.005 \, \text{m}^2/\text{kg}$, the effects of radiation pressure are significantly less than other perturbations. However, for larger values ($> 0.01 \, \text{m}^2/\text{kg}$) it can cause significant perturbation. Indeed, it is now used routinely to help control three-axis controlled spacecraft such as Inmarsat 2, by balancing other longitude influences.

The generalized techniques for examining the influence of perturbations, which were noted in Chapter 4, are not necessarily the optimum methods for analysis of GEO. This is particularly so since the orbit is ideally circular, and hence ω is undefined; the inclination is also zero, resulting in Ω being undefined. A method that is applicable is to linearize the equations of motion for small perturbations of magnitude δa in the semi-major axis a. For small values of δa, i and e, a linearized solution to Kepler's equation yields [13]

$$r = A + \delta a - A e \cos\left((t - t_0)\sqrt{\frac{\mu}{A^3}}\right) \tag{5.27}$$

Table 5.3 Characteristics of the liquid apogee motor for GEO acquisition ($\sqrt{}$ denotes advantage, $\times$ denotes disadvantage) [12]

Attribute	Comments
Multiple engine restarts possible	Burn can be segmented $\sqrt{}$
Longer duration motor firings	Lower acceleration $\sqrt{}$ Reduced efficiency $\times$
Burn duration controllable	Improved reaction to launcher errors $\sqrt{}$ Improved reaction to execution errors $\sqrt{}$ Manoeuvres can be calibrated $\sqrt{}$
Drift orbit not required	Satellite on-station sooner $\sqrt{}$ Reduced propellant requirement $\sqrt{}$
Common fuel supply for primary and secondary propulsion	Fuel savings provide increased life on-station $\sqrt{}$
Extended transfer orbit phase	Increased cost of ground station support $\times$ Power and thermal problems $\times$
More complex propulsion system	Cost $\times$ Control problems induced by 'fuel slosh' $\times$

$$\lambda = \Omega + \omega - t_0 \sqrt{\frac{\mu}{A^3}} - \frac{3}{2}\frac{\delta a}{A}(t - t_0)\sqrt{\frac{\mu}{A^3}} + 2e \sin\left((t - t_0)\sqrt{\frac{\mu}{A^3}}\right) \qquad (5.28)$$

$$\theta = i \sin\left(\omega + (t - t_0)\sqrt{\frac{\mu}{A^3}}\right) \qquad (5.29)$$

where A is the semi-major axis of a truly geostationary orbit (42 164.5 km), λ is the satellite longitude and θ its latitude.

The evolution of r, λ and θ with time, as shown in Figure 5.17, is easily described with the aid of these equations. The geocentric distance r (equation 5.27) oscillates simply about a mean distance $(A + \delta a)$ with an amplitude of Ae. Similarly, the latitudinal variation, from equation (5.29), is a simple oscillation about the equatorial plane with an amplitude equal to the orbital inclination. The longitude, however, manifests an oscillation of amplitude 2e, but also a drift rate. It is worth noting that for circular orbits whose semi-major axes are given by A and whose inclinations are non-zero, the ground track will be a figure of eight of the form shown in Figure 5.20.

Earth triaxiality perturbation is due to the dominant J_{22} Earth tesseral harmonic that produces a force on the spacecraft whose direction is shown in Figure 5.18 [13]. From Section 5.2, it is evident that a positive ΔV will expand an orbit, changing both the

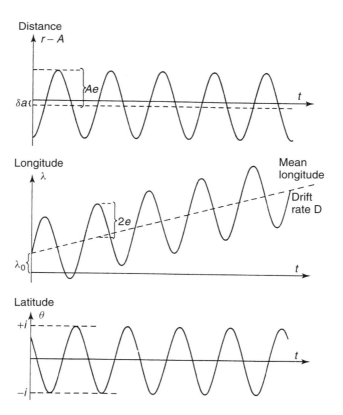

Figure 5.17 Linearized unperturbed spacecraft motion, given as distance (r), longitude (λ) and latitude (θ)

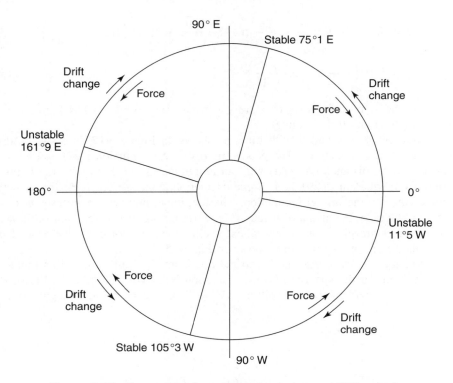

Figure 5.18 Force direction arising from J_{22} on a GEO vehicle

semi-major axis and the eccentricity. Since an expanded orbit relative to GEO has a longer period, a positive ΔV (i.e. an eastward impulse) will lead to westward drift in satellite longitude. Hence, referring to Figure 5.18, the longitude drift rates are opposed to the acting force direction, and it is apparent that 105.3°W and 75.1°E are stable equilibria against equatorial drift. The magnitude of the acceleration as a function of longitude is shown in Table 5.4.

Luni-solar perturbations mainly cause out-of-plane forces acting on the spacecraft, leading to changes of inclination of the orbit. However, since a component of this force necessarily lies in the equatorial plane, then *a*, *e* and hence λ are also influenced. This equatorial component is nearly cyclic with the Earth orbit period, with magnitude less than that caused by the J_{22} term. It therefore does not lead to a significant secular evolution of the orbit. More details may be found in Reference [13].

Returning to the out-of-plane components due to Moon and the Sun, the periodic change of 18.6 years of the lunar orbit plane, plus precession of the Earth's spin axis, results in a 54-year period for the evolution of an uncontrolled geostationary orbit. The net force on the orbit plane evolution, shown in Figure 5.19, must clearly be opposed by an opposite ΔV. The magnitude of the ΔV arising from luni-solar perturbation is typically ~50 m/s/year and thus the propulsion requirements for the control of orbit inclination (i.e. north/south station-keeping) is at least a factor of 20 larger than that required to overcome triaxiality (i.e. east/west station-keeping) effects (see Table 5.4). This results in the need for particularly careful control and planning of manoeuvres in order to avoid

Table 5.4 Acceleration and station-keeping requirements for geostationary vehicles

Longitude (degrees east)	Acceleration (m/s²)	Fuel: ΔV (m/s/year)
−160	-5.3×10^{-8}	1.67
−140	-4.75×10^{-8}	1.50
−120	-2.30×10^{-8}	0.73
−100	$+8.10 \times 10^{-9}$	0.26
−80	$+3.39 \times 10^{-8}$	1.07
−60	$+4.61 \times 10^{-8}$	1.45
−40	$+3.99 \times 10^{-8}$	1.26
−20	$+1.48 \times 10^{-8}$	0.47
0	-2.10×10^{-8}	0.66
20	-5.09×10^{-8}	1.61
40	-5.73×10^{-8}	1.81
60	-3.30×10^{-8}	1.04
80	$+1.17 \times 10^{-8}$	0.37
100	$+5.21 \times 10^{-8}$	1.64
120	$+6.49 \times 10^{-8}$	2.05
140	$+4.44 \times 10^{-8}$	1.40
160	$+4.08 \times 10^{-9}$	0.13
180	-3.40×10^{-8}	1.07

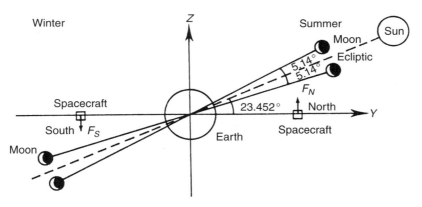

Figure 5.19 Out-of-plane forces F_N, F_S on a geostationary orbit. The Y-axis lies in the equatorial plane and the Z-axis is the Earth's spin axis. The diagram shows positions of the Sun and Moon in summer and winter, relative to the GEO spacecraft

disadvantageous coupling between north/south and east/west control. This interaction is discussed in the following two sections.

In summary, the nature of station-keeping is to locate the spacecraft under the influence of the perturbations described within a specified range of longitude and latitude. This situation is shown in Figure 5.20. Whilst the apparent position of the spacecraft projected on the celestial sphere provides information to the ground control centre indicating whether the spacecraft is within the correct longitude/latitude region, the range and

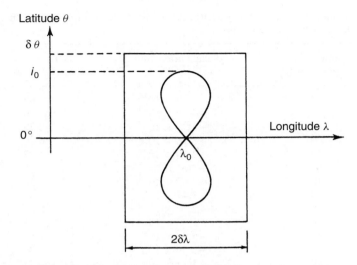

Figure 5.20 Dead band of inclined geosynchronous vehicle. The figure of eight demonstrates the apparent ground track of a synchronized GEO satellite inclined at an angle i_0 to the equator

range rate are also required so that the actual orbit may be determined. This nominal orbit may then be used to schedule station-keeping manoeuvres to preclude departure from the required longitude/latitude location box. The size of this box necessarily impacts on the station-keeping schedule: the tighter the requirements the greater the impact. Requirements for ~0.1° are typical, but this will become less when allowance is made for sensor errors.

5.6.3 Longitude station-keeping

Using the linearized approach noted above, the subsequent change of longitude of a spacecraft following a small impulsive velocity increment ΔV is

$$\lambda = \lambda_0 + \frac{\Delta V}{r} \left\{ -3t + 4\sqrt{\frac{A^3}{\mu}} \sin\left(t\sqrt{\frac{\mu}{A^3}}\right) \right\} \tag{5.30}$$

where t is the time since the impulsive burn. This is shown in Figure 5.21.

Longitudinal control can be effected by using an impulsive along-track burn when the spacecraft reaches one extreme of the required error box. The subsequent longitude drift due to J_{22} (see Figure 5.18) brings the spacecraft back to the same side of the box as shown in Figure 5.22, and the process is repeated as a limit cycle. This neglects the short period fluctuations caused by luni-solar perturbations.

If the ΔV is sized so that the spacecraft just reaches the opposite side of the box following a burn, and a constant acceleration f is assumed (as listed in Table 5.4), then

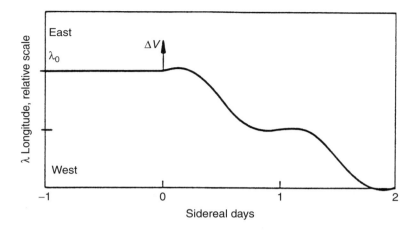

Figure 5.21 Longitude evolution following an east burn

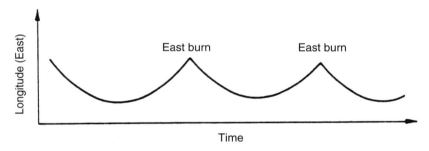

Figure 5.22 Parabolic shape of mean longitude free drift with station-keeping by east burns. In this example, the satellite is stationed where the triaxiality drift is Easterly, and the burns induce a Westerly drift

the ΔV and the corresponding period T of the limit cycle are

$$\Delta V = 4 \sqrt{\frac{rf\lambda_{\max}}{3}} \quad \text{and} \quad T = 4 \sqrt{\frac{r\lambda_{\max}}{3f}} \tag{5.31}$$

where r is the orbit radius ($\sim 42 \times 10^6$ m) and the $\pm\lambda_{\max}$ represents the box size.

Thus, for example, a satellite at a nominal longitude of $-160°$ ($f = -5.3 \times 10^{-8}$ m/s^2) requiring an error box of $\lambda_{\max} = 0.1°$ (remember to convert to radians) would require $\Delta V = 0.14$ m/s and the period of the limit cycle would be $T = 31$ days.

This technique could be referred to as the 'ping-pong method of East–West station-keeping', since a little thought shows that it is entirely analogous to repeatedly tapping a ping-pong ball on a bat. The impact on the bat corresponds to the impulsive burns, the constant gravitational acceleration represents the constant longitudinal acceleration within the dead band, and the constant height amplitude of the ball's motion reflects the size of the longitude dead band. This rather whimsical observation may aid understanding of this method of East–West station-keeping, since the two situations are physically equivalent.

5.6.4   Latitude station-keeping

Latitude drift may usefully be approached by introducing the two-dimensional vectors **i** and **e** given by Reference [13]:

$$\mathbf{e} = \begin{bmatrix} e_x \\ e_y \end{bmatrix} = \begin{bmatrix} e\cos(\Omega + \omega) \\ e\sin(\Omega + \omega) \end{bmatrix}$$

$$\mathbf{i} = \begin{bmatrix} i_x \\ i_y \end{bmatrix} = \begin{bmatrix} i\sin\Omega \\ -i\cos\Omega \end{bmatrix}$$

The natural evolution of inclination during a five-year period is shown in Figure 5.23, corresponding to a number of initial values of Ω. This wavy drift shows the half-yearly effect caused by the motion of the Earth about the Sun. The lunar periodicity of 14 days is smoothed out in this representation.

The objective of latitudinal station-keeping is to maintain the trajectory of the inclination vector within a specified region of the i_x, i_y plane. In the case shown in Figure 5.23, for example, the inclination is to be constrained within a circle of radius 2.5°, which represents a maximum orbit plane inclination of 2.5°.

The philosophy of inclination control is similar to that of longitude, a correcting nodal burn taking place just before the inclination drifts to the maximum permitted by the error box, $\pm i_{max}$, say. If the spacecraft's speed is V, then the directional change needed is $2i_{max}$, and this will require a ΔV of $2V i_{max}$ (see equation 3.21).

For example, if $i_{max} = 0.1°$ and $V = 3075$ m/s, then $\Delta V = 10.7$ m/s.

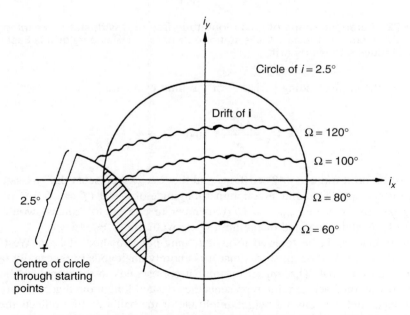

Figure 5.23   Uncontrolled evolution of inclination vector **i** in component form i_x, i_y, where $i_x = i\sin\Omega$ and $i_y = i\cos\Omega$

Table 5.5 GEO burns direction

	North burn	South burn
Spring	Morning	Evening
Summer	Midnight	Noon
Autumn	Evening	Morning
Winter	Noon	Midnight

Clearly, the ΔV impulses for controlling inclination are appreciably greater than those needed for longitude, and their errors due to cold starts will therefore represent a smaller percentage error.

Latitude manoeuvres imply a change of the orbit plane. Such manoeuvres must be performed on the line of nodes, using north/south burns at times of day indicated by Table 5.5. Ideally, the thrust vector should be perpendicular to the plane bisecting the initial and required orbit planes (see equation 3.21), and should produce no change in the spacecraft's speed. The thrust direction should be accurate since a directional error ε will lead to a speed change of approximately $2Vi_{max}\varepsilon$, and this would influence the longitude drift rate; errors of only a few degrees could swamp the natural longitude drift rates noted above.

5.7 HIGHLY ELLIPTIC ORBITS

5.7.1 Space-borne observatories

Spacecraft in elliptic orbits move more rapidly at perigee than at apogee (equation 4.21). This offers the prospect of a pass of increased duration over a ground station if the apogee is situated above it.

This type of orbit has been used to good effect in the past for astronomical observatories such as ISO and the X-ray Multi-mirror Mission (XMM). These types of spacecraft are often used in *observatory mode*, which means the spacecraft instruments are operated as if they were located in a room adjacent to the astronomer's workstation. This requires that for extended periods of time the payload shall be pointed to desired astrophysical targets whilst uninterrupted contact with a ground station is maintained. By placing such a spacecraft into highly eccentric orbits with a low perigee and an apogee substantially higher than the GEO radius, the observatory will spend the majority of its mission time in the apogee region. By optimizing its orbit period with respect to the ground station coverage, operation in observatory mode can be achieved.

Generally, there will be an interruption of observational time while the spacecraft passes through the perigee region. This also corresponds, however, with a traverse of the Van Allen radiation belt (see Chapter 2), which precludes the operation of certain types of payload, such as γ-ray, X-ray and UV detectors, which are particularly sensitive. Observation duty cycles in excess of 90% can be achieved in these orbits.

The alternative of placing this type of spacecraft into GEO has many obvious benefits such as ground station cover and long sensor integration times, but GEO is generally a higher energy orbit than a highly elliptical one, requiring greater ΔV to place it there.

Moreover, the radiation environment in GEO may also preclude the operation of certain types of payload.

5.7.2 Communications spacecraft

In order to maintain the apogee above a ground station for a maximum period, the Earth-oblateness perturbation, which causes the line of apsides to precess can be used (Chapter 4, Section 4.4.1). From equation (4.39), the major axis of the orbit will rotate in the plane of the orbit by an amount, to first order, given by

$$\Delta\omega \approx \frac{3\pi J_2 R_{\mathrm{E}}^2}{p^2} \left(2 - \frac{5}{2} \sin^2 i \right) \quad \text{rad/rev} \tag{5.32}$$

This will cause the apogee of the orbit to move away from above the region where the service is required.

However, there are two solutions of equation (5.32) for which $\Delta\omega = 0$, namely, $i = 63.4°$ (prograde) and $i = 116.6°$ (retrograde). At these inclinations, the line of apsides does not rotate and the apogee is stabilized at its initial position. Using this characteristic, spacecraft in highly elliptic orbits inclined at $63.4°$ have been used to provide communication links between high-latitude ground sites. Spacecraft in GEO cannot provide this service at latitudes in excess of $70°$, North or South, since their elevation at ground sites is then less than $10°$.

Molniya spacecraft

The use of highly elliptic inclined orbits was adopted by the former Soviet Union in April 1965 with the launch of the first Molniya (lightning) spacecraft.

The *Molniya orbit*, shown in Figure 5.24, is a highly elliptic one with a 12-hour period (strictly half a sidereal day, 11 h 58 min), inclined at $63.4°$ to the equator. The initial

Figure 5.24 The Molniya orbit

apogee, at a height of around 39 000 km, is placed above the Northern hemisphere, where it remains for the mission lifetime as a result of the choice of orbit inclination. Because the eccentricity is high, typically about 0.7, the spacecraft moves relatively slowly in the apogee region, and it consequently remains at high elevation over the high-latitude sites for an extended period. Each such spacecraft may then be utilized for communications purposes for typically 8 h of its 12-h orbit period.

To provide 24-h regional services, at least three Molniya spacecraft are needed. One of these resides in each of the three planes whose ascending nodes are spaced 120° apart. Many Molniya spacecraft have been launched; communications traffic in the CIS has grown to warrant the operation of typically eight Molniya I and eight Molniya III spacecraft in eight orbit planes spaced at 45° intervals.

A nominal set of Molniya orbital elements is given in Table 5.6. The ground track is shown in Figure 5.25(a). Since the spacecraft is at near-GEO altitude at apogee, its angular rate is approximately that of the Earth, and so the ground track over the required region closely follows a meridian as shown in Figure 5.25(a).

Table 5.6 Molniya orbit elements

$a = 26\,560$ km (12 h orbit period)
$e = 0.722$ ($h_p = 1000$ km, $h_a = 39\,360$ km)
$i = 63.4°$
$\omega = 270°$ (perigee in southern hemisphere)
Ω arbitrary (dependent upon regional coverage required)

Table 5.7 Principal attributes of highly inclined communications orbits

Advantages	Comments
• Satellite at high elevation at high latitude ground sites	This is an overriding benefit for mobile communication services at high latitude
• No eclipse during communications operation	
Disadvantages	Comments
• Ground stations must track spacecraft	
• More than one satellite required for 24-hour regional coverage	
• Satellite switching protocol required	
• Variation in satellite range and range-rate	This has a number of impacts upon the communication payload design:
	— variation in time propagation
	— frequency variation due to Doppler effect
	— variation in received signal power
	— change of ground coverage pattern during each orbit
• Passage through Van Allen radiation belts each orbit	Accelerated degradation of power and electronic systems
• Orbit perturbations	Third-body forces may perturb the perigee height, causing atmospheric re-entry

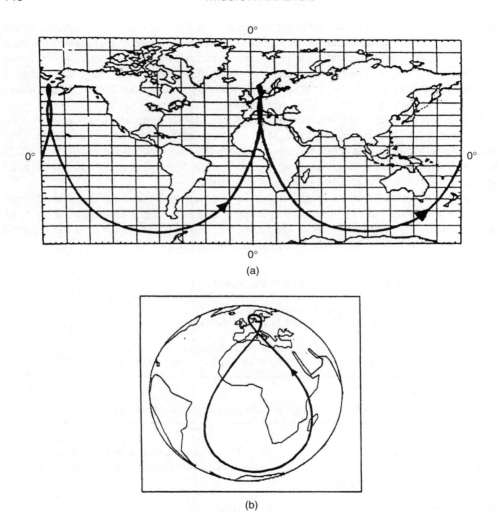

Figure 5.25 Typical global ground tracks for: (a) Molniya orbit and (b) tundra orbit

Interest in Molniya orbits is not confined to the CIS. Other space agencies are considering their use for high-latitude mobile communications and navigation programmes.

Tundra orbit

The tundra orbit is a derivative of the Molniya orbit in that it is elliptic and has an inclination of 63.4°. The principal difference is that its period is one sidereal day—23 h 56 min. An example of a global ground track for a Tundra orbit with an eccentricity of 0.24 is shown in Figure 5.25(b). Its main advantages are that it can provide 24-h coverage with a minimum of only two spacecraft, and that the orbital parameters can be chosen so that the spacecraft does not traverse the Earth's radiation belts. Its main disadvantage is that a higher ΔV is needed for orbit acquisition.

Table 5.7 summarizes the advantages and disadvantages for communications purposes of the inclined highly elliptic orbit compared with GEO.

5.8 INTERPLANETARY MISSIONS

5.8.1 Introduction

It is apparent that in the analysis of interplanetary trajectories, due account must be taken of perturbations caused by the gravitational influence of the Sun and planetary bodies within the solar system, and to a lesser extent to surface forces arising from solar radiation pressure.

For interplanetary spacecraft, the task of trajectory optimization and mission design is one of considerable complexity, requiring software support. Mission specialists at the NASA Jet Propulsion Laboratory in Pasadena have developed such computational tools to analyse complex missions such as that of the Jovian Galileo spacecraft, the Cassini mission to explore the Saturnian system, or the near-Earth asteroid rendezvous (NEAR) spacecraft to orbit a small body. Such a treatment is beyond the scope of this book. However, there is a simplified method, referred to as the *patched conics method*, which is widely used in feasibility studies of interplanetary flight.

The patched conics method assumes, in its simplest form, that at any one time only one body, a central body, is acting on the spacecraft. The region in which a particular body is dominant is called its *sphere of influence*, and in this region the spacecraft is assumed to follow a Keplerian orbit. As it leaves one region, it enters the sphere of influence of another body and its trajectory becomes a new conic with the new body at its focus. Thus, the entire trajectory is formed by patching together the various conic sections. Examples are shown in Section 5.8.2.

Spheres of influence are determined as in the following example of the Earth, which is subject to a disturbing gravitational field because of the Sun. That is to say, a spacecraft at a distance r_s from the Earth, and subject to its dominant gravitational force, is also subject to a weaker force from the Sun, from which its distance is r_d (see Figure 4.14, Chapter 4). Equation (4.45) gives the ratio of the accelerations caused by the Earth as the central body and the Sun as the disturbing one. When this ratio is equal to that which is given by reversing roles (i.e. the Sun is the central body and the Earth is the disturbing one), then the spacecraft is, by definition, situated on the sphere of influence of the Earth. Equation (4.45) can be used again, for the reverse role.

For the Earth–Sun, the acceleration ratios are equal when the Earth–spacecraft distance is

$$r_s = r_d \left(\frac{m_\oplus}{m_\odot}\right)^{2/5} (1 + 3\cos^2 \beta)^{-1/10} \tag{5.33}$$

where $m_\oplus$ and $m_\odot$ are the masses of Earth and Sun, respectively.

This surface approximates to a sphere since $(1 + 3\cos^2 \beta)^{-1/10} \approx 1$. Equation (5.33) then leads to the radius R_{SI} of the sphere of influence about any planetary body of mass m_p with respect to a disturbing body of mass m_d being

$$R_{SI} \approx r_d \left(\frac{m_p}{m_d}\right)^{2/5} \tag{5.34}$$

where now r_d is the distance of the disturbing body from the central planetary body.

This equation shows that the Earth's R_{SI} is about 0.93×10^6 km, assuming the Sun to be the disturbing body. (This compares with the radius of the Moon's orbit of about 0.38×10^6 km.) Reference [14] provides a more detailed analysis.

Table 4.1 shows the radii of the spheres of influence of the planets relative to the Sun.

5.8.2   Patched conic examples

The following three examples show how the initial planning of interplanetary missions may be conducted using the patched conic technique. They focus on the estimation of ΔV. The techniques for calculating a first estimate of transfer times on various trajectories can be found in Chapter 4, Section 4.2. Each phase of the total mission is assumed to have just one central gravitational force leading, for that phase, to a trajectory, which is a conic with the central body at its focus. During each separate phase, the position and velocity of the spacecraft are relative to the phase's central body, and at the handover from one to another they have to be recalculated.

Example 1: interplanetary transfer

The transfer of a spacecraft from an orbit around the Earth to an orbit around Jupiter is considered using a Hohmann minimum-energy transfer (see Section 5.2). The orbits of the Earth and Jupiter round the Sun are taken to be circular and coplanar. Their radii and other relevant data are taken from Tables 2.5 and 4.1.

The sequence of events can be defined by the following mission phases:

- Phase 1: Geocentric—the boost from Earth orbit to a hyperbolic escape trajectory.
- Phase 2: Heliocentric—the cruise in transfer orbit from Earth to Jupiter.
- Phase 3: Jupiter centred—the powered deceleration from hyperbolic approach trajectory to Jupiter orbit.

First step: heliocentric phase

Consider the Hohmann transfer within the Sun's sphere of influence (Figure 5.2). In the case of this interplanetary transfer, equation (5.1) gives the required Earth-relative speed of the spacecraft as it exits the Earth's sphere of influence, as

$$V_\infty^E = \sqrt{\frac{\mu_S}{r_E}} \left(\sqrt{\frac{2r_J}{r_E + r_J}} - 1 \right) = 8.792 \text{ km/s} \qquad (5.35)$$

Parameter values corresponding to the Sun, Earth and Jupiter are denoted by sub- or superscripts S, E or J, respectively.

Similarly, the apoapsis equation (5.2) gives the speed of the spacecraft relative to Jupiter as it enters the Jovian sphere of influence as

$$V_\infty^J = \sqrt{\frac{\mu_S}{r_J}} \left(1 - \sqrt{\frac{2r_E}{r_E + r_J}} \right) = 5.643 \text{ km/s} \qquad (5.36)$$

The transfer time is estimated as half the orbital period of the transfer ellipse, 2.731 years in this case.

Second step: geocentric phase

This phase covers the spacecraft's departure from the Earth, while it is still within its sphere of influence. The geometry of the trajectory is shown in Figure 5.26(a). The space-craft is initially assumed to reside in a circular orbit of height $h = 300\,\text{km}$, perhaps having been deployed by Space Shuttle. An impulsive velocity increment ΔV_1 is provided at point A, in order to inject the vehicle into a hyperbolic escape trajectory having a speed V_∞^E at a great distance from Earth, as given in equation (5.35).

Since the transfer is to be of the Hohmann type, the hyperbola's asymptote must be in the same direction as the Earth's orbital velocity vector. The escape hyperbola is then patched to the Hohmann transfer ellipse at the edge of the Earth's sphere of influence.

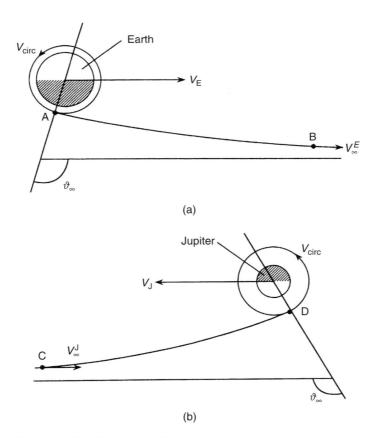

(a)

(b)

Figure 5.26 Geometry of planetocentric trajectories, show-ing: (a) departure from Earth, with rocket burn at A, and hyperbolic path to exit the sphere of influence at B; (b) arrival at Jupiter, entering its sphere of influence at C, hyperbolic path to D, and with a circularizing burn at D

The required velocity increment ΔV_1 is calculated by noting that the spacecraft's speed before the manoeuvre at A, given by equation (4.22), is

$$V_{\text{circ}} = \sqrt{\frac{\mu_E}{R_E + h}} = 7.726 \, \text{km/s}$$

After the manoeuvre, the speed is determined by noting that the hyperbolic path between A and B is at constant energy, and at B the potential energy is zero, so from equation (4.5):

$$\frac{1}{2} V_A^2 - \frac{\mu_E}{R_E + h} = \frac{1}{2} (V_\infty^E)^2$$

from which $V_A = 14.024 \, \text{km/s}$. So

$$\Delta V_1 = V_A - V_{\text{circ}} = 6.298 \, \text{km/s}$$

The location of point A may be identified in terms of the angle θ_∞ (Figure 5.26a). Equation (4.32) may first be used, to find that the eccentricity $e = 2.2950$, and then equation (4.30) shows that $\theta_\infty = 115.8°$.

The final step: arrival at Jupiter

Within the Jovian sphere of influence the trajectory will appear as shown in Figure 5.26(b). The spacecraft enters the sphere at C with speed V_∞^J and follows a hyperbolic path until its conversion into a circular orbit of radius $6R_J$ at D, by means of an impulsive burn with velocity increment ΔV_2. It is assumed that D is the point of closest approach to Jupiter on the hyperbola.

On arrival at C, at the aphelion of the Hohmann ellipse, the spacecraft's heliocentric speed will be less than that of Jupiter; consequently the direction of its arrival will be opposite to Jupiter's direction of travel round the Sun.

The required velocity increment ΔV_2 at D may be calculated by using equation (4.5) to determine the speed V_D before the manoeuvre:

$$\frac{1}{2} V_D^2 - \frac{\mu_J}{6R_J} = \frac{1}{2} (V_\infty^J)^2$$

which leads to $V_D = 24.934 \, \text{km/s}$, and equation (4.22) to determine the circular speed V_{circ} afterwards ($V_{\text{circ}} = 17.173 \, \text{km/s}$). Thus

$$\Delta V_2 = V_D - V_{\text{circ}} = 7.761 \, \text{km/s}$$

For this example, the eccentricity e is 1.1080 and θ_∞ is 154.5°.

Elliptic capture option

It must be noted that ΔV_2 can be reduced by specifying an elliptic capture orbit. For example, such an orbit with a perijove of $6R_J$ and an apojove of $20R_J$ will require a higher post-manoeuvre speed at D, that is, 21.301 km/s, and so ΔV_2 is reduced substantially, to 3.633 km/s.

Using a typical chemical propellant with a specific impulse of 300 seconds, the Tsiolkovsky equation (6.5) shows that the circular capture orbit above requires about 90% of the mass of the spacecraft that arrives at Jupiter to be fuel. For the elliptical capture case, this reduces to about 70%. The choice of capture orbit clearly has a major impact upon the spacecraft's mass budget and upon the mission's feasibility.

Example 2: planetary swing-by manoeuvres

A swing-by manoeuvre is a close pass of a planetary body by a spacecraft. Relative to the planet the trajectory is hyperbolic. If the geometry of the inbound and outbound asymptotes is constructed appropriately with respect to the orbital motion of the planet, then the spacecraft may increase or decrease its heliocentric velocity without expenditure of fuel. This makes the incorporation of these manoeuvres into interplanetary mission design a very attractive and powerful tool.

The Voyager spacecraft missions, launched in 1977 to explore the outer solar system, were of short enough duration to make the mission viable only through the use of this technique. Indeed, the Voyager 2 trajectory design included swing-bys of Jupiter, Saturn, Uranus and Neptune, spanning 12 years. This is to be compared with 30 years for a direct Hohmann-type transfer.

The mission designs for the Galileo Jupiter probe and the Cassini Saturn explorer both incorporate numerous swing-by manoeuvres in the capture and exploration mission phases.

The patched conic method can again be used, in combination with a relative velocity vector diagram, to provide a simple means of analysing spacecraft swing-bys. Figure 5.27(a) shows the swing-by geometry when the spacecraft passes 'behind' the planet (relative to the planet's forward motion), in a planet centred coordinate system. As the spacecraft enters the sphere of influence, the velocity relative to the planet is V_∞^-. If the hyperbola is a constant energy trajectory, then equation (4.5) shows that its outgoing velocity V_∞^+ is of equal magnitude, but its direction is deflected through an angle δ. Relative to the Sun, the planet's velocity is V_p. These vectors can be transferred to a relative velocity diagram shown in Figure 5.27(b), where they appear as dashed-line vectors. Nodes S/C$_1$ and S/C$_2$ represent the spacecraft before and after the encounter. Completion of the diagram as shown gives the Sun-relative velocities V_1 and V_2 before and after the swing-by. The change in the heliocentric velocity is ΔV.

It is clear that V_2 is greater than V_1, and so the spacecraft's speed has been increased by this passage *behind* the planet. A similar construction for a passage *in front of* a planet, shown in Figure 5.28, shows that this produces a reduction in speed. In both cases, the change in speed is of magnitude.

$$\Delta V = 2V_\infty \sin(\delta/2) \qquad (5.37)$$

Example 3: Ulysses-type swing-by

The ESA/NASA probe Ulysses performed a swing-by of the planet Jupiter in February 1992, the purpose being to change the direction of Ulysses, so as to place it into an orbit in a plane that is approximately normal to the ecliptic (see Section 5.1).

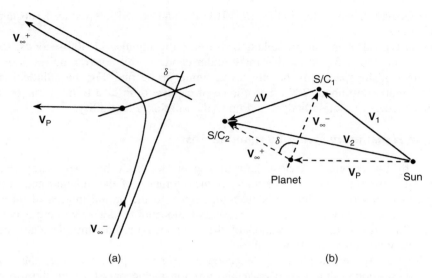

Figure 5.27 Swing-by passage *behind* a planet, showing: (a) hyperbolic trajectory in planet's sphere of influence and (b) relative velocity diagram (all vectors are coplanar)

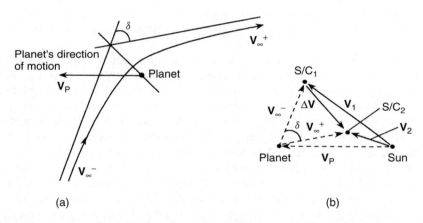

Figure 5.28 Swing-by passage *in front of* a planet, showing: (a) hyperbolic trajectory in planet's sphere of influence and (b) relative velocity diagram (all vectors are coplanar)

After deployment from Space Shuttle in October 1990, the spacecraft was accelerated to a record 15.4 km/s by a stack of upper stages. This gave the probe a heliocentric speed of 41.2 km/s when it left the Earth's sphere of influence, and this became $V_1 = 15.89$ km/s at the end of its journey to Jupiter, which compared with Jupiter's $V_J = 12.6$ km/s. It's direction of travel was at an angle $\gamma_1 = 61.2°$ to Jupiter's. The geometry is shown in Figure 5.29.

Solving the ecliptic plane triangle leads to the conclusion that the probe's velocity relative to Jupiter as it entered its sphere of influence was $V_\infty^- = 14.79$ km/s, with $\gamma_2 = 70.4°$.

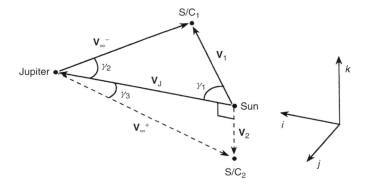

Figure 5.29 Relative velocity diagram for Ulysses-type swing-by. Solid-line vectors lie in the plane of the ecliptic **i** – **j**; dashed-line vectors lie in the orthogonal **i** – **k** plane

The operational orbit for Ulysses was chosen to maximize the time spent over the polar regions, and for this its post-swing-by velocity V_2 was not precisely normal to the ecliptic. However, for simplicity it is here assumed that it *is* normal, in the direction of $-\mathbf{k}$ in Figure 5.29. Noting that $\mathbf{V}_\infty^+ = \mathbf{V}_\infty^-$, this then yields $V_2 = 7.74\,\text{km/s}$, and $\gamma_3 = 31.6°$. The deflection angle δ can be determined from the equation

$$\cos\delta = \frac{\mathbf{V}_\infty^- \cdot \mathbf{V}_\infty^+}{V_\infty^2}$$

when the $\mathbf{V}_\infty$ vectors are expressed in **i, j, k** form using Figure 5.29. This gives $\delta = 73.4°$, and equations (4.31) and (4.32) give e = 1.673 and the closest approach to Jupiter's centre is equal to $5.4R_J$.

The resulting heliocentric orbit is an ellipse with aphelion at Jupiter (5.2 AU) and a perihelion at 1.20 AU.

The mission parameters that have been evaluated by this approximate method are very similar to those of the actual event, and demonstrate the usefulness of a relatively simple method. A more exact solution would be given by using the equations of the restricted three-body problem (Chapter 4, Section 4.5), but these do not have an analytic solution.

5.8.3 Small object missions

Introduction

The majority of interplanetary missions during the twentieth century focussed on exploration of the major planets. There is, however, currently a surge of interest in investigating the properties of minor bodies in the solar system, mainly asteroids and comets, which has led to the proposal of many new spacecraft missions. The principal driver for this activity is the hypothesis that these small objects, particularly comets, hold the key to understanding the composition of the genesis solar nebula, and possibly to the origin of life. The term 'small' in this context has a wide interpretation however, 'small objects' ranging from the satellites of the major planets (some of which are larger than the planets

Mercury and Pluto), to asteroidal chunks of rock perhaps a kilometre across. Missions to bodies of the former type are firmly on the agenda—for example, Jupiter's moon Europa has stirred interest, with the prospect of the discovery of life in a sub-ice ocean fuelled by tidal vulcanism. Objects such as Europa are relatively large, and near-spherical in shape, and methods discussed elsewhere in this chapter can be used to assess a mission profile. For example, a spacecraft orbiting Europa in combination with Jupiter, can be thought of as part of a circular, restricted three-body problem (see Chapter 4, Section 4.5) to a good approximation. A principal concern of a mission designer in this case would be the longevity of the mission orbit around the moon, when subjected to the powerful perturbing influence of the parent planet Jupiter.

The focus of the discussion in this section will be *asteroids* and *cometary bodies*, characterized by an irregular shape, and an overall dimension of up to a few tens of kilometres. Missions to orbit, and to land on such objects pose some interesting challenges to the mission designer. The first such mission, launched in February 1996, was the *NEAR Shoemaker* spacecraft, which was placed into orbit about the asteroid 433 Eros in February 2000. Figure 5.30 is a NEAR image of Eros, showing it to be an irregularly shaped object with a maximum dimension of around 35 km.

Figure 5.30 Image of asteroid 433 Eros, acquired by the NEAR Shoemaker spacecraft in 2000 (Reproduced by permission of NASA/JHUAPL)

Near-body environment

The environment around small objects (particularly comets) is of great scientific interest in itself, but we will focus on those environments that influence the mission and spacecraft system design. Looking at Figure 5.30, perhaps the most obvious of these is that the gravity field does not approximate to that of a sphere, and a method of simulating orbits in such a field is discussed below.

If 433 Eros is to be regarded as typical, the near-body environment of asteroids appears to be 'clean', with minimal dust or debris posing a hazard to spacecraft (although asteroids have been observed to have natural satellites—the *Galileo* probe found the first, christened Dactyl, around the asteroid 243 Ida in 1993).

Cometary bodies are similarly thought to be irregularly shaped, but comprised mainly of ice and dust. As a consequence, they can have a very dynamic near-body environment, particularly when closer than around 3 AU to the Sun, when their surface layers become warm enough to trigger the sublimation of ices. A *nucleus* observed at close quarters is that of Halley's comet, when the European Space Agency (ESA) probe *Giotto* passed within 596 km in 1986. The resulting images revealed a very active nucleus about 16 km by 8 km in size, with bright jets of gas and dust emanating from the sunlit 'hemisphere'. Evidence suggests, however, that cometary bodies more typically have a dimension of roughly 1 or 2 km. When active, the nucleus is the source of all cometary phenomena, which comprise the visible *coma* (extending up to a million km in radius), the *hydrogen cloud* (spanning many millions of km) and of course, the tail. Comets may exhibit *dust tails* up to 10 million km in length, and *plasma tails* extending up to 10 times further. The interaction between the solar environment and these cometary features is complex, and the subject of much research, a review of which can be found in Reference [15]. In terms of mission and spacecraft design, the cometary environment poses greater challenges (and unknowns) than that of asteroids. For example, a near-comet dust and debris environment will cause orbit and attitude perturbations, which must be accounted for in the design. There is an inevitable conflict between the scientists and the engineers, the resolution of which is difficult. The former group would wish to encounter the environment fully, to maximize the scientific return, while the latter group would wish to avoid it to minimize risks to the spacecraft. The difficulty facing the designer, however, is that this environment is not known *a priori*, and therefore a range of environments need to be accommodated in the design.

Motion around small, irregularly shaped bodies

The motion of a spacecraft around a rotating, irregularly shaped body will approximate to the Keplerian orbits discussed in Chapter 4 when the vehicle is at sufficient distance. However, for close orbits where the orbit radius is of the same order as the body's size, the trajectory shape no longer approximates to a conic section, and the resulting equations of motion are sufficiently complex to make the use of a computational solution mandatory.

The gravity field of the body can be modelled using a variety of methods. In terms of an operational scenario, the objective is to determine the gravitational field of the body by observing the close proximity motion of an orbiting spacecraft. On approach, payload imagery will provide quantitative information regarding the size, shape and rotational state of the body. On the basis of the shape assessment, the gravitational potential can be expressed as a series of spherical harmonics (see Chapter 4, Section 4.4.1) or ellipsoidal harmonics. Subsequent precise tracking of the motion of the orbiter will allow the coefficients in these series expansions to be evaluated, so determining the gravitational potential of the object. This fairly complex process provides a general characterization of the body's gravity field, which takes into account its shape and any localized variations in body density. The small sample of evidence currently available suggests that body shapes can best be approximated by triaxial ellipsoids. For example, the asteroid 433 Eros can be approximated by an ellipsoid with semi-axes 17.9 km, 9.2 km and 7.9 km. For such a body, Garmier and Barriot [16] have demonstrated that the performance of the ellipsoidal harmonic expansion is superior to a spherical harmonic representation. Their investigation of lander trajectories showed that an ellipsoidal expansion up to degree 5 gave a landing position error of the order of 1 m, compared to 10 s of m for the spherical harmonic representation.

The above operational methodology concerns itself with determining the gravity field from the analysis of the orbiter's motion. Conversely, given the gravity field of the body, the motion of the orbiter may be determined. An approximate method of determining the orbiter motion, which is adequate for initial study work, is presented here. This is sometimes referred to as the 'mascon' method, and uses the approach of building the body's shape utilizing a collection of spherical masses of uniform size and density. These spherical 'building blocks' can be placed at the vertices of a three-dimensional grid, so that they touch without overlap. The shape of the model of the irregularly shaped body can then be made arbitrarily close to the observed shape of an asteroid or comet, by allowing the size of these mass elements to decrease (within the bounds of computational effort). The overall mass and density of the body must be estimated bearing in mind that ~48% of the model is vacant. The gravitational potential at the spacecraft can then be found by summing the contribution from each element (or equivalently the gravitational force contribution from each may be summed vectorially). The situation is shown in Figure 5.31. The irregularly shaped body is shown rotating with angular velocity $\mathbf{\Omega}$ about the Z axis of an inertial co-ordinate frame (X, Y, Z), the origin of which is at the centre-of-mass of the body. The axis of rotation corresponds to the principal axis of maximum inertia of the body. From equation (4.2), the gravitational force acting on a spacecraft of mass m exerted by the body, comprised of n mass elements each of mass m_i, is

$$\mathbf{F} = \sum_{i=1}^{n} \left(\frac{Gmm_i}{r_i^2} \right) (-\mathbf{r_i}/r_i) \tag{5.38}$$

Noting that

$$\mathbf{r_i} = \mathbf{r} - \mathbf{R_i} = (x - X_i)\mathbf{i} + (y - Y_i)\mathbf{j} + (z - Z_i)\mathbf{k}$$

the equations of motion of the spacecraft may be written as

$$\ddot{x} = Ax + B_x, \ddot{y} = Ay + B_y, \ddot{z} = Az + B_z \tag{5.39}$$

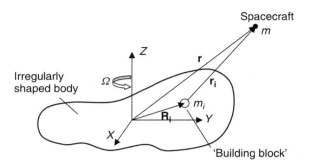

Figure 5.31   Geometry of the spacecraft in orbit around the irregularly shaped body

where

$$A = -\sum_{i=1}^{n} \frac{Gm_i}{r_i^3}, \ B_x = \sum_{i=1}^{n} \left(\frac{Gm_i}{r_i^3}\right) X_i, \ B_y = \sum_{i=1}^{n} \left(\frac{Gm_i}{r_i^3}\right) Y_i, \ B_z = \sum_{i=1}^{n} \left(\frac{Gm_i}{r_i^3}\right) Z_i$$

The equations (5.39) may be easily transformed to a set of six coupled first-order differential equations given by

$$\dot{x} = u, \ \dot{y} = v, \ \dot{z} = w, \ \dot{u} = Ax + B_x, \ \dot{v} = Ay + B_y, \ \dot{w} = Az + B_z \tag{5.40}$$

which may be solved numerically, given appropriate initial values of the spacecraft's position and velocity. To compute the values of the A and $B_{x,y,z}$ functions, the position vector of the elemental masses m_i needs to be known as a function of time. Using the Taylor series for a vector function, we may write

$$\mathbf{R_i}(t + \delta t) = \mathbf{R_i}(t) + \delta t \dot{\mathbf{R}}_i(t) + \tfrac{1}{2}\delta t^2 \ddot{\mathbf{R}}_i(t) + \cdots$$

By noting that the vector $\mathbf{R_i}$ is constant in the co-ordinate frame rotating at angular velocity $\mathbf{\Omega}$ with the asteroid, it is easily shown that

$$\mathbf{R_i}(t + \delta t) = \mathbf{R_i}(t) + \delta t \{\mathbf{\Omega} \times \mathbf{R_i}(t)\} + \tfrac{1}{2}\delta t^2 \{\mathbf{\Omega} \times (\mathbf{\Omega} \times \mathbf{R_i}(t))\} + \cdots \tag{5.41}$$

If using equation (5.41) to update $\mathbf{R_i}(t)$, it should be ensured that the combination of the magnitudes of δt, $\mathbf{R_i}$ and $\mathbf{\Omega}$ provide a convergent series. The above approximate method may be used to simulate close orbits around small irregular bodies. However, the overall accuracy of the method increases with the number of spherical building blocks used (so increasing computational effort), and perturbations because of density anomalies are not modelled.

Mission impact upon orbiter and lander system design

In this section, we discuss briefly the interaction between mission design and system design for a small body orbiter/lander mission. The focus will be on a comet rendezvous mission, since the challenges posed are greater than that of an asteroid mission. Many of the features are common to both. Such a cometary mission, called *Rosetta*, has been implemented by the ESA. This is due for launch in 2003, and rendezvous with comet 46P/Wirtanen in 2011, with a landing planned for 2012. The close orbit phase will occur when the comet is at a heliocentric distance of around 3.25 AU, before the onset of vigorous solar-induced activity.

To plan such a mission, particularly the close orbit and landing phases, the design is very much dependent upon the physical properties of the comet and its environment. Since these are generally unknown prior to the mission, a useful course of action is to propose 'engineering models' of the target comet, which encompass the range of properties expected. This allows the designer to accommodate a range of possibilities in the design. Table 5.8 shows the characteristics of a small, typical and large comet, similar to those considered in the Rosetta mission design. It is assumed that the comet models are spherical during this design phase, to simplify trajectory calculations.

A first examination of the table shows that the operating environment for an orbiter/lander is an extremely 'low-energy' one. Spacecraft generally orbit at a comfortable walking pace—orbital speed, orbital period and escape velocity can be estimated as a function of orbital height from equations (4.22), (4.12) and (4.25), respectively. Landers take a long time to descend to the surface. Care is required when making trajectory manoeuvres not to exceed escape velocity. Landers need to be secured to the surface on touchdown. As we will see, these characteristics have an impact on the design of orbiter, and in particular, lander operations. To highlight some of these impacts, we consider the deployment of a comet lander from an orbiting vehicle.

The first issue is release of the lander from the orbiter, which may be done by a mechanism, possibly in combination with thrusters. The difference between slowing the lander for descent and attaining escape velocity is only a matter of a few cm/s for small comets, so careful thought is needed to ensure the release mechanism is adequately

Table 5.8 'Engineering models' of a small, typical and large comet. The upper part of the table shows their physical characteristics. The lower part gives some dynamical parameters for an orbit radius of 20 km

	Comet size		
	Small	Typical	Large
Radius (m)	1000	2500	5000
Density (kg/m^3)	200	1000	1500
Gravity constant, $\mu = Gm$ (m^3/s^2)	56	4366	52 386
Surface g (m/s^2) (neglecting rotation)	5.59×10^{-5}	6.98×10^{-4}	2.10×10^{-3}
Rotation period (hours)	10	10	250
Equatorial surface speed due to rotation (m/s)	0.17	0.44	0.03
Surface escape velocity (m/s)	0.33	1.87	4.42
Surface weight of 50 kg lander (N)	2.8×10^{-3}	3.5×10^{-2}	1.0×10^{-1}
Orbit speed (m/s)	0.05	0.47	1.62
Escape velocity from orbit (m/s)	0.07	0.66	2.29
Orbit period (hours)	660.4	74.7	21.6

calibrated. If thrusters are used to acquire the correct descent speed, then sizing these is also an issue for the same reason. These manoeuvres require the lander to have knowledge and control of its attitude. Initially, a gyro platform may be established from orbiter data. However, given that the descent time is likely to be large (see below), this may need to be updated using reference sensors (see Chapter 9). To minimize the mass of the lander, payload sensors may be utilized for this attitude-sensing task. The near-body environment of the comet will impact directly on the choice of sensor—an imaging device may not be able to discriminate between intended targets, such as a stellar image, and a particle of debris.

The time to descend to the surface and the surface impact speed is easily estimated if it is assumed that the lander release sequence arrests its orbital motion, so that it free-falls under gravity along a radial path. The time is given by

$$t = -\sqrt{\frac{r_0^3}{2\mu}}\left(\alpha - \frac{\pi}{2} - \frac{1}{2}\sin 2\alpha\right), \quad \alpha = \sin^{-1}\sqrt{\frac{r}{r_0}} \tag{5.42}$$

where r_0 is the initial distance from the comet's centre, and r is the radius of the comet. The impact speed, estimated from the energy equation (4.5), is given by

$$V_{\text{imp}} = \sqrt{2\mu\left(\frac{1}{r} - \frac{1}{r_0}\right)} \tag{5.43}$$

This 'impact speed' is a useful parameter as it is an estimate of the Δv required near the surface to halt the lander for a soft touchdown. Figure 5.32(a) and (b) display values of descent time t and V_{imp} as a function of release altitude, for the three comet types given in Table 5.8. The descent time for small comets is long, which raises issues concerning the attitude stability of the lander during this phase, especially if there are disturbance torques induced by a severe near-body environment. If this disturbance is to be controlled by a thruster limit cycle, then the duration of the descent will govern the amount of propellant used. The use of angular momentum bias may be considered, either in the form of a momentum wheel or by rotation of the lander itself, in order to reduce the sensitivity of the attitude to the disturbances. A sophisticated trade-off analysis is required, involving mass and complexity minimization, to establish the optimum solution. Given these interactions between system and mission design, a constraint on the descent time may be imposed (e.g. $t < 3\,\text{h}$), which in turn will dictate the release altitude. Very low release altitudes may be dangerous with respect to the near-body environment.

The next major event in the sequence is the soft landing. The vehicle may be arrested a few tens of metres above the surface, and may even be required to 'hover' under thruster power. The values of V_{imp} given in Figure 5.32(b) give a measure of the Δv that needs to be imparted by the 'main engine'. The fuel mass required may be approximated by

$$M_{\text{fuel}} = M_0\{1 - \exp[-(\Delta v + gt)/v_{ex}]\} \tag{5.44}$$

where M_0 is the mass of the vehicle before the burn, g is the surface gravitational acceleration and v_{ex} is the effective exhaust velocity of the engine. The thruster sizing again is critical. Very low thrust levels will give good control, with respect to avoiding inadvertently acquiring escape velocity. On the other hand, the gravity loss 'gt' term

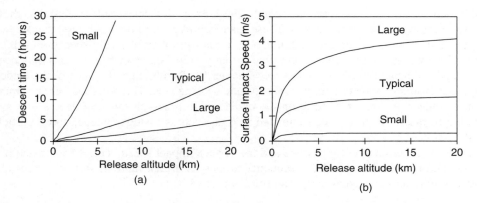

Figure 5.32 (a) descent time t and (b) surface impact speed V_{imp} as a function of release altitude for the three sizes of comet specified in Table 5.8

in equation (5.44), and therefore the propellant mass, is larger for this type of engine. Pollution of the surface by the exhaust products may also be an issue from the point of view of the science objectives. The final phase is touchdown, and in particular, some form of hold-down manoeuvre, which may use an 'upward' firing thruster while the vehicle attaches itself mechanically to the surface.

This brief description of a typical lander mission highlights some of the system issues that pose a challenge to the design engineer.

5.8.4   Planetary atmospheric entry and aeromanoeuvring

Our exploration of the solar system has revealed that not only do several of the major planets have atmospheres, but also some minor bodies, such as Titan, possess a significant detectable atmosphere. Probes sent to explore such worlds by orbiting or landing on these bodies require the ability to actively or passively manoeuvre successfully in such an environment. Indeed, the opportunity is provided by planets having an atmosphere to reduce the total vehicle wet-mass in the execution of a particular mission. Table 2.6 provides data on the atmospheres of those bodies in the Solar System having a significant atmosphere.

It is evident from the discussion in Chapter 4, Section 4.4, that the principal effect of an atmosphere on a satellite's trajectory is to reduce the energy of the orbit. However, it is also possible to utilize the aerodynamic forces, in order to produce both along-track and across-track accelerations without the need for propulsive manoeuvres. The along-track effects, which are generally associated with *aerobraking*, are used and proposed for a variety of mission operations to reduce the translational energy of a spacecraft. These include direct atmospheric entry to reduce the vehicle's speed to facilitate soft landing, orbital *aerocapture* where aeroforces are used to transfer a vehicle's orbital state from hyperbolic to elliptical, and *aero-assisted orbit transfer* where an atmospheric pass is used to modify the orbit. An example of the latter is the transfer of a vehicle from high orbit to low orbit, where an aerobraking manoeuvre is performed in the periapsis region. The across-track aeroforces produce out-of-plane accelerations, and so can be used to modify the orbit plane inclination. It has also been proposed to use them to augment the beneficial effects of gravitational swing-bys (see Section 5.8.2). *Aeromanoeuvring* is a term applied generally to missions utilizing aerodynamic forces. Clearly, savings in propellant mass can be achieved, but this must be weighed in all instances against the additional mass required to protect the vehicle from the dynamic pressure and thermal effects associated with aeromanoeuvring.

A survey of aeroassisted orbit transfer is provided by Walberg [17]. He looked at three specific mission categories, that is, synergetic plane change, planetary mission applications and orbit transfer applications. All these include some degree of aeromanoeuvring which, unlike the simple swing-by missions discussed in Section 5.8.2, have a mass penalty as mentioned above. This arises from the need to incorporate in the design an aeroshield and, in some cases, propellant mass if there is a propelled trajectory between intermediate orbits.

Since Walberg's paper, there have been a number of planetary missions launched, with some missions being completed, which have incorporated a degree of aeromanoeuvring. A review of the mission designs and in-flight performance of several missions: *Mars Global Surveyor, Mars Pathfinder, Galileo, Mars Microprobe, Mars Polar Lander* and *Stardust*, may be found in a special journal issue [18]. The engineering solutions, incorporating the aeroassisted element of the mission design, were in part spurred by the Goldin initiative for NASA of providing faster, better, cheaper missions. This is as a result of aeroassist being a method of both reducing mass of a planetary satellite directly and providing a consequential cost reduction of launch vehicle services. According to Reference [18], 65% of NASA's planetary missions scheduled for the period 1995 to 2005, utilize aeroassist. The following section provides a simplified overview of the key features of the aerodynamic influences of this type of space mission.

Atmospheric entry

As a space vehicle approaches a planet having an atmosphere, it experiences an approximate exponentially increasing atmospheric density. This provides a changing aerodynamic environment for the vehicle. Initially the Mach number (the ratio of the vehicle's speed relative to the local speed of sound in the gas) M_0 may be in the range of 20 to 50. The initial density is so low, however, that the flow field is described as a *free molecular flow*. In this regime, the molecules and atoms that constitute the atmosphere collide so infrequently, that following impact upon a vehicle surface, the molecule will not then collide with the incoming molecules. Under such conditions, shock waves are not formed about the body. However, as the vehicle progresses further into the atmosphere, a *transition flow* commences. This region is difficult to describe analytically, and frequently *bridging functions* are used to describe the aerodynamic properties of the vehicle. Thick viscous shock waves are formed about the vehicle at this time. Eventually, if the vehicle penetrates sufficiently low in the atmosphere, a *continuum flow* region is encountered, which is that type of flow typified by conventional aerodynamics. The velocity, however, is still so great that the flow remains hypersonic.

Key to an understanding of the flow field and the changes that occur within it is the recognition of the amount of energy required to be dissipated. Taking the amount of energy per unit mass to be $\sim 0.5\,V^2$, if a vehicle is approaching the planet on a hyperbolic trajectory with a velocity of tens of km/s, hundreds of MJ must be dissipated. An extreme example is the entry of the Galileo probe into the Jovian atmosphere in 1995. With an approach speed of 47.5 km/s, 3.8×10^5 MJ of translational energy were dissipated in the four minutes before the drogue parachute was deployed. This generated a temperature of 15 000 K, and an estimated 90 kg of ablative material was lost from the probes forward heat shield (out of a total probe initial mass of 340 kg). At such energies, the gas that is incident upon the vehicle undergoes not only chemical reactions, but also excitation of internal energy modes such as vibration, together with dissociation and ionization. Relaxation from these excited states may arise through radiation. The time constants of these processes are large, and hence the flow field is not in equilibrium. As a result, there is great difficulty in analytically predicting the changes that arise in the gas, as the normal relationships of equilibrium thermodynamics cannot be applied. Approximations, such as assuming that the constituents (both in terms of their chemical composition and the degree to which excitation has occurred) have relative number density fractions frozen at some point in the flow, may be used to simplify the analysis. Detailed predictions are made yet more complex by the uncertain role the vehicle surface plays in the chemistry of the reacting flow. The overall net effect however, of the high velocity flow field impinging on a surface is to cause substantial heat transfer to the vehicle. It is also clear from this brief introduction that, because of the chemically reacting flow conditions, the actual heat loads a vehicle will experience will depend upon the constituents of the atmosphere itself. The dominant constituents for each of the planetary atmospheres is shown in Table 2.6. An excellent recent review of the literature concerning planetary entry gas dynamics is given by Gnoffo [19].

Constraints during atmospheric entry

The two principal constraints that occur in the design of an aeromanoeuvring vehicle are the peak dynamic load and the peak thermal load, together with how long these loads exist.

During entry, it is possible to write down the overall governing equations that describe the dynamics. For a ballistic entry, it is assumed that the aerodynamic forces only provide a drag force parallel to the instantaneous direction of motion, with no cross track force (equivalent to a zero for the lift coefficient C_L). It is possible to write down the relationship between the distance to the centre of the planet r at time t, if at that time the flight angle of the vehicle is γ and the velocity V. Assuming the density at some reference height h is given by ρ_s, which decreases exponentially with a scale height of β, then following [20],

$$\frac{\mathrm{d}r}{\mathrm{d}t} = V \sin \gamma \tag{5.45}$$

where

$$\frac{\mathrm{d}V}{\mathrm{d}t} = -\frac{1}{2}\rho_s V^2 \mathrm{e}^{-\beta h} \left(\frac{SC_D}{m} \right) = -\eta \beta V^2 \tag{5.46}$$

Here a dimensionless height variable η has been introduced,

$$\eta = \frac{1}{2}\rho_s \left(\frac{SC_D}{m} \right) \frac{1}{\beta} \mathrm{e}^{-\beta h} \tag{5.47}$$

The ballistic coefficient, (SC_D/m), where S is the wetted surface area for a vehicle of mass m having a drag coefficient C_D, is seen from equation (5.46) to provide a linear influence over the rate at which the vehicle decelerates. However, the maximum deceleration for an initial speed V_0 and entry angle γ_0 is given by δ_{max} and is found to be independent of the ballistic coefficient,

$$\delta_{max} = \frac{\beta V_0^2}{2e} \sin \gamma_0 \qquad (5.48)$$

This function is plotted in Figure 5.33.

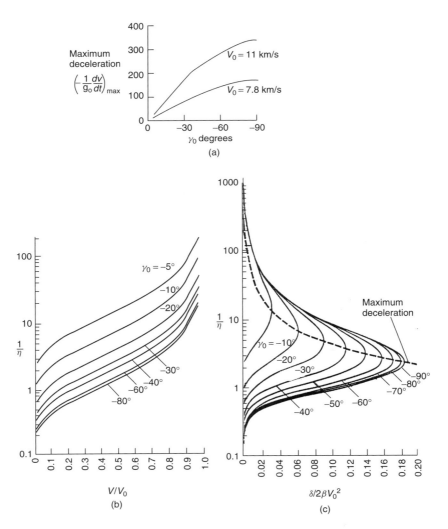

Figure 5.33 Ballistic entry characteristics: (a) variation of the maximum deceleration with entry angle γ_0 and entry speed V_0 for ballistic entry; (b) ballistic entry at large flight path angle: speed-altitude diagram for several values of the initial angle (after Vinh *et al.* [20])(Reproduced by permission of the University of Michigan) and (c) ballistic entry at large flight path angle: acceleration-altitude diagram for several values of the initial angle (after Vinh *et al.* [20])(Reproduced by permission of the University of Michigan)

Turning to the peak-heating load, it is clear from the discussion above that a simple analytic description is not available, if one wishes to describe the real flow situation. Approximations may be used to provide some estimate of the heating profile. These typically omit elements of the various heat transfer processes that take place in the real flow. For example, if only convective heat transfer is considered (or is indeed dominant) [20], then the peak heating rate is given by

$$\dot{q}_{max} \propto V_0^3 \sqrt{\left(\frac{m\beta \sin \gamma_0}{3SC_D}\right)} \tag{5.49}$$

In general, this peak heat flux will occur at a different altitude from that for the peak deceleration load.

Evidently, both the dynamics and heat loads are dependent upon the initial conditions assumed for atmospheric entry [21]. As a result, calculations that are performed must assume an overall mission profile. Thus, the preceding interplanetary manoeuvres will influence the final loads experienced by a vehicle, together with the launch date. Thus, for the Mars Pathfinder mission [22], the inertial arrival velocity could vary by 100 m/s. The worst case (highest entry velocity) trajectory is shown in Figure 5.34.

Various design solutions have been used, or proposed to accommodate both the dynamic loads and thermal heat flux. Simple passive solutions are most appropriate for vehicles that may have been in transit for several years prior to the critical minutes of atmospheric entry. Figure 5.35, taken from Reference [19] shows the range of heat loads experienced

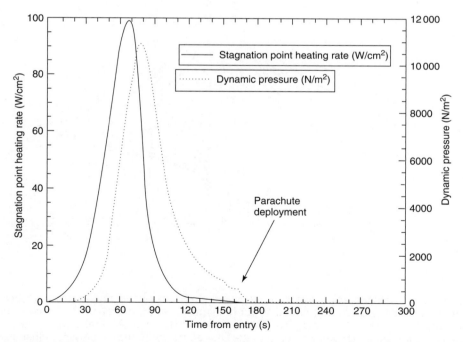

Figure 5.34 Heating rate and dynamic pressure during entry for Mars pathfinder mission (after Spenser *et al.* [22])

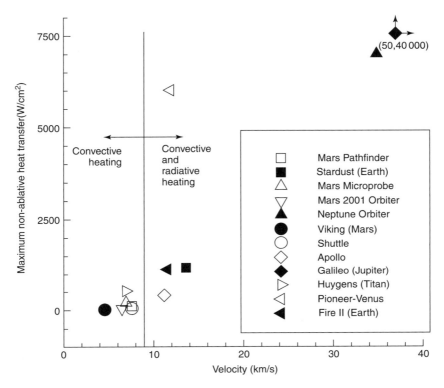

Figure 5.35 Non-ablative peak heating *versus* velocity for past and planned planetary entry vehicles (after Gnoffo [19])

for different vehicles. It should be noted that for some vehicles, such as Apollo, ablative shields are appropriate, since contamination is not an issue. However, where sensitive scientific measurements are to be taken shortly after the main phase of deceleration, the role and final fate of any ablative shield and its constituent materials, need to be carefully examined.

Entry corridor

The foregoing has identified the major constraints arising from the mission profile, which will dictate the preferred engineering solution. There is one other major component to the final mission design that derives from the planet to be encountered. Both maximum heat and dynamic loads are a function of entry angle γ_0 as seen in equations (5.48) and (5.49) above. A design solution will specify the peak value that these loads may have. If the vehicle were to enter at an angle greater than the specified value, then it may be anticipated that the vehicle will either burn-up or break-up. However, there also needs to be consideration of the problem that will arise if the entry angle is shallower than that in the design solution. Chapman [23] defined an entry corridor in terms of the altitude of the first periapsis pass, following atmospheric entry. Undershooting the target height leads to burn-up or break-up, whereas overshooting leads to the vehicle re-emerging from the atmosphere. The entry corridor is then defined as the height difference of periapsis

Table 5.9 Entry corridor widths for entry into planetary atmospheres at parabolic speed $V_0 = (2gr_0)^{1/2}$

Planet	Corridor width (km)					
	$5\,g$ limit			$10\,g$ limit		
	$L/D = 0$	$L/D = 1$	$L/D = 1$ modulated	$L/D = 0$	$L/D = 1$	$L/D = 1$ modulated
Venus	0	43	58	13	84	113
Earth	0	43	55	11	82	105
Mars	338	482	595	644	885	1159
Jupiter	0	55	68	0	84	113

between the acceptable extremes of under- and overshooting. Clearly, this corridor is dependent upon the acceptable limits on both heat load and mechanical deceleration for a specific vehicle.

Table 5.9 provides data for corridor width for several planets.

ACKNOWLEDGEMENT

Thanks is due to Dr Hugh Lewis, for his assistance in compiling material for this chapter.

REFERENCES

[1] Hechler, M. and Van der Ha, J. C. (1981) *J. Spacecraft Rockets*, **18**, 361.
[2] ESA Bulletin, Ulysses Launch Issue, No. 63, August 1990.
[3] De Veubeke, B. F. (ed.) (1969) *Advanced Problems and Methods for Spaceflight Optimization*, Pergamon Press, Oxford.
[4] Lawden, D. F. (1963) *Optimal Trajectories for Space Navigation*, Butterworth, London.
[5] King-Hele, D. G. and Walker, D. M. C. (1987) *Royal Aircraft Establishment Technical Report* 87030, Farnborough, UK.
[6] Lüders, R. D. (1961) Satellite networks for continuous zonal coverage, *Am. Rocket Soc. J.*, **31**, 179–184.
[7] Adams, W. S. and Rider, L. (1987) Circular polar constellations providing continuous single or multiple coverage above a specified latitude, *J. Astron. Sci.*, **35**, 155–192.
[8] Walker, J. G. (1971) Some circular orbit patterns providing continuous whole Earth coverage, *J. Br. Interplanet. Soc.*, **24**, 369–384.
[9] Walker, J. G. (1984) Satellite constellations, *J. Br. Interplanet. Soc.*, **37**, 559–571.
[10] Benedicto, J., Fortuny, J. and Rastrilla, P. (1992) MAGSS-14: A medium-altitude global mobile satellite system for personal communications at L-band, *ESA J.*, **16**, 117–133.
[11] Clark, A. C. (1949) Extraterrestrial relays, *Wireless World*, October, 305.
[12] Smalley, M. (1993) Mission Analysis III: The Geostationary Orbit, Spacecraft Systems Course, University of Southampton, UK.
[13] Soop, E. M. (1983) *Introduction to Geostationary Orbits*, ESA SP-1053.
[14] Cornelisse, J. W., Schoyer, H. F. R. and Wakker, K. F. (1979) *Rocket Propulsion and Spacecraft Dynamics*, Pitman, London.
[15] Beatty, J. K., Collins Peterson, C. and Chaikin, A. (eds) (1999) *The New Solar System*, (4th edn), Cambridge University Press, New York.

[16] Garmier, R. and Barriot, J.-P. (2001) Ellipsoidal harmonic expansions of the gravitational potential: theory and application, *Celestial Mech. Dynamical Astron.*, **79**, 235–275.

[17] Walberg, G. D. (1985) A survey of aeroassisted orbit transfer, *J. Spacecraft Rockets*, **22**, 3–18.

[18] *J. Spacecraft Rockets*, **36**(3), May/June 1999.

[19] Gnoffo, P. A. (1999) Planetary gas dynamics, *Annu. Rev. Fluid Mech.*, **31**, 459–494.

[20] Vinh, N. X., Busemann, A. and Culp, R. D. (1980) *Hypersonic and Planetary Entry Flight Mechanics*, University of Michigan Press, Michigan.

[21] Lees, L., Hastwig, F. W. and Cohen, C. B. (1959) Use of aerodynamic lift during entry into the Earth's atmosphere, *Am. Rocket Soc. J.*, September, 633–641.

[22] Spenser, D. A. and Braun, R. D. (1996) Mars pathfinder atmospheric entry: trajectory design and dispersion analysis, *J. Spacecraft Rockets*, **33**, 670–676.

[23] Chapman, D. R. (1959) An Approximate Analytical Method for Studying Entry into Planetary Atmospheres, NASA Technical Report R-11.

6 *PROPULSION SYSTEMS*

J. Barrie Moss[1] and John P. W. Stark[2]

[1] *School of Engineering, Cranfield University*
[2] *Department of Engineering, Queen Mary, University of London*

6.1 SYSTEMS CLASSIFICATION

The broad classes of propulsion systems for space vehicles are distinguished in Figure 6.1. The opportunities for air-breathing stages in the early phases of an Earth surface launch are briefly reviewed in Section 6.2.4, but we shall focus here on systems that are not reliant on an external oxidizer provision. Of the several alternatives identified, exploitation in terms of practical devices has concentrated largely on thermal and electric rockets. Primary propulsion for launch vehicles is further restricted—currently to solid- or liquid-propelled chemical rockets. The principal options are reviewed in detail in later sections and we refer only briefly to more speculative concepts.

The solar radiation pressure (SRP) at 1 AU from the Sun is approximately $5 \times 10^{-6} \text{ N/m}^2$ and therefore the surface area that must be deployed in order to produce significant thrust for primary solar sailing is extremely large, even in circumstances in which the gravitational force on the sail is small. While missions can be identified for which solar sailing is an attractive option—including the Halley's comet interception mission proposed in the mid-1970s (see Friedman *et al.* [1])—there has been little practical demonstration. Space vehicles of more modest dimensions may, however, be subject to significant perturbing torques, resulting, for example, from asymmetric surface deployments or simply from the cyclic variation of SRP experienced by Earth-oriented spacecraft (see Chapter 9).

Nuclear propulsion has been the subject of very detailed studies over many years, although these have not been pursued to flight demonstration. The NERVA programme of the 1960s (Nuclear Engine for Rocket Vehicle Applications) resulted in a ground-tested solid core (graphite) U235 fission-powered engine delivering approximately 300 kN of thrust with a specific impulse of 825 s. With the emphasis on near-Earth operations, direct thrust nuclear rockets do not now appear to be cost-effective and raise environmental concerns that are not readily assuaged. More recent studies focus on lower-thrust orbit

Spacecraft Systems Engineering (Third Edition). Edited by P. W. Fortescue, J. P. W. Stark and G. G. Swinerd
© 2003 John Wiley & Sons Ltd

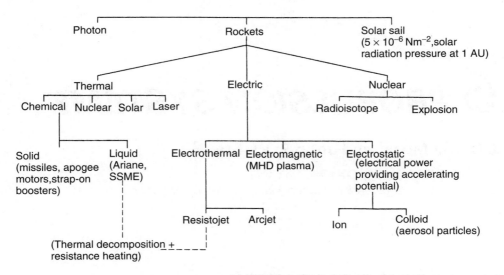

Figure 6.1 Propulsion systems classification

raising and manoeuvring propulsion as aspects of wider programmes for nuclear space power raising (see Buden and Sullivan [2]).

The development of high-power lasers and intermittent interest in the establishment of complex orbiting platforms for power raising and communications in space have also encouraged a number of conceptual studies of laser sustained propulsion (see Caveney [3]).

Crucial parameters that distinguish between the developed systems and introduce important characteristics of systems more generally are readily identified. We show later that the beam (or kinetic) power of the rocket exhaust, P, is given by

$$P = \tfrac{1}{2}\dot{m}V_e^2 \tag{6.1}$$

where $\dot{m}$ is the exhaust mass flow rate and V_e is the exhaust velocity.

Since the thrust delivered by the rocket may be written

$$F = \dot{m}V_e \tag{6.2}$$

then

$$P = \tfrac{1}{2}FV_e \tag{6.3}$$

If we introduce the vehicle acceleration (or vehicle thrust-to-weight ratio), αg_0, where g_0 is the Earth surface acceleration due to gravity, then we may write

$$F \simeq M\alpha g_0$$

(M = vehicle mass), and, from equations (6.2) and (6.3),

$$P/M = \tfrac{1}{2}\alpha V_e g_0 \tag{6.4}$$

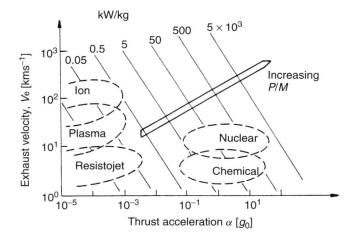

Figure 6.2 Comparative rocket performance

Propulsion systems are readily distinguished by comparisons between their specific power, P/M, exhaust velocity, V_e, and thrust acceleration, α, as illustrated in Figure 6.2. The separately powered electric rocket is characterized by high-exhaust velocity and low specific power or thrust acceleration. Nuclear or chemical rockets, on the other hand, offer high power but with relatively poor propellant utilization through their limited exhaust velocities. Booster operation from planetary surfaces is evidently restricted to these latter systems that are said to be *energy limited*—by the chemical energy stored in the propellants—but which admit high thrust for comparatively modest engine weight. Although the specific impulses of electrically powered systems are high, they are *power limited*. While the energy available from the Sun is unlimited, or that from a radioactive source introduces a negligible fuel mass, the effective energy density is low and the necessary weight of the accompanying systems for electrical conversion is unacceptably large in high-power applications. Electrically propelled space vehicles are thus restricted to very small accelerations. As we shall describe in Section 6.3, the more efficient expellant utilization possible with low-thrust systems thus appears most naturally suited to orbit-raising manoeuvres, interplanetary transfer and spacecraft attitude and orbit control.

While the role of propulsion is most clearly identified in relation to launch vehicles, spacecraft propulsion makes a number of equally important contributions to overall mission success. The range of tasks and performance requirements are illustrated below.

- Launch vehicles—main engines and 'strap-on' boosters—developing continuous high thrust for periods of minutes (approximately 2×10^6 N for 8 minutes in the case of each Space Shuttle Main Engine (SSME), for example).
- Apogee motors for spacecraft orbit circularization and inclination removal (typically, 75 kN for approximately 60 seconds, developing a velocity increment of 2 km/s); perigee motors for orbit raising from low-altitude parking orbit.
- Spacecraft station-keeping, attitude and orbit control (thrust levels ranging from 10^{-3} to 10 N, intermittent and pulsed operation over the complete duration of the mission).

Since propulsive requirements are frequently specified in terms of ΔV (see Chapter 5), it is instructive to compare some typical values:

- ΔV into Low Earth Orbit (LEO) (including drag and gravity losses) $\geq 9.5\,\mathrm{km/s}$
- additional ΔV to equatorial geostationary orbit from a 30° inclined parking orbit $\sim 4.2\,\mathrm{km/s}$
- additional ΔV for a Mars fly-by $\sim 3.4\,\mathrm{km/s}$
- additional ΔV for Solar System escape (without gravitational assist) $\sim 8.5\,\mathrm{km/s}$
- geostationary orbit station keeping (communications satellite)
  North-South (inclination $<0.1°$ throughout a seven-year life) $\sim 0.35\,\mathrm{km/s}$
  East-West ($\pm 0.1°$ of nominal longitude) $\sim 0.03\,\mathrm{km/s}$

6.2   CHEMICAL ROCKETS

The rapid growth in rocket propulsion technology, following World War II, is largely based on chemical rockets. Simple reaction systems, in which the propulsive force exerted on a vehicle arises from changes in system momentum through the discharge of hot products of combustion at high velocity, have proved uniquely successful in high thrust atmospheric and space applications. The burning of chemical propellants, solid or liquid, at high pressure liberates large quantities of energy in a compact volume. The subsequent expansion of these high temperature products of combustion, through a convergent–divergent nozzle, converts thermal energy to directed kinetic energy for rocket propulsion. Since the rocket carries both fuel and oxidizer, the specific fuel consumption is substantially higher than that of an air-breathing reaction system like the turbojet. It is, however, mechanically less complex than an aircraft power plant since moving parts are confined to auxiliary systems such as the propellant feed.

We show later in this section that chemical rockets are conveniently characterized by the *Tsiolkovsky equation* (in field-free space)

$$\Delta V = V_e \ln R \tag{6.5}$$

where ΔV denotes the rocket velocity increment and R is the mass ratio, initial mass to mass at burn-out. ΔV is typically prescribed by the mission, while V_e is essentially fixed by the choice of propellant. Only by increasing the mass ratio in equation (6.5) can the shortfall in propellant energetics be accommodated and mission objectives attained.

6.2.1   Basic principles

We first review briefly those aspects of rocket motor performance, gas dynamics and thermochemistry, which most directly influence design and operation. For further details of the analysis, the reader is referred to the excellent texts by Barrere *et al*. [4] and Sutton [5].

Performance parameters

Consider the rocket illustrated schematically in Figure 6.3. Applying equation (3.18) from Chapter 3 leads to

$$M\frac{\mathrm{d}V}{\mathrm{d}t} = \dot{m}V_e + A_e(p_e - p_a) + F_{ext} \tag{6.6}$$

The rocket thrust F comprises two contributions, from the exhaust momentum flux and the exhaust plane pressure difference

$$F = \dot{m}V_e + A_e(p_e - p_a) \tag{6.7}$$

where $\dot{m}$ is the propellant mass flow rate, V_e is the exhaust velocity, $A_e(p_e - p_a)$ is the resultant force on the rocket caused by the pressure difference between the nozzle exit and the ambient, and F_{ext} denotes the extra force in the direction of motion due to external forces (e.g. aerodynamic drag or gravity).

We show in the following section that nozzle performance, and in particular, the exhaust velocity V_e, is maximized by complete exhaust expansion to ambient pressure, when $p_e - p_a = 0$. In space applications, the ambient pressure is either continuously varying with altitude, or is zero beyond the Earth atmosphere. The effects of any underexpansion, $p_e > p_a$, are thus partially offset by the thrust increase accompanying this pressure difference.

From equation (6.7) the thrust at sea level is

$$F_{SL} = \dot{m}V_e + A_e(p_e - p_{SL})$$

and at altitude h, it may be expressed as

$$F_h = F_{SL} + A_e(p_{SL} - p_h) \tag{6.8}$$

In vacuo, when $p_h = 0$, it becomes

$$F_0 = F_{SL} + A_e\,p_{SL}$$

Typically $A_e p_{SL}/F_{SL}$ approaches 20%.

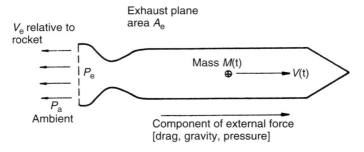

Figure 6.3 Schematic rocket

It is convenient to define an *effective exhaust velocity*

$$V_e^* = V_e + A_e(p_e - p_a)/\dot{m} \equiv I_{SP}\, g_0 \tag{6.9}$$

where g_0 is the acceleration due to gravity at the Earth surface.

I_{SP} is the *specific impulse*, the total impulse per unit propellant weight consumed, and is given by

$$I_{SP} = \frac{I}{M_p g_0} = \frac{\displaystyle\int_0^t F(t)\,\mathrm{d}t}{\displaystyle g_0 \int_0^t \dot{m}(t)\,\mathrm{d}t} \tag{6.10}$$

whence $I_{SP} = F/\dot{m} g_0$ for constant thrust and exhaust mass flow rate.

Equation (6.6) may be expressed in terms of V_e^* and, when this is constant, it may be integrated over the duration of rocket motor firing giving

$$\Delta V = V_b - V_0 = V_e^* \ln\{M_0/M_b\} + \int_0^{t_b} \frac{F_{ext}}{M}\,\mathrm{d}t \tag{6.11}$$

and hence the Tsiolkovsky equation (6.5), where the *mass ratio*

$$R = M_0/M_b$$

is the ratio of the initial to burn-out mass—note that $\dot{M} = -\dot{m}$.

The maximization of V_e^* for a specified velocity increment ΔV is essential for efficient design. We now describe the nozzle flow characteristics necessary to realize high exhaust velocity given a particular propellant selection.

Nozzle flows

We analyse the flow through a convergent–divergent nozzle, downstream from the combustion chamber and as illustrated in Figure 6.4, with the aid of the following simplifying assumptions:

● The combustion products are homogeneous and of constant composition.

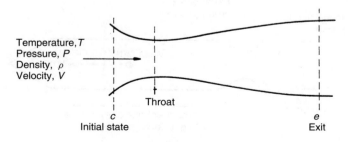

Figure 6.4 Schematic nozzle flow

- The products at temperature T and molecular weight W obey the perfect gas law relating pressure p and density ρ:

$$p = \rho T (R_0 / W) \tag{6.12}$$

where R_0 is the universal gas constant.

- The specific heat of the mixture is invariant with temperature and pressure.
- The flow is one-dimensional, steady and isentropic.

The conservation equations for mass and energy may then be written as

$$\dot{m} = \rho V A \tag{6.13}$$

$$\tfrac{1}{2} V^2 + C_p T = \text{constant} = \tfrac{1}{2} V_c^2 + C_p T_c \tag{6.14}$$

where the subscript c denotes the initial state. For an adiabatic flow process, we have

$$p \rho^{-\gamma} = \text{constant}$$

and
$$\frac{T}{T_c} = \left(\frac{\rho}{\rho_c} \right)^{\gamma - 1} = \left(\frac{p}{p_c} \right)^{(\gamma - 1)/\gamma} \tag{6.15}$$

If the initial velocity V_c is negligibly small, then

$$\frac{\dot{m}}{A} = \left\{ \frac{2\gamma}{\gamma - 1} p_c \rho_c \left(\frac{p}{p_c} \right)^{2/\gamma} \left[1 - \left(\frac{p}{p_c} \right)^{(\gamma - 1)/\gamma} \right] \right\}^{1/2} \tag{6.16}$$

Clearly equation (6.16) exhibits a maximum value for $(\dot{m}/A)$ corresponding to a critical throat condition, subscript t, at which

$$\frac{p_t}{p_c} = \left(\frac{2}{\gamma + 1} \right)^{\gamma/(\gamma - 1)}$$

and
$$\frac{T_t}{T_c} = \frac{2}{\gamma + 1}$$

and
$$\frac{\dot{m}}{A} = (\gamma \rho_t p_t)^{1/2} = \rho_t V_t$$

where the critical throat velocity V_t is given by

$$V_t = \left\{ \gamma \frac{p_t}{\rho_t} \right\}^{1/2} = \left\{ \gamma \frac{R_0}{W} T_t \right\}^{1/2} = a_t$$

which is the speed of sound at the throat.

In convergent–divergent (de Laval) nozzles, the velocity continues to increase downstream from the throat ($p_{\text{exit}}/p_c < p_t/p_c$) but the nozzle is choked—that is, the mass flow is simply determined by throat conditions, independent of the exit flow condition. The choked mass flow rate can be expressed as a function of combustion chamber conditions (p_c, T_c) and throat area A_t,

$$\dot{m} = \sqrt{\gamma} \left\{ \frac{2}{\gamma + 1} \right\}^{(\gamma+1)/[2(\gamma-1)]} \frac{p_c A_t}{\sqrt{(R_0 T_c / W)}} \tag{6.17}$$

We may identify a characteristic velocity

$$c^* = \sqrt{(R_0 T / W)} \Big/ \left\{ \sqrt{\gamma} \left[\frac{2}{\gamma + 1} \right]^{(\gamma+1)/[2(\gamma-1)]} \right\} \tag{6.18}$$

whence $\quad \dot{m} = p_c A_t / c^* \tag{6.19}$

Using the energy conservation equation (6.14) we may determine the exhaust velocity V_e from

$$\tfrac{1}{2} V_e^2 + C_p T_e = C_p T_c \qquad (V_C = 0)$$

whence, after some manipulation

$$V_e = \sqrt{\left\{ \frac{2\gamma R_0 T_c}{(\gamma - 1) W} \left[1 - \left(\frac{p_e}{p_c} \right)^{(\gamma-1)/\gamma} \right] \right\}} \tag{6.20}$$

We note that the exhaust velocity, V_e, increases with the following:

- increasing pressure ratio p_c / p_e—though such benefits are limited by accompanying increases in motor weight;
- increasing combustion temperature T_c—to be set against the adverse effects of higher temperatures on nozzle heat transfer and increased dissociation losses (see the later section on 'Thermochemistry');
- low molecular weight; and
- to a lesser extent, by reducing the ratio of specific heats γ—this is of limited practicality given the other influences.

It is convenient to identify a *characteristic thrust coefficient* C_F^0 such that

$$V_e = c^* C_F^0$$

where $\quad C_F^0 = \sqrt{\left\{ \left[\gamma \left(\frac{2}{\gamma + 1} \right)^{(\gamma+1)/(\gamma-1)} \right] \frac{2\gamma}{\gamma - 1} \left[1 - \left(\frac{p_e}{p_c} \right)^{(\gamma-1)/\gamma} \right] \right\}} \tag{6.21}$

The exit-to-throat area ratio, A_e / A_t, can be determined from the continuity equation (6.13) such that

$$\frac{A_e}{A_t} = \frac{\rho_t V_t}{\rho_e V_e} = \gamma \left(\frac{2}{\gamma + 1} \right)^{(\gamma+1)/(\gamma-1)} \left(\frac{p_c}{p_e} \right)^{1/\gamma} \Big/ C_F^0 \tag{6.22}$$

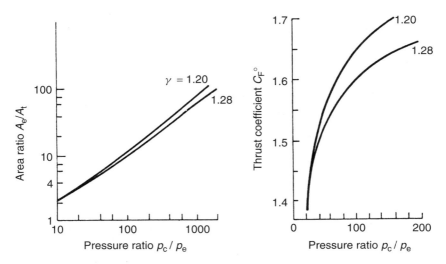

Figure 6.5 The variation of area ratio and thrust coefficient with pressure ratio

The variation of thrust coefficient and of area ratio with pressure ratio are illustrated in Figure 6.5.

Nozzle performance and design

We now combine the rocket motor performance characteristics of Section 6.2 with the gas dynamics of the preceding section.

For a given combustion chamber pressure p_c and mass flow rate $\dot{m}$ (and hence throat area), the motor thrust can be optimized. From equation (6.7)

$$F = \dot{m} V_e + A_e(p_e - p_a)$$

and incremental parameter changes are related by

$$\delta F = \dot{m}\delta V_e + \delta A_e(p_e - p_a) + A_e\delta p_e$$

But conservation of momentum insists that

$$\dot{m}\delta V_e + A_e\delta p_e = 0$$

whence

$$\frac{\mathrm{d}F}{\mathrm{d}A_e} = p_e - p_a = 0 \quad \text{for maximum thrust}$$

The thrust is thereby maximized when $p_e = p_a$ and the nozzle flow is said to be ideally expanded.

The exhaust flow patterns accompanying departures from this ideal behaviour are sketched in Figure 6.6. If the nozzle flow is *overexpanded*, $p_e < p_a$, pressure recovery to ambient conditions is effected through a series of shock waves. Penetration of these

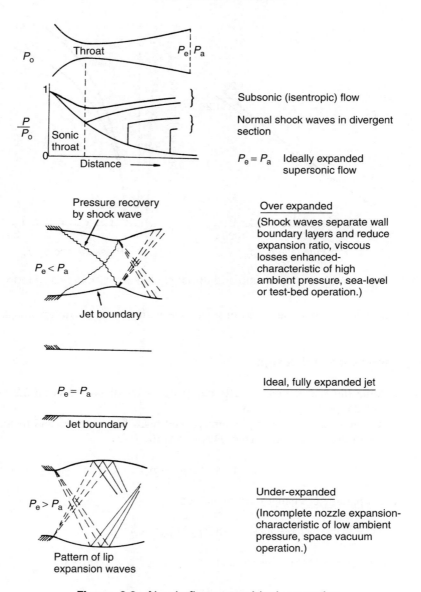

P_o Throat P_e P_a

$\dfrac{P}{P_o}$

Sonic throat

Distance

Subsonic (isentropic) flow

Normal shock waves in divergent section

$P_e = P_a$ Ideally expanded supersonic flow

Pressure recovery by shock wave

$P_e < P_a$

Jet boundary

Over expanded

(Shock waves separate wall boundary layers and reduce expansion ratio, viscous losses enhanced-characteristic of high ambient pressure, sea-level or test-bed operation.)

$P_e = P_a$

Jet boundary

Ideal, fully expanded jet

$P_e > P_a$

Pattern of lip expansion waves

Under-expanded

(Incomplete nozzle expansion-characteristic of low ambient pressure, space vacuum operation.)

Figure 6.6   Nozzle flows: non-ideal expansion

shock waves into the nozzle leads to separation of the wall boundary layers, enhanced viscous losses and reduced expansion ratio. Such behaviour is characteristic of operation at high ambient pressure, typical of sea-level or test-bed firing. *Underexpanded flows* are characterized by incomplete nozzle expansion, $p_e > p_a$, and reduced exhaust velocity. The low ambient pressures that give rise to such flows are typically realized in space-vacuum operation.

Figure 6.7 illustrates the thrust improvement that would result from continuous adaptation of the nozzle geometry, leading to $p_e = p_a$, with increasing altitude, and hence

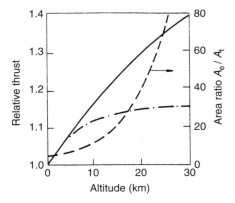

Figure 6.7 Comparative thrust levels:
continuous and sea-level adaptation.
— Continuous adaptation —·— sea
level adaptation

reducing p_a, in comparison with the thrust of an ideally expanded nozzle at sea level. The expansion ratio A_e/A_t of the nozzle becomes very large, however, and significant weight penalties may result. Careful optimization is necessary to reconcile these competing factors.

The thrust coefficient in equation (6.21) can be modified to include the effects of non-ideal expansion in the form

$$C_F = \frac{F}{p_c A_t} = C_F^0 + \frac{A_e}{A_t}\left(\frac{p_e}{p_c} - \frac{p_a}{p_c}\right) \tag{6.23}$$

The *thrust coefficient for ideal expansion*, given p_c, p_a and A_t is, from equation (6.21),

$$\{C_F^0\}_{max} = \sqrt{\left\{\frac{2\gamma^2}{\gamma-1}\left(\frac{2}{\gamma+1}\right)^{(\gamma+1)/(\gamma-1)}\left[1 - \left(\frac{p_a}{p_c}\right)^{(\gamma-1)/\gamma}\right]\right\}}$$

and with equation (6.22) we write

$$\frac{C_F}{\{C_F^0\}_{max}} = \frac{C_F^0}{\{C_F^0\}_{max}} + \frac{\gamma\left(2/(\gamma+1)\right)^{(\gamma+1)/(\gamma-1)}(p_c/p_e)^{1/\gamma}}{C_F^0\{C_F^0\}_{max}}\left(\frac{p_e}{p_c} - \frac{p_a}{p_c}\right) \tag{6.24}$$

As Figure 6.8 indicates, the departure from ideal expansion is less severe with underexpansion than with overexpansion. However, flow separation from the nozzle, which contracts the jet in the overexpanded situation, does lead to an increase in thrust over that which would result in the absence of such separation. This observation also suggests an aerodynamic approach to varying the nozzle geometry that is otherwise mechanically cumbersome—namely, one of controlled fluid injection through the nozzle wall that induces local flow separation and thereby modifies the nozzle contour.

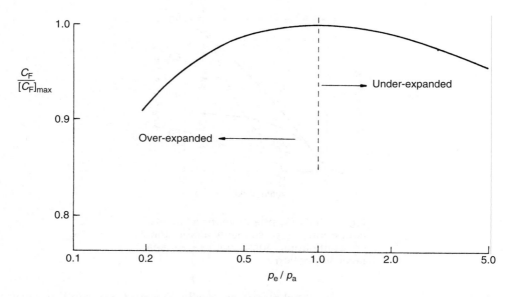

Figure 6.8 The effect on thrust coefficient of departures from ideal expansion

As indicated earlier, nozzle adaptation has important implications for nozzle size and weight. From the manufacturing point of view, convergent–divergent nozzles that are conical represent the simplest designs. Two specific problems then arise, however:

- The exhaust is not directed in the axial direction and the flow divergence implies some loss of thrust.
- In high altitude applications the nozzle tends to be long and correspondingly heavy.

The axial thrust may be shown to be approximately

$$F = \tfrac{1}{2}\dot{m}V_e(1 + \cos\alpha)$$

for cone included angle 2α. For $\alpha = 15°$, the multiplicative factor $\tfrac{1}{2}(1 + \cos\alpha)$ is 0.986 and the effect of divergence is generally small.

Designs to reduce nozzle length, and hence weight, resulting from small divergence angles must expand the flow from the throat more rapidly and subsequently turn the exhaust in the axial direction. Bell-shaped nozzles are designed to achieve this—for example, the established procedure by Rao [6].

The benefits of continuous nozzle adaptation are particularly apparent in relation to single-stage-to-orbit (SSTO) concepts. While the performance compromise introduced by not doing so, and adopting fixed geometry nozzles, may be quite modest in a multi-stage launcher, the SSTO confronts the nozzle design with a progressively reducing ambient pressure over the complete range from sea-level static to vacuum conditions. Unlike the conventional convergent–divergent nozzle, the truncated plug nozzle or aerospike (Figure 6.9) is bounded by a solid wall (ramp) on only one side, with the combustor efflux expanding against the local ambient pressure on the other. The extent of such expansion is thereby adapted to the changing altitude and varies continuously over the

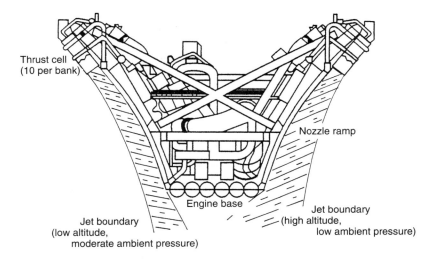

Thrust cell
(10 per bank)

Nozzle ramp

Engine base

Jet boundary
(low altitude,
moderate ambient pressure)

Jet boundary
(high altitude,
low ambient pressure)

Figure 6.9 Schematic linear aerospike nozzle

vehicle trajectory. When allied to a lifting body configuration, as in the Lockheed Martin Venturestar reusable launch vehicle (RLV) concept, or the X-33 demonstrator, the linear aerospike also offers opportunities for propulsion-system integration and base-drag reduction [7].

Thermochemistry

We identified from equation (6.20) that the basic elements in maximizing exhaust velocity (or specific impulse) were a high-combustion chamber temperature and pressure and low molecular weight. The variation of these properties in relation to some representative propellant combinations will be illustrated later.

In developing equations for the ideal rocket motor, we assumed that, in addition to isentropic flow (no viscous or heat losses), the combustion products were of known constant composition throughout the expansion. At the high temperatures of interest, however, a significant proportion of combustion products are dissociated and the extent to which the energy of dissociation can be recovered in the expansion has a substantial impact on nozzle performance.

In the expansion process, the variation of composition with position in the nozzle depends upon both local thermodynamic state—pressure and temperature—and chemical kinetic rates, in particular, upon recombination rates. A complete description is complex and beyond the scope of this chapter but two limiting cases are readily distinguished:

- frozen flow, in which the composition remains constant throughout the expansion irrespective of the variation in pressure and temperature;
- equilibrium flow, in which the equilibrium composition corresponding to local conditions of pressure and temperature prevail along the nozzle.

The former implies that the chemical rates are vanishingly small, whereas the latter implies that they are infinitely fast.

As the temperature decreases along the nozzle, species dissociated in the combustion chamber recombine, releasing energy and changing the composition. While specific impulse (or thrust coefficient) is therefore predicted to be larger for the equilibrium flow at a particular pressure ratio (p_c/p_e), the accompanying area ratio A_e/A_t is also larger.

6.2.2   Propellants

In principle, any chemical system producing heat release in a gas flow through exothermic reaction could be used as a propellant. In practice, consideration of a range of additional factors such as the specific energy content, rate of heat release, ease of storage and handling significantly limits the choice. Heat release in a liquid-propellant rocket may be achieved through the separate injection, mixing and combustion of two liquids—fuel and oxidizer—as in a bi-propellant system, or the exothermic decomposition of a monopropellant such as hydrazine or hydrogen peroxide. A bi-propellant is said to be *hypergolic* if the fuel and oxidizer react spontaneously on contact with each other. More generally, the requirements of separate propellant storage, of pumping and feed to the injector, of mixing and ignition mean that the high specific impulse is achieved at the expense of considerable complexity in design and manufacture. Significant flexibility in operation is, however, gained, including stop-start options and throttling of thrust levels. In contrast, the solid-propellant rocket is of comparatively simple design. The propellant is stored within the combustion chamber in the form of shaped grains bounded by the walls of the chamber. Once ignited, combustion will generally proceed until all the propellant is consumed; the thrust-time relationship is then fixed by the grain configuration. Limited thrust regulation is possible.

Liquid propellants

Table 6.1 illustrates properties of some representative propellants. We recall from equation (6.10) that in order to maximize exhaust velocity, we seek propellants that give high combustion temperatures and low molecular weight products. Peak temperatures accompany the burning of approximately stoichiometric mixtures—that is, mixtures containing just sufficient oxidizer to convert the chemical elements in the fuel to stable combustion products. However, the typical variation of adiabatic flame temperature with mixture strength exhibits only a modest decline in T_c for richer-than-stoichiometric mixtures. It is thus possible to maintain high combustion temperatures with hydrogen as fuel, for example, and capitalize on the low fuel molecular weight by operating fuel-rich.

From the table, the attraction of fluorine as an oxidizer (high combustion temperature, $W_F = 19$) is tempered by its highly corrosive properties that constrain the choice of liner materials for the combustion chamber and nozzle. Both oxygen and fluorine necessitate cryogenic storage as liquids, since they have boiling points of 90 K and 85 K respectively. Long-term storage is therefore difficult and their application is restricted to launch vehicles. Nitrogen tetroxide has found increased application in space propulsion as an oxidizer, despite its high molecular weight ($W_{N_2O_4} = 92$). This is in part because it has a boiling point of 294 K at suitable storage pressures and, though highly toxic, systems for its supply and management are generally simpler and lighter.

Table 6.1 Liquid propellants

Fuel	Oxidizer	Molecular weight of products	Combustion temperature T_c (K)	Ideal specific impulse (s)	Mean density kg/m³
H₂ (hydrogen)	O₂ (oxygen)	10	2980	390	280
	F₂ (fluorine)	12.8	4117	410	460
Kerosine	O₂	23.4	3687	301	1020
	F₂	23.9	3917	320	1230
	RFNA (red fuming nitric acid)	25.7	3156	268	1355
	N₂O₄ (nitrogen tetroxide)	26.2	3460	276	1260
	H₂O₂ (hydrogen peroxide)	22.2	3008	278	1362
N₂H₄ (hydrazine)	O₂	19.4	3410	313	1070
	HNO₃ (nitric acid)*	20	2967	278	1310
UDMH	O₂	21.5	3623	310	970
(CH₃)₂NNH₃ (unsymmetrical dimethyl hydrazine)	HNO₃*	23.7	3222	276	1220
	* hypergolic				
Monopropellants					
N₂H₄		10.3	966	199	1011
H₂O₂		22.7	1267	165	1422

Note: All quoted values are for $p_c = 7$ MPa with an ideal expansion to $p_e = 0.1$ MPa. Higher chamber pressures admit increases in I_{sp}—for example, at 20 MPa, LOX/LH₂ yields a specific impulse of ~460 s.

The traditional, high thrust bi-propellant combinations of LOX/LH$_2$, employed on Saturn V, the SSMEs and the Vulcain motor of Ariane 5, for example, are complemented in apogee motors and orbital manoeuvring systems generally by monomethylhydrazine (MMH)/N$_2$O$_4$ combinations. The latter offers a specific impulse of approximately 310 seconds, a value that is significantly greater than that available from monopropellant hydrazine decomposition, the readily stored—but less stable—alternative (boiling point 387 K). In the context of thermal control during propellant storage, we should note that both hydrazine and nitrogen tetroxide have melting points in the neighbourhood of typical spacecraft ambient temperatures, 275 K and 262 K respectively.

Solid propellants

Solid propellants are typically of two types: either double-base, comprising homogeneous colloidal mixtures of nitrocellulose and nitroglycerine, or composite, comprising mixtures of an organic fuel and crystalline inorganic salt. Ammonium perchlorate, NH$_4$ClO$_4$, is the principal oxidizer used in composite propellants with a polymer fuel binder, typically polyurethane or polybutadiene.

In comparison with the liquid propellants described in Table 6.1, the specific impulses for solid propellants are substantially lower—in the range 200–260 s. These values are not strongly influenced by the particular fuel binder, although high hydrogen to carbon ratios are favoured. The performance of composite propellants is improved by the addition of metals such as aluminium or beryllium in the form of finely ground particles ($\sim$10 μm).

Solid-propellant charges are typically cast or extruded. The components are mixed in the form of a dough, perhaps in the presence of a plasticizer, and cast or pumped into a mould or directly into the combustion chamber. The grain geometry is fixed by a mandrel that is removed after curing and solidification.

Further information on propellant energetics and physical properties is available in Reference [5]; the complexities of solid-propellant combustion are discussed extensively by Kuo and Summerfield [8].

6.2.3   Chemical rocket design

In this section, we seek to identify some of the more important features of motor design in relation to high thrust launch vehicles or booster applications. A detailed discussion of such a major topic is inappropriate in this book that seeks more to familiarize the non-specialist with the diversity of disciplines and technologies at the heart of spacecraft engineering. More specific design issues are addressed in much greater detail in Huzel and Huang [9]. From the standpoint of spacecraft design, the involvement with the launch vehicle lies essentially in the role of procurement and the clear identification of the constraints that are imposed on the spacecraft by the launcher. These form the basis of Chapter 7, while the design of secondary propulsion systems, of more direct concern to the spacecraft engineer, is discussed in Sections 6.3 and 6.4.

If we distinguish the principal rocket components to be

- the thrust chamber
- the propellant feed system
- the propellant storage tanks,

then significant differences are evident between solid- and liquid-propellant devices. For the solid-propellant rocket, offering high thrusts of short duration, key features of the design are the choice of propellant grain to give the appropriate thrust law and of materials selection for the nozzle and casing to combine low weight with reliable operation. The greater complexity of the liquid-propellant rocket requires that consideration be given to thrust chamber design in respect of such components as the fuel injector, cooling system and propellant feed. In the case of the launch vehicle, demanding especially high propellant mass flow rates, both the gas generator and turbopump designs then become critical aspects.

Liquid-propellant rockets

A typical liquid-propellant rocket motor is illustrated in Figure 6.10. Liquid fuel and oxygen are pumped into the main combustion chamber by turbopumps driven by a separate hot gas generator, burning small quantities of the propellants. The starter cartridge is typically a small solid-propellant gas generator.

Two important classes of engine where fuel and oxidant are supplied by turbopumps are differentiated by their description as open or closed cycle designs. In the *open cycle*, the turbine exhaust is discharged into the nozzle downstream from the combustion chamber

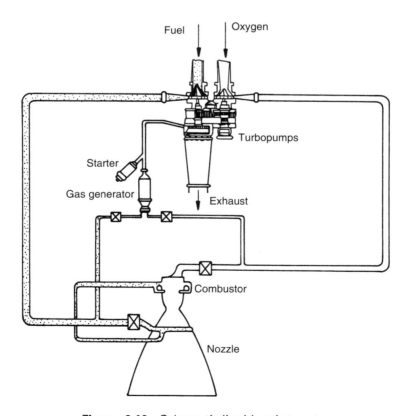

Figure 6.10 Schematic liquid rocket motor

at modest pressure in the expanding section or simply discharged to atmosphere, making no direct thrust contribution. By contrast, in the *closed cycle* the exhaust is injected at high pressure into the combustion chamber, contributing significantly to the energy of the system. The improved performance of the latter cycle must be set against the disadvantage of more complex turbopump design necessary to operate at much greater discharge pressures.

Two open-cycle configurations are illustrated in Figure 6.11: a separate gas generator cycle of the kind employed in the HM-7 and Vulcain engines of Ariane 4 and 5, and a coolant tap-off cycle in which vaporized hydrogen fuel from the nozzle coolant jacket drives the turbine. The turbine power is comparatively low in the latter case but the complexity of the design is much reduced. Also illustrated is the closed cycle, staged-combustion cycle employed in the SSME. It is there that the fuel and part of the oxidizer are supplied to a pre-combustor at high pressure. The high energy fuel-rich exhaust first drives the turbopumps and is then injected into the main combustion chamber with the remaining oxidizer.

An illustrative comparison between open- and closed-cycle engines is presented in Table 6.2, where SSME performance characteristics are summarized together with those of the Ariane 5 Vulcain engine. The gas generator consumes approximately 3.5% of the available propellant and the closed-cycle engine, also incorporating a much greater chamber pressure (approximately twofold higher), yields a vacuum specific impulse some 5% greater. The necessarily higher turbopump discharge-pressures, for both fuel and oxidizer in the staged combustion cycle, imply significantly more extensive design and development in turbines, pumps and ancillary equipment. A simpler closed cycle is the expander cycle in which vaporized fuel drives the turbopumps but, in contrast with the coolant bleed cycle illustrated, all of the fuel is so employed and subsequently passes into the combustion chamber. While particularly suitable for LH_2, the fuel flow rate is now limited by the vaporization rate and hence effectively by heat transfer. Higher chamber pressures and increased thrust are thereby restricted in turn [10].

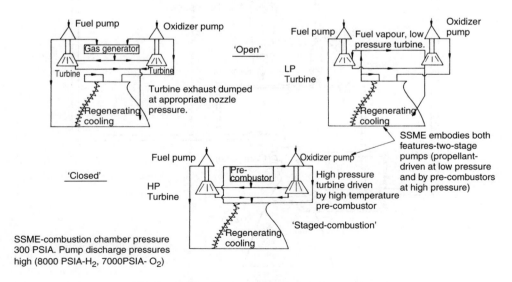

Figure 6.11 Representative engine cycles

Table 6.2   Illustrative comparison of closed- and open-cycle engines

	SSME	Vulcain (Ariane 5)
Thrust (kN):		
Vacuum	2090	1139
Sea level	1700	885
Specific impulse (s):		
Vacuum	455	432
Sea Level	363	
Mixture ratio (stoichiometric 8 : 1 $2H_2 + O_2 \rightarrow 2H_2O$)	6 : 1	5.25 : 1
Chamber pressure (bar)	207	107
Nozzle area ratio	77	45
Flowrates (kg/s)	468 (engine) 248 (pre-combustor)	270 10 (gas generator)
Pump discharge pressure (bar)	309 (LOX) 426 (LH$_2$)	125 (LOX) 150 (LH$_2$)
Burn time (s)	480	590
Mass (kg)	3022	1650

The configurations shown all employ an element of regenerative cooling. Fuel or oxidizer may be used as a coolant, flowing through a jacket surrounding the thrust chamber. The heat absorbed in this way, necessary in prolonged firings, enhances the initial energy content of the propellants prior to injection into the combustion chamber.

Solid-propellant rockets

By contrast, solid-propellant rockets are comparatively inflexible in their design. The gross classification of solid rockets is made on the basis of the propellant grain geometry since, once ignited, erosive burning proceeds to propellant exhaustion. The variation of burning surface area with time then determines the thrust-time history.

A propellant grain is said to be neutral if the thrust remains broadly constant throughout the firing, with a burning surface area that is independent of time. Such behaviour would be characteristic of 'cigarette' burning as illustrated in Figure 6.12. In practice, a convenient cylindrical geometry would insist that the burning surface area be small and hence of limited thrust. An annular grain, on the other hand, that burns from the inner surface outwards is progressive. The exposed area, and hence the thrust, increases with time. A large surface area for burning initially, combined with a period of roughly neutral thrust performance, is provided by the star-shaped cylindrical grain illustrated. The changing internal profile with time will give rise to two-stage burning, combining aspects of both progressive and regressive behaviour. The latter accompanies a decreasing exposed area, as in the case of the external burning of a rod or, with the star-shaped grain, as an initially corrugated boundary is progressively smoothed and simplified as combustion proceeds.

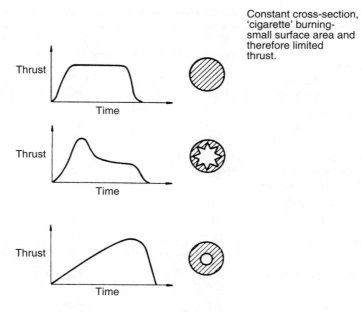

Constant cross-section, 'cigarette' burning- small surface area and therefore limited thrust.

Figure 6.12   Solid-propellant grain geometries

Variations in grain geometry along the length of the propellant charge permit the further tailoring of the thrust-time relationship to the requirements of the mission. The configuration and performance of the solid rocket boosters used on the Space Shuttle are illustrated in Figure 6.13.

Solid-propellant ignition is typically effected by a pyrotechnic or pyrogen igniter. A small quantity of heat-sensitive powdered explosive is ignited electrically and the heat released, in turn, ignites the propellant within the igniter. Typical igniter compounds are aluminium, boron or magnesium, combined with potassium perchlorate or nitrate oxidizers. In lateral burning cylindrical grains, the igniter is placed at the end of the chamber furthest from the nozzle so that the hot products of combustion from the igniter sweep across the whole grain. Main charge ignition then occurs through convective and radiative heat transfer from these products, the balance between these processes being determined by the detailed igniter design.

6.2.4 Alternative high-speed air-breathing propulsion

A review of typical first-stage rocket performance as described in Chapter 7 reveals that approximately 55% of the launch mass is consumed in accelerating the vehicle to 20% of the orbital speed. The bulk of this propellant—consumed within the Earth atmosphere—is the oxidizer; N_2O_4 in the case of Ariane 4. Such bare statistics provide a *prima facie* case for investigation of the contribution that air-breathing propulsion might make to the initial stages of a space launch.

Figure 6.14 identifies the key propulsive options for high-speed air breathing with hydrogen as fuel. Current turbomachinery, familiar from the aircraft jet engine, becomes

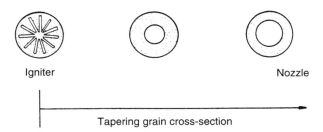

- Propellant mass 5.0×10^5 kg, Inert mass 8.2×10^4 kg
- Vacuum thrust 11.8×10^6 N
- Specific impulse ~ 260 seconds.

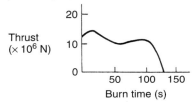

Figure 6.13   Space shuttle solid rocket booster

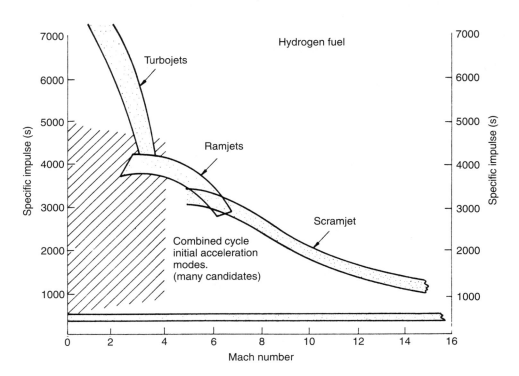

Figure 6.14   High-speed propulsion options in the earth's atmosphere

increasingly unattractive thermodynamically at Mach numbers (ratio of flight speed to sound speed) in excess of 3 and relatively high specific impulses can then be achieved more simply, and efficiently, by capitalizing on ram compression. In the conventional ramjet, however, the ingested air is diffused down to subsonic Mach numbers prior to mixing with fuel and combustion. Isentropic (shock-free) subsonic diffusion leads to a substantial temperature rise, from T to T_0, which increases quadratically with the Mach number, M

$$T_0/T = 1 + \tfrac{1}{2}(\gamma - 1)M^2$$

For $M > 4 - 5$, heat transfer and dissociation losses in the combustor begin to erode subsonic-combustion ramjet performance significantly. However, technologies required by these engines are well established and their range of application may be further extended by the judicious use of high fuel heat capacity in a range of thermal protection and heat exchanger strategies, even embracing air liquefaction. The intimate relationship between vehicle trajectory, kinetic heating and combustion stability is illustrated in Figure 6.15. The target condition for launcher first-stage burn-out identified earlier, at 30 km altitude, however, corresponds to a Mach number of 5.5 and, at such speeds and beyond, the supersonic combustion ramjet (SCRAMJET) is the more attractive prospect. The temperature rise associated with diffusion to low subsonic Mach numbers prior to combustion is avoided by burning the fuel in a supersonic airstream but the necessary technologies for efficient high-speed mixing, ignition and stable burning remain incompletely understood. Residence times in the combustor become very short and comparable with the ignition delay for the fuel.

By comparison with rockets, air-breathing engines offer relatively modest thrust-to-weight ratios and composite engines that seek to use common components over a wide range of flight regimes and operating conditions are crucial to the successful SSTO design.

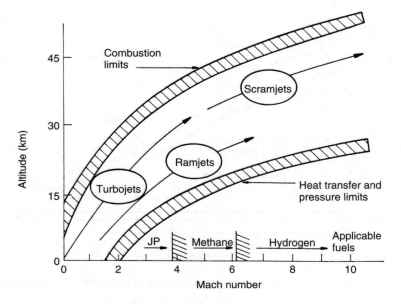

Figure 6.15 Constraints on air-breathing trajectories

Cycle performance analyses and engine simulations have continued to attract recurrent research interest throughout the last two decades as evidenced, for example, by the SANGER [11] and National Aero-Space Plane (NASP) programmes [12]. Current projects again favour more conventional rocket technology, incorporating some measure of reusability (see Reference [13]). Despite improvements in materials technology and simulation tools, advanced design is problematic and the development costs remain intimidating.

6.2.5 Propellant management

The operation of liquid-propellant rockets for space vehicle applications may expose the propulsion system to a dynamical regime not usually encountered in terrestrial applications, namely, that of free-fall or low-residual acceleration. We shall briefly review here some aspects of the problems encountered that do not have a ready analogue in $1g_0$ environments.

The near free-fall situations typical of space manoeuvres or residual drag in the Earth orbit correspond to

$$|g_{\text{local}} - a_{\text{vehicle}}| = O\,(10^{-6}g_0)$$

The very small weights associated with such levels of acceleration invite consideration of other forces, like surface tension, and of equilibrium conditions that might usually be considered to be insignificant except on very small physical scales.

Liquids may be conveniently characterized as wetting (surface spreading) or non-wetting. Liquid to solid surface contact angles may approach zero for perfectly wetting liquids, typical of cryogenics, but may exceed $90°$ for non-wetting liquids such as mercury (employed in some ion engines, as discussed in Section 6.4). The surface tension of LOX (at 90 K) is approximately 13×10^{-3} N/m, compared with 460×10^{-3} N/m for mercury and 72×10^{-3} N/m for water. The range of variation is therefore wide.

It is convenient to review the significance of the several forces prevailing in liquid propellants by introducing the following dimensionless group:

$$\text{Bond number, } B_0 \equiv \frac{\text{gravity}}{\text{surface tension}} \text{ forces}$$

$$= \rho L^2 g / \sigma$$

and $$\text{Weber number, } W_e \equiv \frac{\text{inertia}}{\text{surface tension}} \text{ forces}$$

$$= \rho V^2 L / \sigma$$

We may then distinguish the respective inertia, capillary and gravity dominated regimes as suggested in Figure 6.16. For $B_0 > 1$, gravitational forces predominate and surface tension may be neglected, and *vice versa*. We note that for surface tension to be significant in a $1g_0$ (Earth surface) environment for a liquid such as water, then $B_0 < 1$ implies a characteristic length scale of less than 2 mm and effects such as capillary rise are only significant at such small physical scales. For liquid oxygen in a

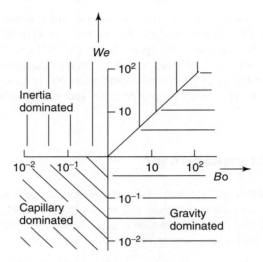

Figure 6.16   Distinctive regimes for forces acting on liquids

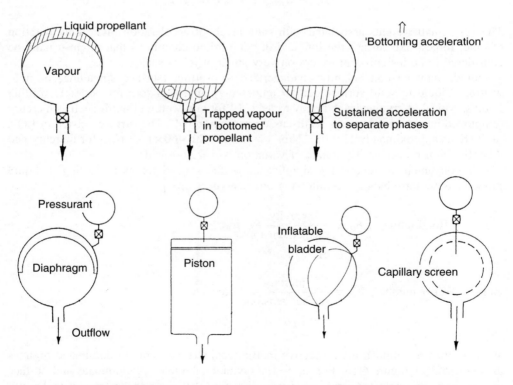

Figure 6.17   Illustrative propellant storage and delivery systems

residual microgravity environment, however, the same B_0 boundary corresponds to a length scale of 1 m.

The equilibrium configuration of a liquid propellant in a partially filled tank, under microgravity conditions, is determined essentially by the minimization of droplet- (or bubble) free energy that is proportional to the surface area. A wetting liquid will therefore preferentially adhere to tank walls, rather than assume a freely suspended droplet configuration. Internal surfaces and screens within tanks can be used to collect the liquid and separate it from the vapour. Active measures must clearly be adopted to ensure that a liquid propellant is available at the tank outlet for rocket motor starting. The principal options for propellant storage and delivery are shown schematically in Figure 6.17 and comprise inertial (or bottoming), positive expulsion and capillary (or surface tension) systems. Supercritical storage is usually ruled out on grounds of tank weight. These options are described in more detail by Ring [14].

In addition to the problem of propellant configuration within the tank, the response to dynamic excitation in flight in the form of propellant sloshing may also be important. The viscosity of cryogenic propellants is in general low, typically by more than an order of magnitude, in comparison with water (or hydrazine). The damping of free-surface oscillation in the fluid, which would otherwise give rise to substantial fluctuating forces and moments on the tanks, may also require active provision in the form of turbulence-generating baffles.

6.3 SECONDARY PROPULSION

The typical functions of spacecraft propulsion, as distinct from launcher operations from the Earth surface, may be summarized, in order of reducing thrust level, as follows:

- final orbit acquisition from the nominal orbit established by the launch vehicle,
- station keeping and orbit control, and
- attitude control

General principles in relation to both solid- and liquid-propellant rockets have been reviewed in earlier sections. Here we shall focus on factors influencing the choice of systems appropriate to the secondary propulsion roles set out above. The principal options are cold gas systems, monopropellant hydrazine, bi-propellant nitrogen tetroxide/mono-methylhydrazine combinations, solid propellants and electric propulsion. We shall briefly consider each of these in turn. Further detail on preliminary design methods appropriate to such devices is provided by Brown [15].

6.3.1 Cold gas systems

These systems simply comprise an inert gas, typically nitrogen, argon, freon, or a hydrocarbon like propane, which is stored at high pressure and fed to a number of small thrusters. In the absence of combustion heat release, the kinetic energy of the nozzle exhaust is then solely determined by the driving pressure in the reservoir. A schematic of a typical system is shown in Figure 6.18. Propellants are selected for the simplicity of

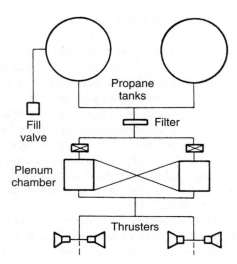

Figure 6.18 Schematic cold gas propulsion system

their storage and compatibility with other facets of spacecraft operation, such as the effect of exhaust-plume impingement on sensitive surfaces, solar cells, sensors and detectors.

Thrust levels are small, typically of the order of 10 mN; levels chosen to provide the small impulse bits required for high pointing accuracy and stable, jitter-free viewing. Minimum impulse bits of approximately 10^{-4} Ns are often necessary for better than 0.1° accuracy in attitude control on some of the larger scientific satellites.

The specific impulse from cold gas systems is comparatively small ($\sim$50 s)—reflecting the reduced reservoir temperature and pressure (T_c and p_c in equation 6.20). The expellant mass is relatively unimportant, however, in the fine-pointing role. Missions requiring larger ΔVs for secondary propulsion do, on the other hand, necessitate higher performance propellants in order to reduce the system mass.

6.3.2 Monopropellant hydrazine

As we described in Section 6.2, the decomposition of anhydrous hydrazine (N_2H_4), either thermally or catalytically, to the products nitrogen, ammonia and hydrogen is exothermic. Expansion of the hot product gas through a nozzle will yield specific impulses in the range 200–250 s. The propellant is readily stored as a liquid, with freezing point 275 K and boiling point 387 K, in tanks under the pressure of an inert gas such as nitrogen or helium.

A representative electrothermal hydrazine thruster configuration is sketched in Figure 6.19. The low temperature monopropellant decomposition is enhanced by a resistively heated metal catalyst—commonly platinum/iridium dispersed on a large surface area or a porous substrate of aluminium oxide. Thruster performance is enhanced by higher temperature operation but the accompanying heat transfer losses and materials compatibility problems also increase.

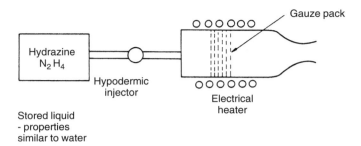

Figure 6.19 Schematic electrothermal hydrazine thruster

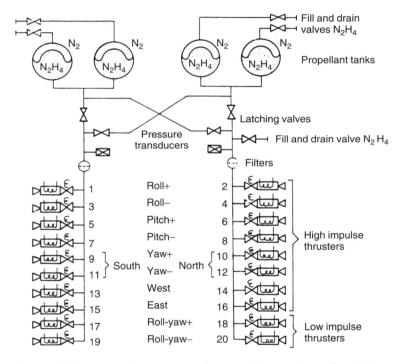

Figure 6.20 Illustrative spacecraft propulsion system using mono-propellant thrusters

Thrust levels ~10 N may be required for orbit control duties and combinations of thrusters are sized accordingly. Figure 6.20 illustrates the configuration employed on an early geostationary demonstrator for which station-keeping (N-S, using the yaw thrusters, and E-W) was a primary consideration. The propellant tanks are of a positive expulsion (elastomeric diaphragm) type, cross-linked between the paired thrusters.

Improved levels of specific impulse are possible if the exothermic decomposition stage is followed by further resistance-heating in a tandem design. Temperature limitations on the nozzle materials, given the corrosive nature of the product gases, restricts the extent of additional heating but specific impulses approaching 300 s have been achieved.

6.3.3   Bi-propellant MMH/nitrogen tetroxide

The increasing size of many spacecraft, particularly into geostationary orbit, implies the allocation of substantial propellant budgets for secondary propulsion systems and greater emphasis on the level of specific impulse delivered. The combination of MMH and N_2O_4 will provide specific impulses in excess of 300 s.

A representative scheme for a geostationary spacecraft, incorporating the functions of both orbit raising and Attitude and Orbit Control System (AOCS), is illustrated in Figure 6.21. The propellants are hypergolic and the layout reflects the additional complexity introduced to ensure safe handling in the propellant storage and feed to the thrusters.

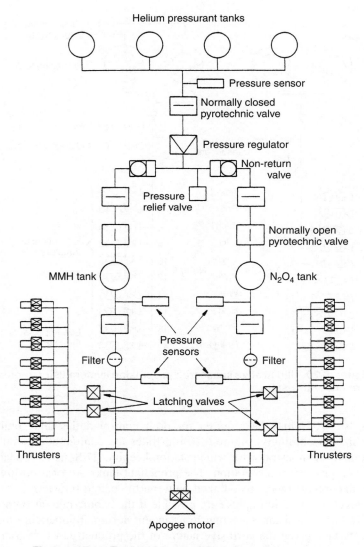

Figure 6.21 Typical bi-propellant propulsion system

The propellants are stored in roughly equal volumes and are both compatible with readily available materials—typically, stainless steel, aluminium or titanium. The accompanying positive expulsion systems employ similar metals in the design of internal bellows since the oxidizer is not compatible with most elastomers.

6.3.4 Solid-propellant apogee motors

In the emplacement sequence for a communications spacecraft into geostationary orbit using an expendable launch vehicle such as Ariane, the final stage injects the satellite into an elliptical transfer orbit with apogee at geostationary height. The circularization manoeuvre is often achieved through a high thrust, short duration burn from a solid propellant apogee boost (or kick) motor. For a satellite with an on-station mass of approximately 1000 kg, the necessary apogee motor fuel approaches 900 kg and the propulsive ΔV is roughly 2 km/s. Payloads launched from the Space Shuttle parking orbit need similar levels of additional orbit raising ΔV (see Chapter 7).

A typical solid rocket motor for secondary propulsion duties is illustrated in Figure 6.22. Table 6.3 refers to the characteristics of the Star 37E motor manufactured by Thiokol. This motor, which weighs 1122 kg, burns for 42 s and delivers an average thrust of 69 kN.

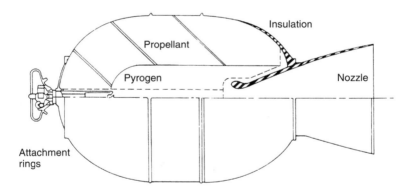

Figure 6.22 Schematic solid-propellant apogee motor

Table 6.3 Performance data for the STAR 37E rocket motor

Total impulse	29.11×10^5 Ns
Motor mass	1122 kg
Propellant mass	1039 kg
Burn-out mass	75.8 kg
I_{SP}	280 s
Burn time	42 s
Propellant mass fraction	0.926
Length	168.4 cm
Basic diameter	109.5 cm
Average thrust	68 800 N
Maximum thrust	75 000 N

Clearly the emplacement roles of the solid-propellant Apogee Boost Motor (ABM), with its near-impulsive ΔV, and the bi-propellant MMH/N_2O_4, with the advantages of higher specific impulse and more controlled burning, are interchangeable. The impulsive burn requires that the spacecraft should also spin for reasons of gyroscopic stability and thrust alignment. It is therefore inherently less accurate than the extended burn, lower thrust level operation of the bi-propellant motor that admits precise spacecraft attitude control throughout the thrusting phase. The trade-off is again that of a propulsion system complexity for improved performance.

6.4 ELECTRIC PROPULSION

The opening section of this chapter revealed some crucial differences in the performance and operation of separately powered rockets. Unlike the chemical systems that we have described thus far, the energy required for expellant acceleration in an electrically propelled rocket is derived from quite a separate source. Whether this source is solar radiation or nuclear fuel, it may be effectively unlimited and the constraints on performance relate to attainable thrust levels and efficient energy conversion.

While electric propulsion has been an active area of development since the earliest space flights, it is only in relatively recent years that electric propulsion has regularly been adopted for commercial, scientific and military missions. Each of the technologies described below have, however, now been operated in space; some as primary systems, although some are still in the technology demonstration stage. In excess of hundred commercial spacecraft in orbit have adopted electric propulsion systems for either attitude control, orbit control or orbit raising. The Deep Space 1 probe, launched in October 1998, saw the use, for the first time, of an electric propulsion system, an ion engine, taking a spacecraft out of Earth orbit to fly-by the near-Earth asteroid 1992 KD. The transfer of technology from the former Soviet Union under the joint National Aeronautics and Space Administration (NASA) / Ballistic Missile Defence Organisation (BMDO) RHETT programme in the early 1990s, has significantly contributed to the increased use of electric propulsion technologies generally. In the following sections we review the key features and the opportunities offered by adopting electric propulsion on a vehicle, together with a brief description and performance available from individual systems.

6.4.1 Electric propulsion fundamentals

In view of the importance of the power plant to the rocket configuration, it is convenient to analyse the performance in terms of the component masses: M_w, the power-plant mass; M_e, the expellant mass; and M_p, the payload mass. The expellant storage and feed system may be assumed to be part of the power plant, while the power-plant fuel mass will be considered to be negligibly small (nuclear fuel) or inappropriate (solar powered). The configuration envisaged is shown schematically in Figure 6.23.

The power plant supplies the exhaust kinetic energy when the jet power is related to the exhaust velocity by an expression of the form.

$$W = \tfrac{1}{2}\dot{m}V_e^2 \tag{6.25}$$

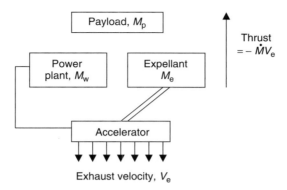

Figure 6.23 Schematic for separately powered electric rocket

where W denotes power-plant output, $\dot{m}$ the expellant mass flow rate and V_e the exhaust
velocity. (If thruster process losses are introduced, then $W_{jet} = \eta W$. For present purposes
we suppose $\eta = 1$.)

We relate the power output W to the power-plant mass M_w linearly and introduce the
inverse specific power α such that

$$M_w = \alpha W \qquad (6.26)$$

If we suppose the exhaust mass flow rate to be constant throughout the burn time, t_b,
such that

$$\dot{m} = M_e/t_b \qquad (6.27)$$

$$M_e = \frac{M_0 - M_p}{1 + \left(V_e^2 / \dfrac{2t_b}{\alpha} \right)} \qquad (6.28)$$

and

$$M_w = \frac{M_0 - M_p}{1 + \left(\dfrac{2t_b}{\alpha} / V_e^2 \right)} \qquad (6.29)$$

where $M_0 = M_p + M_w + M_e$, the total rocket mass.

By identifying $\sqrt{(2t_b/\alpha)}$ as the characteristic velocity V_c and using the above relationships in the Tsiolkovsky rocket equation (6.5) it is possible to show that

$$\frac{\Delta V}{V_e} = \log_e \left[\frac{1 + (V_e/V_c)^2}{\dfrac{M_p}{M_0} + (V_e/V_c)^2} \right] \qquad (6.30)$$

The rocket performance described by equation (6.30) is illustrated in Figure 6.24. For
$V_e/V_c \ll 1$, increases in exhaust velocity at fixed characteristic velocity result in larger

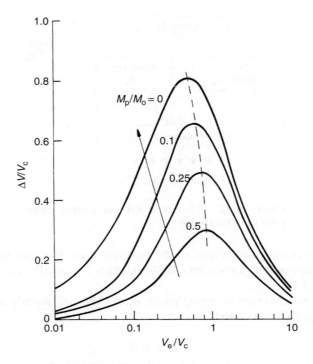

Figure 6.24 Separately powered electric rocket performance

available propulsive ΔV. Such behaviour is essentially that which is observed with chemical propellants. Performance improvements accompany the better propellant utilization reflected in higher exhaust velocities. Unlike chemical propulsion, however, such benefits do not extend without limit and for $V_e/V_c \gg 1$, the ΔV passes through a maximum and then decreases. Further increases in exhaust velocity require higher powers (see equation 6.25) and therefore increased power-plant mass (see equation 6.26). Beyond a certain stage these increases in power-plant mass outweigh the further reductions in expellant mass accompanying the higher exhaust velocities.

The presence of maxima in Figure 6.24 introduces a range of optimization studies. For a particular payload ratio, M_p/M_0, it is desirable to operate in the vicinity of the maximum, broadly corresponding to $V_e/V_c \approx 1$, and high exhaust velocity then requires $2t_b/\alpha \gg 1$. Optimized electric propulsion usage will thus tend to imply lengthy burn times and small inverse specific power, α.

Consider the following illustration: for a payload ratio $M_p/M_0 = 0.5$ and a mission ΔV of, say, 5 km/s, representative of an orbit-raising manoeuvre,

$$V_c \approx V_e \approx 16 \, \text{km/s}$$

For $\alpha = 20$ kg / kW, representative of modern lightweight solar arrays, the burn time t_b would be 30 days. The mean acceleration over the thrust duration, $\Delta V/t_b$, is then $2 \times 10^{-4} \, g_0$. Clearly quite spectacular improvements in α would be necessary for such propulsion systems to operate from planetary surfaces. However, it is quite clear that if long burn times are acceptable from a mission perspective then according to Figure 6.24 substantial mass savings may be achievable. Generally, operational constraints for station keeping of Geostationary Earth orbit (GEO) spacecraft are not a constraint upon long duration thrust arcs. While there may be radiation-induced system problems for substantial orbit-raising manoeuvres for commercial satellites through the Van Allen radiation belts, it is clear that Earth orbit applications may benefit from electric propulsion. Originally, however, it was the opportunity presented by the high specific impulse from some types of devices, particularly the ion engine variants, which encouraged the earliest reviews (e.g. Reference [16]), to focus upon the adoption of electric propulsion for interplanetary missions. We explore these aspects in the following section.

6.4.2 Propulsive roles for electric rockets

If thrust acceleration levels of $2 \times 10^{-4} g_0$ are equivalent to approximately $0.35 \times$ solar gravitational acceleration at 1 AU, then continuous thrust at these levels has a pronounced influence on interplanetary trajectories. Substantial mass savings have been demonstrated to be available for many interplanetary missions, should they adopt electric propulsion. It is perhaps therefore not surprising that papers still regularly appear in journals such as *Journal of Spacecraft and Rockets*, which contribute to the analysis of missions to Solar System bodies. At the time of writing, the possible use of electric propulsion for the European Space Agency (ESA) 2000+ cornerstone mission to Mercury is actively under consideration [17]. Many mission scenarios have been defined, not only to Mars, but also to the other major and minor planets. However, the actual performance of an electric propulsion system during an interplanetary mission has only recently become available, following the success of the Deep Space 1 mission [18]. To date, therefore, the majority of satellites have employed electric propulsion for either station-keeping, orbit control, Earth orbit raising or attitude control.

Substantial benefits in terms of payload into final orbit may be achieved if the potential of high exhaust velocities to reduce propellant mass can be realized. Recalling Figure 6.24, the maxima are distinguished by $V_e/V_c \approx 1$, and therefore high exhaust velocity must accompany extensive burn times, since $V_c^2 = 2t_b/\alpha$. Figure 6.25 presents a simple comparison between a two-impulse Hohmann transfer and low-thrust orbit raising from a LEO of radius r_0, to a higher orbit of radius r. The electric rocket is typically fired continuously and, near-circumferential low thrust leads to a gradual spiralling expansion of the orbit radius. Projected burn times of months would probably be acceptable operationally for an unmanned space tug implying a modest extension of launch procurement schedules. The implications for thruster design are more important. Extended burn times do influence the propulsive ΔV required and low-thrust orbit-raising operations will, in general, require a larger ΔV. Figure 6.26 compares the impulsive and circumferential low-thrust requirements; transfer to geostationary orbit requires approximately 20% more ΔV for the low-thrust mission. The requirements of accompanying manoeuvres for low-thrust orbital operations such as changes in orbit inclination were originally extensively analysed by Burt [19].

The relationship noted in Figure 6.24 between characteristic mission impulse ΔV, the characteristic velocity V_c and the exhaust gas velocity V_e indicates that the type of electric

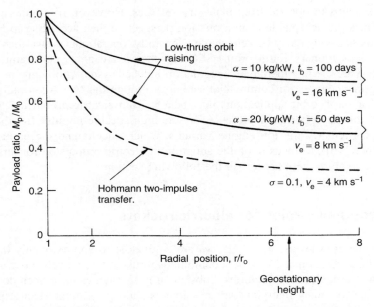

Figure 6.25 Comparative performance: chemical impulse versus low-thrust orbital transfer

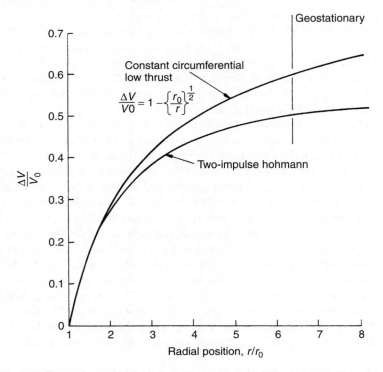

Figure 6.26 Comparative ΔV requirement for transfer between circular orbits r_0 to r

propulsion technology selected for a mission will depend upon the mission itself. In part this demonstrates why in the early years of space flight research was directed to those technologies capable of achieving the highest specific impulse, especially electrostatic and electromagnetic engines.

As an example, for either typical Earth orbit raising in which the ratio of initial to final orbit radius is close to unity, or for N/S station-keeping in GEO, *magneto-plasma dynamic (MPD)* and *thermal arc jets* demonstrate optimum performance [20]. Alternatively for interplanetary trajectories, *electrostatic ion engines* offer good performance. As noted however in Reference [20], care has to be taken in the definition of system performance, specifically whether the satellite power system (SPS) is sized for the propulsion module or for the operational payload or for other on-board subsystems, since this can have an important influence on the selected 'optimum' technology.

In LEO operations, particularly for drag compensation, *resistojets* appear to offer near optimum performance. A variety of propellants have been investigated including bio-waste for space station applications. For very small impulse bit requirements, typical of several missions that have highly demanding orbit control requirements, *field emission electric propulsion (FEEP)* has been developed, principally in Europe.

However, optimum performance is only one issue to be addressed in technology selection. More important is the issue of technology availability. In this respect some technologies offering more modest, sub-optimum improvements in specific impulse, have been developed more rapidly than those with greater potential for saving mass. Thus to date, most electric propulsion systems that have been flown are of the resistojet type. Hence for Motorola's original Iridium constellation, Primex Aerospace's hydrazine resistojets having an I_{sp} of order 300 sec, were used for orbit raising to the operational altitude of 780 km.

Ion engines, Hall effect and MPD engines are however, now available. Xenon propellant is used in variants of both ion and Hall motors. As a result, again far from an optimum propulsion solution, work is also underway on a xenon resistojet with a modest performance (I_{sp} of 50 sec) but obviously with the opportunity for a shared propellant tank.

6.4.3 Electric propulsion systems

The basic principles underlying electric thruster design are well established (see e.g. Stuhlinger [21] Jahn [22] and Sutton [5]). Electrically powered expellant acceleration devices are of essentially three types: *electrothermal*, in which the enthalpy of the expellant is increased and converted into directed kinetic energy via a nozzle; *electrostatic*, in which charged particles, ions or colloids, are accelerated directly in an electric field; and *electrodynamic*, in which crossed electric and magnetic fields induce a Lorentz force in a plasma. We illustrate briefly designs reflecting each of these distinctive approaches.

Electrothermal thrusters

The resistojet is the simplest electrothermal thruster. Typically, the propellant is heated by passing it over a tungsten heating element. Broad features of the thruster are sketched in Figure 6.27. The more common propellants are hydrogen, nitrogen, ammonia and, in decomposition thrusters incorporating electrical heating and chemical heat release,

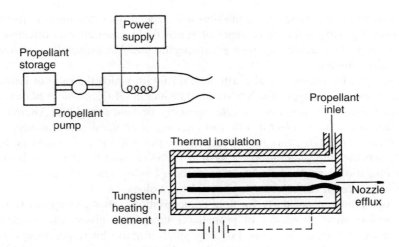

Figure 6.27   The simplest electrothermal thruster: the resistojet

hydrazine (*power-augmented hydrazine thrusters (PAEHT)*). The exhaust velocity is a function of temperature, and therefore materials' integrity and life considerations limit V_e to about 10 km/s. Efficiencies in excess of 70% and thrust levels $\leq 0.5\,N$ have been demonstrated. Hydrogen is particularly attractive as a propellant since it is non-corrosive, with both high specific heat and thermal conductivity. It is, however, difficult to store, requiring cryogenic temperatures, and dissociation to atomic hydrogen at elevated temperature leads to frozen flow losses in the nozzle.

Ammonia is readily stored without refrigeration and is dissociated to lighter species on heating, but it is corrosive in both the heater and nozzle. Higher specific impulses are available from the arc jet ($\sim$2000 s) but at reduced electrical efficiency. The expellant itself is subject to ohmic heating by passing it through an arc discharge, thereby eliminating gas-solid heat transfer. For many years the principal problem in the implementation of arc jet technology arose from the high erosion of the cathode material. This erosion is a function of both the propellant and the thruster configuration. These propulsion units are however, intrinsically simple systems and it is a relatively easy matter to convert the on-board available voltage of most spacecraft buses, to the typical 100 V required for an arc jet. Hydrazine fuelled systems are now space proven, and are being used operationally on many satellites. Performance figures quoted by manufacturers identify an I_{sp} greater than 500 sec, and a thrust greater than 100 mN for an input power of order 1 to 2 kW.

Electrostatic thrusters

Electrostatic thrusters derive their thrust from the direct acceleration of positively charged particles in a static electric field. The stream of positively charged particles must be neutralized to avoid a charge build-up, opposite to that carried away from the spacecraft in the beam, which would lead eventually to stalling of the thruster. Neutralization is generally achieved by a hot cathode electron source, placed in near proximity to the thruster exit plane.

A one-dimensional approximation to the underlying performance of an electrostatic thruster is simply obtained by equating the potential energy of the ion, as it leaves the ionization region in the thruster, to the kinetic energy it has achieved moving through the electrostatic field. Thus assuming a charge of q coulombs exists on an ion of mass m_i, and the potential applied between the ion source and the exit plane of the final accelerator grid is V volts, then

$$\tfrac{1}{2}m_i v^2 = qV$$

If it is assumed that the conversion efficiency from the electrical power-plant output to kinetic energy in the ion stream is unity, then the thrust to power ratio for such a thruster is then simply

$$F/W = \sqrt{\frac{2m_i/q}{V}} \tag{6.31}$$

Thus for a given electrical power and hence power-plant mass (equation 6.26), the thrust is maximized for large mass to charge ratio (m_i/q). While much of the early work therefore considered mercury as a propellant, owing to the relative ease with which it may be ionized and having $m_i/q \sim 2 \times 10^{-6}$, contamination problems have subsequently focused work upon the development of thrusters based upon either argon $m_i/q \sim 4 \times 10^{-7}$ or xenon $m_i/q \sim 1.4 \times 10^{-6}$. For each of these values it is assumed that a single charge only is held on the ion. Each of these propellants have been used in thrusters based upon *electron bombardment* for the production of ions, in the so-called Kaufmann engine. Earlier attempts to adopt the simpler *surface contact ionization* method for ion production have now ceased.

There, the ionization of the vapour of a material having a low ionization potential, for example, caesium, is achieved by passing it through a heated porous tungsten plate. Radiative heat losses limit overall efficiencies. This type of ion production mechanism was studied extensively during the 1970s. Flights on the ATS series of satellites adopted caesium as a propellant, and the engines were used only as experimental payloads. Power input to these thrusters was typically 0.1 kW. The final flight of this technology was on ATS 6 in 1974.

Electron bombardment ion sources are the most extensively developed. Electrons emitted from an axially mounted thermionic cathode are attracted towards a concentric cylindrical anode. A weak, externally applied magnetic field causes the electrons to spiral within the chamber and propellant ionization results from collisions between these electrons and propellant vapour. Although originally developed for use with mercury as propellant, recent research has focused on inert gases, argon and xenon for example, prompted by concern over the environmental impact in near-Earth applications. A representative small thruster design for station keeping is illustrated in Figure 6.28.

Significant mass savings have been predicted to accompany the application of such ion propulsion to North-South station keeping (NSSK) for geostationary spacecraft—for example, in relation to the 25 mN *UK-10 xenon thruster* on Intelsat VII class spacecraft [23].

The first use of a xenon ion engine in space was the *RITA engine* on the ESA Eureca-1 mission in 1992. Since then there has been a steadily increasing adoption of ion engines for NSSK on GEO spacecraft. The PanAmSat organization was the first to have an operational ion engine for one of its satellites PAS-5, which was launched in 1997. This ion engine, similar to one used on most subsequent flights, is the Boeing (formerly Hughes Space and Communications) company's *'XIPS' propulsion unit*. Two variants of the thruster

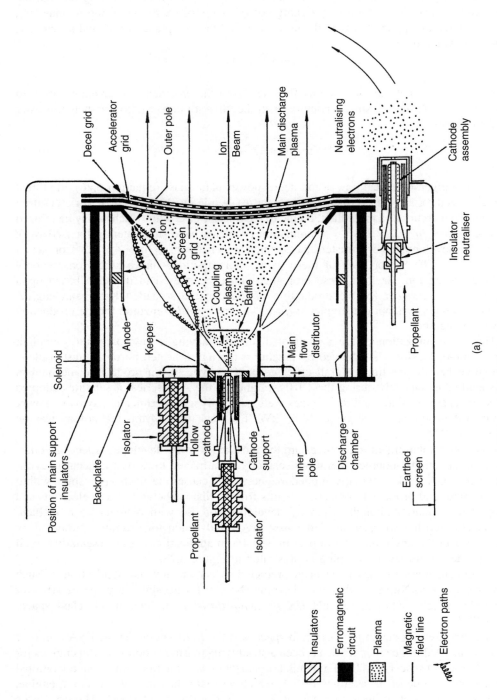

Figure 6.28 Typical ion thruster

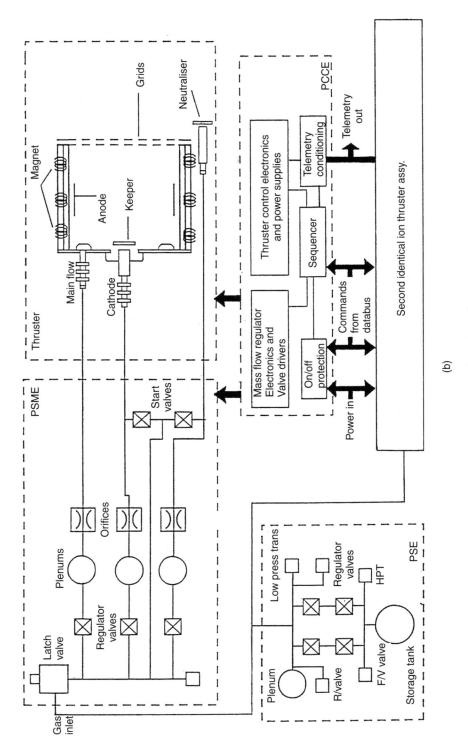

Figure 6.28   (*continued*)

have been flown, one having a 13 cm diameter capable of a thrust of 18 mN at an I_{sp} of 2568 sec, the other with a diameter of 25 cm, with a thrust of 165 mN and an I_{sp} of 3800 sec. The smaller unit has been used on several commercial communications and broadcast satellites including Astra 2A in 1998 for NSSK. In these missions the thrusters are required to fire for approximately 5 hours per day in order to maintain the satellite on station. The larger unit was used on the Deep Space 1 mission.

An alternative form of ion thruster is based upon the extraction of charged fluid droplets directly from a fluid surface. Two forms of this exist: FEEP [24], and colloid propulsion [25]. Both of these depend upon the formation of a 'Taylor cone' [26] at the surface of a fluid that is subject to a strong electric field. The fluid cone forms a capillary jet at its apex that breaks up into a fine spray of positively charged droplets. These droplets in the case of FEEP are sub-nanometre in size, whereas for colloid systems they are sub-micron in size. The colloid droplets, having significantly larger masses than those for the corresponding Kaufmann ion sources, are able to produce higher thrust density, but at the expense of lower I_{sp}. The fluid in the case of a FEEP thruster is a metal, frequently indium or caesium, which is heated so that it becomes liquid. In a colloid system, the fluid is an electrolyte of high electrical conductivity. FEEP thrusters have been developed principally in Europe. Since it is believed that FEEP can produce a highly stable low-thrust beam having a continuous thrust a level of order 10 μN, they have been identified to provide the capability of fine control for several demanding missions that require drag-free operation. This thruster is currently the baseline for the Laser Interferometer Spaceborne Antenna—LISA—a gravitational wave detector that is one of ESA's cornerstone missions, which requires a positional control of order 10 nm. In this mission the thrust must be controlled smoothly in the range 5 to 25 μN with a control resolution of 0.1 μN. The FEEP thruster being developed has an I_{sp} in the range 4000 to 6000 sec. High electrical efficiency is obtained with a value greater than 95%. Power to thrust ratio is of order 60 W/mN.

Hybrid thrusters

Several thrusters rely upon the interaction between electric and magnetic fields for their ability to produce high I_{sp} thrust. Historically in the West, considerable research emphasis was placed upon the magnetoplasmadynamic arc jet. In the former Soviet Union, practical thrusters were produced that utilize the Hall Effect.

As noted above, *Hall effect thrusters* have been the subject of a successful technology transfer programme to the USA. This has resulted in commercially available thrusters that have been adopted for several satellites. In the Hall thruster an externally provided radial magnetic field is required. Within the volume enclosed by this field a continuous axial electric discharge is maintained in the low-pressure xenon propellant gas. The interaction between the axial and radial fields generates a Hall current, perpendicular to the electric field, which is azimuthal within the volume of the thruster. It is this Hall current that interacts with the magnetic field that is responsible for the axial acceleration of the propellant. Performance is dependent upon the size of the thruster. The T-220 system developed by NASA has a thrust of 500 mN at a specific impulse of 2450 sec. The input power for this performance is 10 kW, with an overall electrical efficiency of 59%. Smaller systems have also been flown, with power input of 1.35 kW, for attitude control, and on the STEX spacecraft in 1998, for orbit raising.

The *magnetoplasmadynamic (MPD) arcjet* evolved from the combination of electrothermal arcjet and magnetogasdynamic technologies. A neutral plasma is accelerated by means of both Joule heating and electrodynamic forces. In the full MPD engine, the self-induced magnetic field provides the dominant acceleration mechanism, with thrust being proportional to the current squared. At low power levels the propulsive efficiency is somewhat modest ($\sim$10% for a 1 kW device [27]). As the power level increases into the megawatt range, significantly improved efficiencies result ($\sim$40% at $\sim$MW [28]). For such high-power levels, a pulsed operation is essential and significant developments have taken place in Japan in this area, including flight experiments. As with electrothermal arc jets at present, the major life limitation for these devices is that of cathode erosion. Typical operational characteristics are thrusts of a few newtons, with an $I_{sp} \geq 2000$ sec.

Thrusters using Teflon as a solid propellant have also been developed, generally under the heading of *Pulse Plasma Thrusters*, and are a variant of the MPD thruster. These again rely upon the interaction of an electric field with a self-induced magnetic field. A capacitor is used to initiate a pulse discharge in between two electrodes separated in part by a Teflon bar. The surface of this bar ablates because of the heating caused by the discharge. The induced magnetic field from the high current pulse discharge acts in similar manner to the MPD, to accelerate the ablated material. A very high thrust pulse lasting for some tens of milliseconds can be produced in these thrusters, with a peak thrust value in the region of several hundred newton. These thrusters have been flown on several missions, particularly for station keeping. Specific impulse from these thrusters is similar to that available from Hall Thrusters.

REFERENCES

[1]   Friedman, L. *et al.* (1978) Solar Sailing—The Concept Made Realistic, AIAA Paper 78–82.
[2]   Buden, D. and Sullivan, J. A. (1984) Nuclear space power systems for orbit raising and manoeuvring, *Prog. Astro. Aero.*, **89**, 425–459.
[3]   Caveney, L. H. (ed.) (1984) Orbit Raising and Manoeuvring Propulsion: Research Status and Needs, AIAA, New York.
[4]   Barrere, M., Jaumotte, A., DE Venbeke, B. F. and Vanden Kerckhove, J. (1960) *Rocket Propulsion*, Elsevier, Amsterdam.
[5]   Sutton, G. P. (1992) *Rocket Propulsion Elements* (6th edn), John Wiley & Sons, New York.
[6]   Rao, G. V. R. (1961) Recent developments in rocket nozzle configuration, *ARS J.*, **31**(11), 1488–1494.
[7]   Morel, R. *et al.* (1995) The Clustered Bell Aerospike Engine: Potential, Limitations and Preparation for Experimental Validation, IAF-95-S.2.04.
[8]   Kuo, K. K. and Summerfield, M. (1984) Fundamentals of solid-propellant combustion, *Prog. Astro. Aero.*, **90**.
[9]   Huzel, D. K. and Huang, D. H. (1992) Design of liquid propellant rocket engines, *Prog. Astro. Aero.*, **147**.
[10] Brown, J. R. (1983) Expander Cycle Engines for Shuttle Cryogenic Upper Stages, AIAA-83-1311.
[11] Koelle, D. (1990) Advanced Two-Stage Vehicle Concepts (SANGER) AIAA-90-1933.
[12] Moszee, R. and Snyder, C. D. (1989) A Propulsion Development Strategy for the National Aero-space Plane, AIAA-89-2751.
[13] Caporicci, M. (2000) The Future of European Launchers: The ESA Perspective, ESA Bulletin, **104**, 66–75.
[14] Ring, E. (1964) *Rocket Pressurisation and Propellant Systems*, Prentice-Hall, New York.
[15] Brown, C. D. (1996) Spacecraft Propulsion, AIAA, Washington DC.
[16] Seifert, W. S. (1959) *Space Technology*, John Wiley & Sons, New York.

[17]   Langevin, Yves (2000) Chemical and solar electric propulsion options for a cornerstone mission to mercury, *Acta Astronautica*, **47**, 443–452.

[18]   Rayman, M. D., Varghese, P., Lehman, D. H. and Livesay, L. L. (2000) Results from the Deep Space 1 Technology Validation Mission, *Acta Astronautica*, **47**, 489–502.

[19]   Burt, E. G. C. (1968) The dynamics of low-thrust manoeuvres, *J. Royal Aero. Soc.*, **72**, 925.

[20]   Stark, J. P. W. and Hobbs, L. W. (1989) Optimization of Electric Propulsion for GEO Missions, AIAA-89-2371.

[21]   Stuhlinger, E. (1964) *Ion Propulsion for Spaceflight*, McGraw-Hill, New York.

[22]   Jahn, R. G. (1968) *Physics of Electric Propulsion*, McGraw-Hill, New York.

[23]   Fearn, D. and Smith, P. (1989) The Application of Ion Propulsion to Intelsat VII Class Spacecraft, AIAA-89-2275.

[24]   Marcuccio, S., Genovesse, A. and Andreucci, M. (1998) Experimental performance of field emission micro-thrusters, *J. Propulsion Power*, **14**, 774–781.

[25]   Gamero-Castano, M. and Hruby, V. (2000) Electrospray as an efficient source of nano-particles for efficient colloid thrusters, *AIAA 2000-3265, 36*[th] *Joint Propulsion Conference*, 16–19 July, Huntsville, Alabama.

[26]   Taylor, G. I. (1964) Disintegration of water drops in an electric field, *Proc. Royal Soc. London*, **A280**, 383–397.

[27]   Uematsi, K. (1984) Development of a 1 kW MPD Thruster, AIAA-87-1023.

[28]   Burton, R. L., Clark, K. E. and Jahn, R. G. (1983) Measured performance of a multimegawatt MPD thruster, *J. Spacecraft Rockets*, **20**, 299.

7 *LAUNCH VEHICLES*

J. Barrie Moss[1] and Graham E. Dorrington[2]

[1] *School of Engineering, Cranfield University*
[2] *Department of Engineering, Queen Mary, University of London*

7.1 INTRODUCTION

Although transportation considerations feature prominently in infrastructure provision for a wide variety of terrestrial activities, few of these, if any, pose the difficulties encountered in transporting spacecraft (or payloads) from the Earth's surface to orbit (and back again). While spacecraft design alone introduces considerable technical challenge, the technology involved in the launch will often amplify the complexity—imposing many additional mission and design constraints. The extent to which the user of launch vehicle services is able, in turn, to influence launcher development is, however, still largely unresolved.

Of particular concern to the user, in relation to spacecraft design, are the constraints that the launcher imposes on the mission. These arise especially in terms of payload mass and size, but also include the selection of launch sites and launch windows (see Section 5.3 of Chapter 5), the launch environment (mechanical and electrical—see Chapters 2 and 8) as well as issues of safety and reliability (see Chapter 17).

In this chapter, we seek first to establish the basic principles that determine launch vehicle design and performance and, in so doing, constrain payload and mission. Secondly, we outline the key features of the principal launch vehicle alternatives, as featured in Europe and the United States—notably the expendable Ariane family and the partially reusable Shuttle Space Transportation System (STS)—together with the growing range of smaller vehicles. We address briefly some of the key concerns of the user community in relation to launch costs, operational flexibility and reliability, ending with some speculation about future launch capability.

7.2 BASIC LAUNCH VEHICLE PERFORMANCE AND OPERATION

7.2.1 Vehicle dynamics

For purposes of illustration, we specialize the equations of motion to the vertical plane, parallel and normal to the flight direction, and to the motion of the centre-of-mass and the

Spacecraft Systems Engineering (Third Edition). Edited by P. W. Fortescue, J. P. W. Stark and G. G. Swinerd
© 2003 John Wiley & Sons Ltd

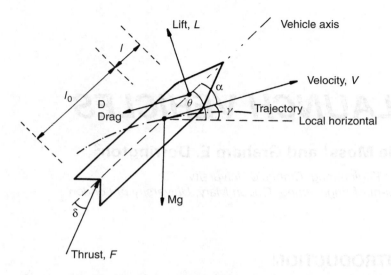

Figure 7.1 Configuration and nomenclature for rocket motion in the vertical plane

pitch rotation. A more comprehensive three-dimensional description of these equations is provided by Vinh [1].

The configuration envisaged and the accompanying nomenclature is shown in Figure 7.1. Parallel to the flight direction, we have

$$M\frac{dV}{dt} = F\cos(\alpha + \delta) - Mg\sin\gamma - D \tag{7.1}$$

and normally,

$$MV\frac{d\gamma}{dt} = F\sin(\alpha + \delta) - Mg\cos\gamma + L + \frac{MV^2}{r}\cos\gamma \tag{7.2}$$

where g is the local gravitational acceleration that varies with radial position of the vehicle r from the centre of the Earth, $g = g_0 r_0^2 / r^2$.

The displacements of the centre of pressure and nozzle from the centre-of-mass give rise to pitching moments. The accompanying angular motion may be written

$$I_P\frac{d^2\theta}{dt^2} = (L\cos\alpha + D\sin\alpha)l - Fl_0\sin\delta \tag{7.3}$$

where I_P is the moment of inertia in pitch and $\theta = \alpha + \gamma$.

For small values of α and δ, equation (7.1) becomes

$$\frac{dV}{dt} = \frac{F}{M} - g\sin\gamma - \frac{D}{M} \tag{7.4}$$

Recall from Chapter 6 that the thrust may be written

$$F = \left(\frac{-\mathrm{d}M}{\mathrm{d}t} \right) I_{\mathrm{SP}} g_0$$

whence equation (7.4) becomes

$$\frac{\mathrm{d}V}{\mathrm{d}t} = -I_{\mathrm{SP}} g_0 \frac{\mathrm{d}}{\mathrm{d}t} (\ln M) - g \sin \gamma - \frac{D}{M} \qquad (7.5)$$

On integration

$$V = V_0 + I_{\mathrm{SP}} g_0 \ln \left(\frac{M_0}{M} \right) - \int_0^t g \sin \gamma \, \mathrm{d}t' - \int_0^t \frac{D}{M} \mathrm{d}t' \qquad (7.6)$$

and in addition to the ideal velocity increment at burn-out, say,

$$\Delta V_{\mathrm{ideal}} = I_{\mathrm{SP}} g_0 \ln \left(\frac{M_0}{M} \right)$$

we now distinguish *propulsive losses* associated with gravity and aerodynamic drag over the burn time t_b

$$\Delta V_g = \int_0^{t_b} g \sin \gamma \, \mathrm{d}t'$$

$$\Delta V_D = \int_0^{t_b} \frac{D}{M} \mathrm{d}t'$$

whence

$$\Delta V = \Delta V_{\mathrm{ideal}} - \Delta V_g - \Delta V_D \qquad (7.7)$$

The relative magnitudes of these terms for an Earth surface launch are sketched in Figure 7.2. We note that the impulse requirement (minimum burn time) that might lead to reduced gravity loss is at variance with the requirement from the standpoint of drag loss, which might suggest a low velocity ascent through the denser atmosphere since $D = \frac{1}{2} \rho V^2 S C_D$. This latter requirement is broadly satisfied. The first stage propulsion of a multi-stage vehicle might typically have an initial thrust-to-weight ratio less than about 1.5, implying an initial vertical acceleration of about 0.5 g_0.

Vehicles are generally launched vertically and minimization of gravity loss, together with the eventual requirement of locally horizontal payload injection, suggests that the flight trajectory be deflected from the vertical as rapidly as possible. The simplest manoeuvre to effect this is the gravity turn.

Neglecting all but the gravitational force in equation (7.2)

$$V \frac{\mathrm{d}\gamma}{\mathrm{d}t} = -g \cos \gamma$$

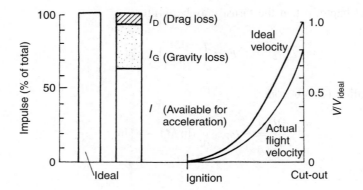

Figure 7.2 Illustration of launch losses due to drag and gravity

and on integration

$$\sin \gamma(t) = \tanh \left\{ \tanh^{-1}(\sin \gamma_0) - \int_{t_0}^{t} \frac{g}{V} dt' \right\} \qquad (7.8)$$

More rapid pitching manoeuvres may be effected by thrust vectoring or the firing of attitude control thrusters. Figure 7.3 shows the full range of flight parameters for a typical Ariane injection into geostationary transfer orbit (GTO). The local pitch angle decreases by 60° in approximately the first two minutes of flight.

Exact analytical trajectory solutions are not available in general since thrust, gravity and drag are complex functions of altitude. However, approximate analytical expressions can be derived when further simplifying assumptions are introduced [2,3]. In particular, when drag and the variation in g are ignored, and the thrust-to-weight ratio is held constant at the initial lift-off value, $a = F/Mg = F_0/M_0 g_0$, the mass ratio of a single-stage vehicle is given by

$$\frac{M_b}{M_0} = \exp \left\{ -\frac{V_b}{g_0 I_{SP}} \frac{a^2}{(a^2 - 1)} \right\}$$

where V_b is the speed at burn-out. In this particular case, the burn-out-to-initial mass ratio M_b/M_0 is lower than the ideal value, $\exp\{-V_b/g_0 I_{SP}\}$ and the gravity loss incurred during the ascent is

$$\Delta V_g = g_0 I_{SP} \ln\{M_0/M_b\} - V_b = V_b/(a^2 - 1)$$

In order to maximize the payload fraction, it is advantageous to increase the vehicle thrust-to-weight ratio—to reduce the gravity loss—provided the gains in mass ratio are not offset by increases in propulsion system mass [3]. From the user's standpoint, this helps to explain why the payload has be designed to withstand final accelerations of about 3 to 4 g_0, although it should be noted that many launch vehicles suffer from higher end-of-burn accelerations resulting from design limitations on engine shut-down conditions and/or the inability to throttle-back.

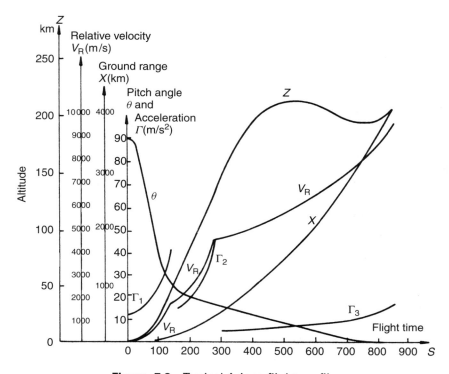

Figure 7.3 Typical Ariane flight profile

When the contribution of aerodynamic drag is introduced, lower accelerations are required to minimize the peak aerodynamic pressures reached in the early phases of launch. Typically, the drag coefficient C_D will be highest in the transonic region and the largest drag losses will occur in the low supersonic region of flight. The actual magnitude of the drag loss will be dependent on the vehicle's aerodynamic reference area S, which is dependent on the vehicle's size—in particular the payload shroud dimensions. For a set of geometrically similar vehicles, the aerodynamic reference area will be proportional to $M^{2/3}$ and the drag loss ΔV_D will be proportional to $M^{-1/3}$ assuming C_D is roughly invariant. Consequently, for small vehicles launching small payloads, drag losses become relatively more important and the user is often forced to accept tighter constraints on payload external dimensions. In general, larger vehicles are able to achieve higher payload fractions since they have lower drag losses.

At this point it is appropriate to review some representative mission requirements for spacecraft launches.

7.2.2 Mission requirements

The standard launch vehicle must provide an efficient means of boosting a spacecraft into the planned trajectory. The payloads may vary from those of Low Earth Orbit (LEO) scientific satellites, through geostationary communications satellites, to Earth escape probes. Figure 7.4 distinguishes some representative launch vehicle burn-out conditions

$$\Delta V_a = V_0 \left\{\frac{R_E + h_0}{R_E + h}\right\}^{1/2} \left\{1 - \left[1 + \left(\frac{R_E + h}{R_E + h_0}\right)\right]^{\frac{2}{R_E + h_0}}\right]^{1/2}\right\},$$

$$V_b = V_0 \left\{\frac{2}{1 + \left(\frac{R_E + h_0}{R_E + h}\right)}\right\}^{1/2}, \quad V_0 = \left\{\frac{\mu}{R_E + h_0}\right\}^{1/2}$$

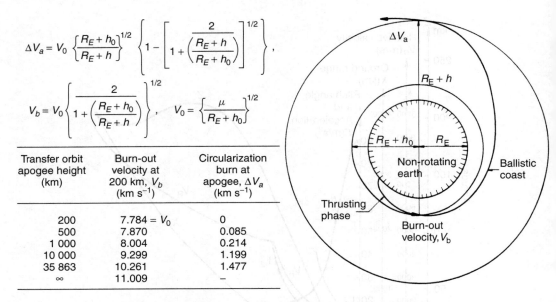

Transfer orbit apogee height (km)	Burn-out velocity at 200 km, V_b (km s^{-1})	Circularization burn at apogee, ΔV_a (km s^{-1})
200	7.784 = V_0	0
500	7.870	0.085
1 000	8.004	0.214
10 000	9.299	1.199
35 863	10.261	1.477
∞	11.009	–

Figure 7.4 Launch vehicle burn-out velocities for spacecraft emplacement

for spacecraft emplacement. As the table of values shows, the minimum velocity at a typical injection height of 200 km is approximately 7.8 km/s, while that for a GTO must be 10.3 km/s (see also Section 5.2 of Chapter 5).

In addition to accelerating the payload to these velocities, the launch vehicle must overcome the effects of aerodynamic drag and gravity. From equation (7.7)

$$\Delta V_{\text{ideal}} = \Delta V + \Delta V_g + \Delta V_D$$

and while detailed determination of the loss terms requires computation, estimates can be made for purposes of preliminary planning using such correlations as those described by White [4]. Representative losses for the first stage burn of Ariane using this approach are

$$\Delta V_g = 1.08 \text{ km/s}$$
$$\Delta V_D = 0.22 \text{ km/s}$$

The values quoted in Figure 7.4 embody a significant simplification relative to Earth rotation. The rotational velocity at the Equator is approximately 0.47 km/s, and an eastward launch from the Earth surface can evidently capitalize on this velocity. The benefit reduces with increasing latitude of the launch site and vanishes completely for launches into polar orbit.

The latitude of the launch site also has important implications for the subsequent inclination of the spacecraft orbit. We have identified some significant benefits in respect to launcher performance accompanying both rapid pitch angle reduction and eastward injection. Figure 7.5 illustrates a typical launch sequence. The spacecraft orbit plane is essentially fixed by the velocity vector at burn-out and, from gravitational considerations, the centre of the Earth. Without a very lengthy dog-leg manoeuvre prior to

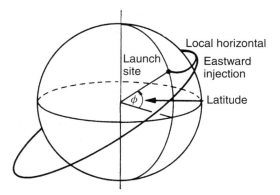

Figure 7.5 Illustration of the link between orbit inclination and the latitude of the launch site

insertion, accompanied by increased losses, both gravitational and aerodynamic, the inclination of the resulting orbit will then be approximately equal to the latitude of the launch site.

As we described in Chapter 5, the correct orbit inclination is a crucial orbital parameter. In the case of the 24-hour synchronous communications satellite, for example, the subsatellite point is only stationary relative to the Earth for equatorial orbits. Inclination leads to the subsatellite point describing a closed figure of eight as illustrated in Figure 7.6 (see also Section 5.6 of Chapter 5). The amplitude of apparent latitude drift of the satellite, north and south, is $\pm i^0$, where i denotes the orbit inclination. The control of orbit inclination, and specifically its reduction to zero in the case of geostationary communications satellites, is a further requirement of the launch phase.

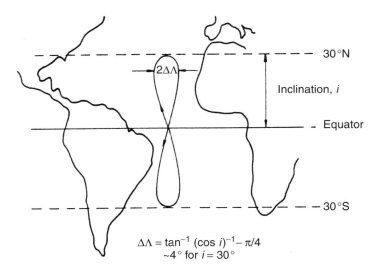

$$\Delta\Lambda = \tan^{-1}(\cos i)^{-1} - \pi/4$$
$$\sim 4° \text{ for } i = 30°$$

Figure 7.6 Drift of the subsatellite point for an inclined geosynchronous orbit

Two guiding principles behind thrusting operations have been identified in Section 3.2.5 of Chapter 3 (see Figure 3.8). The maximum increase in vehicle speed, and hence kinetic energy, is clearly obtained by thrusting in the direction of motion, and the thrust is then used to best effect in orbit expansion. On the other hand, the maximum deviation from the initial direction is obtained if the available increment ΔV is applied at right angles to the required direction. Furthermore, the deviation is larger for a prescribed level of ΔV the smaller the initial velocity. Relative to the problem of orbit raising illustrated in Figure 7.4, directional changes are evidently most economically effected at apogee when orbital speeds are least; for example, 3.07 km/s at geostationary altitude of 36000 km compared with 7.8 km/s at 200 km altitude.

We shall return to these discussions in later sections when reviewing some specific applications but it is appropriate to note at this stage that the geographical position of the launch site may have important mission implications.

7.2.3 Multi-staged launch vehicles

It is evident from Figure 7.4 that the minimum burn-out velocity for orbital operations is approximately 7.8 km/s. Linking this requirement with equation (7.6) and neglecting drag and gravity losses for the moment, then from

$$V_b = I_{SP} g_0 \ln R \tag{7.9}$$

we may estimate the necessary rocket mass ratio R for a particular propellant combination, and hence I_{SP}. Suppose the initial rocket mass, M_0, comprises three components, payload M_P, structure M_S and fuel M_F, whence

$$M_0 = M_P + M_S + M_F \tag{7.10}$$

If we introduce the fractional payload ratio, $p = M_P/M_0$ and the propellant tankage structural efficiency, $\sigma = M_S/M_F$, then we may write

$$R = \frac{M_P + M_S + M_F}{M_P + M_S} = \frac{1 + \sigma}{p + \sigma} \tag{7.11}$$

Substituting equation (7.11) in equation (7.9) gives

$$V_b = I_{SP} g_0 \ln\{(1 + \sigma)/(p + \sigma)\} \tag{7.12}$$

For a typical structural efficiency, $\sigma \approx 0.1$, then the maximum value of the mass ratio, R, corresponding to zero payload is approximately $R = 11$. Rearranging equation (7.12), for given V_b

$$I_{SP} = (V_b/g_0) \ln\{(p + \sigma)/(1 + \sigma)\}$$

and for the maximum value of R together with the minimum burn-out velocity, this implies

$$I_{SP} \geq 331 \, s$$

Thus, even under these ideal circumstances single-stage-to-orbit could only be attained using high-energy propellant (see Chapter 6). If we incorporate a non-vanishing payload and make allowance for performance losses associated with gravity and aerodynamic drag, then LEO is only marginally attainable with the most energetic fuel/oxidizer combinations. Significant contributions, in respect of structural mass reductions, must also come from the exploitation of novel lightweight materials.

Since the mass of propellant tankage is large, significant performance benefits result from the progressive shedding of this mass by multi-staging. In these circumstances, only a small fraction of the initial tankage mass is accelerated to the final speed.

If we identify the operation of each stage by an equation of the form given in equation (7.12), distinguishing the ith-stage by subscript, then the velocity increment produced is

$$\Delta V_i = V_{b,i} - V_{b,i-1} = I_{SP,i} \, g_0 \ln \left[(1 + \sigma_i)/(p_i + \sigma_i) \right] \tag{7.13}$$

We observe that the payload for the ith-stage comprises the vehicle hardware (including the satellite(s) to be launched) 'above' the ith-stage. Summing over n such stages, the final burn-out velocity is then given by

$$V_b = \sum_{i=1}^{n} I_{SP,i} \, g_0 \ln \left[(1 + \sigma_i)/(p_i + \sigma_i) \right] \tag{7.14}$$

An elaborate optimization procedure is required if we seek, for example, to maximize this velocity subject to constraints imposed by way of payload ratio and rocket performance. We should recall that the lower stages will also be subject to drag and gravity losses. The generalized problem has been extensively analysed, for example, by White [4]. It is sufficient for purposes of illustration here to take the simplest case in which the stage specific impulses and structural efficiencies are equal ($\sigma_i = s$), when

$$V_b = \sum_{i=1}^{n} I_{SP} g_0 \ln\{(1 + s)/(p_i + s)\}$$

Using the technique of Lagrange multipliers with the constraint imposed by specification of the overall payload ratio,

$$P = \prod_{i=1}^{n} p_i$$

it can be shown that V_b is maximized when the stage payload ratios are all equal,

$$p_i = P^{1/n} (\text{all } i) \tag{7.15}$$

and
$$V_{b\,max} = I_{SP}g_0 \left\{n \ln(1+s) - n \ln\left(s + P^{1/n}\right)\right\}$$

This expression has the asymptotic value

$$V_{b\,max} \approx I_{SP}g_0 \ln P^{-1} \quad \text{as } n \to \infty \tag{7.16}$$

The benefits introduced by multi-staging do not then increase without bound except for the impractical case of zero overall payload ratio, as shown in Figure 7.7. At the upper end of payload ratios of interest, $P \sim 0.1$, there is little advantage in further partitioning beyond three stages, given the attendant increases in complexity and cost. The inclusion of the propulsive losses described earlier leads typically to a preferential loading of the first stage.

By way of illustration, Figure 7.8 compares the staging of the small air-launched Pegasus vehicle, produced by Orbital Sciences Corporation (OSC), with an equivalent single-stage vehicle having the same overall mass, payload ratio and average stage properties (structure-fuel ratio and specific impulse). Table 7.1 summarizes some of the key features of the propulsive performance of the Pegasus vehicle employed in the comparison.

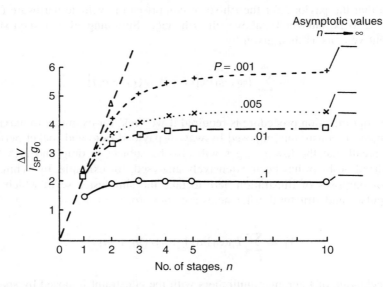

Figure 7.7 Variation of velocity increment with number of stages for fixed overall payload ratio *P*

Table 7.1 Propulsive characteristics: Pegasus launcher

Stage	Specific impulse in vacuum (s)	Average thrust (kN)	Burn time (s)	Gross mass (kg)	Propellant (kg)
1	295.3	486.7	77	14 020	12 150
2	295.5	122.8	75	3400	3025
3	291.1	54.56	65	984	782

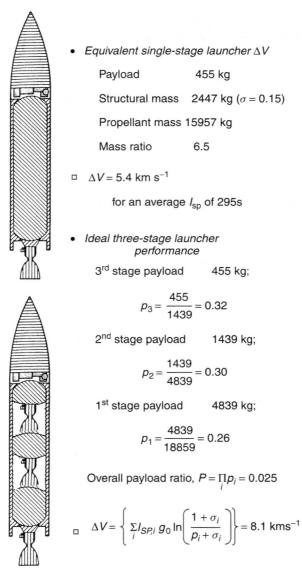

- *Equivalent single-stage launcher ΔV*

  Payload   455 kg

  Structural mass   2447 kg ($\sigma = 0.15$)

  Propellant mass 15957 kg

  Mass ratio   6.5

  □ $\Delta V = 5.4$ km s^{-1}

  for an average I_{sp} of 295s

- *Ideal three-stage launcher performance*

  3rd stage payload   455 kg;

  $$p_3 = \frac{455}{1439} = 0.32$$

  2nd stage payload   1439 kg;

  $$p_2 = \frac{1439}{4839} = 0.30$$

  1st stage payload   4839 kg;

  $$p_1 = \frac{4839}{18859} = 0.26$$

  Overall payload ratio, $P = \prod_i p_i = 0.025$

  □ $\Delta V = \left\{ \sum_i I_{SP_i}\, g_0 \ln\left(\frac{1 + \sigma_i}{p_i + \sigma_i} \right) \right\} = 8.1 \text{ kms}^{-1}$

Figure 7.8 Illustration of the benefits of multi-staging

While the three-stage vehicle shows a substantial improvement over its single-stage equivalent, an available ΔV of only 8.1 km/s would not normally be sufficient to secure LEO once the drag and gravity losses are also incorporated. The vehicle's additional performance is derived from the air launch, - which imparts an initial velocity (0.24 km/s) and the diminished losses accompanying reduced air density and angle of incidence.

7.3 SPACECRAFT LAUNCH PHASES AND MISSION PLANNING

In the preceding sections, we have outlined some basic aspects of flight dynamics, the propulsive requirements for Earth orbit attainment and the response to the intrinsic mismatch between ΔV and chemical propellant performance as reflected in multi-staging. Spacecraft missions invariably require the emplacement of the payload in an orbit having narrowly specified parameters—altitude, period, inclination and inertial orientation. The launch phase therefore embraces each of the propulsive manoeuvres necessary to achieve the initial emplacement and these are necessarily different for each mission.

7.3.1 Geostationary orbit emplacement

A typical three-stage expendable launcher injects the satellite into GTO with a perigee at 200 km altitude and apogee at 36 000 km (Section 5.6.1 of Chapter 5). Upper-stage burn-out occurs at the first equatorial plane crossing. Following a ballistic coast, the transfer orbit apogee then coincides with the second crossing of the equatorial plane or node. This transfer orbit is inclined, the inclination being dependent upon the latitude of the launch site and the launch azimuth. As indicated earlier, a due east launch (launch azimuth $= 90°$) will result in a transfer orbit inclination equal to the latitude of the launch site, while any other launch azimuth must increase the orbit inclination.

Positioning of the geostationary apogee over the equator admits the possibility of emplacement in final equatorial geostationary orbit by a further single motor firing. The satellite is typically fitted with an apogee boost (or kick) motor (ABM/AKM) specifically to effect this combined manoeuvre of orbit circularization and inclination removal. The transfer orbit apogee velocity is approximately 1.60 km/s while the equatorial geostationary orbital velocity is 3.07 km/s. Table 7.2 illustrates the impact of launch site latitude, and hence transfer orbit inclination, on ABM impulse requirement.

Near-equatorial launch sites such as the European Space Agency (ESA) site at Kourou in French Guiana ($\sim$5° N) or the San Marco platform off the Kenyan coast offer significant propulsive advantages over the Eastern Test Range [Kennedy Space Center (KSC)] in Florida (latitude $\sim$28° N) or the Japanese or Russian sites at even higher latitudes.

Injection into the final orbit by the satellite ABM is not usually effected at the first transfer orbit apogee, approximately five hours after launch. The manoeuvre is constrained by the requirement to attain the operational longitude station with a high level of accuracy and within a defined period. Precise determination of the satellite orbit and attitude by ground station telemetry and tracking is necessary in order to correctly orientate the motor prior to commanding its firing. Several apogees will then pass—the elliptical transfer orbit

Table 7.2 ABM impulse requirements for geostationary orbit emplacement

	Transfer orbit inclination (degrees)			
	0	10	30	50
ABM ΔV (km/s)	1.47	1.52	1.86	2.38

period is approximately ten hours—before the ABM burn. The satellite must evidently be ground controlled during this phase and have sufficient electrical power to maintain communications and some on-board systems, for example, attitude and orbit control. If this phase is solely battery powered, then such factors will determine the acceptable length of time spent in transfer orbit.

In view of the possible accumulation of injection errors and the requirement to position the satellite precisely on longitude station, *final positioning* is sometimes achieved from a drift orbit. If the satellite orbit after ABM firing is arranged to be very slightly elliptical, with perigee somewhat less than geostationary height, then the orbit period will be slightly less than the Earth rotational period. Furthermore, as Figure 7.9 illustrates, the subsatellite point at geostationary height will then drift gradually eastwards relative to the Earth. Short duration firing of the satellite's station-keeping thrusters over several days then permits the precise positioning of the satellite. A spacecraft using a relatively low thrust, liquid apogee motor (LAM) can however be precisely stationed without the need for a drift orbit (see Section 5.6.1 of Chapter 5).

7.3.2   Low earth orbit emplacement

Similar, if less elaborate, manoeuvres apply to most other near-Earth missions. The more energetic and accurate is the launch vehicle injection, the smaller is the secondary satellite propulsion requirement and the greater is the true payload fraction. Preliminary mission planning is essentially iterative, however (see Chapter 5). The identification of broad mission objectives—Earth observation, communications, science payload—will all involve the specification of the satellite orbit. Launch vehicle performance summaries that describe the orbits accessible to a particular payload mass introduce the first constraint. Detailed

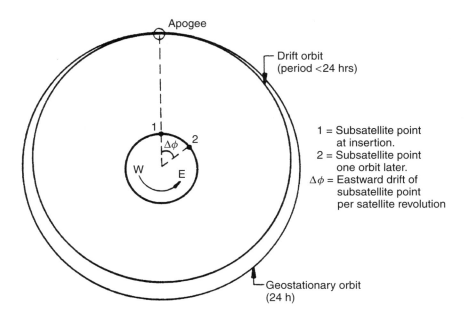

Figure 7.9   Station acquisition from a drift orbit

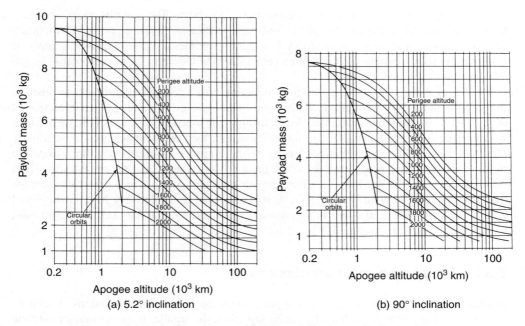

Figure 7.10 Ariane 44L payload capability into elliptical orbits

consideration of the distinctive launch capabilities of Ariane and the Space Shuttle is presented in later sections but we illustrate briefly here some general principles.

Figure 7.10 illustrates typical users' manual information for Ariane 4 launches into near-equatorial (5.2° inclination) and polar orbits (90° inclination). As might be expected, the payload capability into orbits of increasing altitude but fixed inclination (say, 5.2°), decreases steadily −8.5 tonnes for 600 km, 7 tonnes for 1000 km in the case of the most powerful version Ariane 44 L. Further reductions accompany the raising of the apogee height. Changes in inclination also have a substantial impact on payload capability. If the 600 km orbit is inclined at 90°, the payload is further reduced to 6.8 tonnes, reflecting the abandonment of a due east launch and the benefits of Earth rotation. The east coast of South America presents little restriction to eastward launches of any azimuth between zero and 90°, however, but this is not always the case. Figure 7.11 identifies the *launch azimuth limits* that apply at the Western Test Range (WTR) (Vandenberg) in Southern California. Direct injection is here restricted to launches towards the south-west, over the Pacific Ocean, giving access only to retrograde orbits ($i > 90°$). Only along the 164° azimuth limit is there a possibility of an eastwards launch and a small contribution from Earth rotation.

7.3.3 Configuration interactions

While payload mass may be considered to assume an overriding importance in determining the mission parameters, in many circumstances the spacecraft configuration can also be constrained by the size and shape of the available payload volume. Multi-stage

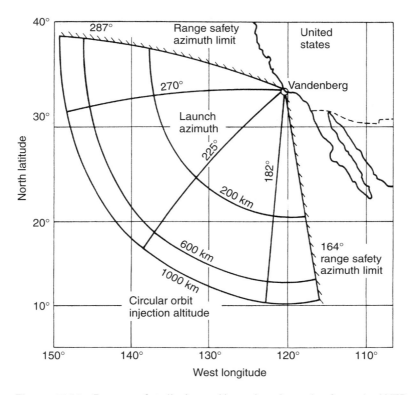

Figure 7.11   Range safety limits and launch trajectories from the WTR

expendable launch vehicles are essentially cylindrical, and aerodynamic considerations naturally restrict the payload fairing (or envelope) to a shape resembling a cone-cylinder combination. Spacecraft designed to be launched by such vehicles tend to be slender—at least in their stowed configuration—and solar arrays, communications antennas and scientific instruments may have to be folded, furled or telescoped to conform to the fairing and then deployed on station. The inclusion of mechanisms necessary to effect such deployment (see Chapter 15) then adds substantially to the complexity and vulnerability of the payload design.

Although the launch phase duration may be measured in hours in a mission with an operational timescale of years, the launch often produces the most demanding environment faced by the mechanical design. The longitudinal acceleration is high; in excess of $4\,g_0$ in the case of Ariane at second stage burn-out. Sensitive elements of the payload and deployable equipment must therefore withstand both the mean acceleration and the structural vibration accompanying motor firing and stage separation (see Chapter 8).

As we indicated in Section 7.3.1, geostationary orbit emplacement can also incorporate quite major satellite thrusting phases at apogee (ABM firing) and, in the case of Shuttle launches (see Section 7.5), a perigee burn in addition. In general, these motors employ solid propellants and deliver relatively high thrusts in firings of short duration (see Chapter 6). In the absence of fine control of thrust, the spacecraft and motor are spun-up to an angular rate of ~100 rpm. This both provides a measure of gyroscopic stiffness for guidance and reduces the effects of any thrust misalignment. In consequence, however, deployment of lightweight, flexible structures such as solar arrays is further delayed and substantial power raising in transfer orbit is often prevented. The enhanced mission flexibility that accompanies lower thrust, more readily controlled and higher specific impulse bi-propellant rocket motors is a major factor in their development as secondary propulsion.

7.4 THE ARIANE LAUNCH VEHICLE

In the preceding section, we have described some general features of launch vehicle performance and operation from the user or payload perspective. Here we describe the Ariane launch vehicle in more detail but from the same standpoint. The reader will be referred to other sources for launch vehicle design information.

7.4.1 Vehicle design summary

While earlier expendable launch vehicles evolved from missile designs and were adapted to a wide range of missions, the Ariane programme was specifically directed towards the task of geostationary communication spacecraft emplacement. Despite much subsequent broadening of the performance envelope through Ariane 1 to 5, each embracing increased payload mass and envelope, the defining mission remains that of injection into GTO. In the presentation here, we concentrate on the capabilities of Ariane 4 and 5, though the removal of the former from the launch vehicle inventory in 2003 has already been announced.

Ariane 4 comprises a three-stage liquid propellant rocket, augmented by a mission-specific selection of strap-on boosters. The first and second stages burn a hydrazine (UH25)-nitrogen tetroxide (N_2O_4) oxidizer combination, while the third stage employs a cryogenic liquid hydrogen-liquid oxygen motor. The mass breakdown for the baseline Ariane 40 is summarized in Table 7.3.

The first stage propulsion system employs 4 Viking 5C motors, each delivering a sea-level thrust of 677 kN. The motor design is based on a gas generator cycle described in

Table 7.3 Ariane 40 mass breakdown (tonnes)

Stage	Structure	Propellants	Stage mass	Cumulative mass
1st stage	18.0		251	251 (83%)
Oxidizer		147		
Fuel		86		
2nd stage	3.3		38.3	289.3 (96%)
Oxidizer		22		
Fuel		13		
3rd stage	1.3		12.1	301.4 (100%)
Oxidizer		6.8		
Fuel		4.0		

Note: Additional mass arise from interstage structure, the vehicle equipment bay, cooling water and propellant pressurants.

Table 7.4   Ariane 4 engine characteristics

	Viking 5C	Viking 4B	HM7B	First stage strap-on boosters	
Vacuum thrust (kN)	759	785	63	140 737	(P) (L)
Vacuum specific impulse (s)	278	294	444	140 278	(P) (L)
Chamber pressure (bar)	58.5	58.5	35	58.5	(L)
Burn time (s)	205	126	725	34 140	(P) (L)
Propellant	N_2O_4—UH25	N_2O_4 —UH25	LOX—LH_2	CTPB (P) N_2O_4—UH25 (L)	

Note: UH25 is a mixture of 75% unsymmetrical dimethylhydrazine (UDMH) and 25% hydrazine hydrate; CTPB is a composite solid propellant. P and L denote solid and liquid propellant, respectively.

more detail in Chapter 6. The second stage employs a single Viking 4B motor, having substantially the same design and delivering a vacuum thrust of 785 kN. The final stage HM7B engine uses LH_2/LOX and is the development of a motor first planned for the ill-fated Europa launcher. The thrust delivered is 63 kN and the motor fires for 725 s.

The component performance is summarized in Table 7.4.

The *Ariane 5* launcher represents a significant departure from the earlier configurations in the family. Most of the lift-off thrust is derived from two large (P230) solid boosters, each delivering a vacuum thrust of 4984 kN at a specific impulse of 275 s. The Vulcain core stage is a cryogenic LH_2-LOX engine, delivering a vacuum thrust of 1114 kN at a specific impulse of 430 s. A small mono-methylhydrazine (MMH)/N_2O_4 upper stage, delivering 29 kN thrust, is available for final orbital injection.

7.4.2   Mission performance

Ariane is launched from the near-equatorial Centre Spatiale Guyanais (CSG) in Kourou, French Guiana, at a latitude of 5.2° N. As described earlier, such a location is particularly favourable for geostationary satellite emplacement but launches can also be made from within azimuth limits −10.5 to 93.5°. A range of performance curves is shown in the appropriate Ariane users' manual [5] for purposes of preliminary mission planning and embraces a variety of elliptical and Sun-synchronous orbits in addition to the primary programme objective of GTO emplacement.

Figure 7.10 illustrates the basic elliptical orbit performance of Ariane 4 for a due east launch ($i = 5.2°$). The maximum single launch payload capability into GTO is identified in Table 7.5 to be 4700 kg. Alternatively this may be configured as two Delta-sized satellites of the following general form:

$$2 \quad \times \quad 1890 \text{ kg satellites}$$
$$\text{SPELDA dual launch system and adaptor } -440 \text{ kg}$$

A brief description of the Système de Lancement Double Ariane (SYLDA)/SPELDA dual-launch systems and payload envelopes is provided later in this chapter.

If we focus specifically on the dual launch into GTO, both satellites are injected into the same orbit although they may be oriented and spun-up (≤ 10 rpm as required) quite independently after third-stage engine cut-off. A minimum relative velocity of 0.5 m/s is established between the two satellites at separation and a safe distance is determined between them before apogee motor firing. The argument of perigee at injection is approximately 180° and the longitude of the first descending node occurs in the vicinity of 10° W, off the coast of West Africa.

The enhanced performance offered into GTO by the addition of alternative combinations of liquid (L) and solid propellant (P) strap-on boosters is illustrated in Table 7.5.

The values quoted in the first part of Table 7.5 include the masses of spacecraft, dual-launch system (if employed) and appropriate adapters.

The injection phase is inertially stabilized and the accuracy is high. Typical values cited for the standard deviations of key injection parameters are: argument of perigee 0.45°, perigee altitude 1 km and apogee altitude 100 km.

While the performance envelope of Ariane 5 has not yet been fully exploited—the first launch of a commercial payload took place on 10 December 1999—Table 7.6 illustrates the substantially improved capability [6].

7.4.3 Dual-launch system

Ariane 4 can simultaneously place two independent satellites into GTO using the SYLDA/ SPELDA (Systeme de Lancement Double Ariane).

The key features of the basic SYLDA are illustrated in Figure 7.12 and comprise a load carrying, carbon-fibre reinforced shell, which encapsulates the lower satellite and supports the upper one.

Table 7.5 Ariane 4 performance into GTO, including mass estimates for dual-launch systems and adaptors

Launcher configuration			Mass (kg)
A40			2070
A42P			2920
A44P			3380
A42L			3450
A44LP			4170
A44L			4700

Dual-launch system	Mass (kg)	Adaptor	Mass (kg)
Short SPELDA	350	'937'	48
Long SPELDA	400	'1194'	43
SYLDA (3990)	185		
SYLDA (4400)	185		

Table 7.6   Ariane 5 performance

Mission	Payload mass (tonnes)
550 km/28.5°	18
Sun-synchronous: 800 km/98.6°	12
GTO: 580 × 35786 km/7°	6.8 (single payload)
	5.9 (dual)
	5.1 (triple)

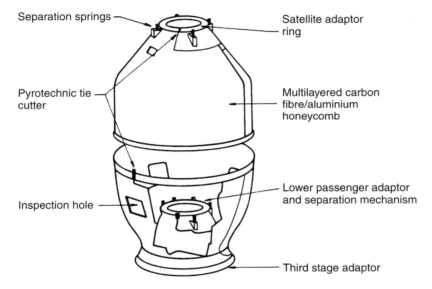

Figure 7.12   SYLDA

The shell incorporates two separable parts, retained in flight by a clamp band, providing satellite adapter rings. The two satellites are quite independent without mechanical or electrical interfaces. The sequence of events leading to the release and injection into orbit of the two satellites is shown in Figure 7.13. Following third-stage engine cut-off, the attitude control system orientates the stage correctly and spins up. Pyrotechnic cutters are fired and release the spring-loaded upper passenger. The upper half of the shell is itself then released pyrotechnically and the lower satellite is spring released. The complete sequence lasts approximately 30 seconds.

The payload accommodation has steadily increased in size over the course of the Ariane programme and Figure 7.14 illustrates the envelopes presently available from the lengthened dual-launch system SPELDA.

Ariane 5 is designed to launch up to three independent payloads using the SPELTRA configuration, providing accommodation up to 7 m in length and 5.4 m in diameter. A modified version of the dual-launch system, SYLDA-5, is also planned as part of the Ariane-5 Evolution Programme, aimed at substantial reductions in system weight and complexity.

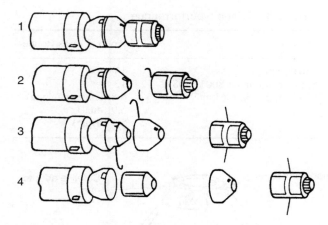

Figure 7.13   Ariane dual-launch separation phases

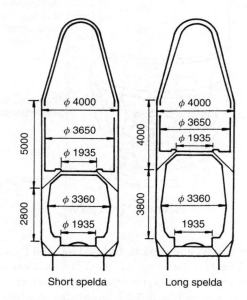

Figure 7.14   Typical payload envelopes
for the SPELDA dual-launch capability

7.5   THE SHUTTLE SPACE TRANSPORTATION SYSTEM (STS)

In contrast to the Ariane launch vehicle described in the previous section, the Space Shuttle is both manned and substantially reusable. As its name implies, its function was always more broadly defined than that of a spacecraft launcher. For the purposes of this chapter, however, we shall neglect its increasingly important role in relation to manned near-Earth operations—for example, in support of the International Space Station (ISS)—and simply focus on its orbital mission performance. Its changed role following the 'Challenger' accident, virtually eliminating from its inventory any launch tasks that could be undertaken by unmanned expendable vehicles, removes any commercial basis for performance comparisons. Some important technical issues remain, however, and these will be highlighted.

7.5.1 Vehicle design summary

The Shuttle primary propulsion comprises two basic elements; an external tank, 47 m in length and 8.4 m in diameter, which carries the LOX/LH2 propellants and feed system. The propellants are then fed to three main engines [Space Shuttle Main Engine (SSME)]. Two solid strap-on boosters augment the available thrust at launch, burning in parallel with the main engines for two minutes.

The solid rocket booster employs a composite propellant; aluminium powder, ammonium perchlorate oxidizer and a polybutadiene acrylic binder, in a tapering eleven-point star grain configuration. Each booster has an inert mass of 8.8 tonnes and propellant charge of 502 tonnes. The performance of the boosters and main engines is summarized in Table 7.7. The spent boosters separate from the external tank and are recovered by parachute approximately 300 km down range.

Table 7.7 Space shuttle main engine and booster performance

	SSME (each of three)	Solid rocket booster (each of two)
Thrust (10^6 N)	2.1 (vacuum)	11.8 (sea level)
Specific impulse (s)	455	260
Burn time (s)	480	120
Propellant	LOX/LH$_2$	PBAN/AP

The SSME employs a staged-combustion closed cycle (see Chapter 6). The hydrogen fuel and a small fraction of the oxidizer are partially burnt in a pre-combustor. The efflux from this drives the high-pressure propellant turbopumps. The hydrogen-rich turbine exhaust, together with the remaining oxidizer, is then injected into the combustion chamber. The latter operates at 190 bar and therefore necessitates substantially higher turbopump discharge pressures—approximately 500 bar. The external tank is jettisoned before final orbit insertion.

Following external tank separation, the orbital manoeuvring system (OMS) of the Orbiter provides the thrust to perform orbit insertion, circularization and such on-station manoeuvres as rendezvous and de-orbit. The OMS engine employs MMH fuel and N$_2$O$_4$ oxidizer, delivering a specific impulse of approximately 316 s and a vacuum thrust of 27 kN. The OMS tankage is sized to provide sufficient propellant for a total ΔV of approximately 305 m/s when the vehicle carries a payload of 24.4 tonnes.

7.5.2 Mission performance

Shuttle launches take place from the Kennedy Space Center (KSC) in Florida, latitude 28.5°N. The payload capability from this site is large (≤ 24.4 tonnes) and therefore factors such as payload volume or in-orbit operations usually impose more significant constraints on payload design than mass alone. Increases in parking orbit altitude can be achieved by a trade-off between payload mass and additional MMH/N_2O_4 OMS propellant (see Figure 7.15). In general, however, satellite orbit-raising is most economically effected by tailoring specialist perigee stages to individual mission requirements.

Shuttle launches due eastwards will characteristically lead to the deployment of satellites into a 204 km circular parking orbit inclined at 28.5° to the Equator.

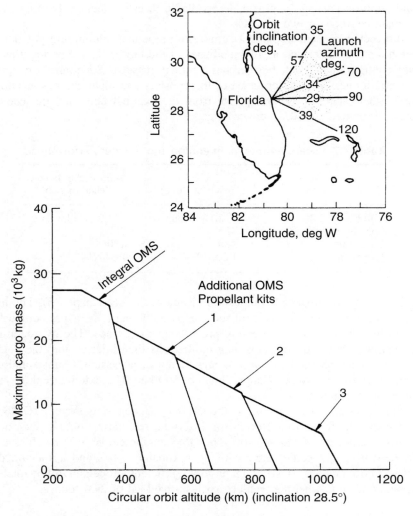

Figure 7.15 Typical STS orbital performance for KSC launch

7.6   SMALL LAUNCHERS

The growing interest in launching small satellites into LEO—including constellations of such satellites, whether for scientific or communications purposes—has prompted a renewed commercial interest in small launchers offering flexible operation and making limited infrastructure demands. A number of such vehicles are now available commercially or are in development. These typically comprise three- or four-stage solid propellant rockets providing a launch capability into Sun-synchronous LEO for payloads of approximately 1 tonne. This class of launchers is typified in the United States of America by the Lockheed Martin Athena or the OSC Taurus, a ground-launched version of the Pegasus discussed earlier [7], and the ESA-developed Vega [8]. As with the earlier, highly successful small launcher Scout, the American vehicles can be launched from Vandenberg AFB, Cape Canaveral or the Wallops Island Facility in Virginia depending on mission requirements—most notably on target orbit inclination. A typical launch cost is approximately 20 M Euros (or US dollars).

7.7   RE-ENTRY INTO EARTH'S ATMOSPHERE

In addition to transporting payloads to orbit, some users also require the recovery of payloads from orbit. With the exception of the Shuttle, recovery from orbit, to date, has been exclusively achieved using ballistic re-entry capsules.

Ballistic recovery essentially involves using aerodynamic drag to decelerate a vehicle from an entry velocity V_0, prior to parachute deployment, and subsequent impact on the Earth's surface with an acceptable velocity. The maximum deceleration rates involved in purely ballistic re-entry are relatively high compared to the launch ascent accelerations, typically exceeding about $-8g_0$, depending on the entry angle (which must be within prescribed maximum and minimum limits). In practice, re-entry capsules are flown at incidence relative to the flight direction to generate a lift force similar in magnitude, though generally less than the drag force. This permits the adoption of trajectories that reduce the peak deceleration and peak heat transfer rates (see Figure 7.16), as well as providing some control on the dimensions of the landing footprint.

The essential features of atmospheric re-entry were first considered by Allen and Eggers [10] and employed extensively in the Apollo programme (see also Section 5.8.4 of Chapter 5). In the absence of lift forces, the deceleration rate of a vehicle re-entering the atmosphere at an entry angle γ is

$$m\frac{dV}{dt} = -\frac{1}{2}\rho V^2 S C_D + W \sin \gamma \qquad (7.17)$$

If atmospheric density is approximated by a simple exponential model,

$$\rho/\rho_s = \exp(-\beta h)$$

where h is the vehicle altitude above some reference height at which the density is ρ_s, γ is assumed to be shallow and held constant at the value γ_0, and the drag force is assumed

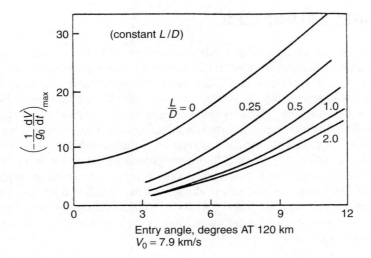

Figure 7.16 Variation of maximum deceleration with (re) entry angle for various L/D ratios (Reproduced by permission from W. H. T. Loh (1968) *Re-entry and Planetary Entry*, Springer-Verlag)

to be much larger than the vehicle weight, then it can be shown that the vehicle velocity will vary as,

$$V = V_0 \exp\left[\frac{-\rho}{2(m/C_\mathrm{D}S)\beta \sin \gamma_0}\right] \quad (7.18)$$

Furthermore, as seen in Section 5.8.4 of Chapter 5, the peak deceleration rate is given by

$$\left|\frac{\mathrm{d}V}{\mathrm{d}t}\right|_{\max} = \frac{\beta V_0^2 \sin \gamma_0}{2e} \quad (7.19)$$

Then for purely ballistic re-entry from a LEO, the peak deceleration rate is essentially dependent only on the entry angle, γ_0.

Lees [11] showed that blunt-shaped bodies are needed in order to reduce re-entry heating, since stagnation point heat transfer rates are inversely proportional to the square root of the nose radius of curvature r_nose in the laminar hypersonic regime,

$$\dot{q} = k\rho^{1/2}V^3/r_\mathrm{nose}^{1/2} \quad (7.20)$$

Even after adopting such features, thermal protection systems are also needed to prevent excessive heating of the payload. In particular, the most prevalent protection schemes employ ablative heat shields in which significant heat is absorbed during vaporization of the surface material (for example, of an epoxy resin), as adopted on the Apollo command module's entry capsule (see Figures 7.17 and 7.18).

Whereas the peak deceleration rate is independent of the so-called 'ballistic coefficient', m/SC_D, Eggers and Allen [10] showed that the peak heating rates are proportional to its

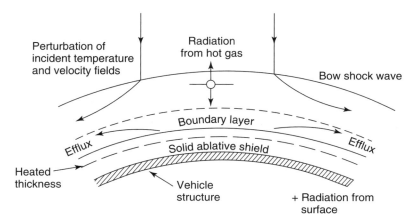

Figure 7.17 Schematic diagram of an ablating surface protection system

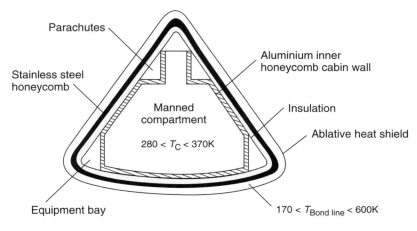

Figure 7.18 Schematic of Apollo command module thermal protection system

square root (see equation 5.49 of Chapter 5) and quadrupling the frontal area of the vehicle will halve the peak heat transfer rate. If the re-entry capsule is constrained in size (for example, by the payload envelope of the launch vehicle), then one possible strategy to reduce heating rate is to install an inflatable 'ballute'. However, such schemes have yet to be tested.

Peak heat transfer rates may also be reduced by modifying the flight trajectory using lift during the re-entry phase (see Figure 7.16). The Shuttle Orbiter re-enters the atmosphere at high angle of attack (about 40°) and at hypersonic speeds it attains a lift/drag ratio of about 2. By also rolling the vehicle about its longitudinal axis, while maintaining the high angle of attack the lift vector can be modulated. In this way, the rate at which the vehicle descends into the atmosphere, and hence the heating rate, can be controlled. This also serves to determine across-track range that can be achieved.

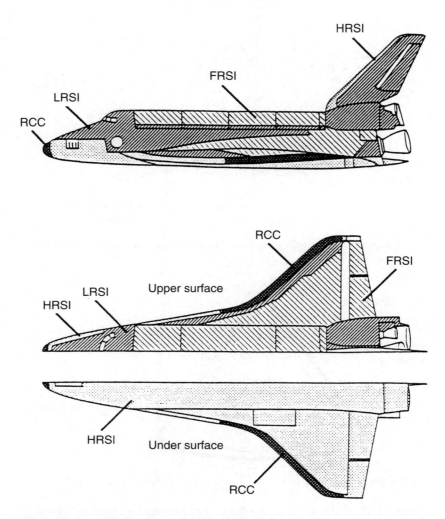

Figure 7.19 Shuttle orbiter thermal protection system. RCC: reinforced carbon-carbon; HRSI: high-temperature reusable surface insulation (silica tiles); LRSI: low-temperature reusable surface insulation; FRSI: flexible reusable surface insulation (coated Nomex)

The Orbiter employs a reusable thermal protection system that is zoned according to the local heating levels (see Figure 7.19). The nose and wing leading edges receive the highest heating rates and are consequently made from Reinforced carbon–carbon (RCC). Most of the body and wing surfaces are covered by pure silica-fibre tiles coated with a boro-silicate glass. On the upper surface of the wing and the payload bay doors where heating rates are much reduced, flexible coated Nomex felt is sufficient.

About 8% of the Orbiter dry mass is taken up by the thermal protection system—which results in a significant payload penalty from the standpoint of a user who does not require recovery. Furthermore, the Orbiter often requires significant post-flight maintenance to check, remove and replace the ceramic tiles.

Second generation shuttle-type vehicles will probably require more advanced ceramic-composite thermal protection systems. Metallic radiation cooled structures only appear to be feasible if and when the lift parameter $m/C_L S$ is sufficiently large.

7.8   SPECIFIC LAUNCH COSTS AND RELIABILITY

The growing commercial exploitation of both low-Earth and geostationary orbits has focussed attention on specific launch cost (cost per kg of payload delivered) as a figure of merit. The launcher and related services such as insurance continue to introduce costs comparable with those of the spacecraft payload itself. Specific launch costs of roughly 25 000 Euros ($) per kg (see Table 7.8) remain high and continue to inhibit much further exploitation of space.

The insurance charges accompanying launch essentially reflect the reliability of the particular vehicle. Mission success levels of about 95% are typical of established systems (see Table 7.9). While these levels can evidently be improved further, the scope is limited and major reductions in overall costs and reliability will only be possible if there are major changes in the design of future launch vehicles.

Given the widespread application of high-energy propellants, notably LH$_2$/LOX, delivering a vacuum specific impulse of 455 s in the case of the SSME, and overall payload

Table 7.8   Illustrative costs for GTO injection (Data from Reference [7])

Launch vehicle	GTO payload (kg)	Launch cost (M)	Specific cost (k$/kg)
Ariane 44L	4770	100–125	~25
Atlas II	3719	90–105	~25
Long March 3C	3800	55–75	~20
Proton K	4350	90–98	~20
Delta 7920	1870	50–60	~30
Shuttle STS	n/a	>300	n/a

Table 7.9   Launch success rate: 1960–1999

Launch vehicle	Number of launches	Successful launches	Reliability (%)
Zenit	32	26	81
Proton K, D	269	234	87
Molniya	308	276	90
Scout	109	98	91
Ariane 4	92	88	96
Delta II	46	44	96
Cosmos	427	409	96
Tsyklon	245	240	98
Shuttle STS	96	95	99

fractions of $\leq 3\%$, there is arguably limited opportunity for substantial improvements in launch costs for expendable vehicles.

The present high cost levels therefore, provide much of the stimulus for feasibility studies and technology demonstration programmes directed to the incorporation of reusability into launch vehicle design. The operational economies that might accompany reusability (and an aircraft-like focus on more autonomous logistical support) have encouraged numerous studies of single- and two-stage-to-orbit concepts in recent years—in some cases incorporating air-breathing elements. However, few have progressed beyond conceptual design [12]. The recent NASA X-33 demonstrator programme [13–15] represents the latest attempt to exploit advanced lightweight materials developments and novel propulsion technologies like the linear aerospike [12] (see Chapter 6).

In most cases, the principal drawback of reusability is that it adds a significant mass penalty to an already performance-stretched system. Furthermore, although the operations of the Shuttle are beginning to show that reusability permits some reliability improvement (e.g. by permitting post-flight subsystem check-out and consistent upgrades or design fault correction), there is no strong evidence that fully reusable systems will offer radically higher reliability and/or permit major cost reductions. In the current market, the costs of insurance are not dominant. Moreover, the costs of the launch itself may be only one third of the payload cost, or less. From a user point-of-view other factors, such as a reduction in the time between commissioning and launch date or the orbit injection accuracy, may be more important launch service considerations.

In this chapter, we have sought to identify the complex tasks involved in transporting a payload from the Earth surface to a designated orbit. We have outlined many of the basic principles that determine launch vehicle design and performance and, in so doing, constrain both payload and mission. On present evidence, rocket technology will continue to underpin this process and the constraints identified seem likely to persist for at least the next decade.

REFERENCES

[1]   Vinh, N. X. (1981) *Optimal Trajectories in Atmospheric Flight*, Elsevier, New York.
[2]   Ruppe, H. O. (1966) *Introduction to Astronautics*, **Vol. 1**, Academic Press, New York.
[3]   Dorrington, G. E. (2000) Optimum thrust-to-weight for gravity-turn trajectories, *J. Spacecraft Rockets*, **37**(4), 543–544.
[4]   White, J. F. (1962) *Flight Performance Handbook for Powered Flight Operations*, John Wiley & Sons, London.
[5]   ARIANE-4 User Manual (1994) Arianespace.
[6]   ARIANE-5 User Manual (1998) Arianespace.
[7]   Isakowitz, S. J., Hopkins, J. P. and Hopkins, J. B. (2000) *International Reference Guide to Space Launch Systems* (3rd edn), AIAA, Reston, VA.
[8]   VEGA User Manual, (1998) NPG/98/016 Iss. 2, ESA.
[9]   Loh, W. H. T. (1968) *Re-entry and Planetary Entry*, Springer-Verlag, Berlin.
[10] Allen, H. J. and Eggers, A. J. (1958) *A Study of the Motion and Aerodynamic Heating of Missiles Entering the Earth's Atmosphere at High Supersonic Speeds*, NACA TR 1381.
[11] Lees, L. (1956) Laminar heat transfer over blunt bodies at hypersonic flight speeds, *Jet Propulsion*, **26**(4) 259–269.
[12] Eldred, C. H. *et al.* (1997) Future space transportation systems and launch vehicles, in *Future Aeronautical and Space Systems*, ed. Noor A. K. and Venneri S. L., Progress in Astronautics and Aeronautics, **Vol. 172**, Chapter 8, pp. 373–408, AIAA, Reston, VA.

[13] Sumrall, J. *et al.* (1998) *VentureStar reaping the benefits of the X-33 program, IAF-98-V.3.03*, *49th Int. Astro. Congress*, Melbourne.

[14] Powell, R. W. and Cook, S. A. (1998) *The road from the NASA access-to-space study to a reusable launch vehicle, IAF-98-V.4.02, 49th Int. Astro. Congress*, Melbourne.

[15] Harmon, T. (1999) *X-33 Linear Aerospike on the Fast Track in Systems Engineering*, AIAA-99-2181, *35th Joint Propulsion Conference*, Los Angeles.

8 SPACECRAFT STRUCTURES

John M. Houghton

Astrium, Stevenage, UK

8.1 INTRODUCTION

The discipline and methods of spacecraft structural design are based upon aircraft design with a stronger emphasis upon minimum weight, vibration interaction and material-selection considerations for space use. The major goals of minimum mass and maximum reliability must be met with minimum cost and schedule. The structure must achieve its goals for the static and dynamic loading of the testing and launch phases, and finally, in the zero gravity operational environment. What makes spacecraft structural engineering perhaps unique is that its goals are strongly dependent on other subsystems such as thermal design, attitude control, communications and power.

Structural design does not only encompass material selection and configuration but must also include analysis and verification as part of the process, with an increasing reliance being placed upon analysis as experience grows.

8.2 DESIGN REQUIREMENTS

The first step in the structures engineering process for any spacecraft project should be the gathering together of all the requirements and relevant information into a specification. This is generally done for all the larger projects and is a worthwhile exercise for even the smallest project. It is good discipline to check whether any preconceived ideas are, in fact, the best solution for the requirements. The following list gives the major points one would expect to find in a specification:

- launch vehicle interface
- load paths
- mounting location for equipment
- launch loads
- on-station loads

- test loads
- safety factors
- stiffness requirements
- environmental protection
- alignment
- thermal and electrical paths
- accessibility
- mass
- fracture control.

The emphasis placed on each of these points can vary considerably from mission to mission. This variation can have a considerable influence upon the design of a spacecraft structure.

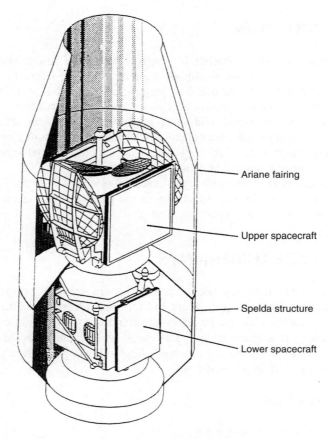

Figure 8.1 Dual launch system showing Ariane 4/Spelda (Reproduced by permission of Astrium Ltd)

For example, an observation spacecraft may require high-pointing accuracy and protection of sensitive instruments from station-keeping momentum wheel microvibration on orbit, while for a high-powered communication spacecraft the emphasis may be on thermal and electrical conductivity. This could require heat pipes to be embedded in structural panels that may be extended to give additional radiator area. Each of the above points will be considered in turn to give a feel for the factors involved.

8.2.1 Launch vehicle interface

The selection of a launch vehicle plays a major role in setting geometric and mass limits on the design. Some launchers have the capability of launching more than one spacecraft at a time. Figure 8.1 shows an arrangement for a dual launch on Ariane. Note that there is no connection between the fairing and the spacecraft. All accelerations and conse-quent inertia loads (except acoustic) are imparted to the spacecraft *via* the launch vehicle interface. There are similar arrangements for triple launches while Figure 8.2 shows that up to eight smaller spacecraft may be launched by adaptors such as the ASAP5 for Ariane 5.

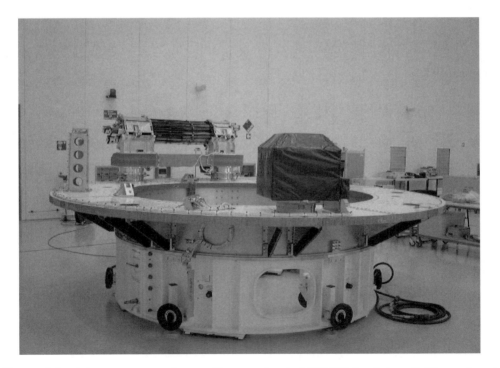

Figure 8.2 Ariane structure for auxiliary payloads (ASAP5) for Ariane 5 (Reproduced by permission of Astrium Ltd)

Expendable launch vehicles

Competition between launch vehicle suppliers encourage the use of standard interfaces giving the customer the option of using alternative launch vehicles with the same interface design. A commonly used adaptor is shown in Figure 8.3. Figure 8.4 shows a sectional view of a manacle clamp used to attach the spacecraft.

Up to 12 accurately machined clamp blocks are placed to form a segmented ring over correspondingly accurately machined wedge shaped lips on both the launch vehicle and the spacecraft side of the interface. The clamp band is then tightened over these blocks by two pyrotechnic bolts. This gives an even distribution of load around the circular launch vehicle interface and a reliable release mechanism. Many launch vehicles also offer discrete pyrotechnic bolt interfaces. These use between three and eight bolts on a fixed pitch circle diameter. This type of interface concentrates the load at these discrete bolt positions and reduces release reliability by increasing the number of pyrotechnics used. It also avoids the need for spacecraft interface ring strengthening to resist crippling under the clamp band tension of the previously described interface. It also allows the use of a non-circular interface if desired. For small spacecraft (less than 100 kg), another alternative is a simple bolted interface on a fixed pitch circle diameter with spacecraft release being provided by a line charge device, which is part of the launch vehicle adaptor.

Space Shuttle

The Shuttle presents a variety of ways in which equipment can be launched. In some cases, the spacecraft is designed solely for a Space Transportation System (STS) launch (see Figure 8.5).

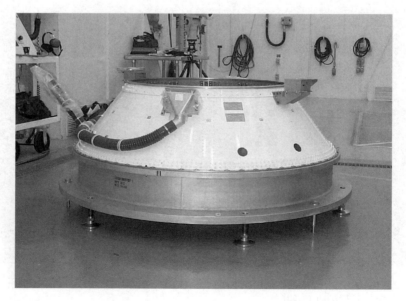

Figure 8.3 Launch vehicle to spacecraft adaptor (Reproduced by permission of Astrium Ltd)

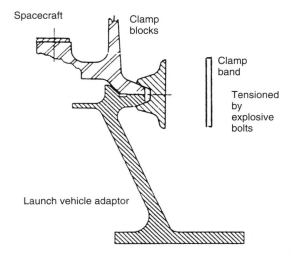

Figure 8.4   Spacecraft attachment clamp band
(Reproduced by permission of Arianespace)

Figure 8.5   Basic structure of UARS (upper atmosphere
research satellite) (Courtesy NASA/GE)

This class of spacecraft will attach to the discrete point standard Shuttle interfaces shown in Figure 8.6.

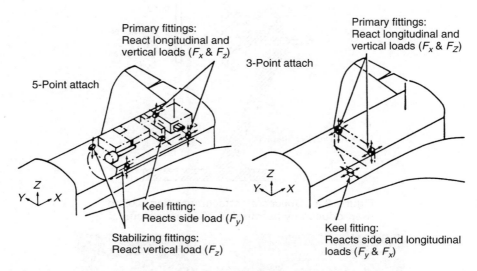

Figure 8.6 Five-point and three-point payload attachment methods for Shuttle (Courtesy NASA)

Alternatively, the spacecraft may be configured for launch by either expendable launch vehicles or the STS. For smaller spacecraft (up to 2 tonnes), the payload adaptor indicated in the upper view of Figure 8.7 is used. This aligns the longitudinal axis of the spacecraft perpendicular to the Shuttle thrust axis thereby generating high lateral loads on the spacecraft, see Table 8.2 and notes. Larger spacecraft (greater than 2 tonnes) will use the larger adaptor as shown in the lower half of Figure 8.7. Shuttle payload adaptors may incorporate a spin table and a boost motor that leaves the Shuttle, with the boost motor subsequently injecting the spacecraft into a higher orbit.

8.2.2 Load paths

Since all launch inertia loads pass through the launch vehicle interface, the choice of interface has a strong influence upon the configuration design of the structure. One route to an efficient structural design is to provide continuity to this circular load path with a circular central thrust structure in the spacecraft. Additional structures must then be designed to carry the launch inertia loads, applied in any direction, from the spacecraft equipment configuration on the most direct and efficient path to the central structure. This may be seen in many of the configuration examples given in Section 8.8.

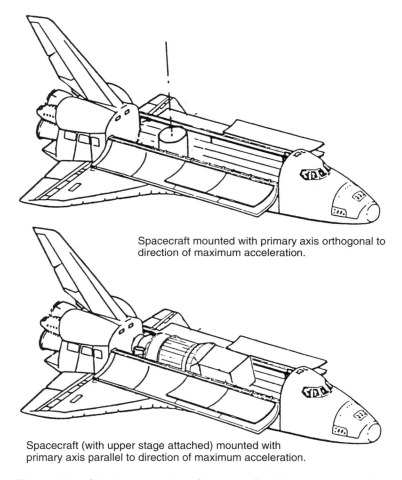

Spacecraft mounted with primary axis orthogonal to
direction of maximum acceleration.

Spacecraft (with upper stage attached) mounted with
primary axis parallel to direction of maximum acceleration.

Figure 8.7 Shuttle mounting of spacecraft, which are compatible
with expendable launchers

8.2.3 Mounting location for equipment

A flat bolted interface is used for most items of equipment dictating the need for large
flat areas on the structure to carry a significant number of equipment units. The flat
panels seen in many of the examples in Section 8.8 are designed primarily for this
purpose. Large areas of panel may need to be supported by perpendicular panels or
struts, not only to strengthen the panel but also to limit the magnitude of vibration
for equipment.

The primary structure may have to be specifically designed to carry large or heavy
equipment efficiently, such as larger propellant tanks, requiring strong and stiff attachments.

8.2.4   Launch loads

Launch loads are divided into the following categories (see also Chapter 2). All are treated slightly differently and all are usually critical for different parts of the structure. They are generated by different flight events.

Quasi-static loads

This term is used to describe loads that occur at the same level throughout the whole spacecraft. They include a vibration or transient component but the frequency content of this will be significantly below the natural frequencies of the spacecraft and therefore, will not cause significant magnification of acceleration through the spacecraft. Maximum quasi-static loads usually occur at a main engine shutdown rather than at lift-off. Just before engine shutdown the thrust is still a maximum, but the launcher fuel mass is at a minimum—accelerations are therefore a maximum. Quasi-static loads are usually critical for the major load bearing parts of the structure such as the central structures shown in the configuration examples. Tables 8.1 and 8.2 give the quasi-static loads for Ariane 4 and a Shuttle payload assist module, respectively.

Table 8.1   Launch quasi-static loads for Ariane 4

Flight event	Acceleration (g) Q.S.L.	
	Longitudinal	Lateral axis
Maximum dynamic pressure	−3.0	±1.5
Before thrust termination	−5.5	±1.0
During thrust tail-off	+2.5	±1.0

Note: The minus sign with longitude axis values indicates compression.

Lateral loads may act in any lateral direction simultaneously with longitudinal loads.
These loads apply uniformly all over the primary structure of a spacecraft complying with

- the frequency requirements and
- the static moments.

The lateral loads for Shuttle are higher than for the expendables represented by Ariane 4. High lateral loads produce higher central structure stresses than a similar level of axial load. The Shuttle cases, therefore, require a stronger structure than the Ariane 4 cases.

Table 8.2   Launch quasi-static loads Shuttle/PAM-D2 (Reproduced by permission of Arianespace)

Spacecraft weight: 1247 to 1950 kg
(2750 to 4300 lb)

Condition	Limit load factor (g's)	
	Lateral	Axial
Maximum lateral	5.2	3.3
Maximum axial compression	3.7	4.7
Maximum axial tension	5.2	−3.3

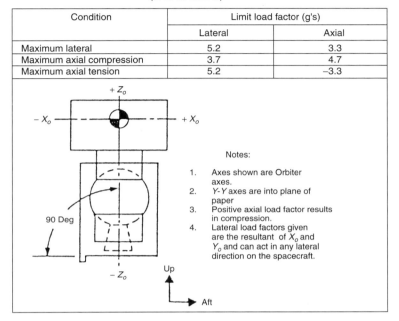

Notes:

1. Axes shown are Orbiter axes.
2. *Y-Y* axes are into plane of paper
3. Positive axial load factor results in compression.
4. Lateral load factors given are the resultant of X_o and Y_o and can act in any lateral direction on the spacecraft.

Sine vibration

The sine vibration environment for Ariane 4 is given in Table 8.3. It is a relatively simply defined envelope of all the low frequency transient vibrations that may occur during launch. For a qualification sine vibration test, a spacecraft in launch configuration is attached to a large vibration test machine or 'shaker'. The shaker starts vibrating at 5 Hz. The frequency of the vibration is increased at a rate of 2 octaves per minute. An octave is a doubling of frequency. Therefore, in one minute the frequency increases from 5, through 10 to 20 Hz. This continues until 100 Hz is reached when the test stops.

Table 8.3   Sine vibration environment for Ariane 4

	Frequency range (Hz)	Qualification levels (0 to peak) (recommended)	Acceptance levels (0 to peak)
Longitudinal	5 to 6	8.6 mm	1 g
	6 to 100	1.25 g	1 g
Lateral	5 to 18	1 g	0.8 g
	18 to 100	0.8 g	0.6 g
Sweep rate		2 oct./min	4 oct./min

Any lightweight spacecraft will have resonant frequencies in this range, which can cause responses in the spacecraft orders of magnitude greater than the specified input. If these responses are in excess of the predicted launch transient responses, the defined test input may be reduced, or *notched* at critical response frequencies by agreement with the launcher agency. Such an agreement usually requires demonstration by the spacecraft and launcher coupled loads transient analysis that the reduced response within the spacecraft in the notched sine test is still greater than that to be expected from launch transients. This coupled loads analysis is a normal part of launch analysis and places a strong emphasis upon the accuracy of the mathematical model of the spacecraft supplied by the spacecraft designer.

Figures 8.8 and 8.9 show two results from a spacecraft's coupled loads analysis. They both show axial force at the spacecraft interface varying with time over 1.5 s from engine thrust cut off.

The decay transient vibration content caused by engine cut-off is smaller and a lower frequency for the first stage compared with the second. This is because the launcher mass is much higher than it is at second stage cut off. The transient magnitude is so great in second stage cut off that the force becomes negative (in this case tensile). This is the reason for the specification of a tensile case in the quasi-static loads (Table 8.1).

These are just two examples of coupled loads analysis recovery. Hundreds of accelerations, displacements and forces may be recovered for many different flight events.

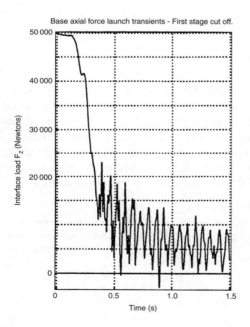

Figure 8.8 Base axial force launch transient for Ariane 4 first stage cut off (Reproduced by permission of Arianespace)

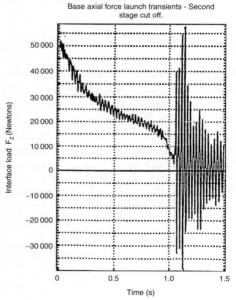

Figure 8.9 Base axial force launch transient for Ariane 4 second stage cut off (Reproduced by permission of Arianespace)

Acoustic noise and random vibration

The largest acoustic noise excitation occurs at the point of lift-off when the reflected noise from the launch pad and ambient air pressure are greatest. An Ariane 4 acoustic noise spectrum is shown in Table 8.4

Table 8.4 Spacecraft acoustic environment for Ariane 4

Octave band (centre frequency Hz)	Qualification level (recommended) (dB)	Acceptance level (flight) (dB)	Test tolerance (dB)
31.5	124	120	−2, +4
63	131	127	−1, +3
125	139	135	−1, +3
250	143	139	−1, +3
500	138	134	−1, +3
1000	132	128	−1, +3
2000	128	124	−1, +3
4000	124	120	−4, +4
8000	120	116	−4, +4
Overall level	146	142	−1, +3
Test duration	2 min	1 min	—

Note: Reference level 0 dB $= 2 \times 10^{-5}$ Pascal.

Octave bands are used to partition the spectrum. Note the doubling of centre band frequency moving down the left-hand column. Acoustic noise can be critical for the design of lightweight structures with large area and low mass such as dish type antenna reflectors and solar array panels. When the resonant frequency of such items is known, the sound pressure level at the corresponding centre band frequency can be used to calculate the magnitude of response. Acoustic test results from similar previous designs are often used to estimate the responses of complex structures.

Structural response to acoustic noise is predicted and measured in terms of random vibration. Random vibration is not completely random. Its magnitude through the frequency domain is expressed in terms of power spectral density. This is the mean square of acceleration in each one Hertz bandwidth of the spectrum. Random vibration is also transmitted to the spacecraft through the launch vehicle structure. Table 8.5 gives a spacecraft random vibration spectrum specified for spacecraft launched on Ariane 4.

Table 8.5 Spacecraft high frequency random vibration for Ariane 4

	Frequency range (Hz)	Power spectral density (g²/Hz)	RMS value (g)
Qualification levels	30–150	+6 dB/oct.	
(recommended)	150–700	0.09	11
	700–2000	3 dB/oct.	
Acceptance levels	30–150	+6 dB/oct.	
	150–700	0.04	7.3
	700–2000	3 dB/oct.	

Note: The levels are identical for longitudinal and lateral vibrations. The test duration is 2 min per axis for qualification, and 1 min per axis for acceptance testing.

Spacecraft random vibration testing is usually only performed on very small spacecraft. For larger spacecraft (greater than 1000 kg), a spacecraft acoustic noise test will generate random responses greater than those of a spacecraft's random vibration test. Random vibration testing is widely used during development and qualification of spacecraft equipment to be sure that no problems will be found at the relatively late stage of integrated spacecraft acoustic noise testing. The required test level is judged from previous test experience, and an example is given in Table 8.6.

Table 8.6 Equipment random vibration spectrum from spacecraft acoustic testing (1 min protoflight, 2 min qualification)

Frequency (Hz)	Level
20–150	+ 6 dB/oct.
150–1000	0.25 g^2/Hz (18.7 g RMS)
1000–2000	−6 dB/oct.

Note: Derived from Ariane 4 experience.

Figure 8.10 superimposes this equipment test level on the response during the eventual spacecraft acoustic test showing how close the acoustic test response is to the equipment random vibration test at 140 Hz, and that an equipment test notch from 600 to 1000 Hz was justified.

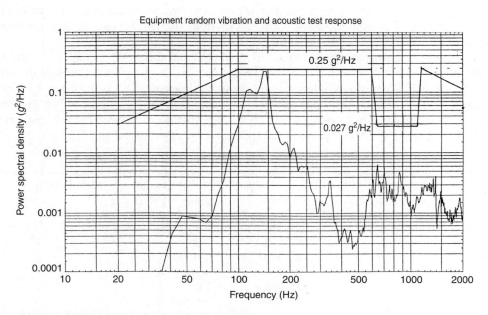

Figure 8.10 Equipment random vibration test and acoustic response, derived from Ariane 4 experience (Reproduced by permission of Arianespace)

Shock loads

Figure 8.11 shows a shock response spectrum for the separation clamp band bolt firing on Ariane 4.

This is a conventional format showing how a simple 'spring and mass' single degree of freedom system with a dynamic magnification factor of 10 would respond if its natural frequency occurred at any frequency on the horizontal axis. The accelerations are alarmingly high but would only happen to a very stiff item with a strong resonance at the given frequencies. High frequency shock energy is attenuated very rapidly with distance from the source and by structural joints between the shock source and the responding item. High frequency shock is usually of more concern to the function of equipment such as relays or glass oscillators rather than structural strength.

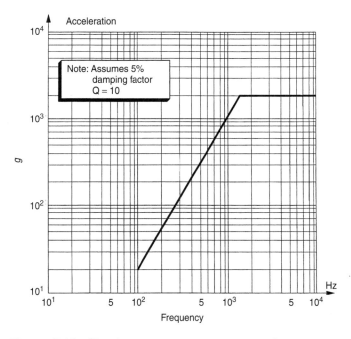

Figure 8.11 Shock response spectrum at the separation plane for Ariane 4 launch (Reproduced by permission of Arianespace)

8.2.5 On-station requirements

On-station loads are of a much smaller magnitude than launch cases but they can have an equally demanding effect upon the structure design. Structural transmission of *microvibration* from sources, such as momentum wheel bearing rumble or thruster firing, to sensitive equipment such as lasers or telescopes may be of critical concern.

Large appendages, such as antenna reflectors or solar array panels that are stowed for launch and then deployed on-station, may have a very low natural frequency when deployed. A minimum deployed natural frequency between 0.5 and 2 Hz is often required to avoid attitude control instability. Although a very low frequency, this requirement may be difficult to meet. Consequently it may be critical for the design of the deployed appendage interface and local backing structure.

A further point is the avoidance of inadvertent pressure vessels. It is easy to forget that items such as strut tubes and honeycomb core cells when manufactured in atmospheric pressure become pressure vessels in the vacuum of space. They must be vented, or if venting is not practicable, designed as a pressure vessel.

Structures are increasingly using non-metallic materials, particularly for thermal finishes. All non-metallic materials must be space-qualified, primarily with respect to outgassing under sun light in vacuum. Release of volatiles is doubly undesirable, since they may degrade the performance of the residual material and they may redeposit on adjacent sensitive equipment.

8.2.6   Test loads

A spacecraft will be tested to be sure that it is adequate for all the previously mentioned loads unless the safety margins are very high or a similar structure has previously been tested. Testing will apply higher magnitude loads than flight to give confidence that minor variations in load, material properties and build differences will not cause failure. Also, the test method may not be able to simulate the conditions of flight exactly, and over-test in some areas may be necessary to ensure adequate testing in others.

It is essential that the structure designer has a clear test plan before the structure design is completed to be sure that the structure is adequate for test as well as flight.

8.2.7   Safety factors

The maximum load that can occur in service with worst weather conditions, roughest launch vehicle vibration and so on is termed the *limit load*. To be sure that structural strength is not a concern at the time of launch, the structure is designed and qualification-tested to levels in excess of the limit load. The first safety level above limit is the *proof level*. This is equal to limit load multiplied by a *proof factor*. The most frequently used proof safety factor is 1.1, but factors between 1.0 and 1.5 are also used. The structure must not sustain any significant permanent deformation once a proof load has been applied and removed. The second safety level above limit is the *ultimate level*. A minimum ultimate safety factor is 1.25 for Ariane 4. Factors up to 4.0 are used for some pressure vessels. The structure must not rupture, collapse or undergo any gross permanent deformation under ultimate load. A commonly used method of expressing the adequacy of a structure under these failure criteria is the *Reserve Factor*. A Reserve Factor at any critical location is equal to the allowable load or stress divided by the applied load or stress. If the Reserve Factor is greater than 1.0, then the structure is adequate for the purpose by this particular criterion. Another widely used term is *Margin of Safety*. A Margin of Safety is equal to the Reserve Factor minus 1.0. A positive Margin of Safety indicates adequacy, and a negative one, inadequacy. On top of the above safety factors some safety agencies require minimum margins on particular failure modes. For example, a minimum margin of 0.25

may be required on first ply failure for composite materials or slender structure buckling. The level of the safety factors may be seen as a measure of the mass criticality of the structure *versus* the schedule and cost risk that a customer is prepared to take.

8.2.8 Stiffness requirements

To be sure that the quasi-static and transient vibration's responses discussed under sine vibration do not exceed the levels indicated in the launch vehicle preliminary design manual, the spacecraft minimum natural frequency requirements must be well separated from those of the launch vehicle forcing functions. Table 8.7 gives the minimum frequency requirements for spacecraft mounted on Ariane 4 or Delta III. If these requirements were not met, then the spacecraft dynamic coupling with the launch vehicle would be stronger, causing the quasi-static loads and dynamic transients to increase. The table also gives notional minimum frequency requirements for major spacecraft appendages and equipment. The spacecraft system designer may change these figures in conjunction with the corresponding quasi-static loads from the results of the spacecraft sine response analysis.

Table 8.7 Minimum natural frequency of vibration requirements (Hz)

Axis	Spacecraft		Appendages	Equipment
	Ariane 4 launch	Delta III launch		
Thrust	31	35	>80	>100
Lateral	10	15	>60	>100

8.2.9 Environmental protection

When the structures designer may consider a framework structure the most appropriate, a requirement for micrometeorite or radiation protection may dictate a *monocoque* (loaded skin) structure with a composite section such as shown in Section 8.7.

8.2.10 Alignment

The accuracy of alignment required can vary widely with mission, from a broad tolerance for a worldwide communication spacecraft to very tight tolerance for accurate location pointing. For missions that require high-pointing accuracy, thermal distortion will require careful attention. Distortion has three main sources. First, all alignment and testing of the spacecraft will be done on earth under a 1 g loading. The structural elastic distortion between critical equipment generated by the 1 g environment must be calculated and possibly verified by mathematical model correlation with static load testing. Secondly, there will be a small movement or hysteresis in the structure as a result of launch vibration. This can usually only be judged from previous test experience. Thirdly, when on-station, there will be temperature variations through the structure and differences in temperature from the time of ground alignment. The distortions these generate through the material expansion coefficient may be calculated using the structure finite element model.

8.2.11   Thermal and electrical paths

Thermal control may require heat pipes to be embedded in structural honeycomb panels or thick conductive doublers added to dissipate heat from local high-powered equipment. Thermal conductance or isolation may be required at structural joints.

The structure may be required to provide a ground return path for electrical circuits. Electromagnetic compatibility (see Chapter 16) may be required to prevent electrostatic discharge. Both these requirements give a preference for metallic structures, but composites may still be a better option, with electrical strapping.

8.2.12   Accessibility

All stages of spacecraft assembly integration and test must be carefully considered with any likely equipment refurbishment to establish the level of accessibility required inside the structure and whether any parts must be interchangeable, testable or transportable with equipment installed.

8.2.13   Mass

The engineering effort expended on a spacecraft's structural design will vary considerably with the mission requirements. When attempting to achieve a minimum mass, it is worth estimating the return on effort, development and manufacturing cost. For the example of a geosynchronous communications spacecraft, the most commonly used Western World expendable launch vehicles cost approximately 25 thousand dollars to launch 1 kg of spacecraft mass to transfer orbit. This can be overridden by absolute mass limits, which may demand high development costs or schedule constraints, which may in turn limit development and manufacturing time.

8.2.14   Fracture control

Fracture control is primarily a Shuttle requirement for safety of a man-rated vehicle. It is also required on pressure vessels for launch range safety of expendable launch vehicles.

The method requires that a structure be subjected to careful crack detection inspection. It is then assumed that cracks may exist, in the most critical locations, at the crack detection limit of the method used. Figure 8.12 gives the detectable limit defined by NASA standards for dye penetrant inspection of a semi-elliptical surface flaw. A *standard initial flaw* is one that can be detected by a competent trained inspector. A *special initial flaw*

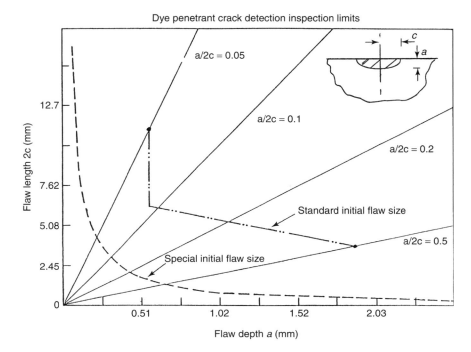

Dye penetrant crack detection inspection limits

Figure 8.12 Dye penetrant crack detection inspection limits (Reproduced by permission of NASA)

is one that can be detected by an experienced inspector who has passed an examination by NASA representatives using samples that contain known flaws.

Fracture-control analysis uses a concept called stress intensity, which is equal to the stress, times a geometric factor for location, multiplied by the square root of the characteristic crack length. The whole essence of fracture mechanics is coupled to strain energy, stress concentration and material state [1].

Crack propagation analysis is performed to predict the crack growth using empirical data, which shows that a crack will grow a tiny amount every time a load or stress is applied and removed. No safety margin factor is applied to the load magnitude, but a safe life factor of 4 is applied to the actual number of load cycles in the service spectrum to be sure that a safe life margin exists. The structure is considered to be *damage tolerant* if the crack does not grow to critical size after application of this load spectrum. A critical crack is one that will give unstable growth when the limit load is applied. Unstable growth will result if the applied stress intensity is greater than the material fracture toughness. ESA has published a computer program with manual [2]. This is based upon a NASA program called FLAGRO that embeds industry standard methods, data and techniques to perform safe life analyses.

8.3 DESIGN PROCESS

Structure design is an iterative process that starts with the initial spacecraft concept and nominally ends at a predetermined 'freeze date' at the start of the construction phase (in practice the active role of design tends to end only shortly before shipment to the launch site). With such multi-disciplinary projects it is convenient to specify a number of formal design reviews to assess the interaction of all the separate design factors and subsystems. Indeed, the design process benefits greatly from concurrent engineering with other subsystems. At the peak of design activity, this can occur on a nearly daily basis. This activity (see Figure 8.13) must proceed rapidly over complex technical (and possibly geographical) interfaces and benefits considerably from the use of sophisticated computer-aided design methods.

The concept of a load path from the interface with the launch vehicle (from which all accelerations are imparted) through the spacecraft structure to the mounting points of individual systems or units is followed. At each mounting point, individual sets of interface requirements (alignment, thermal, field of view, screening, connections and accessibility) will be generated, thus setting some of the constraints to be applied to the design of the structure.

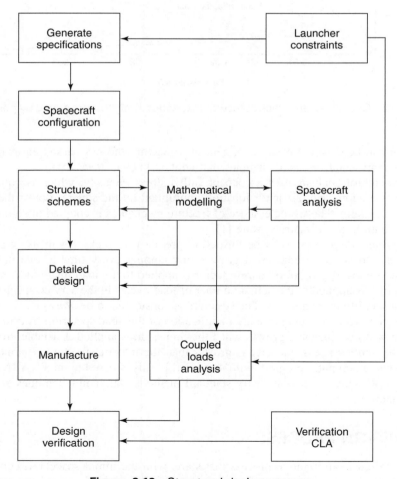

Figure 8.13 Structural design process

Thus far, the problem of providing a structure to meet the specification could be regarded as relatively straightforward. However, it is the need for extreme mass efficiency and reliability, which presents a challenge.

A *scheme design* is produced to give a definition of the concept and materials to be used. The completion of this should coincide with a *preliminary design review*, where the way in which the design will meet its requirements can be seen, together with the extent of development work required to realize the finished product. The first finite element model of the structure will be produced. As soon as there is reasonable confidence that there will be no major changes in configuration, mass or stiffness, a spacecraft mathematical model will be submitted to the launch vehicle authority for couple loads analysis.

Having agreed and frozen the scheme design, the detailed design phase will commence with the production of manufacturing drawings. When all detailed design drawings and analyses have been completed, a *critical design review* will be held to be sure that the design meets its requirements in every detail. A *Structure Strength Summary* will usually be provided for this review, summarizing the strength margins throughout the structure.

When a first flight, or flightlike, model has been produced and subjected to a verification test, the results will be correlated with the finite element model. With a test-verified model, a final coupled loads analysis is performed, usually within six months of the launch, to be sure that the design and qualification testing are within safe limits.

8.4 MATERIAL SELECTION

8.4.1 Material properties

The selection of an appropriate material for an application requires knowledge of the way each property of the material can be best used and where each limitation must be recognized.

Selection criteria can encompass the following:

- specific strength
- specific stiffness
- stress corrosion resistance
- fracture and fatigue resistance
- thermal expansion coefficient and conductivity
- ease of manufacture.

Specific strength

An elementary comparison of materials can be made by examining their proof strengths. This is the stress at which a material will sustain a permanent deformation of 0.1 or 0.2% when the load is removed. Specific proof strength is this stress divided by the material density. Values for a selection of widely used lightweight materials are given in Table 8.8. For metals, titanium shows up well but it is only used where strength is the dominant criteria. Strength is dominant at joints and areas of local load concentration. For all other

Table 8.8 Sample materials properties

C. R. corrosion resistance
S. C. C. stress corrosion cracking

Caution: There is considerable variation of the properties of materials according to conditions (ageing, temper, form and structure orientation). Consult manufacturers' data

	Density (kg/m³)	Young's modulus E (GPa)	Yield strength f (MPa)	Selection criteria				Thermal expansion (μm/m K^{-1})	Fracture toughness (MPa√m)	Fatigue strength (MPa)	Comment
				E/ρ	$E^{1/2}/\rho$	$E^{1/3}/\rho$	f_y/ρ				
Aluminium alloy											
6061.T6	2700	68	276	24	2.9	1.5	98.6	23.6	186	97	Good C.R
7075.T6	2800	71	503	26	3.1	1.5	186.3	23.4	24	159	Prone to SSC in T6 Form
Magnesium alloy											
A2 31B	1700	45	220	26	3.9	2.1	129.4	26			Prone to SCC
ZK 60 A.T5 extrn	1700	45	234	26	3.9	2.1	137.6	26		124	
Titanium alloys											
T1-6A1-4V (annealed)	4400	110	825	25	2.4	1.1	187.5	9	75	500	
(solution treated and aged)			1035						42	690	
Beryllium alloys											
S 65 A	2000	304	207	151	8.7	3.4	103.5	11.5			Hot pressed sheet } Low fracture toughness
SR 200 E			345								
Ferrous alloys											
INVAR		150	275/415					1.66			Low expansion Ferromagnetic
Stainless steel											
AM 350 (SCT850)	7700	200	1034	26	1.84	0.8	134.3	11.9	40/60	550	Austenitic
304L Ann	7800	193	170	25	1.8	0.7	21.8	17.2			
Composites											
KEVLAR 49 0° (Aramid fibre) 90°	1380	76*	1379†	55	6.3	3.1	999.3	−4			Structure members Pressure vessels Rocket casings
	1380	5.5	29.6	4	1.7	1.3	21.4	57			
Graphite epoxy sheets (undirectional) GY70/934	1620	282	586	174	10.4	4.0	361.7	−11.7 (Longitudinal) 29.7 (Transverse)			Sheet
Column ref. (see text)	A	B	C	D	E	F	G	H	I	J	

*Tensile modulus
†Tensile strength

applications of lightweight structures, overall strength is determined by buckling. The material property that determines buckling strength of slender structures is stiffness.

Specific stiffness

For the three common cases shown in Figure 8.14, the stiffness has been expressed in terms of the material properties (Young's modulus and density). This then allows an examination and selection of materials to be made [3]. For example, Figure 8.14 shows the efficiencies of selected materials for three different load cases. The first case, deflection of a beam, is representative also of structures designed to meet a minimum natural frequency of vibration requirement. Notice how similar the efficiencies for titanium and aluminium alloys are despite the greater strength of the former. Ashby and Jones have given an explicit treatment of this subject together with illuminating case studies [4].

Case	Deflection of a beam	Buckling of a strut	Buckling of a panel
			(w fixed) (t and L variable)
Characteristic equations	$\delta = \dfrac{PL^3}{3EI}$ $I = \dfrac{wt^3}{12}$ Weight = $L.W.t.\rho$.	$P_e = \dfrac{\pi^2 EI}{L^2}$ $I = \dfrac{\pi d^4}{64}$ Weight = $\dfrac{\pi d^2}{4} L\rho$	P = const. $\dfrac{E}{1-\sigma^2}\left(\dfrac{t}{W}\right)^2$. tw Weight = $w.t.L.\rho$
Structural efficiency $= \dfrac{\text{load}}{\text{weight}}$	$\dfrac{P}{WT}$ = const. $\left(\dfrac{\delta t^2}{L^4}\right)$. $\dfrac{E}{\rho}$	$\dfrac{P}{WT}$ = const. $\left(\dfrac{P_e}{L^4}\right)^{1/2}\dfrac{E^{1/2}}{\rho}$	$\dfrac{P}{WT}$ = const. $\left(\dfrac{P_e}{L^2}\right)^{2/3}\dfrac{E^{1/3}}{\rho}$
Structure loading coefficient	$\left(\dfrac{\delta t^2}{L^4}\right)$	$\left(\dfrac{P_e}{L^4}\right)^{1/2}$	$\left(\dfrac{P}{L^2}\right)^{2/3}$
Material efficiency criterion	$\dfrac{E}{\rho}$	$\dfrac{E^{1/2}}{\rho}$	$\dfrac{E^{1/3}}{\rho}$
Nomenclature	P = Load δ = Deflection σ = Poisson's ratio (0.3)	P_e = Critical buckling load (Euler theory) L = Length	d = Diameter ρ = density w = Width WT = Weight t = thickness E = Young's modulus

Figure 8.14 Material selection criteria — stiffness

   Materials show variations due to treatment and supply state, and therefore, any application considered should be accompanied by a proper detailed examination of the properties from specialist sources.
   Consideration needs to be given to the availability of materials, difficulties of manufacture, the method by which fasteners are attached and how late modifications would be conducted.

Stress corrosion resistance

In a terrestrial environment, most alloys are susceptible to *stress corrosion cracking* (SCC). Materials under sustained tensile loading, particularly in the short transverse grain direction, may exhibit microcracking at grain boundaries. In the presence of a corrosive medium or moisture, these cracks can develop.

Tensile loading conditions can exist even when in storage, due to weight or residual strain following assembly. The potential effect is a failure occurring at stress levels below values normally considered safe for the material. An ESA design guideline [5] points out that the corrosive environment need not be severe and indeed failed parts may not show visible evidence of corrosion. It also gives a qualitative rating of commonly used metallic materials as having good, moderate or poor resistance.

Weld zones require particular consideration because of microstructure discontinuities in the heat affected areas, particularly if a filler material has been used.

A considerable history of control of SCC has been built up in the aviation industry and although spacecraft experience relatively benign environments and short storage times, with the exception of the Space Shuttle, many of the precautions and inspection procedures are applicable.

Precautions in design and material selection include

- choosing alloys less susceptible to SCC,
- using conservative stress levels,
- specifying the need for close inspection of areas of stress concentration and welds,
- heat-treating components to remove residual stresses due to manufacturing processes,
- avoiding material combinations that promote galvanic corrosion, and
- avoiding exposure to atmospheric conditions that can induce corrosion.

Sublimation/erosion

Normally the sublimation of metals does not pose any major problems in the space environment, although for thin films the rate at which their thickness decreases may be significant. Erosion by atomic oxygen, particularly in respect of polymeric materials in Low Earth Orbit (LEO), can be a significant factor in material selection [6,7]. This is a subject of investigation and research in view of the possible long-term problems that could ensue in the 25-year life of a space station.

Ease of manufacture/modification

Some materials present individual problems during component manufacture. For example, some forms of beryllium and the cutting fluids used present toxic material control problems, which preclude late modification in unsuitable facilities. It is not unusual for additional mounting points or cut outs for cable run clearances to be needed at a very late stage in integration. Thin sections may not support the modifications required. Fibre lay-ups may not allow holes to be cut, which interrupt the homogeneity of the load carrying part.

Advanced materials

The aerospace industry has driven the development of numerous advanced materials in the search for more mass efficient structures. In most cases cost is a significant factor since, particularly in the space sector, designs are very low volume/high quality applications. Many materials developments now fail to reach maturity simply on the grounds of economics. There are exceptions in which the need for a technical solution is paramount. A good example is the thermal protection system for the Space Shuttle, which uses silicon-based, and carbon–carbon materials.

A wide range of materials with developed characteristics is discussed in useful detail with a commentary on fabrication, jointing and applications in [8,9]. These materials can generally be grouped as follows:

- **Polymer composites**
  Continuous fibre reinforcements:
  >   Carbon
  >   Aramid
  >   Glass

  Matrix
  >   Epoxy
  >   Polyimide
  >   Bismaleimide

  Thermoplastic
  >   Carbon in polyether ether ketone (PEEK)
  >   Aramide-based (PEI).
- **Advanced metal matrix and ceramic materials**
  Magnesium alloys and their composites
  Aluminium alloys and their composites
  Titanium alloys and their composites
  Super alloys and their composites
  Intermetallic materials
  Refractory materials
  Beryllium
  Ceramic matrix composites
  Glass and glass-ceramic matrix composites
  Carbon–carbon matrix composites.

These materials offer substantial advantages for some specific requirements provided that their characteristics are well understood. Even so there may be pitfalls in using them in extreme circumstances. The drastic thermal gradients experienced in orbit give rise to concerns about cyclic fatigue and microcracking as a result of differential expansion of fibre and matrix constituents. Also, investigation of the long-term radiation effects upon polymer composites is the subject of current research.

8.4.2 Materials applications

Metals

Ferrous alloys have numerous applications in which their properties of high strength corrosion, resistance and toughness are required.

Austenitic stainless steels are used for propulsion and cryogenic systems because of their excellent low temperature toughness. Other alloys are used in optical and precision structures where properties can be selected to match expansion criteria in a dynamic thermal environment.

Susceptibility to hydrogen embrittlement is a potential hazard for ferrous alloys, particularly where they have been treated in plating solutions. The result is similar to SCC. The corrective treatment is a severe bake-out within a limited time period—observing of course that the materials are not affected.

Some types of stainless steel and invar are magnetic. This can be a problem when a spacecraft carries electromagnetic sensors.

Composite materials (fibre reinforced)

A designer is now able to tailor materials to give properties in chosen directions, much as a marine designer has been doing with wood and resin bonded plywood for a long time. In the use of fibre-reinforced materials (typically with boron or carbon), advantage can be taken of the high strength offered along the fibre. All strands of the chosen fibre are aligned parallel and held in an epoxy matrix. The resulting material exhibits good structural properties along the direction of the fibres but is limited to the properties of the matrix resin in other directions. Additional unidirectional plies can be aligned at an angle to the first and in multi-layered laminates it is possible to create a material with structural properties that are tailored to application. In doing this, it must be recognized that the high unidirectional strength and stiffness, such as is quoted in Table 8.8 will be reduced in any one direction by approximately the ratio of the number of plies in that direction to the number of plies in the complete laminate. Methods for the analysis of individual ply stresses and failure prediction of a laminate have been developed [10,11]. Computer programs are commercially available from agencies such as ESA, which enable much quicker analysis, particularly where optimization is required.

Carbon epoxy materials are used quite extensively in fabrication by filament winding of strut tubes and spacecraft central structures—in which the direction of loading is well defined. It is also widely used for large deployable antenna reflectors in which the combination of high stiffness, lightweight and low coefficient of expansion (to hold shape under temperature extremes) are particularly advantageous.

A note of caution, particularly with carbon fibre-based composites, is that hygroscopic absorption can add up to 2% water by weight in a normal atmosphere. Once exposed to the space environment, they loose the water and exhibit small dimensional changes. Methods of controlling this include total control of the component's environment, baking-out prior to alignment, and even plating to seal the moisture in or out. None of these methods are totally convincing. Careful choice of the resin system can also limit this effect.

Carbon composite materials are not ductile. Consequently, the designer must allow for the fact that stress concentrations produced by features such as holes, sudden changes of section, grooves or fillets will reduce the static strength by the magnitude of the corresponding stress concentration factor. This contrasts with the more generally used high-strength aluminium or titanium alloys, which have sufficient ductility to make them insensitive to most stress concentrations under static loading. For these alloys, stress concentrations are of more concern under cyclic fatigue loading.

For this reason and because complex shapes are more difficult to manufacture in composites, they tend to find application in relatively simple components—strut tubes but not end fittings, and central thrust structures but not attachment fittings and flanges. Complex components such as strut end fittings are sometimes made in composites, but this is usually when stiffness, and not strength, is the design driver.

A further caution should be addressed. Unlike metals, the tension and compression bulk modulus of composite materials may differ significantly, giving a non-linear dynamic characteristic.

Composite materials (metal matrix)

The limiting factors of an epoxy matrix (see above) can be substantially overcome by employing high-strength fibres in a diffusion bonded metal matrix. Although this technology may represent the ultimate direction for 'designed' materials, current costs are high and non-destructive testing (NDT) techniques need development. Further, some materials suffer from low yield strength or elongation to failure, which is an indication of brittleness, notch sensitivity and poor fatigue strength. However, applications are emerging, one particular example being lightweight mirrors [12].

In general, a clear rule to be borne in mind is to choose materials (or their close equivalents) that appear in the NASA or ESA guidelines [13,14]. This will give all the design and approval data necessary, and the confidence that no non-compliances will be discovered later in a programme when any required material qualification testing is completed. When this is not possible, national defence specifications should be used for guidance. In all cases, material tractibility, including treatment history from billet/raw material to finished product, is one of the quality assurance records to be maintained.

8.4.3 Section properties

Hollow or reduced section members such as tubes and 'I' beams give much better mass efficiency than solid bars. Panel deformations such as corrugations can give greatly increased stiffness and resistance to buckling compared with flat sections. It is the art of the design engineer to use materials most efficiently in this way when considering all the duties required of a section.

Honeycomb panels have relatively low weight and high bending stiffness. Figure 8.15 shows the general configuration of a honeycomb sandwich.

A variety of materials and material combinations may be employed with the combination of skin thickness and core depth and density necessary to give the desired overall section properties.

It is important to take care when designing the load attachment points for honeycomb panels. Figure 8.16 shows a widely used blind potted insert. The potting compound transmits bolt tension load to the core while bearing transfers of shear loads from the insert to the hole in the panel skin.

8.4.4 Quality assurance

The final proof of a structure's integrity involves it being tested and then inspected. The NDT of metals is a well-developed field but this is not yet true for composite structures. In the field of low volume production spacecraft structures, considerable care must be taken to ensure that integrity can be assured in the face of variability caused by individual

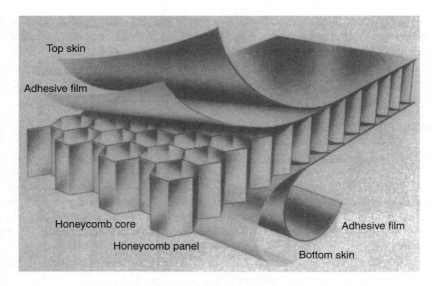

Figure 8.15   Honeycomb panel (Reproduced by permission of Hexcel Corporation)

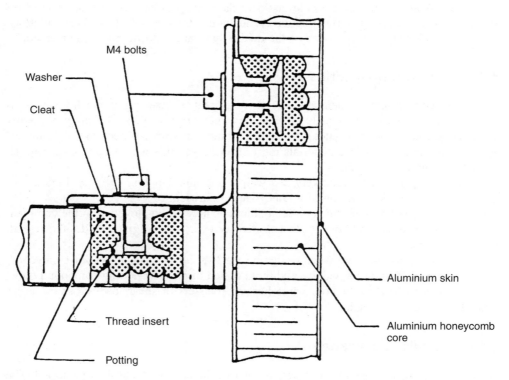

Figure 8.16   Inserts and panel cleat joints (Reproduced by permission of Astrium Ltd)

ply angle errors hidden within a laminate, bonding and complex sections. Added to this is the problem of ensuring that the structure has not degraded during test.

8.5 ANALYSIS

A *finite element model* for analysis of the structure is an essential part of the design development process. It serves the following purposes:

- to predict spacecraft overall natural frequencies;
- to predict the level of spacecraft response to low-frequency vibration;
- to determine load distributions internal to a structure between redundant multiple load paths;
- to be used by the launch vehicle contractor to determine the mission-specific maximum loads, accelerations and deflections and thereby to determine sine test notch levels;
- to apportion stiffness and corresponding strength requirements to sub-structured appendages.

A structure is notionally divided into areas or elements. The size of the elements chosen is important. Consider the modelled area shown in Figure 8.17.

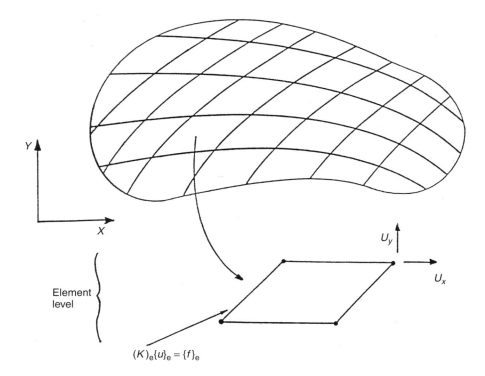

Figure 8.17 Finite element modelled area

The essence of the *Finite Element Analysis* (*FEA*) method is to divide the structure into a large number of discrete elements. Within each of these the load variations, elastic properties and boundary conditions are defined by particular mathematical functions [15]. For each element, appropriate parameters can be fed into the computer, describing material properties, shape, degrees of freedom and connection to the next element. The computer program enables the structure to be evaluated under static loads. It is also possible to conduct a dynamic study (i.e. under fluctuating loads). In a dynamic simulation, its natural frequency can be assessed and relative phase information of mode shapes of the structure can be indicated [16].

The elements chosen here are quadrilateral. Within each element the rate of change of stress is limited by the element formulation. Between any two nodes on the edge of an element there is a constant state of stress and strain. If it is known that stresses are varying very rapidly through the structure, then this mesh may not be fine enough to capture the peaks of stress. The temptation is, therefore, to use a very fine mesh to be sure that peaks are not missed. This approach must be tempered by the size of the complete mathematical model. Every corner point that may be connected to several elements is called a node. Every node has six degrees of freedom for movement—three orthogonal translations and three rotations. The total number of degrees of freedom in a complete model is six times the number of nodes (this may be reduced by introducing constraints that eliminate some unimportant degrees of freedom). It is the total number of degrees of freedom in a model that determines the size of the computation required. This, in turn, is limited by the power of the computer being used.

Models may be composed of a number of element types. A selection of these is shown in Figure 8.18.

The simplest of these are axial elements. These can be made mathematically more complex to carry torsion and moments. The two main types of planar elements are membrane and plate elements. The former cannot sustain out of plane bending and shear, while the latter can. The third major type is solid or three-dimensional elements, which are required to predict complex three-dimensional stresses and deflections. They are not widely used in lightweight structures—only at relatively heavy fittings such as propellant tank attachments and launch vehicle interface brackets.

8.5.1   Building a finite element model

A complete spacecraft finite element model is constructed from submodels, which are ideally provided by the designers of the hardware who have the best knowledge of its structural behaviour. Its construction, therefore, emulates the construction of the hardware. Also, the maturity and accuracy of the model reflects the maturity of the design. At the concept design stage, some gross assumptions may be made. The accuracy of the model should be kept current with the design such that after spacecraft testing the model accuracy should be test verified.

Figure 8.19 shows a model of a telecommunication spacecraft. In this case, *super elements* were used to reduce the computing power required. The complete spacecraft is analyzed in an automatic analysis sequence as a series of subassemblies, which are then automatically assembled as though they were large elements—hence the name super elements.

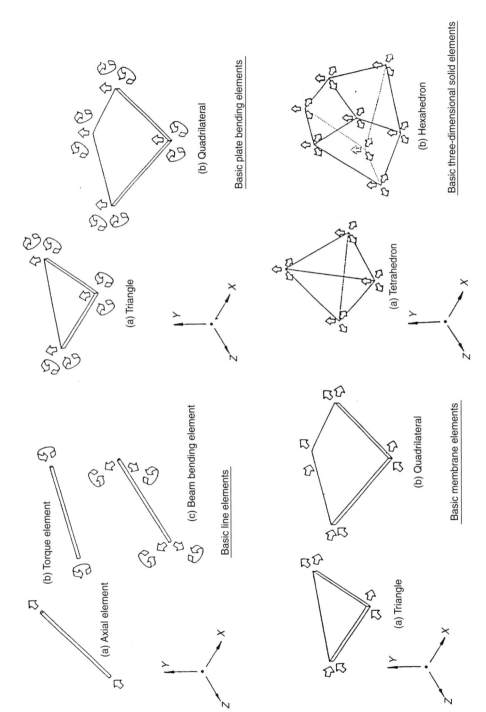

Figure 8.18   Standard finite elements for modelling

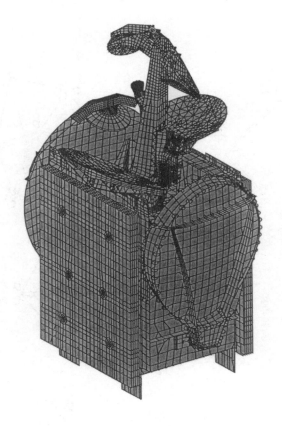

8 Super elements	Degrees of freedom
−X reflector	7145
−X feed	13867
+X reflector	7782
+X feed	13819
−Y solar array	21121
+Y solar array	21121
Top floor antenna	58387
Platform	32498
Total	175740

Element types	
24671	quad 4
7114	tria 3
9371	bar
6936	elas
348	hexa
33302	grids

Figure 8.19 Complete spacecraft finite element model (Reproduced by permission of Astrium Ltd)

8.6 DESIGN VERIFICATION

Test verification that a spacecraft meets its major strength and stiffness requirements will usually be required unless the safety margins are all greater than one or if the design is very similar to a previously tested design. In the latter case, a *qualification by similarity* analysis will demonstrate that the previous test covers the new assembly.

Figure 8.20 Eurostar 3000 structure in the static test rig (Reproduced by permission of Astrium Ltd)

The major requirements to be test-verified are static strength, sine and acoustic vibration responses and strength. It is also important to use a vibration test to prove, or test-verify the accuracy of the mathematical model. Figure 8.20 shows the static test assembly for the *Eurostar 3000* structure. Producing such an arrangement with adequate hydraulic loading jacks, their control systems, recording strain gauges, displacement transducers and data logging systems is a complicated and expensive job. A dedicated *test model* is often used for this test because of the cut-outs required specifically for the test. Load spreader plate attachments are also used to diffuse load from the limited number of jacks into lightweight structures, which in flight will carry a more distributed load.

Sine and acoustic tests may be performed on *engineering models* fitted with mass dummies. They are often performed as protoflight tests on flight hardware with all flight equipment fitted as part of the final *Assembly, Integration and Test* process.

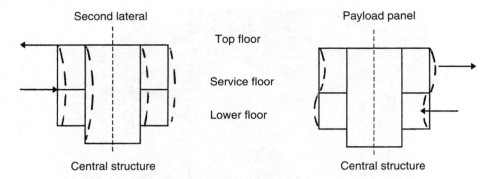

Figure 8.21 Vibration mode isolation by multi-point modal survey testing

Verification that the mathematical model is accurate may be obtained by comparing the sine response analysis with the spacecraft sine test results. There are several aspects to be considered here. First, comparison of natural frequencies and mode shapes will ensure that the model has the correct overall mass and stiffness. Comparing mode shape or the variations of accelerations in any one particular mode of vibration will ensure that the model represents the correct distribution of stiffness through the spacecraft. Comparing magnitude of response will ensure that the correct level of damping has been used.

There may be instances when errors may be present in the model, and modes of similar frequency cannot be separated by base excitation. Figure 8.21 illustrates a possible problem.

The left-hand figure shows a second lateral mode of a spacecraft predicted at 60 Hz. The right-hand figure shows a payload panel mode predicted at 65 Hz. On a base excitation sine test, these could combine into a single mode at 63 Hz making it impossible to isolate and correct the model error. It is in such instances that a multipoint modal survey test would be helpful.

For a modal survey test, the spacecraft is attached to a seismic block. This is a large concrete block mounted on springs or air bags in a pit to isolate the spacecraft from any extraneous ambient ground vibration generated by heavy factory equipment or passing road traffic. The spacecraft modes are excited individually by hand or frame held exciters at the positions indicated by the arrows. The frequency, phase, position and relative magnitude of excitation are varied until a mode is isolated by trial and error. Having isolated the second lateral, the exercise can be repeated for the panel mode, so isolating and correcting the model error [17].

8.7 IMPACT PROTECTION

The predicted growth of the millimetre and sub-millimetre size space debris environment, and the greater understanding of the meteoroid population, have led to a rise in interest in shielding for unmanned spacecraft in LEO. The impacts, typically in the range 5 to 20 km/s are capable of damaging and perforating spacecraft external structures [honeycomb, often covered by multi-layered insulation (MLI)].

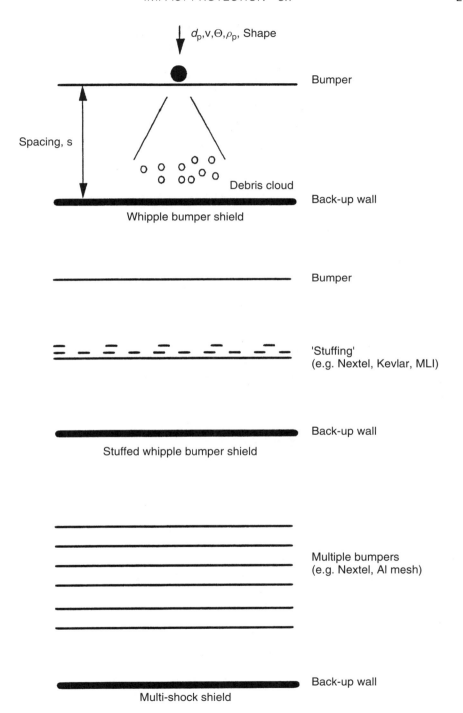

Figure 8.22 Different types of Whipple bumper shields (Reproduced by permission of Astrium Ltd)

A notable early strategy for shielding unmanned spacecraft was that developed for Radarsat [18]. The design changes effected—re-routing of critical harnesses, enhancement of MLI using Nextel—are included in the design recommendations made in the Orbital Debris Handbook [19].

More recently, similar enhanced MLI configurations have been tested for use on the Iridium satellites and on the International Space Station (ISS) [20].

Typically, a space debris and meteoroid shield is based on a Whipple bumper. This consists of a thin bumper shield of aluminium, metallic laminates or composite materials and a thicker aluminium back-up wall separated by a gap (S). The bumper layer disrupts the projectile by either shattering, melting or vaporizing it. The spacing allows the debris cloud to be distributed over a relatively large area, lowering the impact loading on the back-up wall. Placing a range of materials between the bumper and back-up wall (Kevlar, MLI, Spectra, aluminium meshes and Nextel can augment the shield. Multiple thin bumpers ensure that the projectile is repeatedly shocked (and thus fragmented, melted or vaporized) before impacting on the back-up wall. Figure 8.22 shows the three main types of shields.

Different types of bumper material have been tested for their ability to disrupt the projectile into a debris cloud. Metal and ceramic matrix composites were found to give better projectile disruption than the equivalent areal density aluminium bumper, resulting in less damage to the back-up wall. In addition to Kevlar and Nextel, honeycomb and aluminium/titanium laminates have been used as intermediate layers in Whipple bumpers. The ability of laminates to provide good projectile disruption as a result of shock impedance mismatch has been noted [21,22]. Advanced shield designs, Multi-shock shields (MSS), Multiple-mesh bumper shield (MMBS) and Mesh double bumper shields (MDBS) have been investigated [23,24].

8.8   CONFIGURATION EXAMPLES

8.8.1   Eurostar

The Eurostar configuration, shown in Figure 8.23 is a representative example of a three-axis stabilized Geostationary Earth orbit (GEO) communications spacecraft. It bears many of the features common to this class of spacecraft. Loads are carried primarily by a central thrust structure. The nature of the mission requires a large number of electrical components, which are substantial producers of heat and need to be mounted on external walls to promote efficient heat radiation to space. In addition, such units need to be grouped into systems creating the need for large areas. The solution is to employ a box shape using large aluminium alloy honeycomb panels. Load from the panels is carried to the central tube through four shear panels and adjacent box closure panels.

The apogee engine is fitted into the interior of the central tube and propellant tanks are supported symmetrically around the central tube. Main thrust axis inertia loads from the tanks are carried to the launch vehicle interface ring by carbon-fibre tubular struts with titanium end fittings. This is a classical use of carbon fibre, since the load direction is well defined and the construction lends itself to a well-controlled production process. Titanium alloy end fittings are used in this application for their high strength. Steel fittings are used

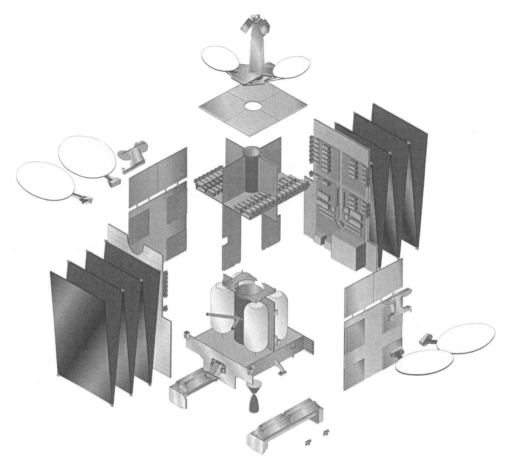

Figure 8.23 Eurostar 3000 structure configuration (Reproduced by permission of Astrium Ltd)

in the highest load missions to achieve the maximum strength in a confined geometric space. Carbon-fibre end fittings would be difficult to produce and would be too weak, with the inevitable stress concentrations located at the lug end fitting.

The flat panel configurations allow good access and logistics for separate panel integration at different sites by the removal of sidewalls carrying whole systems. However, it may be prudent to install non-flight support frames when such a part of the structure is removed. This point emphasizes the need to consider handling at an early stage of design so that appropriate lifting points are incorporated for all stages of assembly, and not just for the complete spacecraft.

The solar cells are fitted to external arrays, leaving the side members of the spacecraft free for heat radiation. The solar arrays are rotated in GEO to ensure that they are always facing the sun. Features of the design that are only used once in a mission include latch release and hinge mechanisms for the solar arrays. Even so they need to be strong enough to withstand launch and shock during deployment.

8.8.2 Boeing 601

The Boeing 601, shown in Figure 8.24 is a similar class of spacecraft to Eurostar that exhibits many similarities. It is a box-shaped construction, with large flat exterior panels to carry many high-powered equipment units. Half the mass at launch is the propellant in the tanks. It does not have the central thrust structure of Eurostar, with the circular launch vehicle interface stopping at the propellant tank's lower interface. A notable effect of this is that on the top floor there is only one central hard point. A structural hard point on a large flat expanse of panel is characterized by a perpendicular supporting structure beneath that panel. On the Eurostar structure top floor, the central cylinder and the four radial shear panels provide this. In the middle area of the Boeing 601 top floor, there is just the single shear wall. Note how all the bracing struts are mounted to the central point. On the Eurostar top floor the central cylinder is often used to support high antennas and feed towers. The importance of factors such as this must be considered when a spacecraft structure configuration is being designed.

Impact protection does not have an influence on the design of GEO spacecraft because of the low debris density currently in GEO.

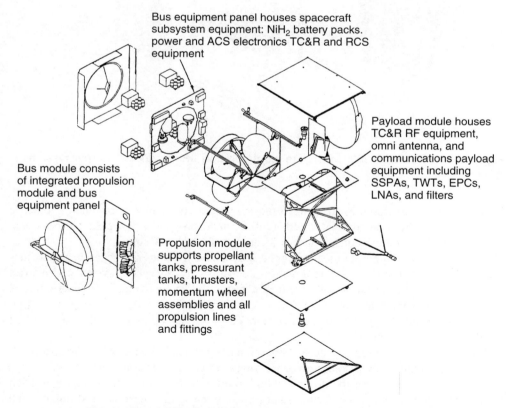

Figure 8.24 Boeing 601 structure configuration (© Boeing)

8.8.3   Metop

The Metop, shown in Figure 8.25 is a LEO meteorological spacecraft. A major configuration design requirement is for a large, flat Earth-facing panel with adjacent side panels to mount a range of relatively low-powered sensing equipment that can also view Earth at 90° to their panel mounting face. The logistics of integration and test also favour modular construction. Although the factors driving the design are different from those of a geosynchronous telecommunications spacecraft, the details of the structural design solution are similar. It is longer and heavier than Eurostar, and a much lower propellant mass is required for the LEO. The central cone and cylinder are of thicker sections to carry greater loads. The flat exterior aluminium alloy sandwich panels mount interior service equipment as well as the exterior experiments, sensors and antennas.

The survivability of the spacecraft with respect to debris impact was considered in the layout of on-board equipment.

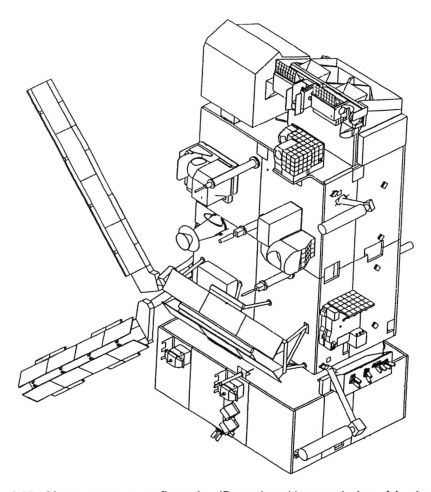

Figure 8.25   Metop structure configuration (Reproduced by permission of Astrium Ltd)

8.8.4 Small spacecraft

Figure 8.26 shows an example of this class of spacecraft, which is being considered more extensively for the launch of individual experiments. This is an alternative to launching many experiments together on a large, complicated and expensive spacecraft, which take many times longer to configure, build and prepare for launch. The small spacecraft's faceted tubular shell is filament wound in carbon fibre reinforced plastic. This relatively simple structure with a minimal number of component parts is cheap and quick to manufacture. Carbon fibre gives minimal thermal distortion. Equipment is mounted on the facets without the need for additional brackets. Launch vehicle attachment may be by an adaptor ring or discrete pyro-bolt brackets. A large flat honeycomb panel (not shown in the figure) is mounted to the cleats on the obliquely cut top face of the tube to carry Earth facing sensors.

With a width of about 1 m, a height from 1 to 2.6 m and a launch mass from 500 to 1500 kg, these spacecraft will take advantage of the multiple launch facilities such as those offered by the ASAP on Ariane 5.

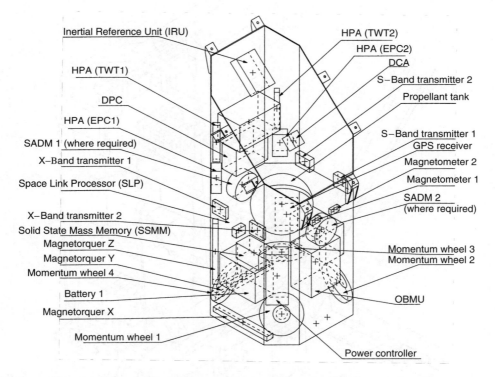

Figure 8.26 Small spacecraft structure configuration (Reproduced by permission of Astrium Ltd)

8.8.5 The international space station (ISS)

The ISS, shown in Figure 8.27 is being assembled in space (at the time of writing) from modules built on Earth. When completed, the ISS will be about 89 m long and 110 m

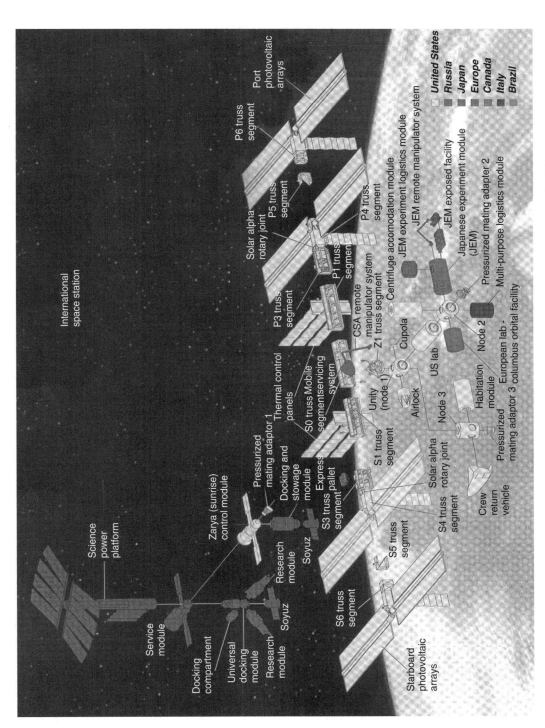

Figure 8.27 International space station (Reproduced by permission of NASA)

wide and have a mass of 402 tonnes. The pressurized volume will be roughly equivalent to that of the passenger cabin of two Boeing 747 jetliners. Structural design requirements of the final assembly are analogous to those of a smaller single-launch spacecraft with complex deployed appendages. The more rigorous launch design requirements must be considered for each individually launched module.

Pressurized manned modules have Whipple style impact protection.

One of the essential building blocks of the ISS is the 'Unity' connecting module shown in Figure 8.28.

Unity has six berthing ports to which future modules will be attached. Essential space station resources such as fluids, environmental control and life support systems are routed through Unity to supply work and living areas. The structure is fabricated from aluminium. Two conical docking adaptors are attached to each end prior to launch to allow the docking systems used by the Shuttle and Russian modules to attach to the node's hatches and berthing mechanisms. Unity is 5.5 m long, has a diameter of 4.5 m and a mass of 11 600 kg.

Figure 8.28 Unity connecting module — international space station (Reproduced by permission of NASA)

8.8.6 Mars express

This is designed to take seven instruments and one lander into Martian orbit (see Figure 8.29).

The mission is a test case for the development of new working methods to speed up spacecraft production and to minimize mission costs. 116 kg is allowed for the instruments and 60 kg for the lander. Off-the-shelf technology or technology developed for the Rosetta comet mission (see Chapter 5) is being used wherever possible.

The instruments will sit inside the spacecraft bus, which is an aluminium honeycomb box 1.5 m long by 1.8 m wide by 1.4 m high. The lander, Beagle 2, is attached to the outside of the bus. Payload, lander, spacecraft and on-board fuel will weigh a maximum of 1070 kg at launch.

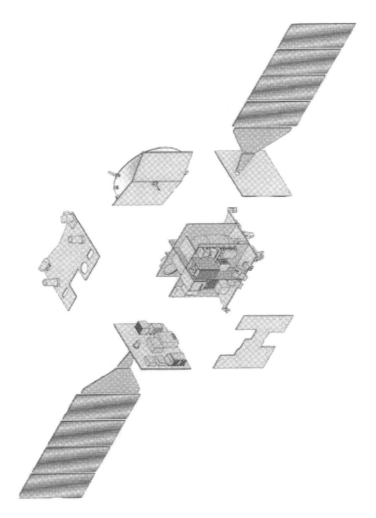

Figure 8.29 Mars express structure configuration (Reproduced by permission of Astrium Ltd)

Most of the power needed to propel Mars Express from Earth to Mars is provided by the Soyuz Fregat launcher, which will separate from the spacecraft after placing it on a Mars-bound trajectory. The spacecraft uses its on-board propulsion to inject into Martian orbit and for orbit corrections. The spacecraft has two Eurostar 2000 bi-propellant tanks carrying a total of 427 kg of propellant at launch.

For up to six hours during the spacecraft's 7.5 h Martian orbit period, the high gain antenna will point towards the Earth. During the remaining 1.5 h, the spacecraft will point towards Mars for observation and communication with Beagle 2 on the Martian surface.

The spacecraft structure uses an arrangement of internal shear panels instead of a central thrust structure. This generates critical joints at the points where the shear walls cross the circumference of the launch vehicle attachment ring. Note the substantial brackets at these positions.

8.8.7 FIRST/Planck

This is a combined launch of two spacecraft, as shown in Figure 8.30. Far Infra-Red Space Telescope (FIRST) is carried by Planck in the launcher. After launch, Planck and FIRST separate and are placed in different orbits around the second Lagrange point of the Earth–Sun System (see Chapter 4). The launch configuration of Planck and FIRST constrains the payload design considerably.

Planck will measure variations in the cosmic microwave background to acquire knowledge of the early universe. A Gregorian Telescope will make observations with detectors at its focal plane. The telescope assembly must be maintained at 60 K. To give dimensional stability, the telescope framework is carbon fibre reinforced plastic sandwich, skins or tubes. The telescope frame will be attached to the Planck service module by a carefully designed arrangement of struts that will provide near isostatic mounting to sustain all inertia loads. It must also prevent thermal distortions of the service module from inducing corresponding distortions of the telescope assembly. The service module will house the cooler units and all other spacecraft functions.

FIRST, the Far Infrared and Sub-millimetre Telescope, is a multi-user observatory, intended to investigate the infrared and sub-millimetre part of the electromagnetic spectrum, in the wavelength range from 85 µm to 600 µm. The detectors of the FIRST instruments have to be cooled to cryogenic temperatures in the range 0.3 to 2 K in order to reach the necessary sensitivity. The satellite cooling system is based on that of a previous spacecraft, the Infrared Space Observatory (ISO). It uses a large superfluid helium dower (liquid helium at 1.6 K), sized for a scientific mission of 4.5 years. This requires

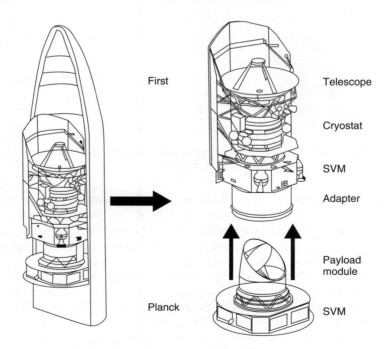

Figure 8.30 FIRST/Planck spacecraft structure configuration (Reproduced by permission of the European Space Agency)

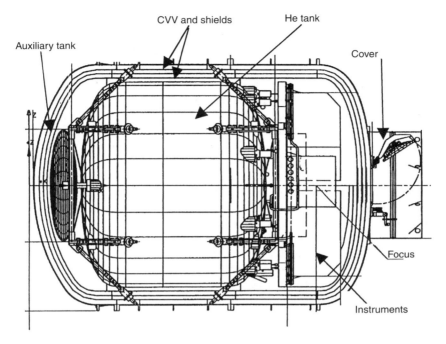

Figure 8.31 The FIRST cryostat (Reproduced by permission of the European Space Agency)

2560 litres of helium. A helium cooler connected directly to the helium tank provides further cooling down to 0.3 K for the bolometers. The cryostat is shown in Figure 8.31. It consists of the following principal items:

- a segmented, cylindrical, main helium tank;
- a lens shaped (ellipsoidal) 79-litre auxiliary helium tank for launch autonomy cooling;
- a carbon fibre reinforced plastic sandwich optical bench to mount three scientific instruments;
- three vapour cooled radiation shields, each consisting of three parts;
- 16-tank support straps, each consisting of four glass fibre composite and two carbon fibre composite chain loops inter-linked by steel bolts that act as thermal anchors and as a mechanical support for the heat shields;
- the cryostat vacuum vessel made in three sections bolted together and
- the cryostat cover.

These examples illustrate the range and scale of projects currently being faced by the spacecraft designer. Most spacecraft structures are designed using flat panels, often honeycomb sandwich or cylindrical and conical outer or inner shells. Mouldings are used when complex shapes are advantageous. When selecting the spacecraft structure-configuration to meet the requirements of any mission, the golden rule is to *keep it simple*. Development of design and realization of detailed requirements as the project progresses invariably increases complexity.

8.9   THE FUTURE OF SPACE STRUCTURES

An examination of the history of structural design shows a progression to the more efficient use of materials. This trend will continue as load path assessments become more refined and application-specific, and tailored high-strength materials are developed. Material and manufacturing process improvements, for composites in particular, generally give increased specific strength and stiffness while improved thermal stability for items such as optical benches is being achieved with carbon–carbon and ceramic materials.

By integrating systems into composite structures, we may approach the concept of the intelligent structure. Already optical fibres can be embedded into a carbon fibre matrix allowing signals to pass along skins. References [25,26] are sources of information on the developing field of electro-active and shape memory materials.

A greater diversity of structural design concepts is emerging. With the interest in small spacecraft for LEO constellations or small single experiments, the designer is required not only to produce an efficient structure to maximize payload but also to respond to the ever-increasing commercial pressure of schedule and cost. Moulded, single component structures that can be stacked for multiple launches are an attractive proposition. The possibility of large numbers of spacecraft for the proposed global constellations (see Chapter 5) requires the design to encompass the needs of mass production, and ease of integration and test. The concept of *six sigma* is being explored to give more robust manufacturing methods. A six-sigma process reduces the probability of defects occurring to less than 3.4 per million or six standard deviations above a 50% chance in a statistical normal distribution. As an example, bolted joints can be six sigma while bonded joints will not be six sigma because there is significant scope for defects in the bond due to a range of possible causes.

Demands are being placed on geosynchronous communications spacecraft for larger unfurlable and inflatable antennae to reduce the size and power of the mobile terrestrial user's equipment. This stretches ingenuity with the requirement for high reliability, simplicity and low cost. The designer is faced with deployment or perhaps assembly after achieving orbit (Figure 8.27).

Potentially, the task of designing for launch can be made easier, since the structure may be stowed compactly within the launch vehicle, and an extremely low mass may be achieved, designed to withstand the zero g environment only when deployed. Careful consideration needs to be given to loads created in hinges or members during automatic or remote deployment, where sections can build up momentum as they are swung into place and are then suddenly stopped when positioned. A similar argument prevails for attitude control or orbit maintenance manoeuvres by the space station. Not only are these likely to excite very low frequency cross-coupled oscillations within the structure, but the designer is faced with a continuously changing situation as mass is added, moved or subtracted from the configuration. This is potentially a fertile area for the application of distributed thrusts and an 'expert system' or 'intelligent' attitude control system to phase the individual firings to counter an undesirable dynamic response.

It can be expected that new families of wire-tensioned structures will emerge that can only be used in a zero g environment.

Much consideration is being given to building sections in orbit directly from materials coiled flat on a roll within a machine to be located in the Shuttle cargo bay. This would roll the stock into a structural section as it emerged. Here, if the concept can be successfully developed, the designer will face the difficult problem of bridge designers on earth—that of ensuring structural safety through all stages, including the stage before the momentum control systems have been installed and the structure is only partly built.

REFERENCES

[1]   Gordon, J. E. (1987) *Structures, or Why Things Don't Fall Down*, Pelican, London.
[2]   ESA *ESACRACK User's Manual*, ESA PSS-03-209.
[3]   Crane, F. A. and Charles, J. A. (1984) *Selection and Use of Engineering Materials*, Butterworths, London.
[4]   Ashby, M. F. and Jones, D. R. H. (1980) *Engineering Materials—An Introduction to Their Properties and Applications*, Pergamon, Oxford.
[5]   ESA *Space Product Assurance* Material selection for controlling stress corrosion cracking ECSS-Q-70-36, January 1998.
[6]   Kinnersley, M. A., Stark, J. and Swingard, B. (1989) Development of, and initial results from a high fluence, high velocity atomic oxygen source, *4th European Symposium on Spacecraft Materials in the Space Environment*, CERT/CNES/ESA, 701.
[7]   *LDEF-69 Months in Space*, Second Post Review Symposium, NASA CP-3914 (1992).
[8]   *Structural Materials Handbook, Vol. 1 Polymer Composites* (1994) ESA-PSS-03-203.
[9]   *Structural Materials Handbook, Vol. 2 New Advanced Materials* (1994) ESA-PSS-03-203.
[10] Jones, R. M. (1975) *Mechanics of Composite Materials*, McGraw-Hill, Washington.
[11] Tsai, S. W. (1988) *Composite Design* Think Composites, Dayton, Ohio.
[12] Mohn, W. R. and Geigel, G. A. (1986) Dimensionally stable metal matrix composites and optics, *Advanced Composites Conference Proceedings*, ASM International and Engineering Society of Detroit.
[13] NASA *Materials Guidelines* (1975) N-75-24848 (SP-3094).
[14] ESA *Materials Guidelines* (1994) ESA PSS-01-701.
[15] Haigh, M. J. (1985) *An Introduction to Computer Aided Design and Manufacture*, Blackwell Scientific, Oxford.
[16] Robinson, J. (1981) *Understanding Finite Element Stress Analysis*, Robinson and Associates.
[17] Zaveri, K. (1984) *Modal Analysis of Large Structures—Multiple Exciter systems*, Technical Note, Bruel and Kjaer Ltd.
[18] Terrillion, F., Warren, H. R. and Yelle, M. J. (1991) *Orbital Debris Shielding Design of the Radarsat Satellite*, AF-91-283, *42nd Congress of the International Astronautical Federation*, 5–11 October, Montreal, Canada.
[19] National Research Council (1995) *Orbital Debris: A Technical Assessment*, National Academy Press, Washington, DC.
[20] Christiansen, E. L. (1998) *Design Practice for Spacecraft Meteoroid/Debris Protection*, Presented at the IAT/NASA Hypervelocity Shielding Workshop, 9–11 March, Galveston, Texas.
[21] McMillan, A. R. (1963) *An Investigation of the Penetration of Hypervelocity Impact Projectiles into Composite Laminates*, Proceedings of the 6th Symposium on Hypervelocity Impact, April–May, Cleveland Ohio, **Vol. III**, pp. 309–356.
[22] Stilp, A. J. and Weber, K. (1997) *Debris clouds behind double layer targets*, Int. J. Impact Eng., **20**, 765–778.
[23] Cour-Palais, B. G. and Crews, J. L. (1990) *A multi-shock concept for spacecraft shielding*, Int. J. Impact Eng., **10**, 135–146.
[24] Cour-Palais, B. G. and Piekutowski, A. J. (1992) *The Multi-shock Hypervelocity Impact Shield—Shock Compression of Condensed Matter 1991*, Schmidt, S. C., Dick, R. D., Forbes, J. W. and Tasker, D. G. (eds) Elsevier.
[25] Gandi, M. V. and Thompson, B. S. (1992) *Smart Materials and Structures*, Chapman and Hall, London.
[26] Spillman, W. B. Jr. (1979) The evolution of smart structures/materials, *1st European Conference on Smart Materials and Structures*, SPIE **Vol. 1777**.

9 *ATTITUDE CONTROL*

Peter W. Fortescue and Graham G. Swinerd

Aeronautics & Astronautics, School of Engineering Sciences, University of Southampton, United Kingdom

9.1 INTRODUCTION

Although the prime purpose of the attitude control system (ACS) is to orientate the main structure of the spacecraft correctly and to the required accuracy, it is worth considering it also as a momentum management system. Angular momentum is a commodity that can be acquired and disposed of, or stored.

In terms of angular momentum management, the ACS designer must decide what his or her best policy is, and provide the hardware to achieve it. He or she will need to assess the momentum implied by the pointing requirements set by the mission objectives, and to specify torquers and storage to provide it.

When one looks at the types of ACS in existing spacecraft, one finds that a profusion of different methods are employed to meet this seemingly simple objective. The structures may be spinning or not, or may be doing so only during certain phases of the mission. The ACS may or may not use momentum bias and/or momentum storage; it may use a variety of torquing methods, in combination or on their own. The number of options open to the designer is large. What then influences the designer to select one solution as opposed to another?

The decisions that he or she takes are not based solely upon the considerations presented here. The designer will be influenced by the experience and the history of the company in which he or she works. A feature of attitude control is that different technical solutions may give very similar performance, and the designer will choose well-trodden paths, the type of solution of which his or her company has experience. And rightly so.

There are complete books on the subject of attitude dynamics and control [1–3], and this solitary chapter cannot present a comprehensive account of the subject. The approach adopted is to look at the fundamentals, to progress from the system level downwards and to identify the options referred to. Detailed design is not covered.

Spacecraft Systems Engineering (Third Edition). Edited by P. W. Fortescue, J. P. W. Stark and G. G. Swinerd
© 2003 John Wiley & Sons Ltd

9.2 ACS OVERVIEW

9.2.1 The design objective

The orientation required of the spacecraft's structure will be determined by the mission. The structure will be seen as the mounting base for the payload(s), and for several 'housekeeping' subsystems that have objects that must be pointed in specific directions. Among the latter will be solar arrays to be pointed at the Sun, thermal radiators to be pointed at deep space, antennas to be pointed at their targets and, prior to firing, thrusters to be pointed in the correct direction. Add to this the pointing requirement of the payload and it will be seen that there is an essentially three-dimensional problem to solve, at the end of which will emerge a proposed layout for the spacecraft with locations for the objects to be pointed and a specification for the orientation of their mounting base, the main structure.

The required orientation will frequently be related to an Earth-based frame of reference, such as, 'One face of the spacecraft must point down the local vertical'. The ACS designer must then analyse the orbit and the mission in order to assess the motion that is required of the structure and the disturbance torques to which it will be subjected. For this objective, the structure requires an angular rate equal to $\dot{\theta}$, where θ is the true anomaly. (See Section 4.2 and Figure 4.3 of Chapter 4.)

There will often be several alternative configurations that will meet the overall objective for a given payload, and these may pose very different problems. The Intelsat series of satellites illustrates this point. Their payloads are all for communications purposes and have antennas to be pointed at locations on the Earth. But their configurations include spinners, dual-spin and three-axis-stabilized types.

In general, pointing mechanisms are to be avoided. However, they are often needed in order to enable an object to remain pointing in one direction whilst the main structure changes its orientation, or *vice versa*. For example, a solar array needs to remain pointing at the Sun whilst the main structure turns to align with the local vertical.

The required accuracy of orientation will be set by the payload. The accuracy with which its direction can be controlled will be less than that to which it can be measured, and this is in line with the requirement. For example, spacecraft astronomers will need to know where a telescope's axis is pointing to great accuracy (typically arc seconds), but the control of its direction may be less accurate, related to its field of view.

A full accuracy specification for both measurement and control of the main structure's attitude may then be determined from the various pointing requirements.

9.2.2 Mission-related system considerations

The ACS designer will need to know the required time-history of the rotational motion, the angular rate ω, which is required of the spacecraft, and of any parts of it that can move independently on bearings. The angular momentum $\mathbf{H}_c$ and the torque $\mathbf{T}$ needed to produce it may then be calculated from the Newtonian law (from equation 3.28 of Chapter 3):

$$\mathrm{d}\mathbf{H}_c/\mathrm{d}t = \mathbf{T} \tag{9.1}$$

where $\mathbf{H}_c$ is the angular momentum referred to the centre-of-mass C, detailed in equations (3.33) to (3.36) of Chapter 3, and $\mathbf{T}$ is the torque.

A circular orbit provides an interesting example. The local vertical will rotate at a constant rate equal to 1 revolution per orbit about the normal to the orbit plane. If one face of a structure with no momentum bias is required to point down the vertical, then no torque is needed to maintain this condition provided that a principal axis is aligned with the normal to the orbit plane (cf. Section 3.4.2 of Chapter 3). If the normal is not a principal axis, then equation (9.1) will indicate that a cyclic torque is needed.

From an elliptical orbit, on the other hand, the local vertical will not have a constant angular velocity. An assessment over one orbit indicates that the angular rate will vary as shown in Figure 9.1. It has a cyclic variation about a mean level of one revolution per orbit.

If one face of the structure is required to point down the vertical and a principal axis is aligned with the normal to the orbit plane, then no torque is needed to maintain the mean angular rate (constant angular momentum). But torques *will* be needed to produce the fluctuations in momentum about the mean value.

If the structure is not rotating, frequently the case for astronomical missions, then again no torque is needed to maintain this condition once it has been achieved. This will be approximately true of solar arrays too, since their angular velocity is likely to be virtually zero as they point towards the Sun, even if the main structure is rotating.

The torque requirements for each part of the spacecraft may be assessed as above, on the assumption that it is rigid, and these will be the *minimum* torques needed. Extra torques will be required in order to combat the uncontrolled (disturbance) torques such as that due to solar radiation pressure (SRP), movement of fuel, mechanisms, friction in bearings and so on, and so these must be evaluated too (see Section 9.4). The assessment must cover *all* phases of the mission, and all three axes of the spacecraft.

It may be expected that the largest torques will be needed during the early stages, between final separation and being on station. The *maximum* torque capability will

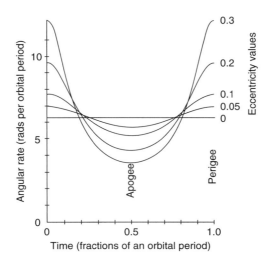

Figure 9.1 Angular rate of the local vertical for satellites in elliptical orbits

govern the time taken to execute manoeuvres such as the repointing manoeuvres shown in Figure 9.7. If manoeuvre times are important, then these may determine the specification for this feature of the torquers.

Momentum storage will normally be provided on spacecraft of medium to large size, especially when there is a requirement for fluctuating torque or momentum or when there are tight tolerances on pointing accuracy (see Table 9.2 and Section 9.4.1). A momentum storage device is basically a wheel fitted with a torque motor, each enabling one component of momentum to be stored (but see Section 9.4.7). On small satellites they are less likely to be used because of their mass and cost, attitude control then being achieved by using external torquers.

In operation, the wheel torque motor will be the prime means of achieving the pointing requirement of the structure, correcting *any* errors, however caused, in a feedback sense. In so doing, it will insert or extract momentum from the wheel. When the wheel reaches an extreme of its permitted range of speeds—its store of momentum has been filled—then its speed must be restored to its normal value, using external torquers to counter the torque on the wheel so as to maintain attitude control. This process is known as *momentum dumping*.

Assessment of the required amount of storage will be based upon the mission, and it will be chosen to achieve infrequent use of the dumping process. Orbit considerations such as its eccentricity, imply rotation as shown in Figure 9.1, for example, and a momentum component ($I\omega$) along the normal to the orbit plane. Dumping will be required during every orbit unless the store can accommodate at least half of the difference between the maximum and the minimum values of $I\omega$. In addition, it must accommodate consequential momentum arising from disturbance torques. Table 9.1 shows a list of the potential sources of these.

Manoeuvres such as the repointing one shown in Figure 9.7 must also be assessed in a similar way.

Table 9.1   Disturbance torques

External torques source	Height range over which it is potentially dominant
Aerodynamic	<about 500 km*
Magnetic	500–35 000 km
Gravity gradient	500–35 000 km
Solar radiation	>700 km*
Thrust misalignment	all heights

Internal torques source	
Mechanisms	
Fuel movement	
Astronaut movement	
Flexible appendages	
General mass movement	

*Values depend upon the level of solar activity.

9.2.3 Momentum bias

The level of momentum that is involved in a storage system will be quite small, and much smaller than will bring any significant benefit from gyroscopic rigidity. This will come from *momentum bias*, which makes the direction of one axis of the spacecraft highly resistant to change. (See Section 3.4 of Chapter 3.)

For example, if the mission requires that one axis of the spacecraft shall always lie in the direction of the normal to the orbit plane, or at right angles to the Sun vector, then the designer might include momentum bias in that direction. The magnitude of the bias, H_b, whilst not critical, is likely to be an order of magnitude greater than the storage system will handle. A torque T at right angles to it will cause the axis to precess at a rate T/H_b (see equation 3.29 of Chapter 3). The attitude response *about* the bias direction will not be altered.

Momentum bias devices may be used for momentum storage too. If more than one bias device is used, and this may well be done in order to enhance the reliability, then it must be remembered that their momenta add vectorially to produce only *one* gyroscopically rigid axis.

9.2.4 The ACS block diagram

The block diagram in Figure 9.2 shows the major components of a general ACS system. The links between components identify major interactions, with arrows indicating that there is a cause–effect relationship; it is convenient to think of them as channels along which information flows. For example, the main structure of the spacecraft is subjected to time-varying torques from torquers, and will respond with attitude motion that will be

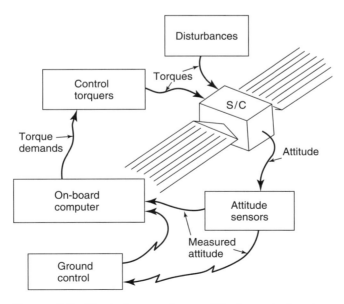

Figure 9.2 Block diagram for an attitude-control system

detected by the sensors. Outputs from these will be sent to computers, on-board and at the ground control station, and their information will be used to determine the torques that should be applied to the structure.

9.3   THE SPACECRAFT ATTITUDE RESPONSE

In the context of the ACS block diagram, the spacecraft may be seen as an object that changes its orientation in response to a torque, which is the simple message conveyed in Figure 9.2. The way in which it responds will depend upon whether it has momentum bias, and the response equations are contained in Sections 3.3 and 3.4 of Chapter 3.

Whilst the decision to include bias and the method of doing so will be decided at system level (see Section 9.2), it is as well to recall the main types of structure and their characteristics with which ACS designer will have to deal. For each type, he or she will be concerned with the following:

- the equilibrium state,
- the response to a steady torque,
- the stability, and the existence of any oscillatory modes.

9.3.1   The three-axis-stabilized spacecraft, with no momentum bias

The dynamic equations for this spacecraft, treated as a rigid body, are covered in Section 3.4.1 of Chapter 3, using principal axes for the analysis. With small angular velocities the responses about these axes are largely uncoupled and may be approximated to

$$I_{xx}\dot{\omega}_x = T_x, \quad I_{yy}\dot{\omega}_y = T_y, \quad I_{zz}\dot{\omega}_z = T_z \tag{9.2}$$

Each of these equations corresponds to a root location diagram[*] having a single root at the origin, as shown in Figure 9.6(a).

Any of the torque components on its own will produce an acceleration about its own axis. Combinations, however, will not do so; they will produce a cross-coupled response except when the moments of inertia are equal. Cross-couplings will increase as the angular velocity increases.

This type of spacecraft usually has flexible solar arrays attached to the main structure as shown in Figure 9.3, with lightly damped flexure modes with a low fundamental frequency. When selecting the algorithms for the ACS computer, up to about 20 of these modes may be mathematically modelled, in order to ensure ideally that they are stabilized, but at least that they are not destabilized.

Another typical configuration for this type is that of the Hubble Space Telescope.

---

[*] A root location diagram is an Argand diagram on which is marked the root(s) of the Characteristic Equation of a linear differential equation.

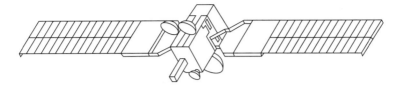

Figure 9.3 Typical configuration of a three-axis stabilized spacecraft

9.3.2 The spinning spacecraft

The dynamics of a spinner are covered in Section 3.4.2 of Chapter 3. It is assumed here that the vehicle is spinning about its z-axis with an angular rate S, and that its mass distribution is axially symmetric so that $I_{xx} = I_{yy}$. The response to a torque will be of interest in one or both of the two sets of axes. A typical 'spinner' is shown in Figure 9.4.

The response in axes fixed in the structure will be of interest when the torque components are in these axes or when the attitude sensors measure components in these axes. The equations, developed from equation (3.40) of Chapter 3, are

$$I_{xx}\dot{\omega}_x + S\omega_y(I_{zz} - I_{xx}) = T_x$$
$$I_{xx}\dot{\omega}_y - S\omega_x(I_{zz} - I_{xx}) = T_y \qquad (9.3)$$
$$I_{zz}\dot{S} = T_z$$

Separating ω_x from ω_y leads to

$$I_{xx}(\ddot{\omega}_x + \omega_{\text{nut}}^2\omega_x) = \dot{T}_x - \omega_{\text{nut}}T_y \quad \text{and} \quad I_{xx}(\ddot{\omega}_y + \omega_{\text{nut}}^2\omega_y) = \dot{T}_y + \omega_{\text{nut}}T_x \qquad (9.4)$$

where $\omega_{\text{nut}} = S[(I_{zz}/I_{xx}) - 1]$ is the frequency (rads/s) of the nutation mode referred to in Section 3.4.2 of Chapter 3, when observed in the spacecraft's axes. This mode will need to be damped, whereupon the eventual constant components ω_x, ω_y in response to constant torques T_x, T_y will become

$$\omega_x = -T_y/S(I_{zz} - I_{xx}) \quad \text{and} \quad \omega_y = T_x/S(I_{zz} - I_{xx}) \qquad (9.5)$$

When these are combined with the spin motion, the result is a coning rotation of the z-axis at the spin frequency S when viewed from outside the spacecraft.

One of the major reasons for spinning a spacecraft is to counter the effect upon the trajectory that is caused by a thrust offset when a high-thrust motor is being used, such as

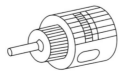

Figure 9.4 Typical configuration of a spinning spacecraft

during orbit-changing manoeuvres. Ideally, it is assumed that the thrust vector will pass through the centre-of-mass along the z-axis. Equations (9.4) indicate that the torque due to an offset will then cause coning and spinning, which is acceptable, with an average thrust in the correct direction.

The z-component of torque T_z produces an acceleration about that axis, which is uncoupled from the other two axes.

The response in non-spinning axes will be of interest when the spin axis has to be redirected and when the reason for spinning the structure is to provide gyroscopic rigidity *via* momentum bias. One of the axes is aligned with the spin axis (z-axis); the x- and y-axes do not rotate with the structure. The response equations become (see equation 3.47 of Chapter 3)

$$I_{xx}\dot{\Omega}_x + I_{zz}S\Omega_y = T_x$$
$$I_{xx}\dot{\Omega}_y - I_{zz}S\Omega_x = T_y \tag{9.6}$$
$$I_{zz}\dot{S} = T_z$$

Constant torque components T_x, T_y produce precessional angular rate components

$$\Omega_x = -T_y/I_{zz}S, \quad \Omega_y = T_x/I_{zz}S \tag{9.7}$$

These are the rotations that will be observed from outside the spacecraft, and they demonstrate gyroscopic rigidity in that as the bias ($I_{zz}S$) increases, so the response to a given torque decreases. The nutation mode, observed in these non-spinning axes, has frequency

$$\omega_{\text{nut}} = S(I_{zz}/I_{xx}) \tag{9.8}$$

and will need to be damped (see Section 9.3.4).

The root location diagrams corresponding to equations (9.3) and (9.6) are shown in Figure 9.6. The response of the spin-rate S corresponds to a single root at the origin, exactly the same as that for all the principal axes in the case when there is no momentum bias (equation 9.2). On the other hand the equations for the components ω_x, ω_y generate a pair of (oscillatory) roots at $\pm\mathrm{j}S((I_{zz}/I_{xx}) - 1)$; the corresponding components Ω_x, Ω_y in *non*-spinning axes generate roots at $\pm\mathrm{j}S(I_{zz}/I_{xx})$. These represent the nutation mode (Section 3.5.1 of Chapter 3).

9.3.3   Hybrid spacecraft

The dynamics of these craft are covered in Section 3.4.3 of Chapter 3. In the dual-spin or partially de-spun versions, momentum bias is provided by rotating a piece of structure on a bearing attached to the non-spinning part; it will be assumed that the spinning part has an axially symmetric mass distribution (see Figure 9.5).

In a three-axis stabilized version with momentum bias, the bias is provided by a momentum wheel (MW) (see Section 9.4.7).

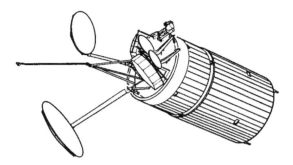

Figure 9.5 Typical configuration of a dual-spin
or partially de-spun hybrid spacecraft

The response equations for the hybrid craft with their bias H_z along the z-axis are
equations (3.57) of Chapter 3. For low body rates, they approximate to

$$I_{xx}\dot{\omega}_x + H_z\omega_y = T_x$$

$$I_{yy}\dot{\omega}_y - H_z\omega_x = T_y \tag{9.9}$$

$$I_{zz}\dot{\omega}_z + H_z = T_z$$

Constant torque components T_x, T_y produce precessional angular rates:

$$\omega_x = -T_y/H_z, \quad \omega_y = T_x/H_z \tag{9.10}$$

The torque component T_z about the bias direction produces an acceleration response about
that direction, which does not cross-couple into the other axes. In the case of MWs, the
total bias may be the vector sum of the biases from three or more wheels in different
directions; nevertheless it is the component of torque in the total bias direction that
produces the acceleration response described above.

The *nutation mode* has a frequency

$$\omega_{\text{nut}} = H_z/\sqrt{(I_{xx}I_{yy})} \tag{9.11}$$

and needs to be damped (see Section 9.3.4).

The root location diagrams for spacecraft with momentum bias, referred to non-spinning
axes (equations 9.9) are shown in Figures 9.6. Motion about the z-axis—the bias direction—in response to T_z, has a single root at the origin as in Figure 9.6(a).

The coupled motion about the x- and y-axes in response to T_x and/or T_y has (in
Figure 9.6b) two roots representing the nutation roots. As the amount of bias decreases
to zero, the roots move to the origin, as in Figure 9.6(a)

Reorientation manoeuvres in response to torques will differ depending upon whether the
manoeuvre changes the direction of the momentum bias, and will depend upon whether
the torquer is an On/Off device or whether its magnitude is controllable.

Torque pulses to produce a manoeuvre about a bias direction—or about any principle
axis if there is no bias—may be achieved by torque pulses as shown in Figure 9.7(b).

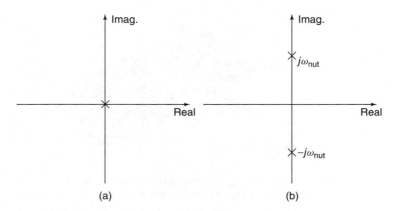

Figure 9.6 Root location diagrams: (a) ω_z in response to T_z; (b) ω_x, ω_y in response to T_x, T_y

Torque pulses used to change a bias direction—the z-axis of the spinner and hybrid spacecraft—will only produce a change of direction whilst the torque acts. At its start and at its termination it will excite the nutation mode as shown in Figure 9.7(a). A torque pulse lasting for an integer number of periods of the nutation oscillation will cancel the oscillation when it ceases.

Torquers whose torque magnitude is controllable admit the possibility of avoiding excitation of the nutation mode. The cross-coupling between pitch and yaw that is characteristic of the mode can be cancelled by introducing cross-coupling in the computer algorithms that control the torquers. Reference [4] shows how the technique of complex summation of the T_x and T_y equations can lead to a suitable algorithm when the spacecraft is axisymmetric.

9.3.4 Oscillatory modes

A characteristic of the space environment is that oscillatory modes have very little damping. The ACS has to avoid undue excitation of these and must include means of damping them. Damping may be enhanced by means of energy dissipation or by active control techniques.

The energy dissipation method is based upon the fact that the energy present in an oscillatory mode is exchanged between kinetic and potential types during each cycle. A constant total amount corresponds to a constant amplitude of oscillation that is proportional to the square root of the energy. By using the oscillation to excite an energy dissipator, the amplitude will decrease and the mode becomes damped. Active damping entails sensing the oscillation and applying a suppressing torque in the correct phase.

Nutation damping may be implemented either way. A passive damper may consist simply of a metallic tube containing a viscous fluid. The fluid is chosen to have a nearly constant viscosity over a wide range of temperatures keeping it from solidifying at low temperatures. A variety of fluids have been used, including alcohol and freon. The geometry of the tube is often a ring, and it is mounted at a location to ensure a good coupling with the nutational motion. When nutation occurs, the frictional force between the fluid

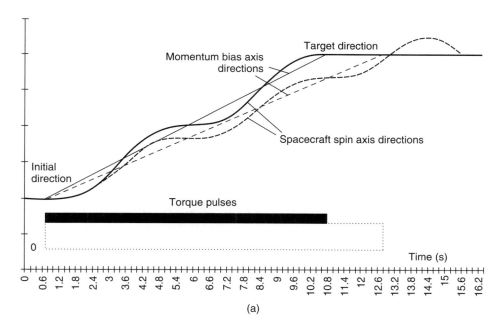

(a)

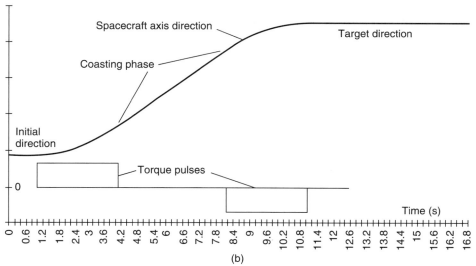

(b)

Figure 9.7 Torque responses: (a) repointing a momentum bias axis and (b) repointing about a principal axis, with no momentum bias

and the tube introduces energy dissipation that dampens the nutational mode. For the long-term stability of the dual-spin spacecraft, it should be mounted in the non-spinning part. (see Section 3.4.3 of Chapter 3.)

Active damping may be achieved by applying a torque component proportional to the angular velocity component along the same axis, that is, T_x proportional to ω_x. The axes may be the spinning (body) axes or non-spinning ones. For the hybrid spacecraft, equations (9.9) then become

$$I_{xx}\dot{\omega}_x + H_z\omega_y = T_x = -K_x\omega_x$$
$$I_{yy}\dot{\omega}_y - H_z\omega_x = T_y = -K_y\omega_y \tag{9.12}$$

where K_x, K_y are the feedback constants expressing the torque per unit angular rate.

The characteristic equation, in terms of the Laplace operator s, is

$$I_{xx}I_{yy}s^2 + (I_{xx}K_y + I_{yy}K_x)s + (K_xK_y + H_z^2) = 0 \tag{9.13}$$

This represents damped harmonic motion when K_x and/or K_y are non-zero and positive.

In practice, there will be other oscillatory modes present, and excited by the same torquers. These may be separated into the so-called rigid-body ones, and those associated with mass movement and flexure, as described in Section 3.5.2 of Chapter 3. The mathematical modelling covered in this chapter is restricted to the rigid body and thus will cover nutation and libration but not the other modes. A more comprehensive coverage is given in References [1] and [2].

9.3.5 Summary of attitude response

The ACS is a feedback system in which the spacecraft is the plant, shown as a block in a block diagram (Figure 9.2) in which torque affects attitude. The dynamic equations linking these are presented in Chapter 3, and above they are shown when the torque components are about the principal axes.

Whilst the torques will excite oscillatory modes, and these must ultimately be stable, their prime purpose is to control the orientation of the axes of the spacecraft. Each component of torque will have an effect, and by resolving into principal axes, their effects are largely uncoupled from each other. These simple responses to each separate component of torque are useful to be preserved as the dominant responses of the block. There are just two types—the precessional and the non-precessional. Figure 9.7 illustrates typical responses to torque pulses in order to achieve repointing in these two cases.

The *non-precessional response* shown in Figure 9.7(b) obeys, for example,

$$I_{xx}\dot{\omega}_x = I_{xx}\ddot{\phi} = T_x \tag{9.14}$$

The rotational response is an angular acceleration about the same principal axis as the torque. This applies to each axis when there is no momentum bias, but applies only to motion about the bias axis when it is present.

The *precessional response*, approximated by omitting the nutation mode, is the steady-state response, which is given by:

$$\Omega_y = \dot{\theta} = T_x/H_z \quad \text{and} \quad \Omega_x = \dot{\phi} = -T_y/H_z \tag{9.15}$$

The response equations with nutation included, in state-space form, are

$$
\mathrm{d}/\mathrm{d}t
\begin{bmatrix} \Omega_x \\ \Omega_y \\ \phi \\ \theta \end{bmatrix}
=
\begin{bmatrix}
0 & -H_z/I_{xx} & 0 & 0 \\
H_z/I_{yy} & 0 & 0 & 0 \\
1 & 0 & 0 & 0 \\
0 & 1 & 0 & 0
\end{bmatrix}
\begin{bmatrix} \Omega_x \\ \Omega_y \\ \phi \\ \theta \end{bmatrix}
+
\begin{bmatrix}
1/I_{xx} & 0 \\
0 & 1/I_{yy} \\
0 & 0 \\
0 & 0
\end{bmatrix}
\begin{bmatrix} T_x \\ T_y \end{bmatrix}
\quad (9.16)
$$

where $\dot{\theta} \approx \Omega_y$ and $\dot{\phi} \approx \Omega_x$.

9.4 TORQUES AND TORQUERS

The torques, arising from moments of forces about the centre-of-mass, must be identified as being *external* or *internal* to the spacecraft. The former will affect its total angular momentum, whereas the latter will affect only the distribution of momentum between its moving parts. The case has already been made that it is *necessary* to include controllable external torquers whereas internal ones, with their momentum storage facility, are *optional* (see Section 3.3.2 of Chapter 3).

The main sources of torques, occurring either naturally or as disturbances, are introduced below and summarized in Table 9.1. The magnitude of torques in space is small when compared with terrestrial standards. Even very small ones become significant when there is no friction to oppose them and when the orientation has to be very accurate.

Some of the phenomena listed as disturbance torques in Table 9.1 may be used as a means of achieving the required orientation of the spacecraft. For this, they will normally need to be controllable; a possible exception is the gravity-gradient torque, which will establish an Earth-facing equilibrium orientation passively, with the axis of least inertia along the local vertical (see Section 9.4.3).

Table 9.2 summarizes the main advantages and disadvantages of various types of torquer.

9.4.1 Thrusters (external type)

Orbit-changing thrusters provide potentially the largest source of force on spacecraft, and potentially the largest source of torque. Being external, the torque will affect the total momentum. Ideally, the thrust vector passes through the centre-of-mass, but inevitably there is a tolerance on this and consequently a disturbance torque arises.

The main means of countering the effects of this torque when large thrust levels are present are either to spin the vehicle about the intended thrust direction (cf. Section 9.3.2) or to provide means of controlling the achieved thrust direction. At lift-off, for example, the latter method must clearly be used. This involves mounting some of the thrusters in gimbals, or using secondary fuel injection into the rocket nozzle, and controlling the thrust direction so as to achieve the required trajectory. Later stages of booster rockets may adopt the alternative method of spinning the vehicle in order to average out the effect of thrust offset upon the trajectory. Thruster firings used for changing the orbit, such as from Low Earth Orbit (LEO) to Transfer Orbit, and again from Transfer Orbit to Geostationary Earth orbit (GEO), are sometimes preceded by a spin-up manoeuvre, followed by de-spin after the orbit changes are complete.

Table 9.2 Types of torquer

Type	Advantages	Disadvantages
External types	Can control momentum build-up	
Gas jets	Insensitive to altitude Suit any orbit Can torque about any axis	Requires fuel On-off operation only Has minimum impulse Exhaust plume contaminants
Magnetic	No fuel required Torque magnitude is controllable	No torque about the local field direction Torque is altitude and latitude sensitive Can cause magnetic interference
Gravity gradient	No fuel or energy needed	No torque about the local vertical Low accuracy Low torque, altitude sensitive Libration mode needs damping
Solar radiation	No fuel required	Needs controllable panels Very low torque
Internal types	No fuel required Can store momentum Torque magnitude is controllable	Cannot control momentum build-up
Reaction wheels (RW)	Continuous, fine-pointing capability	Non-linearity at zero speed
Momentum wheels (MWs)	Provide momentum bias	
Control moment gyroscope (CMG)	Suitable for three-axis control Provides momentum bias	Complicated Potential reliability problem

Thrusters with very much lower levels of thrust are in common use in attitude-control systems for providing controllable external torquing, and hence controlling the total momentum of the spacecraft. For this purpose, they will be mounted in clusters on the surface of the vehicle, pointing in different directions in order to provide three components of torque. They have a number of advantages and disadvantages compared with their main rival, the magnetic torquer.

Their main advantage is that their torque level is independent of altitude and there is potentially no limit to its magnitude. However, the magnitude is not controllable when installed; only the switch-on duration is. This torquing system integrates well with the station-keeping requirement for thrusters, since a common fuel and control system can be used.

Thrusters have a number of disadvantages. If they are used as the prime means of orientating a spacecraft, their restriction to an on–off type of control leads to a limit cycle occurring, a similar process to that occurring in station-keeping manoeuvres (see Sections 5.6.3 and 5.6.4 of Chapter 5) but with a much shorter period. If they are used to achieve accurate pointing, then the torquers will need to provide small and consistent impulses, and their minimum switch-on time of several milliseconds leads to a low maximum torque being needed. Typically a minimum thrust impulse of order 10^{-4} Ns may be required, with a thrust level as low as 10^{-2} N. When the prime means of attitude control is a reaction wheel or momentum wheel (Section 9.4.7), then thrusters may be used for momentum dumping. In this case, their thrust level will be higher than above, and they may be used perhaps only once every several orbits.

A variety of thruster systems may be used, ranging from cold gas to electric propulsion, as indicated in Sections 6.3 and 6.4 of Chapter 6.

The fact that thrusters use fuel is another disadvantage. Although a large amount of fuel is not normally needed for attitude control it will eventually be exhausted, and a number of spacecraft have reached the end of their useful life because of this.

9.4.2 Magnetic torque (external type)

The magnetic field generated by a spacecraft interacts with the local field from the Earth and thereby exerts an external couple on the vehicle. This is a similar effect to that of a compass needle that attempts to align itself with the local direction of the field. If the spacecraft's magnetism is represented as a dipole whose magnetic moment is $\mathbf{m}$, then it reacts with the local flux density $\mathbf{B}$ to produce a torque $\mathbf{T}$ given by

$$\mathbf{T} = \mathbf{m} \times \mathbf{B} \tag{9.17}$$

Care must be taken that electric currents and spurious magnetic effects do not cause a significant disturbance torque (see Chapter 16).

Electromagnets may be used to provide a controllable external torque. Their strength can be controlled by means of the current I. Their reaction with a local field $\mathbf{B}$ leads to the couple:

$$\mathbf{T} = n I A (\hat{\mathbf{c}} \times \mathbf{B}) \tag{9.18}$$

where n is the number of turns

      A is the cross-sectional area of the coil,

      $\hat{\mathbf{c}}$ is the unit vector in the direction of the coil's axis.

Rod-like electromagnets will normally be used, such as those shown in Figure 9.8. A range of strengths is available to suit the mission requirements, and they may be used in an on–off or a proportional control manner, for attitude control or momentum dumping.

Three orthogonal magnets enable the direction and magnitude of the dipole to be controlled. Their mounting locations should be away from instruments that are sensitive to magnetic fields and separate from each other in order to avoid cross-coupling.

Magnetic torquers are in common use in satellites orbiting at altitudes up to geostationary altitude, but their utility decreases at the higher altitudes since the strength of the Earth's field reduces with height. The field's strength and direction also vary with the position of the spacecraft in its orbit in general, and when using magnetic torquers it is common practice to carry a magnetometer to measure the local field.

A feature of magnetic torquers is that they cannot produce a torque component about the local field direction. In a polar orbit, any required torque direction can always be achieved at some point in the orbit since the field direction changes round the orbit. In the equatorial plane, however, the field lines always lie horizontally, north–south.

Figure 9.8 Magnetic torquers. The three torque rods shown are those used on the EURECA spacecraft, and are each in excess of a metre in length (Reproduced by permission of Fokker Space bv)

Consequently a spacecraft whose orbit lies in this plane cannot use magnetic torquers to counteract the north–south component of their disturbance torque, or to dump this component of momentum. A reorientation about the north–south direction is achievable nevertheless, by a rotation of the torque vector about this direction. In the resulting coning motion, the reorientation will slowly take place.

An advantage of magnetic torquers is that they require no fuel and so have virtually unlimited life. They do of course require electrical power. But there is no exhaust pollutant and by providing a couple they are not sensitive to movement of the centre-of-mass.

9.4.3 Gravity-gradient torque (external type)

This source of torque occurs because in a gravitational field that gets weaker with increase in height, a body will only be in stable equilibrium if its axis of minimum inertia is aligned with the local vertical.

The gravitational force dF on an increment of mass dm is

$$dF = \frac{\mu \, dm}{r^2} \tag{9.19}$$

where μ is the Earth's gravitational constant $= 0.3986 \times 10^{15} \, \text{m}^3/\text{s}^2$,
         r is the distance from the Earth's centre.

By summing the moments about the centre-of-mass C, the torque components may be shown to be

$$
\begin{aligned}
T_x &= (3\mu/2r^3)(I_{zz} - I_{yy}) \sin 2\phi \cos^2 \theta \\
T_y &= (3\mu/2r^3)(I_{zz} - I_{xx}) \sin 2\theta \cos \phi \\
T_z &= (3\mu/2r^3)(I_{xx} - I_{yy}) \sin 2\theta \sin \phi
\end{aligned}
\tag{9.20}
$$

where ϕ and θ refer to the roll (about the x-axis) and pitch (about the y-axis) angles, respectively, using the aircraft axis convention (see Figure 3.A2 in the Appendix of Chapter 3).

These torques contribute to the total disturbance torque in general, but an oscillatory 'libration' mode will occur if they govern the motion about the equilibrium state (see Section 3.5.1 of Chapter 3). For small oscillations of an axisymmetric spacecraft ($I_{yy} = I_{xx}$), the motion is like a conical pendulum, whose frequency is

$$\omega_{\text{lib}} = \sqrt{[(3\mu/r^3)(1 - I_{zz}/I_{xx})]} \text{ rad/s} \tag{9.21}$$

Gravity-gradient torque provides a passive self-aligning torque that has been used (e.g. LDEF and UOSAT), but the libration does need damping to be incorporated. The torque levels will be low unless a long thin configuration is used, or in the case of tethered satellites.

9.4.4 Aerodynamic torques (external type)

Aerodynamic torques are dominated by the drag force, which is dependent on frontal area A. Their total moment about the centre-of-mass C may be assessed by considering the projection in the direction of travel. If the spacecraft surface comprises a collection of small incremental areas dA, each with unit normal vectors $\hat{\mathbf{n}}$, such that the position vector of the centre-of-area of each such small area with respect to C is $\mathbf{r}$, then the aerodynamic torque is given by

$$\mathbf{T}_{\text{aero}} = \int_A \mathbf{r} \times d\mathbf{F}_{\text{aero}}, \quad d\mathbf{F}_{\text{aero}} = \frac{1}{2}\rho V_a^2 C_D (\hat{\mathbf{n}}.\hat{\mathbf{V}}_\mathbf{a})(-\hat{\mathbf{V}}_\mathbf{a}) \, dA$$

where ρ is the atmospheric density, $\mathbf{V_a}$ is the air-relative velocity of the vehicle (and $\hat{\mathbf{V}}_\mathbf{a}$ denotes its unit dimensionless vector), and C_D is the drag coefficient (normally taken to be about 2.2). The area A over which the integral is performed is the spacecraft surface for which $\hat{\mathbf{n}}.\hat{\mathbf{V}}_\mathbf{a} \geq 0$—in other words, the surfaces exposed to the incoming flow of atmospheric particles. Usually this integral is not amenable to simple solution for a surface A associated with a complex spacecraft configuration. A commonly used alternative to assess the aero-torques is to represent the spacecraft as a collection of simple geometrical elements (e.g. flat plates over which the integral can be performed easily), and construct a simple summation over all such surfaces exposed to the incoming particles. For a configuration model comprising n flat plates, each with unit normal $\hat{\mathbf{n}}_i$ and area A_i, then

$$T_{\mathrm{aero}} = \sum_{i=1}^{n} \mathbf{r}_i \times \mathbf{F}_{\mathrm{aero},i}, \quad \mathbf{F}_{\mathrm{aero},i} = \frac{1}{2}\rho V_a^2 C_D(\hat{\mathbf{n}}_i.\hat{\mathbf{V}}_\mathbf{a}) A_i(-\hat{\mathbf{V}}_\mathbf{a}) \tag{9.22}$$

for values of i for which $\hat{\mathbf{n}}_i.\hat{\mathbf{V}}_\mathbf{a} \geq 0$. Here $\mathbf{r}_i$ is the position vector of the centre-of-area of the ith plate with respect to C. This technique gives an adequate assessment of the aero-torques, but does not take account of issues such as shadowing of one element by another, or of multiple collisions of particles with the spacecraft.

For zero torque, spacecraft designers will aim to balance the terms in the summation (9.22), but engineering tolerances, shifts of the centre-of-mass and thermal distortion will usually give rise to a residual torque.

The torque is height-dependent, and is not an important effect above about 600 to 700 km, depending upon the spacecraft configuration and the level of solar activity.

9.4.5   Solar radiation pressure (external type)

Solar radiation produces a force on a surface, which depends upon its distance from the Sun; it is independent to first order of the height above the Earth. Large flat surfaces with a significant moment arm about the centre-of-mass, such as solar arrays, may produce a significant torque. Since light carries momentum, when it is reflected at a surface this represents an exchange of momentum with the surface, which gives rise to the SRP. This is not to be confused with the pressure exerted on the spacecraft by the solar wind (the stream of high-energy ionized particles emanating from the Sun), which at Earth orbit can be several orders of magnitude less than light pressure. The force exerted on the surface can be calculated by a vector difference between the incoming and outgoing momentum fluxes. If $\hat{\mathbf{s}}$ is the unit vector from the spacecraft to the Sun, and $\hat{\mathbf{n}}$ is the outward unit normal of an incremental area dA, then the force due to solar radiation is [1]

$$d\mathbf{F}_{\mathrm{SRP}} = -P\cos\theta\, dA\left[(1 - f_s)\hat{\mathbf{s}} + 2\left(f_s\cos\theta + \tfrac{1}{3}f_d\right)\hat{\mathbf{n}}\right], \quad \hat{\mathbf{s}}.\hat{\mathbf{n}} \geq 0$$

where P is the mean momentum flux $\sim 4.67 \times 10^{-6}\,\mathrm{Nm^{-2}}$ at the Earth, $\theta = \cos^{-1}(\hat{\mathbf{s}}.\hat{\mathbf{n}})$ is the angle of incidence of the radiation and f_s, f_d are the coefficients of specular and diffuse reflection, respectively.

The torque may be obtained by integrating the moment of the above force over the spacecraft's illuminated surface area. In practicality, this integration can be problematic, so a similar method to that described for aero-torques (Section 9.4.4) may be adopted.

For example, if the configuration is approximated by a collection of $i = 1$ to n flat plates with area A_i, normal $\hat{\mathbf{n}}_i$ and incidence θ_i, then the SRP torque can be expressed as

$$\mathbf{T}_{\text{SRP}} = \sum_{i=1}^{n} \mathbf{r}_i \times \mathbf{F}_{\text{SRP},i}, \quad \mathbf{F}_{\text{SRP},i} = a_i\hat{\mathbf{s}} + b_i\hat{\mathbf{n}}_i \tag{9.23}$$

where the summation is performed over the plates that are illuminated. Here $\mathbf{r}_i$ is defined as in equation (9.22), and

$$a_i = -PA_i \cos\theta_i(1 - f_{\text{s},i}), \quad b_i = -2PA_i \cos\theta_i(f_{\text{s},i}\cos\theta_i + \tfrac{1}{3}f_{\text{d},i})$$

Both aero and solar radiation pressure torques are in principle usable for counteracting momentum build-up, and for trimming attitude (e.g. INMARSAT 2), but usually the spacecraft is designed such that the forces balance to give zero torque.

9.4.6 Mass movement (internal type)

The movements of masses within a spacecraft may directly exert torques upon the main structure. These are classified as internal torques and do not affect the total momentum. The movements may also alter the location of the centre-of-mass within the spacecraft, and change the inertia matrix.

The centre-of-mass C has been identified as a key reference point for establishing the dynamic behaviour (Chapter 3). Moving the point affects the balance of the vehicle in dynamic ways. It also affects the torques due to forces on the vehicle, but not the couple of the magnetic torquer. In principle, the centre-of-mass location could be controlled in order to balance out the disturbance torques.

A major source of mass movement is that of the fuel. The tanks are normally located in such a way that as their contents are used up the centre-of-mass does not shift. Fuel movement within the tanks causes a different sort of problem in that it moves in a dynamic way in response to the motion of the spacecraft—fuel slosh—affecting its modal characteristics.

Mass movements from one position to another, such as the erection of solar arrays and other appendages and movement of astronauts and so on, have an effect upon attitude, which is best assessed by using the fact that angular momentum is conserved.

9.4.7 Momentum storage torquers (internal type)

Torquers associated with momentum storage such as RWs and MWs are essentially internal torquers, suitable for attitude control but not for controlling the total momentum.

These devices are purpose-built precision-engineered wheels that rotate about a fixed axis, with a built-in torque motor. Figure 9.9 shows a photograph of a MW with its protective cover removed. The basic design comprises a reliable bearing unit (with a life

Figure 9.9 Momentum wheel with cover removed showing
integrated drive electronics. The wheel illustrated is the RDI 68
(Reproduced by permission of TELDIX GmbH)

expectancy of typically 15 years), a spoked flywheel mass and a DC motor in a vacuum-tight, evacuated housing. The stator is controlled by the drive electronics, which may be seen under the wheel. A reaction wheel is shown in Figure 15.14 of Chapter 15.

It is difficult to generalize about the range of wheel characteristics available for use in 'large' spacecraft (mass of the order of a tonne or more), but we may attempt to summarize it as—wheel diameter ~20 to 40 cm; wheel mass (including torque motor) ~3 to 10 kg; momentum storage capacity ~5 to 70 kg m^2/s. With the relatively recent trend to use of wheels (see Chapter 18) to control 'small' satellites (mass of the order of 50 kg), much smaller wheels are required, and these are often designed and manufactured as part of the smallsat programme.

Reaction wheels have a nominally zero speed, and may be rotated in either direction in response to the control torques called for by the spacecraft's ACS. However, it should be noted that at low or zero angular rate, the wheel displays a non-linear response due to 'sticking friction', which can impose an irregular motion on the spacecraft in this region. This problem is often circumvented by setting the nominal operating speed of the wheels above zero rate, at a few rpm.

Momentum wheels on the other hand have a high mean speed (in the range typically 5000 to 10 000 r.p.m.) in order to provide momentum bias. The control torques will then slow down or increase the wheel speed, the permissible amount being about 10% of the mean value.

Both types of wheel provide momentum storage, and need to be used in conjunction with external torquers, as described in Section 9.2.2. (See also Chapter 15.)

For three-axis control, three orthogonal reaction wheels will be the minimum requirement. A redundant fourth is normally added at an equal angle to the other three, in order to avoid a single-point failure. When more than one MW is used, the total bias is the *vector* sum of contributions from the separate wheels.

The principle of MWs has been extended by the development of more advanced forms, such as control moment gyroscopes (CMGs). By mounting the wheel in gimbals fitted with torque motors, all three components of torque may be developed from a single wheel. This can be done to a limited extent with sophisticated wheels mounted on five-degree-of-freedom magnetic bearings (see Chapter 15). There is potential for incorporating attitude sensing with momentum storage and momentum bias in sophisticated devices of this type.

9.5 ATTITUDE MEASUREMENT

9.5.1 Attitude: its meaning and measurement

The meaning of 'attitude' or 'orientation' usually presents no conceptual difficulties. There must be some datum frame of reference, and once this has been chosen then the attitude of a spacecraft refers to its angular departure from this datum. A right-handed set of axes is normally used in order to define a frame of reference, and if both a datum set and a set of spacecraft axes are chosen, then the attitude may be defined in a way that may be quantified.

Specifying attitude may be done in a number of ways such as Euler angles, direction cosines, quaternions and so on [3]. Three pieces of information are needed. A common way is to use the three Euler angles that are defined in the same way as is standard practice for aircraft. These are the angles of yaw ψ, pitch θ and roll ϕ, as measures of the rotations about the z-, y- and x-axes, respectively, in that sequence, which are needed to bring the datum axes into alignment with those of the spacecraft. Figure 3.A2 in the Appendix of Chapter 3 illustrates these rotations.

For a spacecraft in circular orbit, whose z-axis is nominally down the local vertical and whose x-axis is nominally in the direction of travel, the aircraft's standard is frequently used. For other applications, a star-fixed (inertial) set would be better. There is, however, no universally accepted standard for specifying a spacecraft's attitude.

A potential problem when using Euler angles as above is that there is a singularity when the pitch angle θ is 90°. Whilst a set of angles $(\psi, 90°, \phi)$ may be chosen to specify any such attitude, it is not a unique set. For example, if the aircraft's set is used for Shuttle, namely, x is horizontal, z is vertically down and y completes the right-handed set, then when it is on the launch-pad its pitch attitude is 90° and its yaw and roll angles cannot be uniquely specified. Such a problem may be overcome by choosing a more suitable datum set.

It is worth noting that angles, and consequently attitude, are not vector quantities. The combination (ψ, θ, ϕ) should not be thought of as three components of a vector. On the other hand, the rates of change of $\dot{\psi}$, $\dot{\theta}$, $\dot{\phi}$ can be interpreted as vector quantities whose directions are along the (non-orthogonal) axes about which the rotations take place (see Figure 3.A2 in the Appendix of Chapter 3). Resolving $\dot{\psi}$, $\dot{\theta}$, $\dot{\phi}$ along spacecraft axes enables the components of the spacecraft's angular velocity ω relative to the datum axes to be expressed as

$$\omega_x = \dot{\phi} - \dot{\psi} \sin \theta$$
$$\omega_y = \dot{\theta} \cos \phi + \dot{\psi} \cos \theta \sin \phi \qquad (9.24)$$
$$\omega_z = \dot{\psi} \cos \theta \cos \phi - \dot{\theta} \sin \phi$$

The inverse relationship is

$$\dot{\psi} = (\omega_y \sin\phi + \omega_z \cos\phi)/\cos\theta$$
$$\dot{\theta} = \omega_y \cos\phi - \omega_z \sin\phi \tag{9.25}$$
$$\dot{\phi} = \omega_x + (\omega_y \sin\phi + \omega_z \cos\phi)\tan\theta$$

When the angles are small, then $\dot{\psi} \approx \omega_z$, $\dot{\theta} \approx \omega_y$ and $\dot{\phi} \approx \omega_x$.

Equations (9.25) indicate how, by integration, the attitude in the form of the Euler angles (ψ, θ, ϕ) may be obtained from measured components of angular velocity. The singularity at $\theta = 90°$ shows up in the form of $\tan\theta$ and will lead to problems with the integration as θ approaches this value.

9.5.2   Measurement system fundamentals

Fundamentally, the measurement of attitude requires the determination of *three* pieces of information that relate the spacecraft axes to some datum set, whether they are in the form of Euler angles or in other forms. The measurement subsystem must include sufficient sensors to enable the information to be extracted with the necessary accuracy and with reasonable simplicity. This must be done at all phases of the mission.

There are *two categories of sensor*, and they are commonly used to complement each other in a measurement system:

- The *reference sensor* gives a definite 'fix' by measuring the direction of an object such as the Sun or a star, but there are normally periods of eclipse during which its information is not available.
- *Inertial sensors* measure continuously, but they measure only *changes* in attitude, effectively relative to a gyroscope rotor. They therefore need a fix—a calibration from reference sensors. In between fixes, their errors progressively increase because of random drifts.

A *measurement system* may be formed by using reference and inertial sensors to complement each other. In a simple combination, the reference sensors will calibrate the inertial sensor at discrete times and the latter will then effectively 'remember' the reference object's direction until the next calibration. This allows a period in eclipse to be covered. The accuracy of the system will fluctuate, being that of the reference sensor at the calibration instant, and steadily degrading until the next calibration, as shown in Figure 9.10.

It is clear that the achievement of good system accuracy calls for good accuracy from the reference sensors, and a low degradation (drift) rate from the inertial sensors. In practice, it is likely that the mixing will take place in a computational Kalman filter to minimize errors, the design of which is a specialist topic.

Complete attitude information requires *three* pieces of information as explained above. Reference sensors that are based upon detecting the direction of a single vector are incapable of providing all three pieces. A sun sensor cannot detect any rotation of spacecraft about the Sun vector, for example. Two vector directions, ideally orthogonal, are needed for complete attitude information to be obtained from simultaneous measurements.

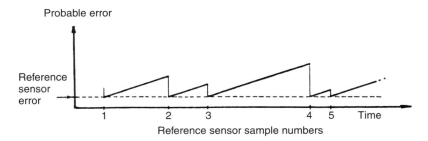

Figure 9.10 Error history for a mixed reference/inertial sensor system

A star sensor may have the ability to track more than one star within its field of view (FOV), thereby monitoring the two or more vector directions that are needed for full attitude information. The angle between these directions will be far from the ideal of 90°, however.

In some cases, the two different vector directions can be monitored by just one sensor, but with the measurements separated by a time interval. Then the change in attitude during the interval must also be measured, by using inertial sensors, and the time interval must be short enough for the build-up of errors from these sensors to be acceptable.

A star scanner is one example. It scans the heavens and so one sensor uses different star directions for complete attitude determination. The time lapse between viewing two stars in orthogonal directions will be short enough to avoid significant build-up of errors.

An Earth sensor (detecting the local vertical) can be used in a similar fashion. The local vertical rotates in space through 90° in a quarter of a circular orbit, a lapse of about 23 minutes in LEO and 6 hours in geostationary orbit. The Sun vector's '90° time' is three months—too long for use in this way! In practice, the Earth vector direction will be monitored continuously; the '90° time' merely gives an indication of the degradation in accuracy due to the sensor system's reliance upon inertial sensors.

The accuracy required will normally be set by the payload, its pointing direction and the required measurement accuracy. There will be an ultimate accuracy of measurement that is determined by the object used by the sensor; stars provide the most accurate sources, with the Sun and Earth being progressively less accurate by virtue of the angle that they subtend at the sensor and the fuzziness of the Earth horizon. A rough guide to accuracies is shown in Table 9.3.

Table 9.3 Potential accuracies of reference sensors

Reference object	Potential accuracy
Stars	1 arc second
Sun	1 arc minute
Earth (horizon)	6 arc minutes
RF beacon	1 arc minute
Magnetometer	30 arc minutes
Narstar Global Positioning System (GPS)	6 arc minutes

Note: This table gives only a guideline. The GPS estimate depends upon the 'baseline' used (see text).

The extent to which the ultimate accuracy is realized in a given instrument depends upon its design, and this will be related to its intended use. Each sensor will have a limited FOV, beyond which it gives no information. In getting a spacecraft to its intended attitude it will normally be necessary to include very wide-angle low-accuracy sensors for use when steering the craft towards the state in which the accurate sensor has its objective in its field of view.

Each phase of the mission must be addressed when the list of sensors is being compiled. In the early stages, the visibility conditions will be quite different from those when it is on station, partly because the orbit is different, but also because of stowed arrays and so on. The sensor list must also cover the possibility that the spacecraft attitude may need to be recaptured following a failure of some sort.

The *datum axes* that are used when defining the attitude of the spacecraft are normally related to its payload. A telescope is likely to need star-based datum axes for specifying its attitude, for example. For the control system, it is the error measured from the intended attitude that is important and in many systems, sensors will be chosen to measure this directly. For example, the pointing of the solar arrays towards the Sun will use a Sun sensor. Earth-facing hardware will possibly use the local vertical as a datum, or maybe beacons on the ground that have been set up for the purpose. The error in measuring the pointing direction of the payload will be reduced if the sensor uses a related reference object.

Errors in measuring the payload's pointing direction will also depend upon the physical separation between it and the sensor. If each is mounted on the same base, then the mounting tolerances of both instruments will be sources of error, and so will distortions of the base due to thermal or other effects. Error paths of this type should be kept to a minimum and must be carefully assessed when the location of sensors measuring payload-pointing directions is being considered. For extreme accuracy, the payload pointing will be calibrated in space.

9.5.3 Types of reference sensor

There are numerous different engineered forms of sensor. Only a brief review of the selection on offer is given here.

Sun sensors

The Sun subtends an angle of about 30 arc minutes at Earth, and provides a well-defined vector, which is unambiguous because of the intensity of the radiation. Sensors range from mere presence detectors that determine whether the Sun is in a specified FOV, to instruments that measure its direction to an accuracy of better than one arc minute. An example of the latter is the *reticle slits mask detector*, shown schematically in Figure 9.11. The sensor comprises a transparent block of known refractive index, coated in a thin opaque film. An entry slit is etched into the upper surface, and reticle slits in the base, and the block mounted on a sequence of photo-cell detectors. When exposed to the Sun, a plane of sunlight intersects the base so that some of the photo-cells receive photons giving a '1' output, whereas others do not and so are assigned a '0'. The digital output can be mapped onto a unique entry angle, using the Gray code, allowing the plane in which

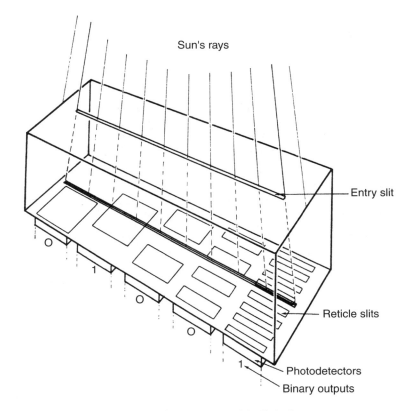

Figure 9.11   Sun sensor with digital output

the Sun lies to be determined. Two such detectors, mounted on a three-axis stabilized spacecraft with their entry slits orthogonal to each other, give sufficient information to determine the spacecraft-Sun vector in the body axes. One detector mounted on a spinning spacecraft is sufficient to determine the solar aspect angle—between the Sun vector and the spin axis.

Another type of digital sun sensor again has orthogonal entry slits, which permits photons to be incident upon two orthogonally-mounted linear charge coupled device (CCD) arrays—again allowing the spacecraft-Sun vector to be determined.

An alternative method to determine the Sun vector on a spinning vehicle is to use a '*V slit*' *sensor*, an example of which is illustrated in Figure 9.12. This shows the sensor array on the *STRV-1* (Space Technology Research Vehicle), which is a spinning microsatellite (~50 kg mass) built by QinetiQ at Farnborough, UK. The V slit Sun sensor is indicated by the arrow. The Sun is viewed through the two slits, set at an angle as illustrated. Detection of the Sun through the first, 'vertical' slit gives a measure of the rotational 'azimuth' of the spacecraft relative to the Sun. With knowledge of the spin rate of the vehicle, the time taken for the Sun to appear in the second slit, gives a measure of the 'elevation' of the Sun above or below the plane in which the instrument rotates. Other types of optical/detector combinations are available.

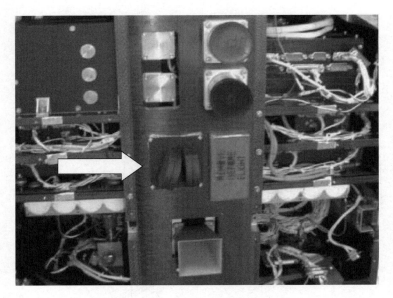

Figure 9.12   The V slit Sun sensor on the STRV-1 microsatellite
(Reproduced by permission of QinetiQ)

Earth sensors

The Earth, radius R_e, subtends an angle $2\sin^{-1}[R_e/(R_e + h)]$ at a spacecraft at height
h. At $500\,km$ altitude this is about $135°$ (so that an Earth-presence detector is not very
useful!), and this falls to about $17.5°$ at geostationary height. The determination of the
nadir vector (the spacecraft-to-geocentre direction) can be achieved, however, by sensing
the position of the Earth's horizon, and Earth-horizon sensors provide the means of doing
this. For example, the nadir vector can be determined by bisecting the directions to the
horizon at the ends of a diameter of the Earth's disc.

Earth sensors are usually designed to operate in the infra-red, often in the $\sim15\,\mu m$ CO_2
absorption band due to the following reasons:

- There is less variation between maximum and minimum radiance compared to the
  visible band.
- The *terminator* (the dividing line between night and day on the Earth's disc) disappears in the infra-red. Although the night-time is generally cooler than the day,
  the variation is small compared to absolute zero. The modelling of the complex and
  variable geometry of the visible terminator is also avoided.
- The 'infra-red Earth' is always present as a reference object, even when the spacecraft is in eclipse.
- However, *bolometers* (heat detectors) are required, which generally have a slow
  response.

The *static Earth-horizon sensor*, which is illustrated schematically in Figure 9.13, is a
conceptually simple device used on three-axis stabilized spacecraft in high circular orbits,
in particular, GEO. The infra-red image of the Earth is focused on the focal plane of the

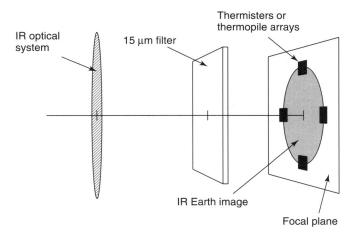

Figure 9.13   Schematic of a static Earth-horizon sensor

instrument. With nominal attitude, the image falls equally upon heat-detecting arrays, as shown. However, if the spacecraft attitude has errors in pitch or roll, the 'signal' will decrease in some arrays and increase in others. The instrument cannot detect errors in yaw (rotations about the local vertical).

A *scanning Earth-horizon sensor* is generally used in LEO to determine the nadir vector. Typically, a narrow FOV ('pencil beam') is swept across the Earth horizon to detect the abrupt change in infra-red signal between viewing 'deep space' and Earth. The beam scans the horizon by means of an internal mechanism or by using the rotation of a spinning spacecraft. Figure 9.14 shows schematically a typical scan geometry. The pencil beam traces out a cone, of known semi-angle γ, that intersects the Earth horizon at two locations, one corresponding to acquisition of the infra-red signal and the other to the loss of signal. With knowledge of the scan rate, the measurement made is an estimate of the fraction of the scan period between acquisition and loss of signal. This, combined with knowledge of γ and the Earth's angular size, can be used to estimate the angle between the spin axis of the scan mechanism and the nadir vector. Figure 9.15 shows a scanning infra-red Earth-horizon sensor, which has two such scan cones on opposite sides of the spacecraft. If their pulses are of equal duration, then the local vertical lies in the plane bisecting the cone angles, and this can form the basis of a nulling control system for roll motion. Pitch sensing uses the fact that the local vertical lies in the plane defined by the bisector of the pulse from each conical scan.

For this system, one cone angle will be suitable for only a limited range of altitudes. In some sensors, the angle may be adjusted to accommodate the changes in altitude as the spacecraft moves from LEO to geostationary height.

Star sensors

Star sensors are the most accurate reference sensors in common use for measuring attitude. Accuracies of 1 arc second or better may be obtained. But the large number of stars means that sophisticated techniques are needed in the instrument and its associated computer in order to identify any particular star in its FOV. This problem is usually overcome by

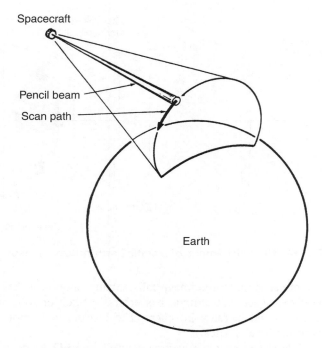

Figure 9.14 Scanning technique for an Earth horizon sensor

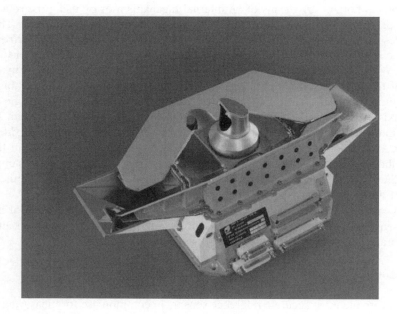

Figure 9.15 A scanning infra-red Earth-horizon sensor for LEO. The unit illustrated is the STD16 sensor (Reproduced by permission of SODERN)

relying on a degree of *a priori* knowledge of the attitude of the spacecraft, or by adjusting the detection threshold of the sensor to reduce the number of 'targets'. Star sensors once had the reputation of being massive, power-hungry and expensive, but the advent of CCD technology has led to considerable improvements in their system characteristics.

Star sensors may be classified as

- *Star scanners* for mounting on a spinning spacecraft. The spin provides a means for the instrument to scan the sky. The characteristics of stars passing through the FOV can be compared with a star directory in order to determine attitude. A 'V slit screen' is often used (see section on Sun sensors) to determine the spacecraft-star vector.
- *Star trackers* for mounting on a three-axis stabilized spacecraft. Their detectors/ controllers enable them to select, locate and track one stellar image with precision. Clearly one such fix is insufficient to determine the spacecraft attitude uniquely. Two such trackers 'staring' in orthogonal directions, as used on the US Space Shuttle, will provide an optimal, unique attitude estimate.
- *Star mappers* for mounting on a three-axis stabilized spacecraft. Their FOV is suffi- cient to include several stars. They locate and record the position of each, so allowing the spacecraft's orientation about the sensor's axis to be determined.

Considerable improvements in star sensors, referred to above, have been achieved with the use of solid-state detectors. The CCD sensor has led to lower mass, volume and power characteristics for the same performance in terms of accuracy. Figure 9.16 shows two SED16 star sensors, each with a 1024×1024 CCD matrix array detector. This particular unit can track up to 10 stars simultaneously, and has a fully autonomous ('lost-in-space') capability. Its mass is 2.55 kg (plus baffle at 0.7 kg), and the power consumption around 10 W.

Figure 9.16 CCD star sensors. The units shown are SED16 sensors (Reproduced by permission of SODERN)

Radio frequency beacons

Direction-finding techniques may be used to detect the direction of an RF source, with an accuracy of order 1 arc minute. There are several techniques by which this can be done.

For example, Ulysses, rotating at a nominal 5 rpm, carries an antenna whose axis is offset from the spin axis. The intensity of the signal that it receives from a ground station is thereby modulated at the spin frequency. The actual spin rate and its phase, and the angle between the spin axis and the ground station direction, can then be derived respectively from the frequency, the phase and the depth of modulation.

Magnetometers

The magnetometer is a robust instrument but with an accuracy that is limited to about 0.5°. It measures the direction and possibly the strength of the local magnetic field. But the field is not well mapped and has abnormalities that make the sensor of limited use for attitude sensing. To gain attitude information, the measured field is compared to a magnetic field model held in the on-board processor. The magnetometer is also used in conjunction with magnetic torquers as described in Section 9.4.2.

Attitude determination using global navigation satellite systems (GNSS)

GNSS, such as the Navstar GPS system, is commonly used for the determination of orbital position, but it can also be used to determine spacecraft attitude. Figure 9.17 illustrates the principle. A set of GPS patch antennae (at least three), slaved to a GPS receiver, are located on an 'upward looking' face of the spacecraft. If **L** is the baseline vector between two such antennae, and **e** is the unit vector along the line of sight to a GPS satellite P, then the range from P to each of the antennae differs by an amount

$$\delta\rho = \mathbf{L}.\mathbf{e} = L\cos\theta$$

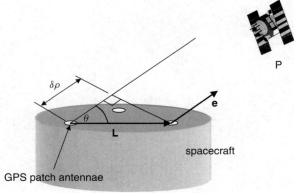

Figure 9.17 The principle of attitude determination using GNSS

The measurement made is the phase difference between the two antennae, which allows $\delta\rho$ and therefore θ to be estimated. A similar procedure using each antenna baseline and multiple GPS satellites, allows the attitude of the spacecraft to be constructed. Accuracies of the order of $\sim 0.1°$ are possible, although this will depend on the length of the baseline. Increasing this is not without its problems, however, as there is more flexure in long baselines that will compromise the accuracy. The phase difference measurement is also inherently ambiguous because of the unknown number of GPS signal cycles received between antennae. Finally *multi-path*—unwanted local reflections of the signal—can also pose problems. Despite these difficulties, however, attitude determination using GNSS is an established technique, which has been demonstrated in orbit [5].

9.5.4 Inertial sensors

Gyroscopes form the basis of the inertial sensing system for attitude. The conventional, mechanical gyro has a rotor mounted in a single gimbal in an environment that is very carefully controlled. In the rate- and rate-integrating types, the gimbal is torqued so that it follows the motion of the spacecraft. The torque is then a measure of the angular rate about the instrument's sensitive axis (see equation 3.29).

A set of three orthogonal rate-gyros will measure the components $(\omega_x, \omega_y, \omega_z)$ of the spacecraft's inertial angular velocity; a fourth at a skew angle is normally carried to avoid a single-point failure. The output of a rate-integrating gyro (RIG) is the integral of the angular velocity component, such as $\int \omega_x \, dt$ and so on. Only when the direction of a RIG axis remains fixed in space does its output represent the angular displacement about the axis. In more general motion equations (9.25) apply.

In high-quality sensors, drift rates of less than 0.01 degree/h are obtainable. Recent years have seen the development and introduction of gyroscopic sensors without moving mechanisms. The best known of these is perhaps the *Ring Laser Gyroscope (RLG)*. This comprises a small triangular prism of ceramic glass, the vertices of which are truncated to form optically flat surfaces. Laser light is then introduced at one vertex, so that two beams are internally reflected around the prism in opposing directions. If the prism is rotated around the axis normal to the triangle, then the path length of light traversing the prism in the same sense as the rotation is longer than that for the beam in the opposite direction. This difference, detected by examining the interference between the two beams, gives a measure of the inertial angular rate about the sensitive axis. Figure 9.18 shows a montage of the Quasar 3000 unit used in the guidance of the Ariane 5 launch vehicle. This comprises linear accelerometers, and a RLG to give angular rates about three axes. Such triaxial information is usually achieved using three independent laser prisms as described above, but in this case is achieved by a single prism as shown. The drift of the Quasar 3000 RLG is quoted as less than 3×10^{-2} degrees per hour.

A device that uses a similar principle is the *Fibre Optic Gyroscope (FOG)*, which is generally less massive than the RLG, the need for the prismatic block having been eliminated. In this instrument, laser light is fed simultaneously into both ends of a long ($\sim$ km) fibre optic coil. When the coil rotates about its axis, the counter-rotating beams again travel different distances before they reach the detector, and the phase difference gives a measure of the rotation rate.

Another class of gyroscopic device is based upon the response of a mechanical resonator to the Coriolis force generated by rotation, first analysed by Bryan in 1890 [6].

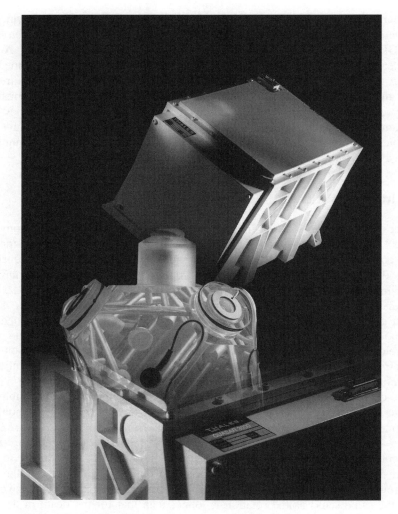

Figure 9.18 Ring laser gyroscope as used in the Ariane 5 launcher inertial reference system (Reproduced by permission of Thales Avionics)

The principle of operation of one such device, the *Hemispherical Resonator Gyroscope (HRG)*, can be thought of in terms of a resonating wine glass. When the glass is made to 'sing' (resonate), a standing wave is produced on its surface. When the glass is rotated about the stem (the sensitive gyro axis), the nodes and anti-nodes of this wave precess at a rate dependent upon the inertial angular velocity. The HRG therefore comprises a hemispherical fused silica or quartz shell, which is electrostatically actuated and sensed to determine the shell's response and measure the wave's precession. Such devices have been used in flight (e.g. the *Cassini* mission to Saturn), and have a performance that matches advanced RLGs. The principle of the HRG has been extended to other resonator geometries, to simplify manufacture. Gyroscopic devices, sometimes referred to as *Piezoelectric Vibratory Gyroscopes (PVGs)*, are available (such as the Murata 'Gyrostar'). These units are relatively inexpensive, with an overall size smaller than a matchbox. An example of this type of gyro is based upon the measurement of the vibrations of a small steel bar with a triangular cross-section [7]. Piezoelectric transducers are attached to the bar to actuate and sense the motion, to determine its response to Coriolis effects, so estimating the inertial angular rate. Furthermore, very small devices based upon this principle can be produced using advanced microfabrication techniques. At the time of writing, however, the performance of these microgyros is insufficient for many space applications, due largely to manufacturing imperfections and their temperature sensitivity.

9.6   ACS COMPUTATION

9.6.1   The computer

The development of digital computers for use in spacecraft has proceeded rapidly. They must perform reliably in the radiation environment of space, and many space-qualified processors exist. Further development is providing more power and speed, and the capability of being programmed in higher-level languages. These on-board computers (OBCs) link with ground control computers, which will normally host their software development tools. The availability of powerful computers means that spacecraft are given greater autonomy, and many of the sophisticated control techniques that find applications in ground-based systems may be used on spacecraft.

Robustness is a requirement for ACS and other on-board systems. For example, the ACS must potentially operate with large flexible structures such as solar arrays, whose natural frequencies cannot be established accurately before launch. Fixed algorithms will tolerate only limited variation from their expected value. The ability to reprogram the OBC from Ground Control permits any necessary adjustment of the control algorithms to be made following calibration of the spacecraft's parameters after launch. For full autonomy or immediate response to any changes that occur such as hardware failures, adaptive control techniques may be used. Indeed, the uploading of mission or control software is routinely performed to adapt to changing requirements or to enhance performance.

Computer power will also benefit the attitude measurement subsystem. The fusion of sensor data to achieve maximum accuracy *via* the Kalman type of filter requires computer modelling. In addition, they can provide the substantial data backup, which is needed when star mappers and scanners are used.

9.6.2   Active attitude control using a PID algorithm

The sophisticated techniques mentioned above are beyond the scope of this text, and for these the reader is referred to more specialized texts [3,8]. However, there are many control situations that are satisfactorily achieved by using a simple *Proportional, integral and differential (PID)* algorithm, and the following discussion addresses the question of controlling a spacecraft's attitude using this.

The principle of active control involves the measurement of the spacecraft's attitude using installed sensors, and comparing this with the desired attitude as explained in Section 9.2.4. The *error signals*—the difference between the measured and the desired attitude—are then used in appropriate algorithms within the OBC to determine corrective torques (see Figure 9.2). Euler angles ϕ, θ, ψ are commonly used as measures of the errors when they are small and may also be used when only one of them is large, but alternative measures are direction cosines or quaternions.

A torque acting upon a spacecraft may cause an angular acceleration or, when momentum bias is used, an angular velocity. A PID algorithm can be used in both cases, but it is discussed here in the context of zero-momentum-bias when the torque causes an acceleration of the existing error towards zero—its desired value. Equation (9.14) shows that $I_{xx}\ddot{\phi} \approx T_x$, and similarly $I_{yy}\ddot{\theta} \approx T_y$ and $I_{zz}\ddot{\psi} \approx T_z$ under the constraints mentioned in the previous paragraph.

The algorithm attempts to make the pointing error—ε say—obey a differential equation of form

$$\ddot{\varepsilon} + 2\zeta\omega_n\dot{\varepsilon} + \omega_n^2\varepsilon = 0 \tag{9.26}$$

In this, the undamped natural frequency ω_n (rads/sec) governs the speed of the response to a disturbance, and the damping ratio ζ, ideally of order 0.5, affects how rapidly the (typically) oscillatory response dies away. For example, the overshoot in the response is about 16% when $\zeta = 0.5$ (see Figure 9.19).

The implementation of a PID control algorithm for the roll error ϕ

Although we treat the roll angle ϕ in the remainder of this section, the equations for the pitch θ and yaw ψ take the same form.

The total roll torque T_x will in general include a disturbance torque T_{xd} as well as the control torque T_{xc} from the actuator, leading to

$$I_{xx}\ddot{\phi} = T_x = T_{xd} + T_{xc} \tag{9.27}$$

Using a control torque T_{xc} that involves only the proportional and differential parts of the PID algorithm, namely,

$$T_{xc} = -K_{xp}\phi - K_{xd}\dot{\phi} \tag{9.28}$$

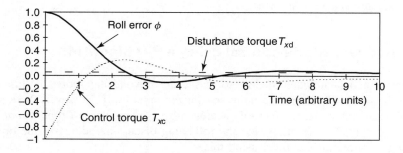

Figure 9.19 The response of the roll error ϕ and the control torque T_{xc} due to a PD algorithm (no integral term) in the presence of a disturbance torque T_{xd}

and inserting this into equation (9.27) leads to

$$\ddot{\phi} + (K_{xd}/I_{xx})\dot{\phi} + (K_{xp}/I_{xx})\phi = T_{xd}/I_{xx} \tag{9.29}$$

This has the same form as equation (9.26) except that its right-hand side is not zero.

The response of equation (9.29) to an initial error, caused by a command for a new orientation say, is shown in Figure 9.19. When the transient has died away, then the roll attitude will have an error $\phi = T_{xd}/K_{xp}$. This will reduce to zero when the integral term in the PID controller is included. The full algorithm, in general form, is

$$T_{xc} = -K_{xp}\phi - K_{xd}\dot{\phi} - K_{xi}\int\phi\,\mathrm{d}t \tag{9.30}$$

Some practical issues are involved:

- The torquers above were assumed to be ones whose torque magnitude is controllable. RWs answer this description (Table 9.2). There will, however, be a limit to the amount of torque that these devices can provide. Therefore, high values of the PID constants will lead to a call for large torques to achieve a rapid response.
- The spacecraft will not be entirely rigid as was assumed in the theory above. Its structure will have oscillatory flexure modes, the natural frequencies of which will be very low if there are large flexible appendages such as solar arrays. Their damping ratios may be only of order 0.015—definitely stable, but only just. The output of all attitude sensors will include a measure of modal excitation, the amount depending upon their location within the modal shapes. It is important that the control algorithm does not lead to a destabilizing feedback to these modes. In large spacecraft, it will be necessary to include many modes in the mathematical model when designing the final form of the algorithms, but the modal data in this model is likely to be imperfect prior to launch. Some calibration and adjustment of the controller parameters may be necessary in orbit.
- The angular rate term $\dot{\phi}$ may be provided by a rate gyroscope. It may alternatively be calculated by differentiating the signal from a roll *angle* sensor. Any sudden repointing demand, however, will create an immediate change in the roll *error*, and differentiating this step function should be avoided if possible. In general, differentiation will amplify any high-frequency 'noise' and also the content of any high-frequency flexure modes in the signal. Some attenuation of this must be included in the differentiation process.
- When large repointing manoeuvres are required, cross-coupling between the roll, pitch and yaw motions may become detrimental. Lower values of the PID constants will alleviate this, but an alternative is to carry out these large manoeuvres as a sequence of roll followed by pitch and then yaw, say, so that only one of these angles is large at any one time.

In summary, although the preceding theory indicated no limit to the magnitude of the PID parameters K_{xp}, K_{xi}, K_{xd}, they will nevertheless be limited by the practical considerations, and possibly severely so.

On/Off control

Thrusters used for applying a torque for attitude control can only be used in an on/off manner, with different jets being used for positive and negative torques about each axis. The pulses required for repointing a momentum bias direction as opposed to a zero-momentum bias axis are very different, as illustrated in Figure 9.7. The zero-momentum bias case used above will also be discussed here.

A PID algorithm controlling the level of torque is theoretically capable of producing zero attitude error, even when there is a constant disturbance torque, as has been shown above. With on/off control the limitation of having only three torque levels—clockwise, zero or anticlockwise—means that the control objective becomes 'the maintenance of an attitude error within acceptable bounds'. Furthermore, this must be done with a minimum of fuel usage.

With on/off control the error ϕ is expected to settle into a *limit cycle* under steady conditions as shown in Figure 9.20. A simple 'P' control law has been used, the appropriate torquer being switched on when the roll error exceeds a specified threshold value, $|\phi| \geq \phi_0$. With no momentum bias there is an angular acceleration while $|\phi| \geq \phi_0$, and constant angular rate otherwise. Under these conditions, the speed at which the error crosses between the boundaries where $\phi = \pm\phi_0$ is the same in each direction, and so the period and hence the fuel consumption will depend on its speed when the thrusters are first switched off. Figure 9.20 illustrates this. The control strategy shown in Figure 9.20(a) has twice the initial roll-rate as that of Figure 9.20(b), and consequently twice the fuel consumption per cycle, and a shorter period.

This shows that a switching signal that is proportional to the error alone is not satisfactory. However, it will be used to demonstrate some other aspects of on/off control.

The effect of a disturbance torque can be beneficial, as shown in Figure 9.21. Its optimum value is that which returns the error to the same switching level, $\phi = +\phi_0$ in this case, without needing to switch on the reverse thruster. Greater disturbance torques will increase fuel expenditure.

The PID controller for an On/Off system

Figures 9.20 and 9.21 involve the torquer switching ON when $|\phi| \geq \phi_0$. Implementing this will be achieved by a computer-generated signal ε that switches on the appropriate

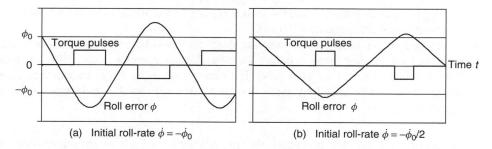

(a) Initial roll-rate $\dot{\phi} = -\dot{\phi}_0$ (b) Initial roll-rate $\dot{\phi} = -\dot{\phi}_0/2$

Figure 9.20 Roll error limit cycle for a simple P controller, in the absence of a disturbance torque

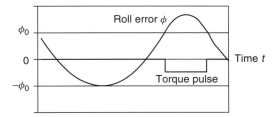

Figure 9.21 Roll error limit cycle with optimum disturbance torque

torquer when $|\varepsilon| \geq \phi_0$. The equation for the switching signal above was $\varepsilon = \phi$, a PID control signal with the proportional 'P' term only.

The inclusion of a differential term to form a 'PD' controller means that as ϕ approaches the switching level $\phi = \phi_0$ say, the torquer will switch ON *before* reaching it, and as it returns it will again switch OFF early since the sign of $\dot{\phi}$ will have changed. This progressively reduces the speed of entry into the zero torque 'corridor' as shown in Figure 9.22. In theory, this could result in an indefinite zero angular rate within the desired corridor, but a disturbance torque will not allow this to happen.

The inclusion of an integral term, to create a full PID controller, can cause a response such as is shown in Figure 9.23 under a constant disturbance. It has the effect of lowering the switching boundary progressively under steady disturbance torque conditions, until the average error over a limit cycle is zero.

Some practical considerations

- Torquer pulses will not be the instantaneous sharp-edged ones that have been assumed above. Furthermore, there will be a minimum switch-on time.
- Fuel consumption is an important issue, and there may be a compromise between achieving a short period limit cycle with small average error—'bouncing along close to a switching value' as in Figures 9.22 and 9.23—and making relatively large excursions across the switching 'corridor' as in Figure 9.21 in order to achieve as long a period as possible. Furthermore, the parameters of the PID controller can be modified if a change of mission emphasis occurs, say from 'high pointing accuracy' to 'fuel conservation'.

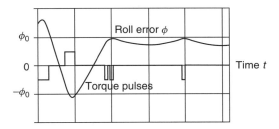

Figure 9.22 Roll error response with a PD controller, in the presence of a disturbance torque

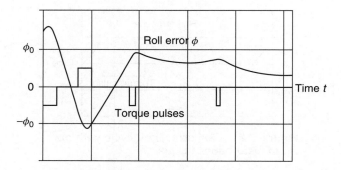

Figure 9.23 Roll error with full PID control, with a disturbance torque

As we have indicated, there are a number of limitations to the brief treatment presented here. For a text in which these are addressed in the context of spacecraft control, the reader is referred to Reference [3]. More generally, the topic of automatic control has a vast literature. Reference texts include [9] and [10], the latter being a brief and approachable text for those wishing to acquire the basics rapidly.

REFERENCES

[1]   Wertz, J. R. (1978) *Spacecraft Attitude Determination and Control*, Kluwer, Dordrecht, The Netherlands.
[2]   Hughes, P. C. (1985) *Spacecraft Attitude Dynamics*, John Wiley & Sons, New York.
[3]   Sidi, M. J. (1997) *Spacecraft Dynamics and Control—A Practical Engineering Approach*, Cambridge University Press, New York.
[4]   Fortescue, P. W. and Belo, E. M. (1989) Control decoupling analysis for gyroscopic effects in rolling missiles, *J. Guidance, Control Dynam.*, **12** (6), 798–805.
[5]   Chu, Q. P. and van Woerkom, P. Th. L. M. (1997) GPS for low-cost attitude determination, IAF-97-A.2.03, 48th *International Astronautical Congress*, Turin.
[6]   Bryan, G. H. (1890) On the beats in the vibrations of a revolving cylinder or bell, *Proceedings of the Cambridge Philosophical Society*, Vol. 7, pp. 101–111.
[7]   Fujishima, S., Nakamura, T. and Fujimoto, K. (1991) Piezoelectric Vibratory Gyroscope Using Flexural Vibration of a Triangular Bar, 45th *Annual Symposium on Frequency Control*, pp. 261–265.
[8]   Gelb, A. (1974) *Applied Optimal Estimation*, MIT Press, Cambridge, MA.
[9]   Nise, N. S. (2000) *Control Systems Engineering* (3rd edn), John Wiley & Sons, New York.
[10] Schwarzenbach, J. (1996) *Essentials of Control*, Longman, UK.

10 ELECTRICAL POWER SYSTEMS

John P. W. Stark

Department of Engineering, Queen Mary, University of London

10.1 INTRODUCTION

Provision of electrical power for space vehicles is, perhaps, the most fundamental requirement for the satellite payload. Power-system failure necessarily results in the loss of a space mission, and it is interesting to note that many of the early satellite systems failed due to such a loss. The demand for power has increased and is characterized by enhanced spacecraft operational complexity and sophistication. The earliest spacecraft, such as Vanguard 1, typically required a power raising capability of only $\sim 1\,W$, whereas current communications satellites typically require three orders of magnitude greater than this. Evolving trends suggest that a further two orders of magnitude may still be needed. However, prediction of future power demand must always be treated with a great deal of caution. In the early 1980s there was great focus upon large systems, particularly with the infrastructure elements associated with space stations, containing both manned and unmanned elements. The potential for Solar Power Satellite systems (SPS), large orbiting power stations generating GW of electrical power for terrestrial use, was investigated in many studies, initially supported in the United States of America by both NASA and DoE, and in Europe by the European Space Agency (ESA) [1]. Political and technical issues have slowed down such developments, and large is no longer necessarily more beautiful. Manned systems do require higher power levels, perhaps of the order of 300 kW. Commercial communication services and broadcast systems are still seeing a slowly evolving trend to higher power levels, but enhanced system performance is reducing power demands. However, much focus is now also upon smaller satellite systems (see Chapter 18) where cost and system efficiency are critical system drivers.

The best methods of raising power can be broadly related to power level and mission time as shown in Figure 10.1. It is apparent that photovoltaic (solar cells) or radioisotope thermoelectric generators (RTGs) are appropriate for the power requirements typical of present generation spacecraft, namely, a few kilowatts for missions of several years. For shorter periods, fuel cells are advantageous, and for periods of less than a few days

Spacecraft Systems Engineering (Third Edition). Edited by P. W. Fortescue, J. P. W. Stark and G. G. Swinerd
© 2003 John Wiley & Sons Ltd

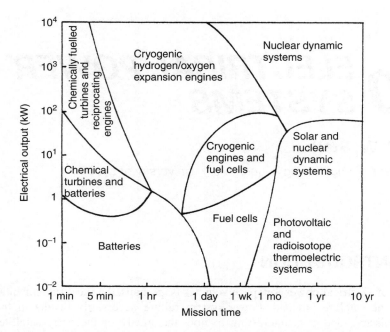

Figure 10.1 Power outputs: mission duration relationship between energy source and appropriate operational scenario [2] (From Angrist, S. W. (1982) *Direct Energy Conversion*, 4th edn, Copyright Allyn and Bacon, New York)

batteries come to the fore. It is not surprising, therefore, to discover that batteries are used in launch vehicles to provide the primary energy source, fuel cells are used in the Shuttle and both photovoltaic devices and RTG are used for general spacecraft operation, dependent upon the mission. It should be noted that nuclear sources of power such as RTG and nuclear dynamic systems are used for military applications, but are not generally acceptable for civilian vehicles in Earth orbit. Indeed, during the 1990s, an increasing public awareness of environmental issues resulted in protests concerning the launch of nuclear-based sources for scientific purposes—for example, the launch of the Cassini–Huygens mission. Before the individual elements of a spacecraft power system are considered, the overall power system configuration will be described briefly.

10.2 POWER SYSTEM ELEMENTS

In general a spacecraft power system consists of three main elements: primary and secondary energy sources, and a power control/distribution network. These are shown schematically in Figure 10.2.

The *primary energy source* converts a fuel into electrical power. On early space flights and on launch vehicles, batteries have provided this. Strictly these systems do not have a fuel element, in that a battery is a device that stores energy rather than performing a direct energy conversion process.

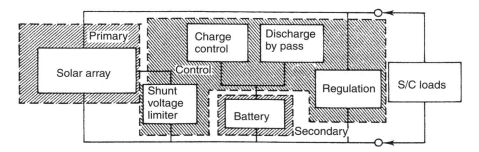

Figure 10.2   Schematic of typical spacecraft power system block elements

The majority of present-day spacecraft use a solar array as the primary energy source. The 'fuel' in this case is solar radiant energy, which is converted *via* the photovoltaic effect (see Section 10.3.1) into electrical energy. On manned missions of short duration, fuel cells have been used most frequently as the primary source; for longer duration flights associated with space stations such as Mir and ISS, the combination of solar arrays and fuel cells has been adopted. These electrochemical devices perform a controlled chemical reaction, in such a way that electrical energy may be derived rather than heat energy. The fuels used for space operation are hydrogen and oxygen yielding water as the reaction product: this may then be drunk by astronauts. Nuclear systems utilize either a radioactive decay process (RTG uses this) or a nuclear fission process as the energy source. RTG makes use of the thermoelectric effect, whereas fission reactors operate in a manner similar to terrestrial nuclear power plants [3].

The *secondary energy source* is required to store energy and subsequently deliver electrical power to the satellite system and its payload, when the primary system's energy is not available. The most usual situation when this condition arises is during an eclipse period when the primary system is a solar array. The eclipse's duration depends on the spacecraft orbit (see Section 5.3.2 of Chapter 5). Typically for Low Earth Orbit (LEO), a 35-minute eclipse occurs in each orbit for low-inclination satellites: in Geostationary Earth orbit (GEO), eclipses occur only during equinoctial periods, with a maximum duration of 1.2 h in a 24-hour period. For such short times, batteries demonstrate the highest efficiency. However, for systems that require high-power levels, typically 100 kW, a solar array/regenerative fuel–cell combination has improved characteristics over a solar array/battery combination. Regenerative fuel cells operate in a closed fuel cycle: H_2/O_2 fuel is consumed to form water on the 'discharge' cycle and electrolysis of water is performed during the 'charge' cycle, with power for this being derived from the solar array [4]. Whilst the net efficiency is low, only 50 to 60% compared to nearly 90% for a battery, it is possible by judicious sizing of the fuel (H_2/O_2) component to reduce the size of the solar array required for primary power raising. For LEO operations in which aerodynamic drag is significant, the reduction in array area reduces the mass of propellant required for orbit control leading to a lower wet mass of the system at launch compared with the conventional array/battery configuration.

The *power control and distribution network* is required to deliver appropriate voltage–current levels to all spacecraft loads when required. Several salient features should be noted.

The primary power source always degrades during the mission. Thus, at its start an excess of power will be generated, and it is necessary to provide an ancillary load to dissipate it. The simplest way to do this is to use a resistive load, generally external to the main spacecraft structure in order to simplify the thermal design.

Both primary and secondary power system characteristics will change during the mission, leading to a requirement for voltage and/or current regulation. The changes arise due to both degradation effects, such as cell failures, and also illumination variations caused by changing solar array aspect angles with respect to the Sun. The customary approach is to use a voltage shunt regulator across the array.

Charge control of a battery system is particularly important to maintain the lifetime and reliability of battery units. It generally necessitates both current and voltage control. A variety of techniques may be used to sample the state of a battery and these will be discussed in Section 10.5. Discharge control is also required in order to limit current output.

10.3   PRIMARY POWER SYSTEMS

10.3.1   Solar arrays

A solar array is an assembly of many thousand individual solar cells, connected in a suitable way to provide dc power levels from a few watts to tens of kilowatts. For a detailed description of both terrestrial and space solar arrays, Rauschenbach [5] is recommended.

Each *solar cell* assembly has a semiconductor $p - n$ junction as shown schematically in Figure 10.3. For spacecraft applications, the base material typically has a resistivity of between 10^{-3} and $10^2 \, \Omega\,\text{cm}$. Using silicon, for example, it might be doped with boron to form the p-type material (electron deficient), and with phosphorous for the n-type material (electron excess).

With no illumination, the junction achieves an equilibrium state in which no current flows. But when it is illuminated with suitable radiation, photons with sufficient energy will create electron–hole pairs, and the radiation is converted to a potential across the cell with usable electrical power. The incident photon energy needed for this must exceed a *band gap* that depends upon the material, as shown in Table 10.1. Photons with excess energy dissipate it as heat within the cell, leading to reduced efficiency.

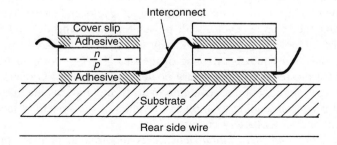

Figure 10.3   Schematic of a typical solar cell assembly

Table 10.1 Properties of semiconductor
materials

Material	Band gap (eV)	Maximum wavelength (μm)
Si	1.12	1.12
CdS	1.2	1.03
GaAs	1.35	0.92
GaP	2.24	0.554
CdTe	2.1	0.59

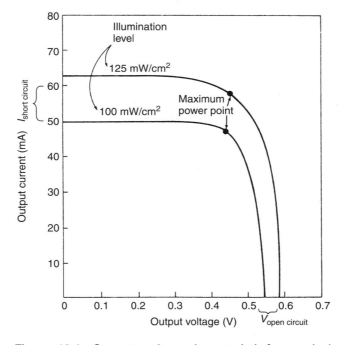

Figure 10.4 Current–voltage characteristic for a typical
solar cell. The short-circuit current is dependent upon
both the illumination level and the size (area) of the cell
(From Angrist, S. W. (1982) *Direct Energy Conversion*,
4th edn, Copyright Allyn and Bacon, New York)

Characteristic voltage-current curves for cells are shown in Figure 10.4. Typically open-circuit voltages for silicon cells lie between 0.5 and 0.6 V under solar illumination. The plot of power against voltage (Figure 10.5) has a clear maximum, with a particularly rapid fall once the optimum voltage is exceeded. On some spacecraft maximum-power-point tracking is used to operate the array most efficiently. Increase in cell temperature results in decreasing open-circuit voltage with only a modest increase in short-circuit current. The theoretical maximum efficiency of both silicon (Si) and gallium arsenide (GaAs) cells

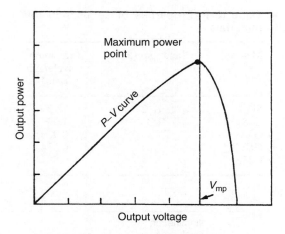

Figure 10.5 Power–voltage characteristic for a typical solar cell (From *Solar cell array design handbook* by Rauschenbach, H. S. Copyright © 1980 by Van Nostrand Reinhold. All rights reserved)

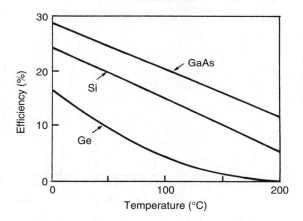

Figure 10.6 Theoretical cell efficiency as a function of temperature for three semiconductor materials (From *Solar Cell Array Design Handbook* by Rauschenbach, H. S. Copyright © 1980 by Van Nostrand Reinhold. All rights reserved)

is shown as a function of temperature in Figure 10.6. It shows the particular sensitivity of Si to temperature and also the improved performance of GaAs at high temperatures. This means that it is theoretically possible to use a focusing optical arrangement for GaAs cells such that these cells may be illuminated by an intensity greater than the nominal radiation intensity of $1.4\,kW/m^2$ in Earth orbit. Fewer cells would then be required to provide a given power level, which could result in array cost reductions since the cell cost is a large proportion of the total cost. Concentration ratios as high as 100 : 1 for GaAs cells have been investigated in such studies, but at present the benefits of these systems are not conclusive. The majority of commercial satellites, together with ISS have utilized silicon for solar cells. Perhaps two of the most notable civilian uses for GaAs have been on the Mir space station and the satellites used in the Iridium constellation.

The *n-type contact* on the upper surface of the cell is in the form of a multiple-finger arrangement. These fingers are required for efficient current collection, whilst maintaining good optical transparency (typically ~60%). They are connected at a bar, along one edge of the cell. Frequently titanium/silver (Ti/Ag) is used for this.

Radiation damage is a problem with solar cells. In general, Si cells having higher base resistivity (~10 Ω cm) are the most tolerant of radiation. Furthermore, cells have the *n*-type material uppermost, since on early space flights it was discovered that cells having the *p*-type material as the upper region rapidly suffered from radiation damage. Thin cells suffer less than thicker ones, but at present they have a lower conversion efficiency. GaAs cells are more radiation tolerant than Si and for this reason there is considerable interest and effort in their development.

The *cover glass* provides environmental and radiation protection. For design purposes, the particle fluence of a spacecraft's radiation environment may be expressed as an equivalent fluence of monoenergetic 1 MeV electrons (see Figure 10.7). Degradation of cell output to this irradiation is generally available from manufacturers' data; Figure 10.8 shows typical degradation curves for cells with a variety of thicknesses.

The effectiveness of the cover glass depends on its density and thickness. Suitable glass microsheet is commercially available in several thicknesses from 50 μm to 500 μm [6]. Their absorption of radiation follows approximately an exponential law, so that the intensity of radiation after traversing a depth x into the glass is

$$I \sim I_0 e^{-k\rho x} \tag{10.1}$$

where I_0 is the initial radiation fluence at $x = 0$, ρ is the density of material and k is an energy-dependent absorption coefficient. For fused silica, Figure 10.9 shows the effect of changing glass thickness. It should be noted that in order to evaluate the total radiation fluence absorbed by the solar cell, both front-side and rear-side fluences must be calculated. On rigid panel arrays mounted on honeycomb structure, and on solar arrays bonded to the spacecraft wall (e.g. spinning spacecraft) the rear side is effectively screened from radiation, whilst on lightweight arrays this is not the case. Using the data supplied in Figures 10.7 to 10.9, it is possible to derive the area of active solar cells required to meet a specific mission requirement of end of life (EOL) performance. Suppose that an output of 1 kW is required at EOL for a satellite in a circular, equatorial orbit at 1000 km altitude. Assume that the cell to be used is made of silicon, 150 μm thick, having the properties shown in Figure 10.8. The mission duration is to be 5 years. Note that in the discussion that follows, the numerical values are approximate.

The starting point is to use Figure 10.7 to find the total damage equivalent 1 MeV electron fluence for a cell protected by a 150 μm cover slip. At 1000 km, this Figure shows that the damage equivalent due to protons is 1.7×10^{14} electrons/cm/year and that due to electrons is 2×10^{12} electrons/cm/year. The total flux in 5 years is therefore $5(1.72 \times 10^{14})$ electrons/cm/year, or 8.6×10^{14} electrons/cm/year. It is evident that the damage due to protons is much greater than that from electrons, as noted in Section 2.3

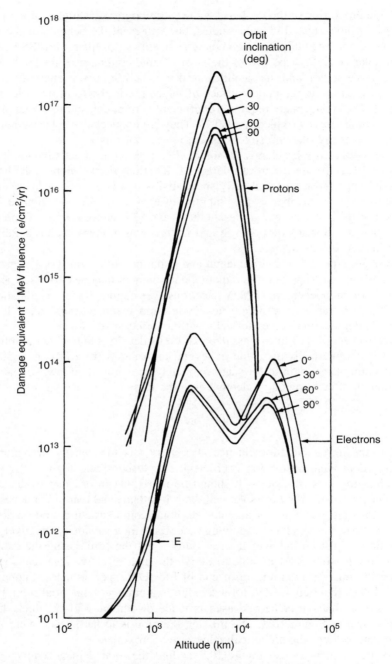

Figure 10.7 Damage equivalent 1 MeV fluence caused by electrons and protons due to trapped particles, to silicon cells protected by 150 μm fused silica covers and infinitely thick rear shielding [5] (From *Solar cell array design handbook* by Rauschenbach, H. S. Copyright © 1980 by Van Nostrand Reinhold. All rights reserved)

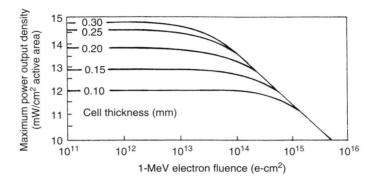

Figure 10.8 Effects of thickness and fluence on conventional non-p^+ silicon solar cell performance [5] (From *Solar Cell Array Design Handbook* by Rauschenbach, H. S. Copyright © 1980 Van Nostrand Reinhold. All rights reserved)

of Chapter 2. From Figure 10.8, the power per unit area is $11.5\,\mathrm{mW/cm^2}$. It is however, noticeable from this Figure that a significant deterioration in the performance of the cell is evident at such a high radiation dose. A reduction in dose may be achieved by increasing the cover slip thickness. For example, if a cover slip of $500\,\mu\mathrm{m}$ were to be used, then the flux will be reduced according to the data presented in Figure 10.9 by a factor of 0.6. Thus the expected radiation dose absorbed in the cell after 5 years would then only be 5.2×10^{14} electrons/cm, yielding an EOL performance of $12\,\mathrm{mW/cm^2}$, an improvement of approximately 4%. The mass increase associated with the use of this cover slip, assuming as a first approximation that the density of the cover slip is the same as that of the cell, will be (mass of $500\,\mu\mathrm{m}$ cover slip plus $150\,\mu\mathrm{m}$ cell)/(mass of $150\,\mu\mathrm{m}$ cover slip plus $150\,\mu\mathrm{m}$ cell), equivalent to a factor of nearly 2.2. This increase in mass needs to be considered however alongside the cost increase associated with the alternative of a 4% increase in the number of cells. The thinner cover slips result in an active area of $(1000/0.0115)\,\mathrm{cm^2}$, or $8.7\,\mathrm{m^2}$, whilst the thicker protection requires an area of only $8.3\,\mathrm{m^2}$.

Additional features required of the cover glass are that it provides good optical coupling between free space and glass and also between glass and adhesive, and that it provides suitable wavelength selection, limiting the UV flux to the adhesive layer and the cell. These features are achieved using an anti-reflection coating, such as magnesium fluoride on the upper surface, perhaps with an additional indium oxide conductive coating as described in Chapter 2. A UV-filter coating may be applied to the underside of the cell, to reflect UV radiation. For a cover glass with cerium doping, additional UV filtering is unnecessary.

For *efficient cell operation* and insensitivity to radiation, a shallow junction depth (typically less than $10\,\mu\mathrm{m}$) is required. Various Si cell configurations have been investigated to improve conversion efficiency. These include the following:

- Back-surface reflectors (BSRs), used, for example, on Spot and Orion, to reflect unabsorbed radiation from the rear side of the (p)-region back through the cell. This reduces cell heating.

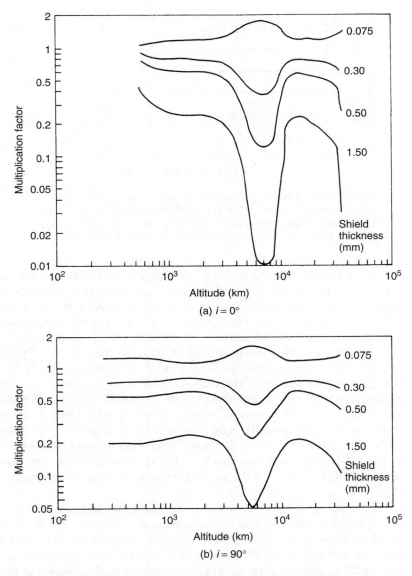

Figure 10.9 Multiplicative factors to be applied to damage fluence on a solar cell as a function of cover slip (shield) thickness, and operational orbit height (From *Solar Array Design Handbook* by Rauschenbach, H. S. Copyright © 1980 Van Nostrand Reinhold. All rights reserved)

- The introduction of a p$^+$-region or back-surface field (BSF) at the rear of the *p*-regions. This exhibits higher output due to enhanced carrier collection efficiency, but the improvement is lost under high fluence damage. This technology has been used on the array for the Hubble Space Telescope (HST) and Envisat.
- The use of a textured front surface of the cell reduces reflection from the cell surface.

Solar arrays using Si cells are made from individual cells that at present are generally rectangular, 2 cm × 4 cm, having a conversion efficiency of ∼12 to 14%. To improve packing efficiency, much work on high-efficiency 5 cm × 5 cm cells is under way. The cells have a thickness of between 50 and 250 μm, ∼200 μm being used for the majority of arrays. Silicon cells having a thickness of 50 μm are now able to convert with a power density of 120 W/m^2; thinner GaAs cells ∼5 μm with an efficiency of 21% have also been developed, but are not available for space systems at the time of writing.

Interconnections between cells represent a major array failure hazard. This arises because of the thermal cycling inherent upon entry/departure from sunlight to eclipse. Since the materials used for cell and substrate are different, differential expansion takes place during the rapid temperature change (∼100 °C in a few minutes). Thermal stress-relieving loops are required to reduce such failure mechanisms as interconnect lift-off and fracture.

Atomic oxygen effects on exposed interconnects have been mentioned earlier, in Chapter 2. A particular problem of the solar cell interconnection is that historically these have always been made of thin silver foil. Silver has a high capture efficiency for atomic oxygen, resulting in the formation of a variety of silver oxides. The process results in thinning due to flake-off of the oxides and hence an increase in interconnection resistivity. This leads to a loss of power. In the case of the HST array, the interconnects were originally designed to be pure silver. The delay in the original launch date (because of the Challenger accident) permitted a redesign in which silver was used only as a surface layer on a molybdenum interconnect. During flight, the silver eroded revealing the molybdenum, which is oxidation resistant. The power output from the array has thus been maintained at the predicted value for the life of the mission.

A variety of *substrate materials* have been used and proven in space. Frequently Kapton with glass- or carbon- fibre reinforcement, ∼100 μm thick, forms the immediate interface with the cell, which may be mounted on a honeycomb panel for rigidity (e.g. Tracking and Data Relay Satellite (TDRS) solar array). Flexible cell blankets have also been used to reduce mass. The HST solar array is a typical example, with glass fibre reinforcing the Kapton. In general, for the largest arrays, flexible substrate materials offer mass savings. But it is interesting to note that early work [7] has shown that for power levels up to 6 kW, mass savings occur when using advanced rigid arrays, wherein Kapton is reinforced with carbon fibre. Fokker Space has continued the development of advanced rigid solar arrays using both Si and GaAs, with in-orbit power (beginning of life (BOL)) in excess of 8 kW [8].

As noted earlier, individual cells produce power at a voltage of ∼0.5 V, and it is necessary to connect many cells in series. Reliability is then achieved by additional parallel coupling at each cell; typically three or four cells form a parallel combination. This series–parallel arrangement is called a *solar cell string*. Further protection is afforded using shunt diodes that provide current bypass paths should individual cells become shadowed. Shadowing can cause cell failures since if a cell is unable to generate power because of loss of illumination, then the entire string voltage may appear as a reverse bias voltage across the cell.

System level interactions

Now consider the *system level interactions* between the solar array design and the vehicle itself. The relatively low conversion efficiency of an array results in the need for large areas of solar cells to intercept sufficient solar radiation for the power demand. Examples of typical array configurations, associated with generic stabilization types are illustrated in Chapter 9.

For a spinning satellite, using either dual spin or simple spin ACS, the 'drum' size evidently limits the power that can be generated. The drum itself is limited by the launch volume. A possible solution is to introduce an additional mechanism such as the 'drop skirt' as used on Intelsat VI, where a larger-diameter hollow cylinder is deployed to expose additional cell area. Thruster plume impingement in such a configuration can, however, cause both disturbance torques and contamination.

Other aspects to note for the spinning satellite solar array are coupled with the thermal environment. Assuming a typical spin rate of $\sim$50 rpm, the average temperature of the array can be maintained at a lower value than for the three-axis configuration. This leads to an increased efficiency of the individual cells, and thus a decrease in the required active cell area. However, since not all the array is instantaneously illuminated, a factor of $\sim\pi$ times the number of cells is required for the same collected power. Since also on the spinner, the array is mounted essentially on the body of the spacecraft, the temperature excursions noted on the array (between sunlit and eclipse phases) are reduced, having a significant impact on reducing the thermal shock characteristic of a three-axis deployed solar array.

Considering the three-axis-stabilized satellite, the solar array requires a mechanism to deploy the stowed array following launch and then orientate it appropriately to track the Sun. These deployment mechanisms may be of a simple extending telescopic construction, or of the 'Coilable' variety. (See Chapter 15). In the Astromast, the carbon-fibre members of the mast are deformed by coiling them into a stowage container. They may then be driven out by a screw mechanism to the underformed (extended) configuration.

Tensioning wires are then required to achieve an acceptable minimum fundamental frequency of the array largely because of AOCS requirements. On three-axis-stabilized vehicles, power take-off from the array generally, but not always, requires a rotary degree of freedom between the satellite and the array, in order that pointing requirements of the array and the payload may be met. This requires two elements—the mechanical rotation device to allow the body to move relative to the array (with appropriate sensing systems) and an electrical power take-off device. (See Chapter 15) The provision of power using a solar array clearly has many design interactions with the rest of the spacecraft system.

Table 10.2 summarizes performance characteristics for several solar arrays. A method for array sizing is given in Section 10.6.

Table 10.2 Solar array performance figures

Array	Type	BOL power (kW)	Specific mass (W/kg)	Power density (W/m^2)
XMM	Rigid	2.5	32	215
Astra 2B	Rigid	9.2	52	409
Comets	Flexible	6.3	34	146

Note: XMM: X-ray multi-mirror mission.

10.3.2   Fuel cells

Fuel cells provide the primary power source for the Shuttle orbiter. Originally they were designed as part of the Mercury, Gemini and Apollo US manned missions. Table 10.3 shows how their performance has evolved since the earliest days of manned space flight. A fuel cell converts the chemical energy of an oxidation reaction directly into electrical energy, with minimal thermal changes. From a system viewpoint, a major advantage is its flexibility. For example, it provides power during both sunlit and eclipse periods, and the fuel has a high-energy density and thus provides a compact solution compared with a solar array. The evident disadvantage is the need to carry fuel.

The *hydrogen/oxygen fuel cell* has been used for space applications, a product of the reaction being water. This is clearly useful for manned missions. A schematic diagram of such a cell is shown in Figure 10.10. This technology has also been proposed for lunar rover missions [9].

The voltage that appears at the terminals of an ideal cell is given by

$$E_r = \frac{-\Delta G}{nF} \tag{10.2}$$

where ΔG is the change of Gibbs free energy occurring in the reaction, n is the number of electrons transferred and F is the Faraday constant (product of Avogadro number and elementary charge) equal to 9.65×10^4 C/mol. For the reaction of the H_2/O_2 cell, two electrons are transferred per mole of water formed and ΔG has the value of -237.2 kJ/mole at $25\,^\circ$C. The reaction takes place spontaneously.

Thus the reversible voltage of the ideal cell is $237.2 \times 10^3 / (2 \times 9.65 \times 10^4) = 1.229$ V.

In practice this is not realized because there are various irreversibilities, termed polarization losses. Figure 10.11 shows a typical current–voltage curve for a hydrogen/oxygen fuel cell. Initially, as soon as a current is drawn from the cell, a rapid drop in voltage occurs. This is associated with the energy required to activate the electrode reactions. For the H_2/O_2 fuel cell these are three-phase: gas (fuel), solid (electrode) and liquid (electrolyte—this can also be a solid). It is necessary for the reactants to be chemisorbed onto the electrode, a process that requires breaking and forming new chemical bonds, and requires energy; hence the voltage drops. The process is called activation polarization.

Table 10.3   Performance summary of fuel cells for space use

System	Specific power (W/kg)	Operation
Gemini	33	
Apollo	25	
Shuttle	275	2500 h at P_{ave}
SPE technology	110–146	>40 000 h
Alkaline technology	367	>3000 h
Alkaline technology	110	>40 000 h
Goal (lightweight cell)	550	

Note: SPE solid polymer electrolyte.

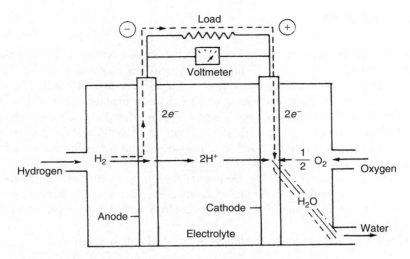

Figure 10.10 Schematic of a hydrogen/oxygen fuel cell. At the anode–electrolyte interface, hydrogen dissociates into hydrogen ions and electrons. The hydrogen ions migrate through the electrolyte to the cathode interface where they combine with the electrons that have traversed the load [2] (From Angrist, S. W. (1982) *Direct Energy Conversion*, 4th edn, Copyright Allyn and Bacon, New York)

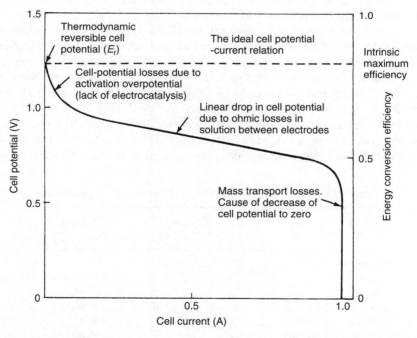

Figure 10.11 Typical cell potential and efficiency–current relation of an electrochemical electricity producer showing regions of major influence of various types of overpotential losses (Source [10])

The magnitude of voltage drop is given by the Tafel equation:

$$\Delta V_{ACT} = a + b \ln J \tag{10.3}$$

where J is the current density at the electrodes, and a and b are temperature-dependent constants for the reaction/surface description.

As the current drawn increases, a linear voltage–current relation is noted. This is simply because of the resistive nature of the electrolyte.

At high current values problems arise because of the transport of reactants to the reaction sites, a feature that is particularly important at porous electrodes since pressure gradients are set up that limit flow rates. Additionally, species concentrations are not uniform and thus ionic species can create a back emf. This concentration polarization provides the ultimate limit on current density that may be achieved by a fuel cell. All of these electrochemical polarization processes are common to both batteries and fuel cells, and their voltage–current characteristics are very similar.

Early fuel-cell systems were primarily based upon the technology of solid polymer electrolyte (SPE). For the Gemini series, 1 kW was produced at a specific power of 33 W/kg, within a volume of 0.05 m^3, and the objective to extend missions to greater than four days was achieved. However, the water produced was not of drinking quality because of degradation of the fuel-cell membrane.

The Apollo system, also used for Skylab, was based upon matrix aqueous alkaline technology and achieved a power level of 1.5 kW at a specific power of 25 W/kg. It had to operate whilst the vehicle was on the lunar surface, at a temperature greater than 394 K. The selected system, a Bacon fuel cell, operated at 505 K.

Shuttle developments, also based upon the alkaline technology, have improved the specific power by an order of magnitude, $\sim$12 kW, 275 W/kg. Further, the start-up time for this cell is 15 min with shutdown being instantaneous, whereas for Apollo, 24-hour start-up periods were required with 17-hour shutdown. Table 10.3 summarizes past and present fuel-cell status. Regenerative fuel cells wherein water is also electrolysed are not yet space-proven.

10.3.3   Radioisotope thermoelectric generators (RTG)

For deep-space missions, the use of fuel cells is precluded by their long duration. Solar arrays produce less power as they move away from the Sun, by a factor of approximately $(r_E/r_{S/C})^{1.5}$, where $r_{S/C}$ and r_E are the distances from the Sun to the spacecraft and to the Earth respectively. This factor comes about from the combined effect of the reduction in the intensity of the illumination from the Sun as the spacecraft moves away from it, partly offset by the beneficial effect of the reduction in the temperature of the solar cells.

For spacecraft travelling further than Mars, solar arrays show disadvantages from a system viewpoint, compared with radioisotope generators.

The operation of a RTG is based on the thermoelectric effect noted by Seebeck, that it is possible to generate a voltage between two materials, A and B (either conductors or semiconductors) if a temperature difference is maintained (see Figure 10.12). This is analogous to a thermocouple. Practical RTG space systems utilize two semiconductor materials—one p-type, the other n-type—in order to exploit the effect.

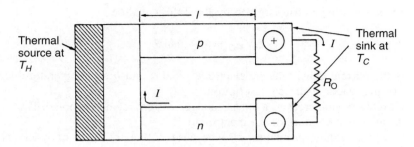

Figure 10.12   Schematic diagram of a semiconductor radioisotope generator (From Angrist, S. W. (1982) *Direct energy conversion*, 4th edn, Copyright Allyn and Bacon, New York)

The power output from such a device is a function of the absolute temperature of the hot junction, the temperature difference that may be maintained between the junctions and also the properties of the materials. Because such devices are relatively inefficient (less than 10%), one major problem in their design is removing waste heat.

The heat source used in space systems is derived from the spontaneous decay of a radioactive material. As this decays, it emits high-energy particles that can lose part of their energy in heating absorbing materials. Suitable fuels are listed in Table 10.4, and shows the half-life ($\tau_{1/2}$) for each of the fuels, namely, the time required for the amount of a given radioactive isotope in a sample to halve. Thus over a period of time t, the power available from such a fuel decreases by an amount given by

$$P_t = P_0 \exp\left(\frac{-0.693}{\tau_{1/2}}t\right) \tag{10.4}$$

where P_t is the power at time t after some initial time t_0.

Table 10.4 indicates that high specific power levels are available from sources with shorter half-lives (and hence shorter duration missions). For deep-space missions a long life isotope is essential; for example, the design life for the Cassini–Saturn orbiter is 11 years, after which time the electrical power source is required to be 628 W. For these missions Plutonium is used exclusively.

Table 10.4   Possible fuels and their performance for radioisotope generators (From Angrist, S. W. (1982) *Direct energy conversion*, 4th edn, Copyright Allyn and Bacon, New York)

Isotope	Fuel form	Decay	Power density (W/g)	$\tau_{1/2}$(yr)
Polonium 210	GdPo	α	82	0.38
Plutonium 238	PuO$_2$	α	0.41	86.4
Curium 242	Cm$_2$O$_3$	α	98	0.4
Strontium 90	SrO	β	0.24	28.0

The advantages of RTGs over other systems include the following:

1. They provide independence of power production from spacecraft orientation and shadowing.
2. They provide independence of distance from the Sun (deep-space missions are possible).
3. They can provide low power levels for long periods of time.
4. They are not susceptible to radiation damage in the Van Allen belts.
5. They are suitable for missions with long eclipse periods, for example, lunar landers.

The disadvantages of RTG systems need also be considered, and include

1. They adversely affect the radiation environment of the satellite whilst in orbit. This will influence the spacecraft configuration significantly as may be seen from Figure 10.13, which shows the Galileo spacecraft. In this instance, the RTG needs to be deployed on a lengthy boom away from the main satellite bus.
2. Careful handling procedures are required during satellite integration owing to the radiation hazard posed by the radioactive source.
3. High temperature operation is required for efficient energy conversion. This impacts upon the thermal environment of the vehicle, and again on vehicle configuration.
4. RTGs are a source of interference for plasma diagnostic equipment that may be carried as part of the scientific objectives of the mission.
5. At the political level there has been increasing concern expressed at the inclusion of radioactive material on board a satellite. This is principally of concern because of the potential for such a source to be dispersed in the atmosphere, should there be a launch failure.

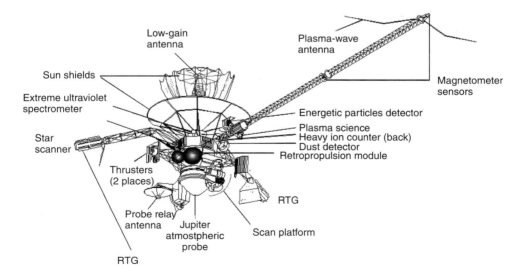

Figure 10.13 The Galileo spacecraft configuration, showing the position of the RTG sources (Courtesy of NASA/JPL/Caltech)

Table 10.5 RTG system performance

Name	Power (W)	kg/kW
Cassini (1997 launch)	628	195
Galileo probe/Ulysses (GPHS RTG, late 1980s)	285	195
Nimbus/Viking/Pioneer (SNAP 19, mid 1970s)	35	457
Apollo lunar surface experiment:		
SNAP-27, early 1970s	25	490
SNAP 9A, 1960s	73	261

System for Nuclear Auxiliary Power (SNAP-19), which powered the Viking lander vehicle to Mars, had a specific power of 2.2 W/kg, with a thermal/electric efficiency of ~5%. The output electrical power was 35 W. Table 10.5 summarizes data on RTG systems.

10.3.4 Other primary power systems

Two other primary power systems have been developed for operation—nuclear fission and solar heat. The former has been extensively used in the former Soviet space programme for military purposes. The latter has not flown in space yet, but a substantial effort has taken place in recent years in developing such systems for potential use on the International Space Station.

Nuclear fission systems

These systems operate in a similar way to conventional ground-based nuclear power stations, in that fissile material such as uranium-235 is used as a heat source. In space systems, this is used to drive a thermoelectric converter as noted in the preceding section. Specific features of space-based systems relate to the fail-safe requirement, particularly during launch.

Whilst the USA has invested substantially in the SNAP systems (even number SNAP designates a reactor system), these have not been used regularly. The main focus of their activity in recent years has been on the SP100 system [11]. It is notable, however, that the purchase of Russian technology by the US suggests that this US programme may not be pursued vigorously in the future.

Solar heat systems

The use of solar energy directly in the form of heat can provide system advantages. The heat energy can be used to drive a heat engine and then a rotary converter to electricity (solar dynamic), or directly be used as a heat source for a thermoelectric converter (solar thermoelectric).

Solar dynamic systems have had the greatest concentration of effort for the ISS. Design studies show that their end-to-end conversion efficiency is approximately 25% greater than for photovoltaics. This results in a reduced need for deployed collection area by

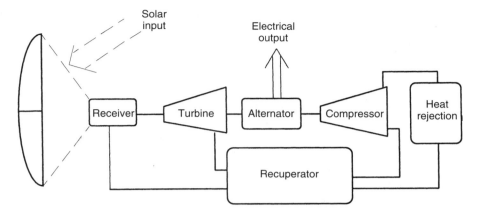

Figure 10.14 Solar dynamic Brayton cycle

about 25%, and consequently in reduced aerodynamic drag for LEO satellites. In the original concept for the ISS, primary power for the initial in-orbit capability was to be 75 kW, derived from photovoltaics. Power expansion was then assumed to be provided by solar dynamics in two units of 25 kW.

Solar dynamic systems, resulting in less drag, lead to lower fuel usage for orbit maintenance. This reduces the cost of station operation, principally by reducing the demand for Shuttle refuelling flights. An additional cost benefit arises from the lower maintenance costs envisaged for solar dynamic systems compared with photovoltaic systems. Over the lifetime of the Space Station, cost savings of several billions of dollars have been identified [12].

The concept studied by NASA, and which was the original baseline for the ISS solar dynamic system, is based upon the Brayton Cycle engine [12]. A block diagram of this is shown in Figure 10.14. The working fluid for this all-gas phase cycle is helium and xenon in such a proportion that the equivalent molecular weight is 40. This all-gas phase cycle minimizes problems of handling wet vapours (leading to erosion) and gravitational effects in transporting fluids. Storage of power within the concept was to be performed thermally, using the latent heat of fusion for a lithium fluoride/calcium fluoride mixture. This phase change occurs at 1042 K. Storing power in this way provides mass savings compared to battery technology, because of the high quality (temperature) heat energy.

Funding problems for the ISS have led to a cancellation of the power extension requirements. This has resulted in the cessation of most technology development work in this field. In the longer term, however, it is apparent from the activities that have been performed so far, that solar dynamics will play a role in large LEO space stations.

10.4 SECONDARY POWER SYSTEMS: BATTERIES

Batteries have been used extensively for the secondary power system, providing power during periods when the primary one is not available. As a back-up for a solar array this means that the batteries must provide power during eclipses, and that the array must recharge the batteries in sunlight.

In GEO operations, eclipses only occur during the two equinoctial periods producing eclipse seasons for the spacecraft. These last about 45 days at each equinox. Initially they are short (~minutes), but lengthen to a maximum of 1.2 h before decreasing again. The total of ~90 eclipses, thus, occur irregularly with significant periods of time when no battery operation is required. With more than 22 h of sunlight available in each orbit, a trickle charge solution is possible.

In LEO, on the other hand, the spacecraft may be in eclipse and thus require battery power for 40% of each orbit. Although the precise duration will depend on orbit inclination, it is fairly regular, and the eclipse cycle results in typically 5000 to 6000 charge/discharge cycles of the battery per year. This results in the array-power sizing needing to be nearly twice the nominal load requirement (see Section 10.6).

In summary LEO operations require a large number of low-depth discharges, whereas in GEO a few deep discharges suffice. This inevitably influences battery type, resulting in the present trend of using nickel–cadmium (Ni–Cd) or silver–zinc (Ag–Zn) cells for LEO operation and nickel–hydrogen (Ni–H_2) cells for GEO operations. Cell cycle life, specific weight (kW h/kg) and volume (kW h/m^3) all influence the acceptability of a particular battery technology. However, work on more exotic materials, for example, Li–SO_2, is continuing [13] and alternative technologies continue to be implemented on spacecraft (see e.g. the annual Goddard Space Flight Center (GSFC) battery workshop proceedings [14]). One of the more recently adopted technologies is the use of a Li–SO_2 battery to provide power to the Huygens probe, after separation from the Cassini Saturn orbiter, an event scheduled to occur in 2004. For this rather specialized mission, the probe is in hibernation for 7 years. The battery is then required to provide power during a low power coast for 22 h, and then a high-power load for 2.5 h during descent. For the final 30 min of this period the probe operates from the surface of Titan. For this, the overall battery contains five individual battery units. Each battery consists of two modules of 13 Li–SO_2 cells in series. Each cell has a capacity of 7.5 A-h. Table 10.6 summarizes the performance characteristics of a number of cell technologies. Table 10.7 provides specific data for the Ni–H_2 batteries for the HST and Intelsat VII.

The detailed electrochemistry of batteries is covered in References [15] and [16]. The main function of battery operation, which is of importance to spacecraft design, is the way in which the reliability and charge efficiency are related to charge control. Parameters of critical importance are the charge/discharge rate, the depth of discharge (DOD), the extent of overcharging and the thermal sensitivity to each of these parameters. Figure 10.15 summarizes some available data on Ni–Cd batteries. A further feature not indicated

Table 10.6 Performance of battery technologies for space use [14]

Type	Specific energy (W h/kg)
Ni–Cd	39
Ni–H_2	52
Ag–Zn	60
Ni–MH	60
Li–Ion	80
Li–TiS_2	125
Na–S	150

Table 10.7   Hubble space telescope and Intelsat VII
Ni–H$_2$ battery summary

Parameter	HST	Intelsat VII
Specific energy (W h/kg)	57.14	61.26
Capacity (A h)	96	91.5
Cell dimensions:		
Diameter (cm)	9.03	8.89
Length (cm)	23.62	23.67
Terminal/terminal (cm)	24.66	29.67
Cell mass (kg)	2.1	1.867

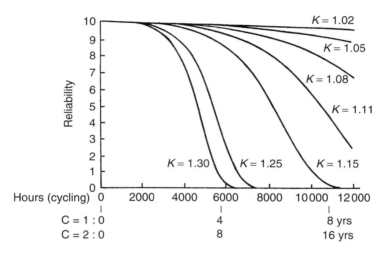

Figure 10.15   Ni–Cd cell reliability as a function of overcharge factor. Hours cycling is related to operation for two cases: $c = 1$, charge rate is battery capacity $\frac{1}{20}$ A/A h; $c = 2$, charge rate is battery capacity $\frac{1}{10}$ A/A h [17] (Reproduced by permission of European Space Agency and P. Montalenti)

in these graphs is the changing performance of a battery after cycling, specifically the change in voltage–current characteristics. The predominant effect is that the charge control system must be flexible if long missions are to be successfully executed. One other notable feature of battery ageing is the effect of hysteresis on the battery capacity. Figure 10.16 demonstrates the loss of this charge capacity over several cycles. It has been noted that if a battery is completely discharged, then capacity may be regained. Whilst this process may result in reverse polarization problems, battery reconditioning before an eclipse season is regularly used for GEO spacecraft. For a given reliability, Ni–H$_2$ batteries may be operated at a greater DOD compared with a Ni–Cd battery, thus requiring for the Ni–H$_2$ a lower installed capacity. As an approximation, for a given number of cycles, a Ni–H$_2$ will be capable of operating at an extra 15% DOD above what is safe to operate a Ni–Cd battery.

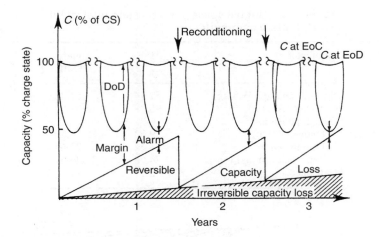

Figure 10.16 Battery reconditioning via complete discharge to improve battery capacity. Both reversible and irreversible capacity loss occurs [17] (Reproduced by permission of European Space Agency and P. Montalenti)

10.5 POWER MANAGEMENT, DISTRIBUTION AND CONTROL

The basic features of power control were outlined in Section 10.2. A key aspect of the power management system is that it must be designed to operate with both a primary and a secondary power system whose characteristics are changing with time as outlined in the previous sections. Several philosophies for power management are outlined in Figure 10.17.

The *electrical 'bus'* may be required to provide a variety of voltages to meet the needs of the various equipment. Generally within Europe, the trend has been to have a regulated dc power bus, typically at 28 or 50 V. For example, the bus of the second generation Meteosat system has a bus voltage of 28.2 V ± 1%, and for Artemis it is 42.5 ± 0.5 V. Both of these satellites were designed in Europe. In contrast, US spacecraft generally use unregulated buses; for example, the NASA standard for unmanned spacecraft provides a voltage in the range 21 to 35 V dc. Present spacecraft are being designed with higher bus voltages (∼150 V) to reduce resistive losses and harness mass. For both regulated and unregulated systems, dc–dc converters are required to provide the variety of voltages needed; this conversion frequently takes place at equipment level rather than centrally.

An ac bus is sometimes used to augment the dc one. The hybrid system can provide mass savings due to both the simplicity of conversion from ac to a variety of dc levels, and also the fact that it is possible to run the power distribution harness at higher voltages if an ac supply is used. Indeed it has been noted [18], that the equivalent wiring cross-section is six times smaller for a three-phase ac network than for a dc one. An ac distribution is mainly applicable to high-power spacecraft and where a large number of dc voltages are required at equipment level. The voltages in ac buses generally have a square waveform, an example being Hipparcos. A notable exception to this is on the Shuttle Space-lab, where the ac bus is more sophisticated, providing a three-phase sinusoidal voltage at 400 Hz.

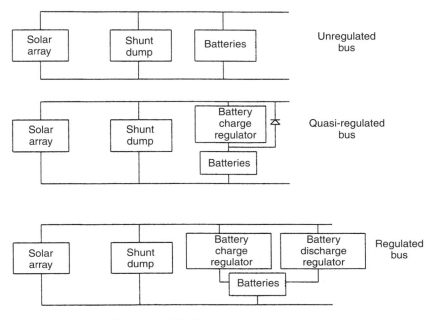

Figure 10.17   Schematic bus concepts

As noted above, the bus can be either regulated or unregulated; in some circumstances it can be quasi-regulated. The fully regulated bus provides voltage regulation during both normal sunlight operation (battery charge cycle) and during an eclipse (discharge cycle). With quasi-regulation, the regulation occurs only during sunlight operation.

The primary units

The main units used in a power system are described below for a typical configuration used on European communications spacecraft. The terminology may differ slightly for US spacecraft. Figure 10.18 shows the overall power system layout.

● *Array regulator.* As the power available from an array varies during a mission, or the power demanded by the payload varies, it is possible to switch in or out segments of the solar array. Switching out is achieved by grounding the individual segments in the shunt dump module. The array itself can be structured into various modules, sometimes called *solar power assemblies* (SPA). The regulation of power output from the array bus is typically achieved by using either a pulse-width modulation scheme, or a sequential switching-shunt regulation, known as S^3R [19]. A considerable number of satellites have used this latter approach. The voltage sensing that is used to control the shunt dump module is termed the *mode control unit* (MCU). An additional strategy used on some spacecraft is maximum power-point tracking. As can be seen from Figure 10.5, the power output from an individual cell shows a distinct maximum. Maximum power-point trackers control the operating point of a string of cells by varying the voltage at which the string operates. As an example [20] of a solar array regulator, the unit on the Artemis satellite, which has a

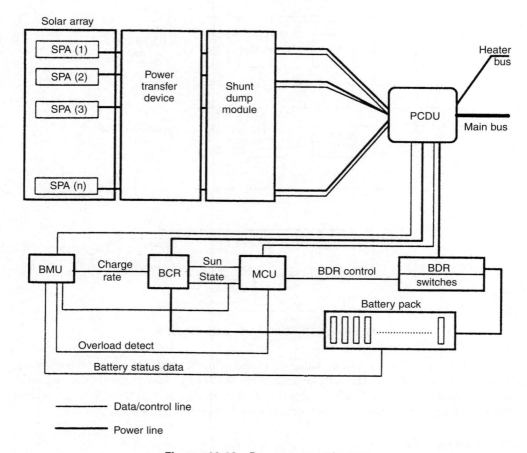

Figure 10.18 Power system layout

sunlight load of 3.3 kW and an eclipse load of 2 kW, provides shunt regulation. This regulator has a mass of 6.5 kg and dissipates 117 W at a nominal load of 2.65 kW. During eclipse operations the regulator requires 18 W.

- *Battery control.* Three units are typically associated with battery control. These are the *battery management unit* (BMU), the *battery charge regulator* (BCR) and the *battery discharge regulator* (BDR). The BMU's functions are to monitor the battery's temperature and voltage as well as individual cell voltages, pressures and temperatures. It is the interface between the power subsystem and the data-handling subsystem, and also provides control inputs to the charge regulation of the batteries, carried out by the BCR. The principal function of the BCR is to provide a constant current charge of the battery during sunlight operation, whilst that of the BDR is to supply a constant current to the spacecraft bus during eclipse operation. Control of this current is derived from the MCU, typically with further protection from the BMU. Whilst the charge/discharge rate of the battery may be controlled in a fairly simple manner through current regulation, the monitoring of the state of charge in a battery is more complex. The principal methods that may be used to sense charge-state are cell voltage, cell temperature, or cell pressure. It should be noted that the level of full charge noted by each of these methods results in a different level of over-charging. Pressure and temperature sensing results in overcharging by 20 to 30%, whereas voltage sensing may indicate 10 to 20% overcharge. A principal problem with voltage sensing arises because of the voltage–charge–temperature characteristics that may cause significant errors in determining the state of charge of the battery. Again taking the Artemis spacecraft as an example, the overall battery control unit has a mass of 25.5 kg. In sunlight, the power required to operate this is 63.5 W and in eclipse, 282 W. The efficiency of the BDR is 89%, and the BCR is 91%.

- *Power control and distribution unit (PCDU)*. This unit provides monitoring and protection for the bus current. Protection is normally achieved either by current limiting or by fusing, the latter generally requiring a redundant path to be switched into operation, normally by command from ground-control.
- *Power conversion unit (PCU)*. This unit supplies the individual voltage/current characteristics required for loads. The typical low voltage outputs (e.g. ± 15 V and 5 V) will be regulated using solid-state switches that are pulse-width modulated. This unit must also be able to cope with transient protection for over- and under-voltage and in-rush current limiting when units are switched on or off.

10.6 POWER BUDGET

The foregoing sections have outlined the principles of the technology behind a power system for use on a spacecraft system. In this section the methodology used to provide the size of a power system is outlined. It is noteworthy that the power subsystem is often the most massive.

10.6.1 Mission specific design issues

The starting point for any power system is in the definition of spacecraft electrical loads. In general these will not be constant throughout the mission, or even throughout a single orbit. The initial analysis must therefore take into account the mission profile and hence the power demand. The three critical issues that need be considered are the orbit parameters, the nature of the mission (communications, science, or other) and the mission duration.

The *orbit selection* has a major influence upon the radiation environment experienced, and hence the degradation anticipated in any solar array-based solution. Further, the orbit will define the duration of eclipse periods (see Section 5.3.2 of Chapter 5), which together with the number of eclipses anticipated, will define battery requirements and the degradation that might be expected to occur during the mission. Clearly in deep-space missions an investigation of the most appropriate technology for the primary power source will be influenced by the orbit.

The *nature of the mission* will have significant impact on the type of loads expected. Thus for a communications satellite, independently of the orbit specification, it can be anticipated that the primary payload will be required at all times. Further, the power demand in eclipse may well exceed that in sunlight, owing to the need both to operate the payload and to meet the additional burden from active, or power augmented elements, in the Thermal Control System (TCS). Navigation and broadcast satellites will have similar requirements to these. In contrast, a remote sensing spacecraft during eclipse may well not require the whole payload to be operational, particularly if there are passive optical instruments. For such satellites, there may be a very great range of power demands owing to the mission requirements, especially if there are active microwave instruments. These instruments may operate for only limited portions of the orbit, perhaps when communication of data to a ground station is available; this can provide a very high peak load for the power system. Scientific satellites may have very specific payload operational requirements, with additional complexity being provided by the need to download scientific data only when the vehicle is within sight of a ground station. Deep-space missions will inevitably face the temperature extremes that may require additional power for environmental control if the spacecraft either goes very close to or very far from the Sun.

The *mission duration* will provide a major influence on the degradation of the power system. The two most significant influences are

1. The total radiation dose expected which may determine, for a satellite carrying a solar array, the amount of shielding required, and hence influence the specific mass of the power system.
2. The number of eclipse cycles, which will influence the system reliability and its degradation. Solar cell failure through open-circuit losses will clearly increase with increased number of thermal cycles, driven by the entry to and from eclipse. Battery degradation will progress with number of eclipse cycles; indeed for any given technology there is only a maximum number of charge/discharge cycles that a battery can sustain before failure, as shown in Figure 10.16.

10.6.2   Power budget evaluation

The format of a typical power budget is shown in Table 10.8. In this each of the subsystems are identified, with, in this example, power shown simply in terms of eclipse and sunlight loads and peak loads. It is notable that the power subsystem itself also places a load on the spacecraft.

The method adopted for deriving suitable values to insert in this table is as follows: Suppose data is available for each subsystem. Initially such data will exclude the power subsystem. For each subsystem, data must be provided for both the sunlit orbit phase, having, say an average value of P_{sun} and the eclipse phase having an average value of $P_{eclipse}$. Since the satellite must be provided with power throughout the mission, this specifies the EOL requirement.

In reality this approach is highly simplified, since there will be specific events that may place a high transient load on the power system. One such event is the firing of a pyrotechnic device followed by the actuation of some form of deployment mechanism. As an example, the firing of a pyrotechnic release, followed by the deployment of a boom on the Ulysses spacecraft, provided a transient load of nearly two thirds of the total available power from the RTG at the BOL.

In the principal operational phase of a mission, generally not all the spacecraft equipment will be operating at one time. As a result the potential power demand, identified by summing all the spacecraft loads, will never be a realistic value for the peak demand load.

Table 10.8 Typical structure of a power budget

Subsystem	Peak power	Sunlight power P_{sun}	Eclipse power $P_{eclipse}$	Intelsat VIIa (%)	Average GEO comms. satellite (%)
AOCS				5.0	3.6
Power				10.4	11.2
Thermal control				4.9	6.4
Comms.				n/a	n/a
Data handling				0.6	1.6
Payload				79.1	77.2
Average total power				100	100

Indeed in many cases it will be found that such a summation exceeds the total power availability from the power bus. One subsystem that has widely varying requirements during a mission is the thermal system. This subsystem must meet both a hot and cold case, which may require very different levels of heater input. Again taking the Ulysses mission as an example, the 'hot' case heater power (8.7 W) was only one third that of the 'cold' case heater load (24.7 W).

For telecommunications spacecraft operating in GEO there is a fairly well-defined power profile between subsystems. Increasingly, as noted in Chapter 6, electric propulsion is being used on such missions for station keeping control, which results in an increase in the power required for the propulsion subsystem. Power for propulsion in Table 10.8 is included in the AOCS subsystem in the power profile. This is shown as a percentage for each of the subsystems for recently launched GEO communications satellites. The specific profile for Intelsat VIIa is also shown.

10.6.3 Power system sizing

A simplified block diagram for the power system is shown in Figure 10.19, in which the efficiencies of various components are also identified. Representative values for the efficiencies may be found in Sections 10.3, 10.4 and 10.5.

Consider a general case for sizing a power system. Assuming that the orbit period is τ, with the time spent in sunlight τ_{sun}, and the time spent in eclipse $\tau_{eclipse}$, then the power required from the array to meet the eclipse load is clearly given by P_{charge}, where

$$P_{charge}\tau_{sun} = \frac{1}{\eta}P_{eclipse}\tau_{eclipse} \qquad (10.5)$$

and η is the product of the efficiency terms shown in Figure 10.19,

$$\eta = \eta_{BDR}\eta_{BCR}\eta_{AR}$$

The total power required to be available from the array is thus approximately given by

$$P_{array} = P_{sun} + P_{charge} \qquad (10.6)$$

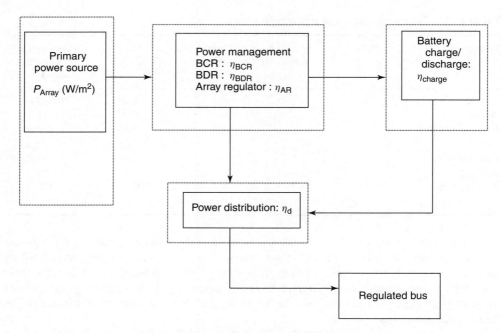

Figure 10.19 Definition of system efficiencies

In the limit, if all the efficiencies are equal to unity, it is apparent from equations (10.5) and (10.6) that if the eclipse power demand is the same as that in sunlight then the array power is simply given by

$$P_{\text{array}} = P_{\text{sun}} \left(\frac{\tau}{\tau_{\text{sun}}} \right) \tag{10.7}$$

Thus for LEO orbits where the fraction of the orbit in eclipse is large, typically of the order of 30 min out of 90 min, the array needs to be oversized relative to the sunlight provision by a factor τ/τ_{sun} of approximately 1.5. Hence, in this approximate scenario, the array power in LEO is required to be 50% in excess of the bus load. This may be contrasted with the GEO case where a maximum eclipse duration of $\sim$70 min in the 24-hour orbit arises. In this case the 'oversizing' of the array amounts to only 5%. This feature clearly has significant influence over the design of a solar array.

If the orbit period is τ (hours), the battery-stored energy will be given approximately by E_{B} (W-hrs)

$$E_B = P_{\text{eclipse}}(\tau - \tau_{\text{sun}})/(\eta_{\text{charge}} DOD) \tag{10.8}$$

where the *DOD* is the depth of discharge of the battery. The battery mass can be estimated by dividing the stored energy (W-hrs) by the energy density (W-hrs/kg) for the chosen battery technology. For example, for Ni–Cd the energy density is approximately 30 to 40 W-hrs/kg.

In any particular case, the individual loads that will be switched in and out need to be considered. We can then modify equation (10.5) to obtain the total energy required from the array. This may be written in the form $\varepsilon_{\text{array}}$ where

$$\varepsilon_{\text{array}} = P_{\text{array}}\tau_{\text{sun}} = \frac{1}{\eta_{\text{sun}}}\left(\sum_{i=1}^{k} P_i t_i\right) + \frac{1}{\eta_{\text{ecl}}}\left(\sum_{i=k+1}^{n} P_i t_i\right) \qquad (10.9)$$

in which the P_i, $i = 1, \ldots, n$, gives the typical power profile for payload and subsystem operation throughout the orbit. A typical profile may be as illustrated in Figure 10.20. The battery charge requirement during sunlight is explicitly excluded from this profile, since the eclipse profile (second term in equation 10.9) is equivalent to the battery charge energy (see equation 10.5). For the purposes of a first estimate of the array size, the efficiency factors have typical values of $\eta_{\text{sun}}\sim 0.8$, from array to loads, and $\eta_{\text{ecl}}\sim 0.6$, from batteries to loads.

In terms of calculating the array size, allowance needs to be made for any pointing-angle offset of the array relative to the sun line. This results in the array area being given by A_{array} where

$$A_{\text{array}} = P_{\text{array}}/(S\cos\delta\theta\,\eta_{\text{cell}}\eta_{\text{packing}}(1 - D)) \qquad (10.10)$$

Here S is the solar flux ($\sim 1400\,\text{W/m}^2$ in a near-Earth orbit); $\delta\theta$ is the array pointing error with respect to the Sun, which will typically be of order $1°$, but is highly dependent on the mission; η_{cell} is the solar cell efficiency; η_{packing} is the cell packing efficiency that is typically 0.90; D is the array degradation factor over the spacecraft lifetime, which may be calculated in the manner described in Section 10.3.1.

In a specific design, the sizing of the array given by the equivalent of equation (10.7) will be modified to include specific mission profiles for 'hot' and 'cold' cases. These differing cases arise during the year, as a result of variation in solar insolation, which will influence the array temperature (see Chapter 11); this variation occurs due to the influence of the angle between the orbit plane and the ecliptic. As a result, the final EOL array and battery capacity can be specified. To define the BOL characteristics, loss factors due to radiation damage, micrometeorite damage and battery hysteresis losses must included. Thus this process evidently requires the array and battery capacity to be oversized initially to meet the EOL power demands. Typically these loss factors may rise to 25% of the

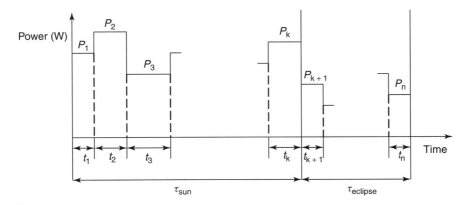

Figure 10.20 Approximate power profile for payload and subsystem operation through-out the orbit

initial provision, although this figure is highly dependent upon both mission duration and the type of orbit, with Highly Elliptical Orbits (HEOs) being particularly affected by radiation damage.

REFERENCES

[1]   *SPS Concept Development and Evaluation Programme Reference System Report* (1978) US DOE and NASA DOE/ER 0023.

[2]   Angrist, S. W. (1982) *Direct Energy Conversion*, Allyn and Bacon, New York.

[3]   Bennett, G. (1995) Summary of the US Use of Space Nuclear Power, ESA SP 369, 163–169.

[4]   Sheibley, D. W. (1983) Regenerative H_2-O_2 Fuel Cell—Electrolyser Systems for Orbital Energy Storage, NASA CP-2331, 23–38.

[5]   Rauschenbach, H. S. (1980) *Solar Cell Array Design Handbook*, Van Nostrand Reinhold, New York.

[6]   Taylor, H., Simpson, A. F. and Dollery, A. A. (1984) CMX-50: A New Ultra-thin Solar Cell Cover for Lightweight Arrays, ESA SP-173, 211–214.

[7]   Mawira, D. (1982) Advanced Rigid Array, ESA SP-173, 9–14.

[8]   van Hassel, R. (1999) Achievements and Prospects for ARA Mark III Solar Array Product Range, IAF-99-R.2.07.

[9]   Knorr, W., Theurer, G. and Schwartz, M. (1995) A Regenerative Fuel Cell System for a Lunar Rover, ESA SP-369, 21–26.

[10] Bockris, J. O.'M. and Srinivason, S. (1969) *Fuel Cells and Their Electrochemistry*, McGraw-Hill, New York.

[11] Terrill, W. and Haley, V. (1986) Thermoelectric Converter for SP-100, *21st IECEC*, 1950–1955.

[12] *Solar Dynamic Power System Development for Space Station Freedom*, NASA RP1310 (1993).

[13] Dudley, G. J. (1998) Lithium Ion Batteries for Space, ESA SP-416, 17–24.

[14] Halpert, G. and Surampudi, S. (1993) Advanced Energy Storage for Space Applications, NASA CP 3254, 91–110.

[15] NASA RP 1052 (1979).

[16] NASA SP 172 (1968).

[17] Montalenti, P. (1977) Software/Hardware Interface in Control and Protection of Space Batteries, ESA SP-126, 271–278.

[18] Eggers, G. (1985) AC Buses for LEO—A Viable Alternative, ESA SP-230, 17.

[19] Knorr, W. (1998) Power System for 2nd Generation Meteosat, ESA SP-416, 11–16.

[20] ESA SP-369 4th European Space Power Conference (1995).

11 THERMAL CONTROL OF SPACECRAFT

Chris J. Savage

Head of Thermal and Environmental Control Section, European Space Research and Technology Centre, European Space Agency

11.1 INTRODUCTION

Spacecraft thermal control—that is the control of spacecraft equipment and structural temperatures—is required for two main reasons: (1) electronic and mechanical equipment usually operate efficiently and reliably only within relatively narrow temperature ranges and (2) most materials have non-zero coefficients of thermal expansion and hence temperature changes imply thermal distortion.

Spacecraft equipment is designed to operate most effectively at or around room temperature. The main reason for this is that most of the components used in spacecraft equipment, whether electronic or mechanical, were originally designed for terrestrial use. It is also much easier and cheaper to perform equipment development and, eventually, qualification and flight acceptance testing at room temperature. Typically, operating electronic equipment requires to be maintained in a temperature range between about $-15\,°C$ and $+50\,°C$, rechargeable batteries between about $0\,°C$ and $+20\,°C$ and mechanisms (solar array drives, momentum wheels, gyroscopes etc.) between about $0\,°C$ and $+50\,°C$. There are, of course, exceptions to this—for example, some detectors within astronomical telescopes that need to be cooled to very low temperatures.

Many spacecraft payloads require very high structural stability, and therefore thermally induced distortion must be minimized or strictly controlled. For example, the search for ever-higher resolution from space-based telescopes means that temperature stability of a fraction of one degree is often required within telescope systems several metres in size.

Heat is generated both within the spacecraft and by the environment. Components producing heat include rocket motors, electronic devices and batteries. Initial ascent heating effects are minimized by the launch vehicle's nose fairings or, in the case of launch by the Space Shuttle, by the cargo-bay doors. Heat from the space environment is largely the result of solar radiation. Heat is lost from the spacecraft by radiation, mainly to deep space. The balance between heat gained and heat lost will determine the spacecraft temperatures.

Spacecraft Systems Engineering (Third Edition). Edited by P. W. Fortescue, J. P. W. Stark and G. G. Swinerd
© 2003 John Wiley & Sons Ltd

The configuration of a spacecraft is dictated by many factors and 'thermal control' is only one of them. The task for the thermal control engineer consists, in fact, of three main parts. Firstly, *analysis*—he or she must be able to analyse a given spacecraft configuration and predict equipment and structural temperatures for all phases of the mission. Secondly, *design*—in the rather-likely circumstance that the results of the analysis show temperatures falling outside allowed limits, the engineer must devise suitable solutions, for example, by modifying heat-flow paths or implementing heaters, radiators and so on. Finally, *testing*—the engineer must perform sufficient and appropriate testing to confirm the accuracy of the analysis and of the thermal predictions for the mission.

11.2   THE THERMAL ENVIRONMENT

An important characteristic of the space environment is its high vacuum. Spacecraft are generally launched into orbits where the residual atmospheric pressure, and hence drag, is very small (although often not negligible—the International Space Station (ISS) will require re-boosting a few times per year to compensate for air drag). Fortunately for the thermal control engineer, the very low level of drag implies also the absence of any significant *aerodynamic heating*. For an orbiting spacecraft, aerodynamic heating and indeed any *convective interaction* between spacecraft and environment can be ignored.

The rate at which the Earth's atmospheric pressure falls with altitude [1] is shown in Table 11.1 for conditions of moderate solar activity. Spacecraft in orbit around the Earth usually orbit at altitudes higher than 300 km where the residual atmospheric pressure is typically less than 10^{-7} mb. During the launch phase, the transition from being fully protected within the launch vehicle to autonomous operation in space is the result of a compromise. The sooner the nose fairings can be jettisoned, the more payload a given vehicle can launch. However, if they are jettisoned too early, dynamic pressure and aerodynamic heating will damage the spacecraft. Fairings are normally jettisoned at the point at which residual aerodynamic heating is the same or less than the incident solar heating, which occurs at around 100-km altitude, depending on the characteristics of the

Table 11.1 Atmospheric pressure as a function of altitude (moderate solar activity)

Altitude (km)	Pressure (mb)
0	1013
50	7.98×10^{-1}
100	3.20×10^{-4}
150	4.54×10^{-6}
200	8.47×10^{-7}
250	2.55×10^{-7}
300	8.77×10^{-8}
350	3.51×10^{-8}
400	1.45×10^{-8}
450	6.45×10^{-9}
500	3.02×10^{-9}

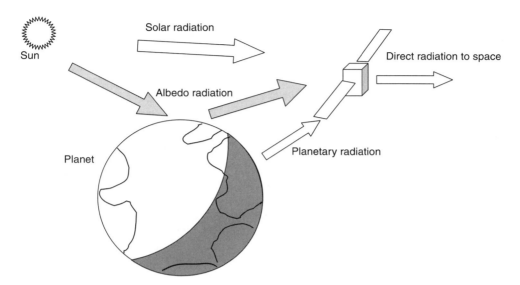

Figure 11.1 Typical spacecraft thermal environment

launch vehicle and trajectory. Hence, once again, aerodynamic heating effects can be ignored by the spacecraft thermal designer. During re-entry or aerobraking manoeuvres (see Chapter 5), specific protection is provided, which ensures that these phases do not drive the spacecraft thermal design.

A spacecraft in space can interact with its environment only by radiation and this interaction is characterized by the exchange of energy by means of the following (see Figure 11.1):

- direct solar radiation;
- solar radiation reflected from nearby planets (albedo radiation);
- thermal energy radiated from nearby planets (planetary radiation);
- radiation from the spacecraft to deep space.

The spacecraft will experience thermal equilibrium when the sum of the radiant energy received from the first three sources listed above, together with any thermal dissipation within the spacecraft, is equal to the energy radiated to deep space. It is this balance that will determine the physical temperature of the spacecraft.

11.2.1 Solar radiation

The solar radiation parameters of interest to the thermal design engineer are (1) spectral distribution, (2) intensity and (3) degree of collimation. The spectral distribution can be considered constant throughout the solar system and the solar irradiance, or spectral energy distribution, resembles a Plank curve with an effective temperature of 5800 K (see Chapter 2). This means that the bulk of the solar energy (99%) lies between 150 nm and 10 μm, with a maximum near 450 nm (in the yellow part of the visible spectrum).

The solar radiation intensity outside the Earth's atmosphere and at the Earth's average distance from the Sun (1 AU) is called the solar constant and is about $1371 \pm 5 \, \text{W/m}^2$. The solar radiation intensity J_s at any other distance d from the Sun can be found from the simple relationship

$$J_s = \frac{P}{4\pi d^2} \tag{11.1}$$

where P is the total power output from the Sun, $3.856 \times 10^{26} \, \text{W}$. Table 11.2 shows the resulting variation in solar intensity that can be expected at the average distance from the Sun of each of the planets in the solar system.

The angle subtended by the Sun in the vicinity of the Earth (at 1 AU from the Sun) is about $0.5°$. This means that the sunlight incident on a spacecraft can, for thermal control purposes, be regarded as a parallel beam emanating from a point source. This is not true, however, for spacecraft whose mission takes them very close to the Sun.

The fraction of the solar radiation that is reflected from the surface and/or atmosphere of a planet is known as the *planetary albedo*. Its value is highly dependent on local surface and atmospheric properties. For example, for the Earth, it varies from as high as 0.8 from clouds to as low as 0.05 over surface features such as water and forest [2,3]. Fortunately for the thermal engineer, such changes occur rapidly in relation to the thermal inertia of most spacecraft, and an orbital average value can be used for thermal design purposes. For the Earth, this is in the range 0.31 to 0.39. Table 11.2 lists the albedo values [4] for the planets of the solar system. The reader should be aware that measuring the albedo of the more distant planets is not an easy task and that the quoted figures should be treated with caution. Although the spectral distribution of albedo radiation is not identical to that of the Sun, as is evidenced by the diverse colours of planetary surface features, the differences are insignificant for thermal engineering purposes and can be ignored.

The intensity of the albedo radiation, J_a, incident on a spacecraft is a complex function of planet size and reflective characteristics, spacecraft altitude and the angle β between the local vertical and the Sun's rays. This can be expressed in terms of a *visibility factor* F as follows:

$$J_a = J_s a F \tag{11.2}$$

Table 11.2 Planetary solar constants and albedo values [2,3]

Planet	Solar radiation intensity, J_s (percentage of solar intensity at 1 AU)	Planetary albedo, a
Mercury	667	0.06–0.10
Venus	191	0.60–0.76
Earth	100	0.31–0.39
Moon	100	0.07
Mars	43.1	0.15
Jupiter	3.69	0.41–0.52
Saturn	1.10	0.42–0.76
Uranus	0.27	0.45–0.66
Neptune	0.11	0.35–0.62
Pluto	0.064	0.16–0.40

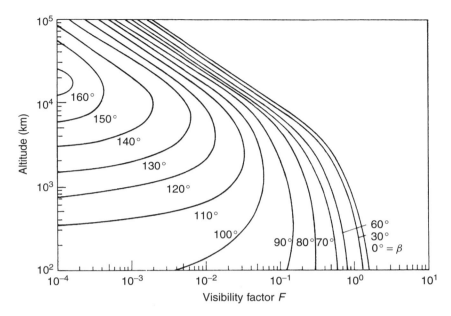

Figure 11.2 Spacecraft albedo irradiation. β is the angle between the local vertical and the Sun's rays

For the purpose of calculating albedo radiation inputs, the Earth can be regarded as a diffuse reflecting sphere, in which case the visibility factor varies approximately as shown in Figure 11.2.

It is emphasized that the above treatment is approximate. For complex spacecraft, particularly in low orbits, accurate calculation of albedo inputs may need to be performed as a function of orbital position for each external surface element. These are complicated calculations for which specific software tools are available.

11.2.2 Planetary radiation

Since the planets of the solar system all have non-zero temperatures, they all radiate heat. Because of its relatively low temperature, the Earth radiates all of its heat at infrared wavelengths, effectively between about 2 and 50 μm with peak intensity around 10 μm. For this reason, the radiation is often referred to as *thermal radiation*. The spectral distribution of the Earth's thermal radiation is shown in Figure 11.3 [2]. The atmosphere is essentially opaque over much of the infrared spectrum, with important transparent windows at around 8 and 13 μm. The radiation that a spacecraft sees is hence composed of radiation from the upper atmosphere, radiating with an effective black-body temperature of 218 K. Superimposed on this is radiation from the Earth's surface passing through the infrared windows. Since terrestrial temperatures vary with time and geographical location, the intensity J_p of the thermal radiation incident on orbiting spacecraft can also be expected to vary with time and position around the orbit. In fact, because of the Earth's large thermal inertia with respect to diurnal and seasonal changes and the spacecraft's

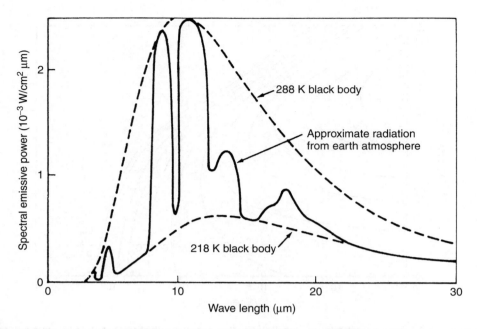

Figure 11.3 Typical spectral emissive power for the thermal radiation from Earth (*Note*: The 288-K black-body curve approximates the radiation from the Earth's surface, and the 218-K black-body curve approximates the radiation from the atmosphere in those spectral regions where the atmosphere is opaque)

large thermal inertia with respect to its orbital period, only very small errors occur if averaged values are used. For most practical purposes, the thermal engineer can assume that the Earth radiates with an intensity of 237 W/m^2 and that the thermal radiation emanates uniformly from the whole cross-sectional area of the Earth.

Since the intensity falls with altitude according to the inverse-square law, the approximate value of J_p in W/m^2 at a given altitude can be found from the following expression

$$J_\mathrm{p} = 237 \left(\frac{R_\mathrm{rad}}{R_\mathrm{orbit}} \right)^2$$

where R_rad is the radius of the Earth's effective radiating surface and R_orbit is the orbit radius. The precise value of R_rad is not easy to determine and for most practical purposes it can be assumed equal to the radius of the Earth's surface, R_E. For other planets, care needs to be taken to verify the validity of these assumptions on a case-by-case basis. For example, Mercury, with a sidereal day of the same order of magnitude as its year (59 and 88 Earth days, respectively), sustains temperature differences of hundreds of degrees between sunlit and shadowed sides, with a slow-moving terminator. Its orbit is also sufficiently eccentric for its solar constant to more than double between apogee and perigee, giving rise to large seasonal variations.

11.2.3 Spacecraft heat emission

The spacecraft itself has a finite temperature, so it will also radiate heat to space. Since the spacecraft temperature will be similar to that of the Earth (if the thermal engineers have done their job properly), it too will radiate all its heat in the infrared region of the spectrum.

11.3 THERMAL BALANCE

As already noted, the temperature of a spacecraft depends on the balance between the heat received from external and internal sources, and the heat radiated to space. In order to control spacecraft temperatures, it is necessary to control the heat absorbed, the heat radiated or (usually) both.

If spacecraft were *black bodies*, that is, radiated as black bodies and absorbed all the radiation that fell on them, they would acquire a certain temperature and that would be the end of the story. Thermal control would be impossible, except perhaps by varying the internal heat dissipation. However, spacecraft are not black bodies but absorb only a fraction α of incident energy. They also emit as a *grey body*, radiating a fraction ε of the radiation of a black body at the same temperature. This may be expressed as

$$J_{\text{absorbed}} = \alpha J_{\text{incident}} \tag{11.3}$$

$$J_{\text{radiated}} = \varepsilon \sigma T^4 \tag{11.4}$$

where α and ε are known as the *absorptance* and the *emittance*, respectively, and σ is the *Stefan–Boltzmann constant* equal to $5.67 \times 10^{-8} \, \text{W/m}^2 \, \text{K}^4$.

For a spacecraft with no internal heat dissipation, an effective absorbing area (projected area facing the Sun) A_α and emitting area A_ε, the equilibrium temperature, T, is given by

$$A_\alpha J_{\text{absorbed}} = A_\varepsilon J_{\text{radiated}} \tag{11.5}$$

which from equations (11.3) and (11.4) gives

$$A_\alpha \alpha J_{\text{incident}} = A_\varepsilon \varepsilon \sigma T^4$$

so that

$$T^4 = \frac{A_\alpha}{A_\varepsilon} \frac{J_{\text{incident}}}{\sigma} \left(\frac{\alpha}{\varepsilon} \right) \tag{11.6}$$

Since A_α, A_ε and σ are constants, and for a given value of J_{incident}, the value of T can be controlled by varying the value of α/ε.

In fact, things are not quite as simple as that, as α and ε are not independent variables. Over any given wavelength range, the laws of thermodynamics require that $\alpha = \varepsilon$ (Kirchhoff's law). Thus for any surface, its *absorptivity* at a given wavelength is equal to its *emissivity* at the same wavelength.

Fortunately for the thermal control engineer, absorptivity and emissivity generally vary with wavelength and we have already learnt that the radiation environment of a spacecraft is basically composed of radiation either at 'visible' wavelengths or in the infrared. It is this feature that makes spacecraft thermal control possible. For the spacecraft thermal control engineer-

- α means the absorptance of a surface to solar radiation (peak intensity at about $0.45\,\mu$m)—it is therefore often referred to as the *'solar absorptance'*.
- ε means the emittance of a surface radiating in the infrared region (peak intensity at about $10\,\mu$m)—it is therefore often referred to as the *'infrared emittance'*.

Note that, according to Kirchhoff's law, the absorptance of a spacecraft to planetary radiation is equal to its infrared emittance, ε.

By way of an example, let us consider a simple spacecraft in Low Earth Orbit (LEO). For the sake of convenience, let us assume a polar orbit that does not suffer from eclipses (a so-called dawn–dusk orbit—see Chapter 5) and let us furthermore assume that our spacecraft has a high thermal inertia and is isothermal. We have

$$
\begin{aligned}
\text{heat received directly from the Sun} &= J_s \alpha A_{\text{solar}} \\
\text{albedo contribution} &= J_a \alpha A_{\text{albedo}} \\
\text{planetary radiation contribution} &= J_p \varepsilon A_{\text{planetary}} \\
\text{heat radiated to space} &= \sigma T^4 \varepsilon A_{\text{surface}} \\
\text{internally dissipated power} &= Q
\end{aligned}
$$

where A_{solar}, A_{albedo} and $A_{\text{planetary}}$ are the projected areas receiving, respectively, solar, albedo and planetary radiation, and A_{surface} is the spacecraft total surface area. If we assume that J_s, J_a, J_p and Q remain constant, our spacecraft will acquire an equilibrium temperature T given by

$$(A_{\text{solar}} J_s + A_{\text{albedo}} J_a)\alpha + A_{\text{planetary}} J_p \varepsilon + Q = A_{\text{surface}} \sigma T^4 \varepsilon$$

Hence

$$T^4 = \frac{A_{\text{planetary}} J_p}{A_{\text{surface}} \sigma} + \frac{Q}{A_{\text{surface}} \sigma \varepsilon} + \frac{(A_{\text{solar}} J_s + A_{\text{albedo}} J_a)}{A_{\text{surface}} \sigma}\left(\frac{\alpha}{\varepsilon}\right) \qquad (11.7)$$

So once again we see that the spacecraft temperature is dependent on the ratio α/ε, particularly for spacecraft for which Q is small. For simplicity, let us assume a spherical spacecraft, radius r, for which

$$A_{\text{surface}} = 4\pi r^2, \; A_{\text{solar}} = A_{\text{albedo}} = A_{\text{planetary}} = \pi r^2, \; Q = 0$$

$$J_a = 0.33 F \; J_s (\text{Earth albedo } a = 0.33),$$

$$J_p = 220 \,\text{W/m}^2 (\text{corresponding to an orbit altitude of around } 240\,\text{km})$$

Then equation (11.7) reduces to

$$T^4 = 9.70 \times 10^8 + 4.41 \times 10^6 (1 + 0.33 F) J_s \left(\frac{\alpha}{\varepsilon}\right)$$

For $J_s = 1371 \,\text{W/m}^2$, F = 0.15 (from Figure 11.2) and a black paint finish for which $\alpha/\varepsilon = 1$, our spacecraft equilibrium temperature is about 293 K or 20 °C.

If we now turn the orbit plane until the spacecraft passes through the Earth's shadow, the heat absorbed from the Sun (directly and as albedo) will be reduced. Assuming a sufficiently high thermal inertia, a new equilibrium temperature will be obtained. Let us consider the case in which the Earth–Sun vector lies in the plane of the orbit. This will evidently give the minimum time in sunlight, which, for a LEO spacecraft at an altitude of 240 km, is about 59% of its orbit period. Under these conditions, an average albedo visibility factor can be estimated from Figure 11.2 for the illuminated part of the orbit, F ∼ 0.7. Note that the albedo radiation is zero during eclipse. The new equilibrium temperature will then be obtained from

$$T^4 = 9.70 \times 10^8 + 5.43 \times 10^6 J_s \left(\frac{\alpha}{\varepsilon}\right) f$$

Table 11.3 Equilibrium temperatures for a simple spacecraft in LEO

Surface finish	White paint $\alpha = 0.15$ $\varepsilon = 0.9$	Black paint $\alpha = 0.9$ $\varepsilon = 0.9$	Electroplated gold $\alpha = 0.25$ $\varepsilon = 0.04$
No eclipse	−61 °C	+20 °C	+176 °C
Maximum eclipse	−70 °C	−2 °C	+138 °C

Table 11.4 α and ε values for several surfaces and finishes [5,6,7]

Surface	Absorptance (α)	Emittance (ε)	α/ε
Polished beryllium	0.44	0.01	44.00
Goldized kapton (gold outside)	0.25	0.02	12.5
Gold	0.25	0.04	6.25
Aluminium tape	0.21	0.04	5.25
Polished aluminium	0.24	0.08	3.00
Aluminized kapton (aluminium outside)	0.14	0.05	2.80
Polished titanium	0.60	0.60	1.00
Black paint (epoxy)	0.95	0.85	1.12
Black paint (polyurethane)	0.95	0.90	1.06
—electrically conducting	0.95	0.80–0.85	1.12–1.19
Silver paint (electrically conducting)	0.37	0.44	0.84
White paint (silicone)	0.26	0.83	0.31
—after 1000 hours UV radiation	0.29	0.83	0.35
White paint (silicate)	0.12	0.90	0.13
—after 1000 hours UV radiation	0.14	0.90	0.16
Solar cells, GaAs (typical values)	0.88	0.80	1.10
Solar cells, Silicon (typical values)	0.75	0.82	0.91
Aluminized kapton (kapton outside)	0.40	0.63	0.63
Aluminized FEP	0.16	0.47	0.34
Silver coated FEP (SSM)	0.08	0.78	0.10
(OSR)	0.07	0.74	0.09

Note: SSM, Second Surface Mirror.
OSR, Optical Solar Reflector.

where f is equal to the fraction of the orbit that is illuminated by the Sun, 0.59 in this case. It will be seen that our black-painted spacecraft has fallen in temperature to about $-2\,°C$. Table 11.3 shows the results of similar calculations for a white paint finish (low α/ε) and electroplated gold (high α/ε). It is evident that, by using different surface finishes in different ratios, spacecraft temperatures can be controlled over quite large ranges. Table 11.4 lists α and ε values for a number of common spacecraft surface finishes [5,6,7]. It should be realized that, particularly for metal surfaces, the values of α and ε may be very dependent on preparation and surface treatment.

Real spacecraft are, of course, far more complicated than the one discussed above. They are certainly not isothermal and often contain components (e.g. the solar arrays) with a relatively low thermal inertia, which will change temperature significantly around an orbit (particularly when entering or leaving an eclipse). Whilst the overall spacecraft thermal balance is determined by its external surface characteristics and the radiative environment, the internal thermal balance determines equipment temperatures and is hence of crucial importance to the thermal engineer. Calculation of the internal thermal balance, involving radiative and conductive exchanges between all the spacecraft components, is complex and is covered in some detail in the next section.

11.4 THERMAL ANALYSIS

11.4.1 Thermal mathematical model (TMM)

Spacecraft are generally very complex structures within which temperatures are varying continuously as a function of location and time. Calculating these temperature fields in rigorous detail is, for all practical purposes, impossible. In order to progress further, it is first necessary to simplify the problem. This is done by generating an approximate representation of the spacecraft that is amenable to mathematical treatment. Such a representation is known as a *thermal mathematical model* (TMM).

In order to construct a TMM, the spacecraft is considered as being composed of a number of discrete regions within which temperature gradients can be neglected. These regions are known as *isothermal nodes*. Each node is characterized by a temperature, thermal capacity, heat dissipation (if any) and radiative and conductive interfaces with the surrounding nodes. Nodes that can 'see' space directly will also have radiative interfaces with the external environment.

11.4.2 Conductive heat exchange

It will be recalled that the conductive heat flow rate is given by

$$Q_c = \frac{\lambda A}{l}\Delta T \tag{11.8}$$

where λ is the thermal conductivity, A the cross-sectional area, l the conductive path length and ΔT the temperature difference. The term $\lambda A/l$ is known as the *thermal conductance*, h_c, and hence the temperature difference can be written as

$$\Delta T = Q_c \frac{1}{h_c} \tag{11.9}$$

In most engineering applications, A and possibly λ may vary significantly along the path length. If the conductive path is considered as a number of discrete conductive paths connected in series, the temperature difference can be rewritten as

$$\Delta T = Q_c \left(\frac{1}{h_1} + \frac{1}{h_2} + \frac{1}{h_3} + \cdots \right) = Q_c \frac{1}{h_c}$$

and hence the effective thermal conductance, h_c, for the path can be found from

$$\frac{1}{h_c} = \frac{1}{h_1} + \frac{1}{h_2} + \frac{1}{h_3} + \cdots \tag{11.10}$$

For a spacecraft composed of n isothermal nodes, the heat conducted from the ith to the jth node is given by

$$Q_{c_{ij}} = h_{ij}(T_i - T_j) \tag{11.11}$$

where h_{ij} is the effective conductance between nodes i and j and T_i and T_j are the temperatures of the ith and jth nodes, respectively.

11.4.3   Radiative heat exchange

Radiative heat exchange between two surfaces is determined by three important parameters—the surface temperatures, the radiative view factors and the surface properties. For diffuse surfaces, the amount of radiation leaving a surface i and absorbed by a surface j can be shown [8] to be of the form

$$Q_{r_{ij}} = A_i F_{ij} \varepsilon_{ij} \sigma (T_i^4 - T_j^4) \tag{11.12}$$

where A_i is the area of the surface i, F_{ij} is the view factor of surface j as seen from surface i and ε_{ij} is a parameter known as the *effective emittance*. Note that it is tacitly assumed in the above that the value of the view factor F_{ij} remains constant over the surface i.

View factors

The *radiative view factor* F_{ij} is defined as the fraction of the radiation leaving one surface that is intercepted by another. It follows that from any node i inside a spacecraft, the sum of the view factors to surrounding equipment must be unity,

$$\sum_{j=1}^{k} F_{ij} = 1 \tag{11.13}$$

where k is the number of surrounding surfaces.

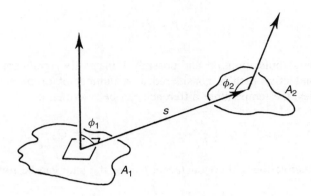

Figure 11.4 View factor geometry between surfaces A_1 and A_2

Consider two surfaces A_1 and A_2 as indicated in Figure 11.4 and let us assume that they are both diffuse surfaces. The radiation emitted from A_1 in the direction of A_2, per unit solid angle and per unit area of A_1, is given by

$$I_1 = I_0 \cos \phi_1 \tag{11.14}$$

where I_0 is the radiation intensity normal to A_1.

For the elementary surfaces δA_1 and δA_2, the total radiation leaving δA_1 and falling on δA_2 is given by

$$\delta Q_{r_{12}} = I_0 \delta A_1 \cos \phi_1 \frac{\delta A_2 \cos \phi_2}{s^2}$$

Hence, the total radiation reaching A_2 from A_1 is given by

$$Q_{r_{12}} = I_0 \int_{A_1} \int_{A_2} \frac{\cos \phi_1 \cos \phi_2}{s^2} \, dA_1 \, dA_2$$

However, from equation (11.14) it can be easily shown that the total radiation leaving A_1 is given by

$$Q_{r_{tot}} = A_1 I_0 \int_0^{\pi/2} 2\pi \sin \phi \cos \phi \, d\phi = A_1 \pi I_0$$

The view factor, F_{12}, is then

$$F_{12} = \frac{Q_{r_{12}}}{Q_{r_{tot}}} = \frac{1}{A_1} \int_{A_1} \int_{A_2} \frac{\cos \phi_1 \cos \phi_2}{\pi s^2} \, dA_1 \, dA_2 \tag{11.15}$$

which can be rewritten more generally in the form

$$A_i F_{ij} = \int_{A_i} \int_{A_j} \frac{\cos \phi_i \cos \phi_j}{\pi s^2} \, dA_i \, dA_j \tag{11.16}$$

From an examination of the symmetry of this equation, the important reciprocity relationship can be deduced as

$$A_i F_{ij} = A_j F_{ji} \qquad (11.17)$$

The calculation of view factors is a tedious process and many examples of standard configurations are available in the literature [6] to make the task easier.

For simple spacecraft models, involving relatively few isothermal nodes, view factors may be obtained by a method originally attributed to Eckert [9]. With reference to Figure 11.5, the surface from which the view factor is to be measured is placed in the plane of the table and a scale model of the second surface is constructed in the correct relative position as shown. A milk-glass hemisphere is placed over the model, with its base centred at the point P on the first surface from which the view factor will be measured. If a light source is placed at P, a shadow of the second surface will be obtained on the milk-glass hemisphere. The ratio of the projected area of the shadow on the table A and the area of the base of the hemisphere gives the view factor. A practical method for complex shapes would involve observing the model from above, perhaps with the aid of a camera, from sufficiently far away to avoid parallax. Alternatively, for simple shapes, the result can be obtained by scale drawing without the need to construct a physical model.

Another variant of the above approach [10] uses a convex paraboloid mirror that is graduated in such a way that, when viewed from above, the reflections in each segment represent equal view factors. The instrument, known as a 'factometer' and shown schematically in Figure 11.6, is used by placing the base of the mirror flat against the first surface and then counting the segments covered by the image of the second surface. For example, for an instrument with 100 segments, if the image of the second surface covers 15 segments, the view factor is 15/100, that is, 0.15. The advantage of this instrument is that it can be used with either a scale model of the spacecraft or a full-sized development or even a flight-standard version. The factometer was used extensively in the past, for example, for developing the thermal models for the Orbital Test Satellite (OTS) and European Communications Satellite (ECS), but has now been largely superseded by sophisticated software tools, capable of handling large and complex spacecraft models, such as ESARAD [11,12].

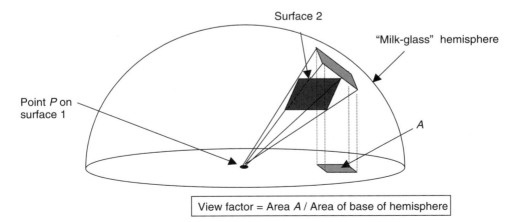

Figure 11.5 Experimental method for view factors, using milk-glass hemisphere

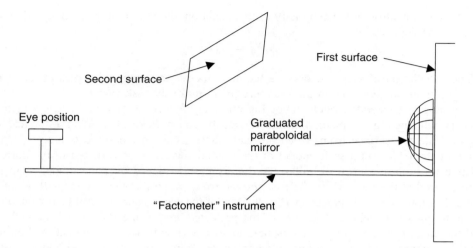

Figure 11.6 Practical instrument for view factor measurement, using paraboloid mirror

The above experimental techniques are useful for simple spacecraft (or, more correctly, simplified spacecraft models), for complex local configurations that are not amenable to easy mathematical description and for trouble-shooting during thermal balance testing.

Effective emittance

The effective emittance between two surfaces has a complicated dependence on surface optical properties, mutual reflections and reflections via other nearby surfaces. Its detailed treatment is beyond the scope of this book but can be found in Reference [8]. In the simple case in which all surfaces are perfectly black (i.e. $\varepsilon = 1$ for all surfaces, not just surfaces i and j), the effective emittance reduces to the trivial result $\varepsilon_{ij} = 1$. Specular surfaces and/or emittance values less than unity give rise to more complicated expressions for ε_{ij}. For the relatively simple case of two parallel, diffuse surfaces, separated by a distance that is small compared with their surface areas, it can be shown that

$$\varepsilon_{ij} = \frac{\varepsilon_i \varepsilon_j}{\varepsilon_i + \varepsilon_j - \varepsilon_i \varepsilon_j} \tag{11.18}$$

In spite of the fact that this is a special case, seldom encountered in practice, errors introduced by its use are small, provided the surfaces involved are diffuse and have relatively high ε values. This expression is commonly used for the calculation of practical effective emittance values and is acceptable for normal purposes in which the spacecraft interior is painted ($\varepsilon_i = \varepsilon_j \approx 0.9, \Rightarrow \varepsilon_{ij} \approx 0.82$). The use of specular and/or low-emittance surfaces will involve complicated analysis, which will need to be repeated for every change in the spacecraft internal configuration, and should be avoided where possible.

11.4.4 Calculation of nodal temperatures

Let us consider, at a particular time, the heat balance of the ith node of a TMM consisting of n nodes. The net heat absorbed by node i per unit time is given by

$$Q_{\text{external},i} + Q_i - \sigma \varepsilon_i A_{\text{space},i} T_i^4 - \sum_{j=1}^n h_{ij}(T_i - T_j) - \sigma \sum_{j=1}^n A_i F_{ij} \varepsilon_{ij}(T_i^4 - T_j^4)$$

where

$$Q_{\text{external},i} = J_s \alpha_i A_{\text{solar},i} + J_a \alpha_i A_{\text{albedo},i} + J_p \varepsilon_i A_{\text{planetary},i}$$

is the external heat input. $A_{\text{solar},i}$, $A_{\text{albedo},i}$ and $A_{\text{planetary},i}$ are the effective areas receiving, respectively, direct solar, albedo and planetary radiation, $A_{\text{space},i}$ is the effective area with an unobstructed view of space and Q_i is the internal heat dissipation. If the mass and specific heat of the node are m_i and C_i, respectively, the heat balance equation for node i can be written as

$$m_i C_i \frac{dT_i}{dt} = Q_{\text{external},i} + Q_i - \sigma \varepsilon_i A_{\text{space},i} T_i^4 - \sum_{j=1}^n h_{ij}(T_i - T_j) - \sigma \sum_{j=1}^n A_i F_{ij} \varepsilon_{ij}(T_i^4 - T_j^4)$$

(11.19)

The thermal behaviour of the complete spacecraft is thus described by the set of n simultaneous non-linear differential equations as above, with i varying from 1 to n. It should be noted that, in most practical cases, $Q_{\text{external},i}$ and Q_i will also vary with time.

Generally speaking, the above set of equations is not amenable to analytical solution, and numerical methods must be used. Perhaps the simplest approach to a solution is to assume that the spacecraft is launched at time t_0 from an air-conditioned launch vehicle such that all the temperatures are known initially. Substituting these initial temperatures allows the values of $dT_{i,0}/dt$ to be calculated for $i = 1$ to n. The temperature at a small increment of time, δt, later can then be calculated from

$$T_i = T_{i,0} + \delta T_{i,0}, \qquad \delta T_{i,0} = \frac{dT_{i,0}}{dt} \delta t$$

(11.20)

These new temperature values, together with the new values for Q_{external} and/or Q, can then be used to recalculate the values of dT/dt in a second iteration, and so on. Eventually, provided that δt is small enough, the evolving temperature history will settle into a profile determined by Q_{external} and Q rather than by the assumed initial temperatures. The problems with this simple approach are mainly a tendency to instability (particularly if δt is too large or thermal capacities are small) and a large amount of computer time required. It is, therefore, seldom used in practice.

A more common approach relies on first linearizing the equations as follows:
From equation (11.20)

$$T_i^4 = (T_{i,0} + \delta T_{i,0})^4$$

Assuming that $\delta T_{i,0}$ is small compared with $T_{i,0}$ and expanding we can write that

$$T_i^4 \approx T_{i,0}^4 + 4T_{i,0}^3 \delta T_{i,0}$$

and hence that

$$T_i^4 \approx T_i(4T_{i,0}^3) - 3T_{i,0}^4$$

(11.21)

Steady-state calculations

In the special case of steady-state calculations, where $Q_{\text{external},i}$ and Q_i are constant and dT/dt is zero, use of equation (11.21) allows us to write equation (11.19) in the linearized form as follows:

$$T_i \left[\sum_{j=1}^{n} h_{ij} + 4\sigma T_{i,0}^3 \left(A_{\text{space},i}\varepsilon_i + \sum_{j=1}^{n} A_i F_{ij} \varepsilon_{ij} \right) \right]$$

$$- \sum_{j=1}^{n} T_j \left[h_{ij} + 4\sigma T_{j,0}^3 A_i F_{ij} \varepsilon_{ij} \right]$$

$$= Q_{\text{external},i} + Q_i + 3\sigma T_{i,0}^4 A_{\text{space},i}\varepsilon_i + 3\sigma \left(T_{i,0}^4 - T_{j,0}^4 \right) \sum_{j=1}^{n} A_i F_{ij} \varepsilon_{ij} \quad (11.22)$$

This is a set of linear equations in n unknown T_i, where $i = 1$ to n, and can be solved by standard matrix inversion techniques. The calculated temperatures can then be substituted for the original values of $T_{i,0}$ and the calculation repeated until the difference between newly calculated and previously calculated temperatures is sufficiently small for steady-state conditions to be assumed, that is, dT_i/dt is sensibly zero. It should be noted that, although the criterion that $\delta T_{i,0}$ should be small may not be met initially, it will be so for the final iterations and hence the final calculated temperatures will be numerically correct.

Transient calculations

A similar approach can be used to analyse non-steady-state conditions. However, the left-hand side of equation (11.19) must be replaced by the term

$$m_i C_i \frac{(T_i - T_{i,0})}{\delta t}$$

and the temperatures and heat inputs by their average values over the time interval δt, $(T_{i,0} + T_i)/2$, $(T_{j,0} + T_j)/2$, $(Q_{\text{external},i,0} + Q_{\text{external},i})/2$ and $(Q_{i,0} + Q_i)/2$, respectively.

Replacing the values of $T_{i,0}$ by the newly calculated values T_i and repeating the calculation will provide the temperature history at successive intervals of time δt. It should be noted that, for this to be accurate, $\delta T_{i,0}$ must be small since, at a given time, the temperatures are the result of a single calculation and not the result of multiple iterations as in the case of the steady-state calculation. In practical terms, this means that δt should not be too large and that the initial temperature values $T_{i,0}$ should be as realistic as possible. In the latter context, the temperature distribution resulting from a steady-state calculation using orbital-average heat inputs is often used as the starting point for a transient calculation.

Many software codes have been developed over the years to perform the above calculations. The most modern and comprehensive, and the one most commonly used in Europe, is the ESA-sponsored ESATAN package [13,14]. For small TMMs, a spreadsheet-based programme is also available, THERMXL [15].

11.5 THERMAL DESIGN

11.5.1 Requirements and constraints

The first task is to examine the proposed spacecraft concept and assemble the thermal parameters for the on-board equipment. The equipment designer should provide upper and lower safe operating temperatures for his equipment and should also specify the operating power dissipation or dissipation ranges. Detailed drawings and materials lists will be required in order to calculate nodal thermal capacitances, conductance paths and view factors. Orbital and other operating constraints must also be defined. For example, an Earth observation spacecraft may well require to be in a near-polar orbit with the plane of the orbit perpendicular to the solar vector in order to emphasize ground features (shadow effects). There may also be constraints on allowable attitudes. For example, an Earth observation spacecraft must turn so that its instruments always face the Earth, a solar observatory will be oriented such that its instruments always face the Sun and an astronomical spacecraft must ensure that its telescopes never come too close to the solar vector.

As an example, let us look again at our spherical spacecraft. For simplicity, let us assume that its mission is to enable upper atmospheric density to be inferred from measurements of the orbit decay over a period of a few weeks. The example chosen is relevant, and yet simple enough to show thermal design methodology in action. The principal investigator requires the external geometry to be a clean spherical surface that is as inert as possible to try to minimize any possible chemical effects (the spacecraft will be flying through a residual atmosphere that is nearly 100% atomic oxygen—see Chapter 2). The ideal would be a gold-plated sphere with a diameter of 1 m. Launch site and launch vehicle constraints require that the spacecraft be spinning and that it will be in a polar orbit, but the angle between the orbit and solar vector is unknown. The angle between the spin axis and solar vector is also unknown and not controllable. The spacecraft systems engineers determine that the spacecraft will be tracked via a battery-powered beacon, located at the centre of the sphere, which dissipates 5 W. This is mounted through a circular panel across the spacecraft's 'equator' and is supported by two tubes along the spin axis, as shown schematically in Figure 11.7. The two hemispheres on each side of the equatorial panel act as the beacon's antenna. The battery supplier confirms that, for the short lifetime of the spacecraft, the batteries can tolerate temperatures between $-15\,°C$ and $+60\,°C$. We have already seen that the thermal analysis process involves approximations, particularly concerning the establishment of the TMM. We should therefore take an appropriate margin here, and design to stay within the range $-5\,°C$ to $+50\,°C$.

11.5.2 Definition of worst-case conditions

The task of the thermal designer is not usually to achieve a specific temperature but rather to ensure that equipment stays within certain acceptable limits. To do this, *worst-case conditions* should be defined. These would typically be the orbits with maximum and minimum periods of sunlight, combined with certain extreme spacecraft attitudes and operational modes. In most cases, these worst-case conditions are established by inspection and experience. In the case of our air density spacecraft, we must design for

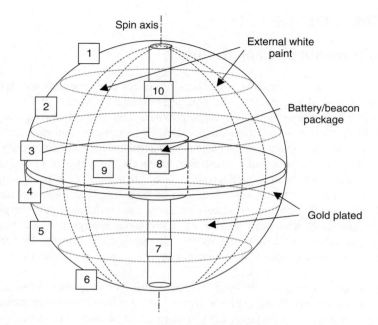

Figure 11.7   Spacecraft configuration and nodal breakdown

both full-sun (hot case) and 59% sun (cold case) orbits, and for any attitude with respect to the solar vector.

From an inspection of Table 11.3, it is clear that our principal investigator will have to accept some compromise concerning the external surface of the sphere. A gold-plated sphere will have an average temperature between 138 °C and 176 °C, neglecting the internal heat dissipation. It is evidently necessary to apply a low-α/ε surface, say a white paint, to part of the external surface. However, this should be the minimum area consistent with keeping the battery–beacon package below 50 °C. A paint pattern consisting of several segments (see Figure 11.7) is relatively easy to apply and gives the necessary degree of surface uniformity.

11.5.3   The TMM

Figure 11.7 shows a simple nodal breakdown for the air density-measuring spacecraft. It consists essentially of an aluminium sphere with a honeycomb panel across its equator, which supports the battery–beacon package at its centre. Two aluminium tubes are used to provide additional support along the spacecraft spin axis. Since the spacecraft is spinning and the only source of internal dissipation is at the centre, we may assume that temperatures vary only with spacecraft 'latitude'. Each hemisphere can therefore be modelled as a number of isothermal nodes (zones) generated by planes parallel to the spacecraft equator. For our example, each hemisphere is divided into three such nodes as shown in Figure 11.7. The various conductance terms, view factors and nodal thermal capacitances can now be calculated as indicated previously. Table 11.5 lists the resulting view factor–area products and conductance terms.

Table 11.5 View factors and conductance terms

View factor × area products (m²)

i \ j	1	2	3	4	5	6	7	8	9	10
1		0.068	0.079	0	0	0	0	0.013	0.215	0.058
2	1.03		0.089	0	0	0	0	0.018	0.241	0.031
3	0	2.10		0	0	0	0	0.016	0.246	0.018
4	0	0	2.54		0.089	0.079	0.018	0.016	0.246	0
5	0	0	0	2.10		0.068	0.031	0.018	0.241	0
6	0	0	0	0	1.03		0.058	0.013	0.215	0
7	0	0	0	0	0	0.18		0.0027	0.031	0
8	0	0	0	0	0	0	0.16		0.054	0.0027
9	0	0	1.41	1.41	0	0	0	0.94		0.031
10	0.18	0	0	0	0	0	0	0.16	0	

Conductance terms (W/K)

11.5.4 Results and discussion

For the analysis, the incident heat flux densities were the same as those used for the example worked in Section 11.3. That is,

$J_s = 1371\,W/m^2$, $J_p = 220\,W/m^2$

$J_a = 67.9\,W/m^2$ [~0.33 (0.15) J_s] for the 100% sunlit orbit and 316.7 W/m² (~0.33 (0.7) J_s) for the sunlit part of the 'cold' orbit.

The internal surfaces of the spacecraft were assumed painted with a matt white paint with $\varepsilon = 0.9$, whence $\varepsilon_{ij} = 0.82$. The paint pattern on the external surface consists of several (the number is not important) segments, each of which is defined by planes containing the spin axis. This means that the percentage painted area of each node is the same. For this example, a silicate-based white paint with $\alpha/\varepsilon = 0.16$ has been used.

Table 11.6 and Figures 11.8 and 11.9 show the results (obtained using the ThermXL [15] software) for the case in which 15% of the external surface is covered with white paint, giving an effective $\alpha/\varepsilon = 1.38$ for the external surfaces of nodes 1 to 6.

It can be seen that the maximum predicted temperature of the battery–beacon package (Table 11.6) is comfortably inside the 50 °C maximum. If required, the design can be trimmed, for example, by modifying the external paint coverage, and the analysis re-run until the desired maximum predicted temperature has been attained.

For the cold cases (Figures 11.8 and 11.9), the battery–beacon package remains well above the −5 °C lower allowed limit. We may therefore conclude that our thermal design is adequate for this spacecraft and its mission.

More complex spacecraft with more complex missions and more variable environments will usually require more complicated thermal design. Reliance on thermal conduction, radiation exchange and insulation systems is known as *passive thermal control* and is the initial starting point for most spacecraft thermal design. However, where this proves inadequate, *active thermal control* techniques can be used. These are predominantly concerned with the active control of internal dissipation, for example, by the use of heaters, and of

Table 11.6   Calculated temperatures for 100% sunlit orbit

Node number	Location	Temperature (°C)	
		Side-to-Sun case	Top-to-Sun case
1	External surface	40.6	72.3
2	External surface	44.6	57.2
3	External surface	45.8	43.8
4	External surface	45.8	32.1
5	External surface	44.6	25.3
6	External surface	40.6	22.1
7	Support tube	43.3	30.7
8	**Battery/beacon**	**46.**7	**45.**5
9	Support platform	44.6	42.3
10	Support tube	43.3	59.2

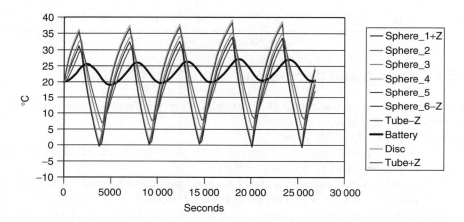

Figure 11.8   Temperature evolution, side-to-Sun, maximum eclipse

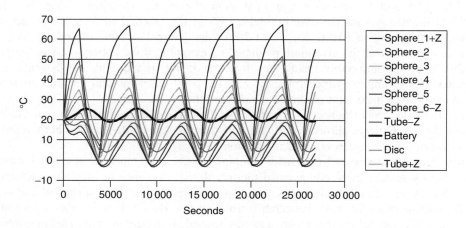

Figure 11.9   Temperature evolution, top-to-Sun, maximum eclipse

radiative, conductive and forced convective heat flows as a function of temperature. These active techniques are described in more detail in the next section.

In the event that temperature requirements cannot be met, or can be met only at great expense or with significant risk, it will be necessary to revisit the requirements. This may involve re-qualifying hardware for different temperature ranges or changes in spacecraft system design or mission profile. Project managers will expect such feedback earlier rather than later. Consequently, thermal design should start early with simplified TMMs, only moving to more detailed models once the overall concept has been shown to be sound.

11.6 THERMAL TECHNOLOGY

11.6.1 Passive control

Passive thermal control techniques available to the engineer consist essentially of the selection of surface properties, the control of conduction paths and thermal capacities and the use of insulation systems.

Surface finishes

As indicated in Table 11.4, surface finishes are available covering a wide range of α/ε values. Care needs to be taken, however, concerning the ageing characteristics of these surfaces. Contamination of low-α surfaces (white paint, polished or electroplated metal surfaces) will increase the α value and should be avoided. Paints generally consist of a binder that is transparent to visible light and a filler material that gives it its colour. Often the binder will be opaque in the infrared, which implies that most paints have high ε values. Unfortunately, many binders degrade and discolour under the influence of solar ultraviolet radiation, becoming less transparent to visible light. White paint on the outside of a spacecraft will suffer an increase in its α/ε value with time and care must be taken to select the most suitable formulation [5,6,7]. The presence of contamination (grease, oil etc.) will exacerbate the situation. Reflectors such as *Second Surface Mirrors* (*SSM*) (plastic film, aluminized or silver coated on the back side) or *Optical Solar Reflectors* (*OSRs*) (similar to SSM but using thin sheets of glass instead of plastic film) are less sensitive to solar radiation and are easier to clean. Compared to paints, however, SSMs and OSRs are more difficult to apply to spacecraft surfaces, particularly those with complicated geometry. Suitable allowance must be made for any likely changes in surface optical properties during the design lifetime of the spacecraft.

Conduction paths

Solid conduction needs little discussion. However, care needs to be taken at joints, particularly under vacuum conditions. During ground testing, the thin air gap between non-contacting areas of a simple bolted joint contributes significantly to the effective conductance of the joint. Under vacuum conditions in space, this contribution disappears. The conductance of such a joint under vacuum conditions is strongly dependent on the quality of the mating surfaces and on the bolt tension. These uncertainties can be minimized by the use of interface fillers such as soft metals (e.g. indium foil) or loaded polymers (e.g. silver-loaded silicone) [6].

Heat pipes and two-phase systems

Two-phase heat transport systems, such as *heat pipes*, *loop heat pipes* (*LHPs*) and *capillary-pumped loops* (*CPLs*), are devices that transfer heat in the form of the latent heat of vaporization, using a volatile working fluid that is circulated by capillary action in a porous wick structure. They are capable of large heat transport rates at very low temperature gradients (i.e. very large effective conductance).

A basic heat pipe is shown schematically in Figure 11.10. It consists essentially of a sealed tube possessing a porous structure (the wick) on its inside surface and containing a volatile liquid (the working fluid) in thermodynamic equilibrium with its vapour. The quantity of liquid is just sufficient to saturate the wick structure. The working fluid and wick material are chosen to ensure mutual wetting (i.e. a small contact angle), which guarantees retention of the working fluid in the wick. The pressure of the vapour is everywhere equal to the saturation vapour pressure of the working fluid, which is a strong (exponential) function of the temperature at the liquid–vapour interface (as described by the Clausius–Clapeyron equation). If heat is applied to the evaporator section, the local temperature will increase and the working fluid will evaporate, causing a local pressure rise. Vapour will then flow to all other parts of the pipe and, since the pressure will then exceed the local saturation vapour pressure, condensation will occur. If we neglect the pressure drop associated with the vapour movement, it can be seen that the vapour space and internal surface of the wick are always sensibly isothermal and the heat pipe heats up uniformly along its whole length. Similarly, if heat is removed from the condenser section, the heat pipe cools uniformly over its whole length. If we simultaneously heat the evaporator section and cool the condenser section, heat will flow as long as there is sufficient liquid available in the evaporator section of the wick.

As the working fluid in the evaporator section evaporates, the liquid surface withdraws into the wick and the surface tension forces generate a pressure drop across the (curved) liquid–vapour interface. It is this pressure drop across the liquid–vapour interface that generates the pressure gradient between the condenser and evaporator sections to drive the condensed working fluid back to the evaporator. Assuming that the working fluid perfectly wets the wick material (i.e. contact angle is zero), the maximum capillary pressure that can be generated is given by

$$\Delta P_{max} = \frac{2\sigma}{r_0}$$

where σ is the surface tension and r_0 is the effective pore radius of the wick.

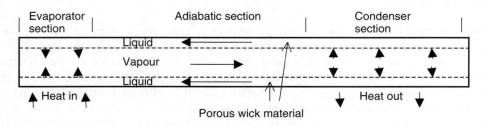

Figure 11.10　Schematic illustration of a basic heat pipe

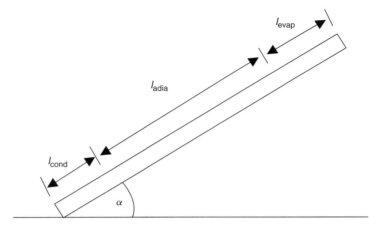

Figure 11.11 Heat pipe oriented with evaporator above condenser

With reference to Figure 11.11, the maximum rate dm/dt at which the working fluid can be recycled is then governed by

$$\frac{2\sigma}{r_0} = (l_{\text{eff}} \sin \alpha) \rho g + \frac{l_{\text{eff}}}{A_{\text{wick}}} \frac{\eta}{\phi \rho} \frac{dm}{dt} + \Delta P_{\text{vap}} \tag{11.23}$$

where ρ = density of the liquid phase, g = acceleration due to gravity,
   η = dynamic viscosity of the liquid phase, ϕ = permeability of the wick structure,
   l_{eff} = heat pipe effective length = $l_{\text{adia}} + 1/2(l_{\text{evap}} + l_{\text{cond}})$,
   A_{wick} = cross-sectional area of wick, ΔP_{vap} = pressure drop in the vapour flow.

In most practical applications, the pressure drop in the vapour is small compared to that in the liquid phase. Equation (11.23) can then be rearranged as follows:

$$\frac{dm}{dt} \approx \frac{A_{\text{wick}}}{l_{\text{eff}}} \frac{\phi \rho}{\eta} \left(\frac{2\sigma}{r_0} - \rho g l_{\text{eff}} \sin \alpha \right)$$

Hence, if H_v is the latent heat of vapourization, then the maximum rate of heat transport Q_{max} is given by

$$Q_{\text{max}} \approx \frac{A_{\text{wick}}}{l_{\text{eff}}} \frac{\phi \rho H_v}{\eta} \left(\frac{2\sigma}{r_0} - \rho g l_{\text{eff}} \sin \alpha \right) \tag{11.24}$$

When operating in space ($g = 0$), it should be noted that the product $Q_{\text{max}} l_{\text{eff}}$ is a constant for a given heat pipe design—the longer the heat pipe, the smaller its heat transport capability. For this reason, heat pipe performance is often specified in units of watt-metres. The second thing to note in equation (11.24) is the dependence on the angle of tilt α when operating on the ground. Capillary forces are relatively weak forces and a heat pipe with a performance of several hundreds of watt-metres under zero-gravity conditions may cease to operate on the ground if its evaporator is raised more than a few millimetres above its condenser section. On the ground, a heat pipe only approximates to its zero-gravity performance when tested horizontally. The effect of gravity on performance must be fully taken into account during system design, particularly if a system of several heat pipes (e.g. in a radiator) is used or heat pipes are required to be bent into complex shapes. It is easy to design heat-pipe-based thermal control systems that prove to be unverifiable on the ground!

In analysing the effect of a heat pipe in a TMM, it should be recalled that the vapour is essentially isothermal. The effective conductance h_{eff} is given by

$$\frac{1}{h_{eff}} = \frac{1}{h_{evap}} + \frac{1}{h_{cond}}$$

where h_{evap} and h_{cond} are the radial conductance values through the wall and wick in the evaporator and condenser zones, respectively. Unlike normal thermal conduction, the effective conductance of a heat pipe is not a function of its length.

LHP and CPL are variations on the basic heat pipe, designed to improve ultimate performance and/or to improve flexibility of application. In an LHP, the working fluid is returned to the evaporator via an external pipe. The CPL takes the process a step further and several evaporators, operating in parallel, may be attached to the same liquid return line. Heat pipes, LHPs and CPLs often use anhydrous ammonia as the working fluid, since it permits operation over a wide temperature range from approximately $-50\,^\circ$C to about $+80\,^\circ$C. Working fluids are available, however, allowing heat pipes to operate from cryogenic temperatures up to several hundred degrees.

The subject of two-phase heat transport systems is, unfortunately, outside the scope of this text. The interested reader is referred to References [16,17,18].

Phase change materials

Phase change materials (*PCMs*) can be used where increased thermal capacity is required, for example, to minimize temperature excursions during eclipse or other transient conditions such as occasional high equipment power dissipation. They work by absorbing or releasing latent heat during solid–liquid phase change. Care needs to be taken to ensure that there is sufficient free volume (ullage) to accommodate the volume changes accompanying phase change. There should also be sufficient internal structure to ensure adequate thermal conduction throughout the PCM material, especially under zero-gravity conditions. Typically, PCM consists of a hydrocarbon wax but many substances are available [19] covering a wide range of temperatures.

Insulation systems

Extremely effective insulation systems can be designed to minimize radiative exchanges in the vacuum of space. These are normally referred to as *super-insulation systems* or multi-layer insulation (MLI) blankets. They consist typically of several layers of aluminized plastic film (e.g. Mylar of Kapton) acting as radiation shields, each separated by a low-conductance spacer (e.g. silk, Nylon or glass-fibre net). In practice, in many applications the spacers are omitted and instead alternate layers of aluminized film are crinkled or mechanically dimpled to minimize contact between adjacent layers. A typical insulation

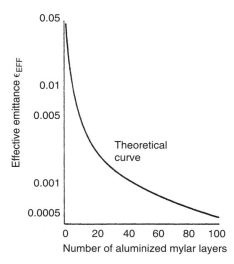

Figure 11.12 Effective emittance versus number of aluminized Mylar layers (theoretical)

blanket might consist of 40 or more layers, each about $10\,\mu m$ thick and aluminized on both sides. This would have an effective emittance value of about 0.001. The theoretical values of effective emittance as a function of number of layers, neglecting the effects of conduction between layers, are shown in Figure 11.12.

It should be remembered that, in practical systems, full account must also be taken of edge effects, joints and fasteners. The design of MLI blankets and their method of installation must also ensure adequate venting, particularly during the rapid depressurization accompanying the launch phase. Further extensive information concerning insulation systems can be found in the literature [6].

11.6.2   Active control

Active thermal control systems are generally more complex than passive systems and often consume power and sometimes telemetry resources. Such systems are typically less reliable and often heavier. As a general rule, active systems should be used only when it has proved impossible to meet requirements by passive means alone. Active systems will typically be used for very temperature-sensitive equipment (telescopes, scientific instruments, atomic clocks etc.), in which environmental conditions are very variable (missions to Mercury, Mars, Europa etc.) or in which heat dissipations are variable/random (e.g. in the cabin of a manned space vehicle).

Heaters

Heaters constitute, probably, the simplest and most obvious active thermal-control device. A passive thermal-control design may lead to minimum predicted temperatures falling below permissible limits for a given equipment or subsystem. A thermostatically controlled heater can be used to prevent this. In view of the high reliability of modern thermostats, such a solution may well be considered competitive with alternative passive approaches, such as use of PCM units. Other typical applications for heaters include the propulsion subsystem (thrusters, fuel lines and valves, tanks etc.), batteries, experiment sensors and mechanical actuators. Heaters may be in the form of compact metal-mounted resistors, metal co-axial cables in which the heating element forms the core of the cable (thermo-coax) or adhesive sheets in which the element is configured rather like a Printed Circuit Board (PCB) between two layers of Kapton foil.

Variable conductance heat pipes (VCHPs) and diodes

A *variable conductance heat pipe* (*VCHP*), shown schematically in Figure 11.13, is a variant of the simple heat pipe described previously. A non-condensable gas, typically nitrogen, is used to progressively block the condenser section as a function of evaporator temperature.

There are several forms of VCHP—the one illustrated is known as a cold, wicked reservoir design. During operation, the vapour always streams from evaporator to condenser, so driving the non-condensable gas always towards the reservoir. The reservoir is cold and hence the partial vapour pressure of the working fluid is low. Most of the gas in the reservoir is hence composed of the non-condensable gas. Since the vapour can only travel through this gas by diffusion, the vapour flow is effectively stopped when it meets the non-condensable gas. Hence, the heat flow to regions of the condenser section that are downstream of the vapour-gas front is very much reduced. The reservoir is equipped with a wick structure that connects with the main heat pipe wick. This ensures that any vapour that does enter the reservoir and condenses will be removed by capillary action. In operation, any increase in evaporator temperature will cause the internal pressure to rise. This will compress the non-condensable gas, causing the vapour-gas front to retreat down the pipe and exposing a greater length of condenser to the vapour flow. The device thus acts as a thermostat and, configured as described, operates completely passively. The ultimate temperature stability that can be obtained depends on the ratio of the reservoir and condenser volumes—the larger the reservoir, the better the stability. If even greater temperature stability is required, a heater can be attached to the reservoir to control the partial vapour pressure of the working fluid and hence control the position of the vapour-gas front. This enables the vapour-gas front to be moved as a function of the temperature of the equipment being cooled, allowing temperature-control precision to fractions of a degree.

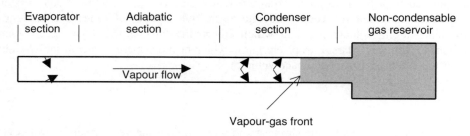

Figure 11.13   Schematic drawing of a cold, wicked VCHP

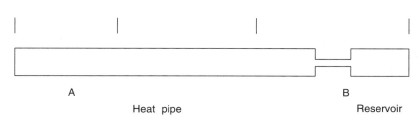

Figure 11.14 Schematic drawing of a liquid-trap heat pipe diode

Consider now the effect of removing the connection between the heat pipe and reservoir wicks and omitting the non-condensable gas. Such a device, known as a *liquid-trap heat pipe diode* [20], is shown schematically in Figure 11.14. When section 'A' is hotter than section 'B', liquid accumulates by condensation in the reservoir until the heat pipe dries out and stops conducting. When 'B' is hotter than 'A', the liquid evaporates from the reservoir, re-condenses in the heat pipe and the heat pipe works normally. Such a device can be useful, for example, for switching between radiators to prevent backflow of heat from a temporarily sunlit radiator. Use of a heater on the reservoir enables the heat pipe to be operated as a thermal switch.

Mechanically pumped two-phase loops

Mechanically pumped two-phase loops are similar the CPLs with the addition of a mechanical pump in the liquid return line. This increases the heat transport capability and also renders the loop less sensitive to instabilities caused, for example, by gas bubbles in the liquid lines or nucleation within the evaporators. Such systems should be considered for complex thermal control applications involving large heat transport requirements such as might be encountered on a large space station.

Liquid loops

Perhaps the simplest concept for the active transport of heat is to use *liquid loops*, such as is used for cooling automobile engines or in the domestic central heating system. The liquid coolant is pumped between the various heat sources (dissipating equipment) and heat sinks (e.g. radiators or inter-loop heat-exchangers) by, typically, an *electrically driven centrifugal pump*. Temperature control is effected by using bypass valves to divert a controllable fraction of the liquid flow-rate around active elements such as radiators and heat exchangers. The pressure in the loop must be maintained within a range that ensures that no cavitation or structural damage can occur anywhere in the loop. This can be achieved by including an *accumulator* in the loop, in which a flexible diaphragm separates the coolant from a gas, usually nitrogen, the pressure of which is controllable. *Gas traps* should also be included to minimize the risk of damage from gas bubbles, particularly in the pump. Since the pump is both a single-point failure risk and the most vulnerable item in the loop, the pump package will usually consist of two pump units in cold redundancy.

Liquid loops are inherently rather massive. They would normally only be used where their particular attribute of system and operational flexibility is required. They are sometimes, but infrequently, used in unmanned spacecraft, such as the *European Retrievable*

Carrier (*EURECA*) platform, which was designed to be launched from and retrieved by the Shuttle and to accommodate a wide range of payloads. They find most application, however, in manned spacecraft such as the Shuttle, Spacelab and now the ISS. The liquid-loop-based active thermal-control system for the Columbus pressurized module, a European contribution to the ISS, is shown schematically in Figure 11.15.

The choice of liquid should be made with some care. In order to minimize power consumption, the specific heat should be high (low mass flow rate) and dynamic viscosity low (low drag forces). In order to minimize system mass, the boiling point should be high (to minimize containment pressure). However, there must be no risk of freezing during any phase of operation, particularly if water is considered, which expands on freezing. For use in cabins containing a human crew, toxicity in the event of leaks must be taken into account. The coolant used on EURECA was Freon-114. Water is used in all the manned volumes of the Shuttle and ISS, and was used in Spacelab. Because of its undesirable freezing characteristics and high freezing temperature, water is not used outside the pressurized modules. Instead, liquid–liquid heat exchangers interface between the water loops and a second external loop, containing a low freezing point liquid, which transports the heat to the radiator assemblies. The Shuttle uses Freon-21 in its external loop, while the ISS uses ammonia.

Louvres and shutters

A *louvre* [6] is a device that varies the effective emittance of a radiator in response to temperature. It is usually—but not always—mounted on the outside of a radiator

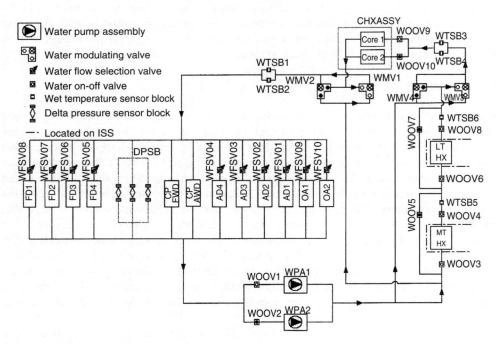

Figure 11.15 Schematic diagram of the Columbus water loop (courtesy of ASTRIUM GmbH)

panel and typically consists of an array of metallic blades rather like a venetian blind. The concept is shown schematically in Figure 11.16. When the blades are open (perpendicular to the radiator surface), the radiator has a good view of space and radiates accordingly. When they are closed, however, the radiator sees a low-emittance surface and most of its radiated heat is reflected back. The blades may be activated individually, for example, by bimetallic springs or by liquid-filled Bourden spiral actuators thermally connected to the radiator or as a complete assembly by a single actuator. Actuators can be bimetallic, hydraulic or electrical, and operate in response to equipment temperatures, radiator temperatures or computer- or ground-generated commands. An example of a hydraulically activated louvre, used in the thermal-control system on the NIMBUS series of spacecraft [21], is shown schematically in Figure 11.17. Figure 11.18 shows a photograph of a lightweight louvre assembly driven by bimetallic springs, which has recently been developed for the ROSETTA spacecraft (see Chapter 5). Louvres operate most effectively when mounted on radiators that are not illuminated by the Sun. In sunlit locations, complex reflections occur that make the prediction of louvre performance complicated and unreliable. Under these circumstances, an alternative approach, which was used on the GIOTTO spacecraft during its mission to Halley's comet, is to partially cover the radiator surface with a motor-driven low-emittance roller-blind made from a sheet of metallized Kapton film.

Although louvres have traditionally been regarded as devices for controlling heat rejection from radiators, they are currently under active consideration as devices to control solar heat input for missions close to the Sun, such as the BepiColombo mission to Mercury.

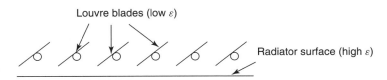

Figure 11.16 Schematic illustration of a louvre

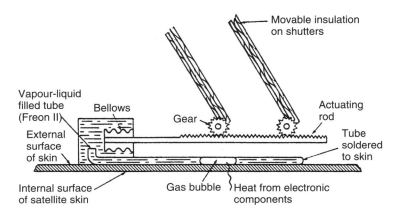

Figure 11.17 Schematic of NIMBUS louvre system (General Electric)

Figure 11.18 Louvre assembly developed for
ROSETTA (courtesy of SENER)

Refrigerators and heat pumps

A *heat pump* may be used to increase the temperature of a radiator. Since the amount of heat radiated from a surface increases with the fourth power of its temperature, quite small temperature rises are sufficient for a significant increase in heat dissipation rate. The penalties for this are the mass and power consumption of the heat pump and any ancillary interfacing equipment such as liquid loops. Under normal circumstances, the engineer will find that it is usually better to try to find more radiator area. Exceptions would be where the amount of radiator area is fundamentally limited or where heat must be rejected in a relatively warm environment such as might occur on the lunar or Martian surface during daytime. The more usual application for heat pumps and refrigerators is to cool sensors or other equipment in order to reduce the signal-to-noise ratio. In the specific case of manned space activities, refrigeration is also needed for storing astronaut food and biological specimens resulting from scientific research.

Refrigeration requirements tend to fall within one of four categories, corresponding to temperature ranges

1. $+4\,°C$ *to* $-20\,°C$. This range is basically concerned with food storage and is normally provided by *thermoelectric cooling* or mechanical two-phase heat pumps based on the *Rankine cycle*. Thermoelectric refrigerators are simple, rugged and reliable (no moving parts). They are not very efficient thermodynamically, however, and become very power-hungry when heat loads or the temperature-lift (i.e. the difference between the cooled equipment and heat sink) become large. Two-phase heat pumps, such as those used in domestic refrigerators/freezers, are more efficient but much more complicated [22]. They work by absorbing heat as the working fluid evaporates at low temperature and pressure, and rejecting heat as the working fluid condenses at higher temperature and pressure. Since the temperatures directly determine the pressures in the evaporator and condenser (governed by the Clausius–Clapeyron relation), the temperature-lift that can be achieved is directly limited by the compression ratio of the vapour compressor. In addition to the need for moving parts, further complications arise from the need to ensure that damage cannot occur as a result of liquid accidentally entering the vapour compressor under

zero-gravity conditions. The crew 'fridge-freezer' developed in Europe for the ISS uses thermoelectric refrigeration.

2. $-80\,°C$. This temperature is used for long-duration storage of biological specimens. It is too cold for thermoelectric cooling and a two-phase refrigerator would require several stages, involving substantial mass, power and reliability penalties. Instead, single-phase gas cycles such as the *Stirling* or *Brayton cycles* are used [23,24]. The Brayton cycle is a continuous loop, whereas the Stirling cycle is a reciprocating one. The Brayton cycle is inherently easier to interface with a large refrigerator, although it is not as thermodynamically efficient as the Stirling cycle. The MELFI specimen storage freezer for the ISS uses the Brayton cycle.

3. $80\,K\ (-193\,°C)$. This is an important temperature for Earth observation spacecraft since it is the temperature needed by the infrared detectors of instruments designed to look through, for example, the 8-μm atmospheric window (see Section 11.2.2). If heat loads are low and there is an unrestricted view of space, carefully designed radiators can achieve these temperatures. The METEOSAT spacecraft, orbiting at geosynchronous height, uses such a radiator. In LEO, however, the effect of the Earth's planetary radiation is difficult to avoid, and mechanical coolers using the Stirling cycle are now common. They consist essentially of reciprocating compressor and displacer assemblies. The compressor alternately compresses and expands the working gas, typically helium. The displacer, moving approximately 90° out of phase with the compressor, ensures that the gas is in the compressor when it is compressed and in the displacer when it expands and cools. In passing between compressor and displacer, the gas moves through a regenerative heat exchanger (the regenerator). The displacer can be actively driven, run passively (driven by the pressure wave, with the phase shift determined by the impedance between compressor and displacer, and displacer inertia etc.) or replaced completely by a pulse tube that operates passively containing no moving parts.

4. $<4\,K$. Temperatures in this range are predominantly of interest to astronomers wishing, for example, to measure the cosmic infrared background or detect the infrared emission from planets orbiting other stars. Depending on the application, cooling may be required down to the mK range, although it is an arguable point as to whether the achievement of such low ultimate temperatures is legitimately the responsibility of thermal control or more properly the province of the instrument developer. Techniques available for these temperatures [23,24] include *Joule–Thomson*, *Brayton* and related *Claude cycles, sorption cooling, dilution refrigeration* and *adiabatic demagnetization (ADR)*.

11.7 THERMAL DESIGN VERIFICATION

Thermal design verification consists of two fundamentally different activities: (1) qualification of the hardware making up the thermal-control subsystem and (2) verification of the accuracy of the TMM. The latter enables the TMM to be used to make reliable temperature predictions for any particular mission scenario (orbit, attitude, payload operational timelining etc.).

11.7.1 Hardware verification

As with any other spacecraft components, thermal-control hardware must be shown to be suitable for its purpose. The optical properties of thermal-control surfaces, the effective conductance of joints, and the performance of insulation systems, must be measured. The performance of heat transport loops, heat pipes, cryogenic systems and so on must be verified. It is also necessary to verify that the hardware will operate once in space (launch vibration tests, vacuum tests, temperature exposure tests) and will continue to operate correctly for the life of the mission (life tests). This is achieved by exposing qualification samples or units to conditions more severe than will be encountered in flight, to verify that the design is suitably robust. The flight hardware will then be tested to limits that also exceed expected flight conditions but are less severe than the qualification values. In many cases the thermal engineer will be using hardware that has a proven track record in space (paints, insulation etc.), in which case qualification can be established by similarity with past applications. However, care should be taken on a case-by-case basis to ensure that previously established qualification limits remain valid for the current case. For example, external thermal-control paints that have good performance in an Earth orbit will probably prove unacceptable in the vicinity of Mercury, because of the high temperatures and intense solar ionizing radiation encountered. It should also be noted that, although the atmosphere is extremely thin at LEO altitudes, it is composed almost entirely of atomic oxygen with a very high kinetic temperature (because of the satellite velocity). This environment, particularly when combined with solar UV radiation, can be very damaging for some thin film materials [5].

11.7.2   Thermal model verification

It will be recalled that the TMM is an approximation to the real situation, devised by the thermal engineer to enable temperature predictions to be made. It is essential to verify the accuracy of these models and, where inaccuracies are found, to amend the TMM accordingly. This is done by performing thermal balance tests that make use of subscale or full-size models of spacecraft, or parts of spacecraft (e.g. a payload module may be tested on its own if the spacecraft service module to which it is to be attached is already a well-established design).

A spacecraft thermal balance test requires high vacuum conditions to minimize air conduction/convection, a heat sink to simulate the cold radiative environment of space and heat source(s) to simulate the external heat inputs (Sun, albedo, planetary infrared radiation). Facilities to test even small spacecraft are extremely expensive, both to manufacture and to operate. There are therefore relatively few of them and, in Europe, their use is coordinated by ESA/ESTEC [25]. The thermal engineer needs to consider carefully just how accurately the environment needs to be simulated, in order to verify his TMM, since the higher the required accuracy the higher will be the costs. It may be necessary to use a well-collimated and calibrated solar simulator to verify the external thermal interface between spacecraft and environment (including shadowing effects and multiple reflections). The absolute value of the simulated solar input is not so important, however, provided it is known. The same applies to the surrounding heat sink—most facilities use matt-black-painted shrouds cooled by liquid nitrogen for this purpose. The shroud temperatures are monitored and suitable allowance made in subsequent analysis. Albedo and planetary infrared radiation are seldom simulated, although a slightly higher solar constant may be used to ensure that the test article (particularly if it is a real spacecraft) does not run too cold. A typical test sequence will consist of several steady-state tests at different spacecraft attitudes, together with a transient test (e.g. simulating eclipses). Such a test on a full-size development model of a spacecraft (or, indeed, on the proto-flight spacecraft) should provide sufficient information to enable the TMM to be verified and, if necessary, 'trimmed'.

These full-size solar simulation tests are expensive, however, and for large modern spacecraft problematic owing to the difficulty of getting a very large diameter solar simulation beam of the required quality. In such cases, adequate verification of the TMM can usually be obtained by a combination of (1) a full-size spacecraft test using infrared heaters instead of a solar simulator [26] and (2) verification of the external thermal interface using a subscale model specially designed for the purpose and sized to fit within the beam of available solar simulators.

As a general rule, the more testing that can be done, the better. However, in practice the amount of testing that a project manager will sanction will be based on compromise and will depend on such parameters as criticality of the thermal design, size of current temperature margins, complexity of external geometry, mission operational complexity and the project budget. However, the availability of reliable active thermal-control techniques and increasingly sophisticated thermal analysis tools, and the tendency to base new spacecraft on past well-proven designs (the spacecraft bus approach), is steadily reducing the need for detailed extensive thermal-balance testing. Experience indicates that the thermal engineer will always request more testing than the project manager is prepared to pay for.

11.8 EXAMPLE OF SATELLITE THERMAL DESIGN – XMM/NEWTON

In this chapter, we have attempted to describe the techniques that are used to devise, analyse and verify a spacecraft thermal design. In practice, every spacecraft is different and each thermal design tends to be unique. It will be a compromise between the thermal engineer's desire for perfection and other constraints such as mass, cost and available power and telemetry resources. Although designers will attempt to make use of the legacy of past successful designs, there is no 'standard thermal-control subsystem'. As a rule of thumb, the thermal control system will usually constitute between 2 and 5% both of spacecraft mass and development cost [27]. An example helps to illustrate the way in which these design techniques are used on modern spacecraft.

The X-ray Multi-mirror Mission (XMM) spacecraft [28], subsequently renamed 'Newton' and shown schematically in Figure 11.19, is an advanced X-ray observatory that uses three grazing-incidence mirror modules, each containing 58 concentric mirror assemblies, to focus X-ray images on the charge coupled device (CCD) detector arrays located in the focal plane. The spacecraft 'housekeeping' subsystems are located in the service module that surrounds the mirror modules. The solar arrays are attached to the service module. The mirror modules and focal plane instrumentation are separated by the telescope tube that is about 6.8 m in length. XMM has an overall length of around 10 m and a mass of about 10 tonnes.

In order to avoid interference from the Earth's trapped radiation, XMM is in a highly elliptical orbit and only performs its observations when well outside the radiation belts. Specifically, it is a 48-h orbit with apogee at 114 000 km and perigee at 7000 km.

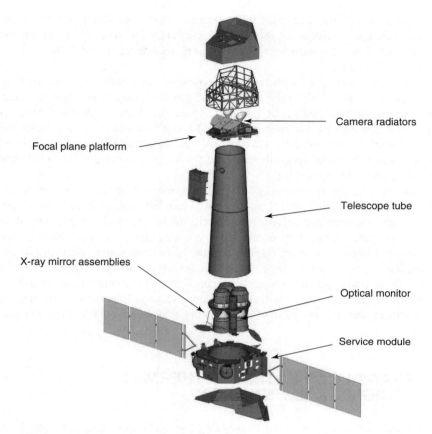

Camera radiators

Focal plane platform

Telescope tube

X-ray mirror assemblies

Optical monitor

Service module

Figure 11.19 'Exploded view' of the XMM/Newton spacecraft (Reproduced by permission of European Space Agency)

Observations are made during the 40 h of each orbit when it is above 40 000 km. Spacecraft attitude is controlled to within ±20° of 'side-to-Sun'. The inclination of the orbit is 40°, so it suffers from periodic eclipse seasons. The eclipses always occur well below the altitude at which observations are made, however, and never exceed 1.7 h. Under these circumstances, the external thermal environment is rather stable. There is little influence from the Earth, even during an eclipse season, and none during observing periods, and this has been taken advantage of in the thermal design.

The most critical requirement of the thermal design [29] was to ensure that thermal distortions would not jeopardize the alignment of the X-ray telescope (which has a 7.5-m focal length) or the relative alignment between this telescope and, for example, the star trackers. This translated into a requirement to control the temperature of the mirror modules and mirror-support platform (which also supports the star trackers) at an average temperature of 20 °C, and to limit temperature gradients to less than 2 °C. This was achieved by equipping each mirror module, the mirror-support platform and the entry and exit baffles with heaters controlled by the mirror thermal control unit (MTCU). The heaters are pulse-modulated with a duty cycle that is updated periodically by command from the ground. This approach was made possible by the large thermal mass and stable external

Figure 11.20   XMM/Newton lower module showing mirror entry baffles and surrounding service module (Reproduced by permission of European Space Agency)

environment. The surrounding service module is maintained at an average temperature of 15 °C by a combination of equipment dissipation and thermostatically controlled heaters. The service module (Figure 11.20) is insulated from the environment by multi-layer super-insulation blankets, except for the panel radiators. At the other end of the spacecraft, the focal plane instrumentation compartment is controlled passively under normal operating conditions. If an instrument is turned off, its thermal dissipation is automatically replaced by switching on equivalent heaters. Contingency thermostats are also provided, which will switch these heaters off if an instrument approaches its non-operating temperature limit for any reason. The sunlit side of the compartment is fully insulated—thermal balance is achieved by adjusting the heat rejected from the shadowed sides. Special cone-shaped radiators are provided to cool the camera detectors to cryogenic temperatures. As shown in Figure 11.21, these are inclined away from the Sun at an angle of 20° to avoid any risk of sunlight entering them, as well as being located behind the fully insulated Sun-shield. The telescope tube is constructed from carbon fibre (high-ε surface) and is fully insulated externally. This results in a very stable environment with gradients across the tube diameter not exceeding 3 °C. No heaters were implemented on the telescope tube.

As can be seen from the photographs, the thermal design of XMM involves extensive use of MLI blankets. These were typically constructed from 20 double-sided aluminized layers, separated by Dacron net spacers, together with an outer layer of carbon-impregnated Kapton. This black outer layer, which gives XMM its rather sinister black appearance,

Figure 11.21 XMM/Newton focal plane instrumentation compartment (Reproduced by permission of European Space Agency)

is electrically conducting and is intended to prevent the build-up of static electricity and consequent electrostatic discharges. It has the added advantage that its thermo-optical properties will not change during the 10-year life of the mission, so helping to maintain spacecraft temperature stability.

11.9   ACKNOWLEDGEMENT

The assistance of the author's ESTEC colleagues is gratefully acknowledged. Particular thanks are due to Mr. Olivier Pin who ran the ThermXL analysis for the example in Section 11.5.

REFERENCES

[1] Lide, D. R. and Frederikse, H. P. R. (eds) (1995) *CRC Handbook of Chemistry and Physics*, CRC Press, Boca Raton, FL.

[2] NASA (1983) *Space and Planetary Environment Criteria Guidelines for Use in Space Vehicle Development* (1982 revision), TM82478, Volume 1.

[3] NASA (1983) *Space and Planetary Environment Criteria Guidelines for use in Space Vehicle Development* (1982 revision), TM82501, Volume 2.

[4] ESA (1995) *Space and Planetary Reference Data—Thermal Radiation Environment*, ESA PSS-03-101.

[5] ESA (1994) *Data for the Selection of Space Materials*, ESA PSS-01-701, Issue 1, Revision 3.

[6] ESA (1989) *Spacecraft Thermal Control Design Data*, ESA PSS-03-108, Issue 1.

[7] ESA (1993) *Outgassing and Thermo-optical Data for Spacecraft Materials*, ESA RD-01, Revision 4.

[8] Redor, J.-F. (1990) *Introduction to Spacecraft Thermal Control*, ESA EWP1599, Version 1.10.

[9] Eckert, E. R. G. (1959) *Heat and Mass Transfer*, McGraw-Hill, New York.

[10] Semple, E. C. (1967) *Principles and Techniques in the Passive Thermal Control of Spacecraft*, Royal Aircraft Establishment Technical Report No. TR 67100.

[11] ESA (1999) *ESARAD Getting Started Guide*, UM-ESARAD-113.

[12] ESA (2000) *ESARAD User Manual*, UM-ESARAD-024, ESARAD Version 4.2.

[13] ESA (2000) *ESATAN Training Manual*, TM-ESATAN-008, ESATAN Version 8.5.

[14] ESA (2000) *ESATAN User Manual*, UM-ESATAN-004, ESATAN Version 8.5.

[15] Knight, R., Pin, O. and Thomas, J. (2000) *ThermXL: A Thermal Modelling Tool Integrated Within Microsoft Excel*, 30th International Conference on Environmental Systems and 7th European Symposium on Space Environmental Control Systems, Toulouse (France), 9–13 July.

[16] Dunn, P. D. and Reay, D. A. (1994) *Heat Pipes* (4th edn), Pergamon Press, Elmsford, New York

[17] Peterson, G. P. (1994) *An Introduction to Heat Pipes—Modelling, Testing and Applications*, John Wiley and Sons.

[18] Faghri, A. (1995) *Heat Pipe Science and Technology*, Taylor and Francis, Washington, DC.

[19] Humphries, W. R. and Grieggs, E. I. (1977) *A Design Handbook for Phase Change Thermal Control and Energy Storage Devices*, NASA TP-1074, November.

[20] Groll, M., Muenzel, W. D., Supper, W. and Savage, C. J. (1980) *Transient Behaviour of Liquid Trap Heat-pipe Thermal Diodes*, in Heat Transfer, Thermal Control and Heat Pipes, Progress in Astronautics and Aeronautics, volume 70, edited by W. B. Olstad, published by AIAA.

[21] London, A. (1965) *Thermal Control of the NIMBUS Satellite System*, General Electric MSD, AIAA Unmanned Space Meeting.

[22] Berner, F. and Savage, C. J. (1984) *Design and Characteristics of a Dynamic Cooler for Space Use*, Paper No. 840964, 14th Intersociety Conference on Environmental Systems, San Diego, CA (USA), 16–19 June.

[23] Walker, G. (1983) *Cryocoolers, Part 1: Fundamentals and Part 2: Applications*, Plenum Press, New York.

[24] Ros, R. G. (ed.) (1994) *Cryocoolers 8*, Proceedings of the 8th International Cryocooler Conference, Vail, Colorado (USA), 28–30 June, published in 1995 by Plenum Press, New York (USA).

[25] ESA (1993) *Europe Qualifies for Space—The European Coordinated Test Centres*, ESA BR-46, ISBN 92-9092-035-1 (2nd edn), December 1993.

[26] Tan, G. B. T. and Walker, J. B. (1982) *Spacecraft Thermal Balance Testing Using Infrared Sources*, 12th Space Simulation Conference, NASA CP 2229.

[27] Wertz, J. R. and Larson, W. J. (1999) *Space Mission Analysis and Design* (3rd edn), Microcosm Press and Kluwer Academic Publishers.

[28] Barré, H., Nye, H. and Janin, G. (1999) *An Overview of the XMM Observatory System*, ESA Bulletin No. 1000, December.

[29] Van Katwijk, K., Van Der Laan, T. and Stramaccioni, D. (1999) *Mechanical and Thermal Design of XMM*, ESA Bulletin No. 100, December.

12 TELE-COMMUNICATIONS

Howard Smith[1] and Ray E. Sheriff[2]

[1]*Astrium Space, Portsmouth, UK*
[2]*Department of Electronics and Telecommunications, University of Bradford, UK*

12.1 INTRODUCTION

12.1.1 The development of telecommunications satellites

Long before artificial Earth satellites became a reality, their potential in the field of telecommunications had been appreciated by visionaries such as Arthur C. Clarke [1]. In the years following the launch of Sputnik 1, the demand for global communications systems was one of the main driving forces—along with military and political considerations—in the rapid development of space technology.

The birth of telecommunications by satellite can be seen, perhaps, in the launch of Telstar I in 1962. Telstar I permitted, for the first time, transoceanic communications by satellite. By today's standards it was, of course, a very modest affair. Just under a metre in diameter and weighing 77 kg at launch, the satellite had one channel with a 50 MHz bandwidth, providing about 12 telephony circuits. But it was the start of a revolution in international communications.

The low elliptical orbit of Telstar I limited usage for transatlantic communications to three or four half-hour periods in each day. The first successful geosynchronous communications satellite was Syncom II, used for experimental transmissions between America and Japan. This again had a rather limited capacity, but the design after further improvements, became the basis for the first commercial communications satellite, Early Bird or, as it was later called, Intelsat I. This satellite, launched in 1965, was similar in size and mass to Telstar I but was capable of providing 240 telephony circuits or a high-quality television channel.

The first worldwide satellite communications system was not established, however, until 1969 when Intelsat III satellites were in position over the Atlantic, Pacific and Indian Oceans, giving coverage of all parts of the globe other than the polar regions.

The subsequent development of satellite communications has been a response to a dramatic increase in the demand for international telephone, video and Internet traffic. The

Spacecraft Systems Engineering (Third Edition). Edited by P. W. Fortescue, J. P. W. Stark and G. G. Swinerd
© 2003 John Wiley & Sons Ltd

Table 12.1   The growth of the Intelsat series of satellites

Satellite	First successful launch	Approximate mass in orbit (kg) (BOL)	Primary power (W) (BOL)	Approximate capacity (voice circuits)	Number in GEO
Intelsat I	1965	40	45	240 (or 1 TV)	1
Intelsat II	1967	85	83	240 (or 120 + 1 TV)	3
Intelsat III	1968	150	160	1 200 (+1 TV)	5
Intelsat IV	1971	730	550	3 500 (+2 TV)	7
Intelsat IV A	1975	825	700	6 250 (+2 TV)	5
Intelsat V	1980	1010	1800	12 500 (+2 TV)	8
Intelsat V A	1985	1190	2100	15 000 (+2 TV)	6
Intelsat VI	1989	2550	2600	33 000 (+4 TV)	5
Intelsat K	1992	1550	4800	32 TV	1
Intelsat VII/VII A	1993	~1800	3900/4800	18 000 (+3 TV)/ 22 500 (+3 TV)	8
Intelsat VIII/VIIIA	1998	~1900	N/A	22 000 (+3 TV)	6

Note: The channel capacities shown are typical values. These are generally less than the theoretical maximum signalling rates, particularly for the larger satellites. The number of telephone circuits depends not only on the bandwidth allocated to television channels but also on the distribution of telephone channel allocations between various countries or spot beams.
Note: BOL, Beginning of life GEO, Geostationary Earth orbit.

need to provide services to the mobile user has also resulted in a number of initiatives over the last decade. This extraordinary increase in demand has been matched by a rapid increase in the size, power capability and traffic capacity of the spacecraft, as well as in the number of satellites in simultaneous use. Table 12.1 shows this development for the Intelsat series (see also Chapter 19). A satellite of the principal current series (Intelsat VIII) has a capacity of about 22 000 telephone circuits and three colour television channels.

Geostationary satellites were first used to provide mobile communications to the maritime sector in 1981. The initial service provided by the International Maritime Satellite Organisation (*INMARSAT*) was based on frequency modulation (FM) telephony and was known as the Standard-A service, which was later renamed Inmarsat-A. Over the last twenty years, Inmarsat has evolved its family of services such that aeronautical and land environments are also served by its network of nine satellites.

Driven by the huge potential market offered by the introduction of mobile telephones, the last decade was notable for a new direction in satellite communications, with the introduction of non-geostationary satellites in Low Earth Orbit (LEO). Unfortunately, by the time these multi-satellite systems were introduced, terrestrial cellular networks had already substantially eroded the anticipated markets. As a consequence, potential subscribers have proved allusive, placing severe financial difficulties on the backers of such systems. The use of non-geostationary satellites for the provision of fixed-satellite services may also become a reality within the next few years.

The success of satellite navigation has demonstrated the viability of non-geostationary satellite technology. In particular, the US Global Positioning System (*GPS*) and the Russian Global Navigation Satellite System (*GLONASS*) networks, both of which have their origin in military applications, have successfully been applied to civilian use. With the backing of the ESA and the EC, Europe is currently developing its own navigation system, which will culminate with the deployment of the *GALILEO* network of non-geostationary satellites in 2008.

12.1.2   The role of the communications payload

Some early experimental systems used orbiting objects as passive reflectors or scatterers in order to achieve intercontinental communications. It quickly became apparent, however, that in order to meet the demand for high traffic capacities it would be necessary to use active satellites containing transponders that receive the signals transmitted from the ground, change their frequencies and amplify them before re-transmitting them to Earth. Figure 12.1 is a very much simplified block diagram of such a link. The power

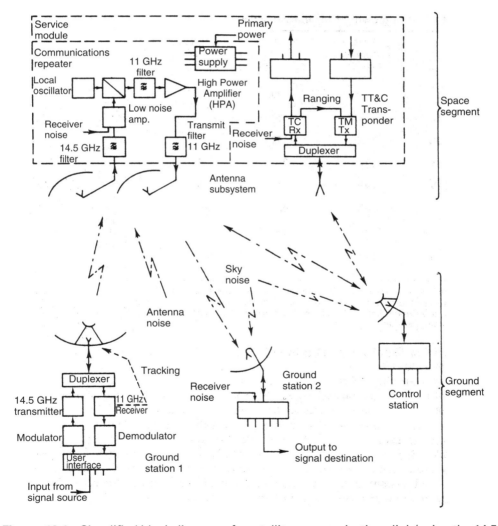

Figure 12.1   Simplified block diagram of a satellite communications link (using the 14.5 and 11 GHz bands)

amplification factor (or *gain*) provided by the satellite is typically in the region of 10^{12}–10^{15} (120–150 dB).*

The most striking difference between a satellite link and a terrestrial link is in the length of the transmission path. This is typically about 50 km for a terrestrial microwave link and 36 000 km from an Earth station to a satellite in geostationary orbit, or 10 000 km to 20 000 km for a Medium Earth Orbit (MEO) or 700 km to 2000 km for a LEO. The immediate result of this enormous increase in path length is that the most critical parameter in a satellite system is normally the available transmitter power—particularly on the downlink, in which the transmitter power is in turn one of the major demands on the primary satellite power. For broadcast or multicast services, the huge coverage area offered by a geostationary satellite, combined with its fixed position in the sky, makes it particularly suitable for this mode of service delivery. For mobile services, the selection of one particular orbit over another is not so straightforward. The smaller coverage area offered by a non-geostationary satellite, which continuously moves over the Earth, necessitates the need for multi-satellite constellations to guarantee continuous coverage on a regional or global basis. This, in turn, increases the complexity of the network infrastructure. However, the reduced transmission distance in comparison with a geostationary orbit, allows the power requirements to be relaxed and results in a reduction in the latency of the link.

Satellite communications is not restricted to the use of circular orbits. For instance, satellites of the Russian Molniya series are in elliptical, high-inclination, zero-drift orbits that allow good coverage of high latitudes. During the slow apogee transit, near GEO altitude, the satellite appears almost stationary in the sky, allowing continuous operation for up to about two-thirds of the orbital period (see Chapter 5).

The simple system illustrated in Figure 12.1 is but one of a large variety of telecommunications applications of artificial satellites. The main types of systems are as follows:

- trunk telephone and television services;
- broadcast services;
- videoconferencing and other business services;
- communication with mobiles (hand-held, ships/aircraft/land vehicles);
- data transmission (between ground stations);
- data relay (between other satellites and ground stations);
- internet access;
- navigation;
- remote monitoring and meter reading;
- fleet management and tracking.

A single satellite may include transponders for more than one of these functions.

12.1.3   System constraints

The design of a satellite communications system involves considerations not normally required for a terrestrial system. In general these design constraints can be divided into

---

* In telecommunications engineering, a power ratio (in this case the ratio of output power to input power) is often expressed on a logarithmic scale with decibels (dB) as units. On this scale the ratio P_1/P_2 is represented by $10 \log_{10}(P_1/P_2)$ dB. Thus, for instance, a power ratio of 10 is 10 dB and a factor of 2 is about 3 dB.

three broad categories: customer requirements, technical requirements and international regulations.

Customer requirements form the starting point of the design, and are as many and as varied as the customers themselves. Among the parameters that one might expect the customer to specify would be

- type of signals (voice, television, data and so on);
- capacity (i.e. number of channels of each type)—or bandwidths and frequencies;
- coverage area served by the satellite and the site(s) of the control station(s);
- uplink and downlink signal strength and quality—or types of ground terminals to be served;
- connectivity between different channels and traffic routes;
- availability (i.e. times of day/year, permitted outage times and so on);
- lifetime (typically 7 years for LEO, 12 years for MEO and 12 to 15 years for GEO satellites currently in service, but likely to be about 20 years or more for future designs).

There may be a further requirement that some of these parameters can be altered in orbit by command from a ground station and there may also be other limitations such as a need for a high level of security, both against illicit interception and decoding of signals and against interference, jamming and illegal telecommands.

As with all major engineering projects, however, the most important constraints imposed by the customer and market are usually those of cost and timescale.

The *technical constraints* (such as transmitter power, receiver sensitivity, interference, environment, available components etc.), which apply to a terrestrial link generally apply also to a satellite link, but with some additional restrictions. Of those factors that are peculiar to satellite systems, the most significant are those imposed by the payload mass, the available power and the need for equipment to survive the launch phase and operate unattended in the space environment.

The interplay between technical and cost factors is often crucial. In most terrestrial systems, it is possible to guarantee adequate overall performance by allowing generous margins in the design. Such extravagance in the case of a satellite system will result in an increase in mass and power consumption and hence increased launcher costs. On the other hand, the cost of developing and space-qualifying new tailor-made equipments or of elaborate integration and test procedures may mean that the best technical design is not the most cost effective.

International regulations for communications systems are required to control possible interference between different systems and to ensure compatibility between the various national systems that may need to be connected end to end.

The International Telecommunications Union (ITU), which is now part of the United Nations Organization, exists for the purpose of defining and enforcing international standards. More than 180 nations are signatories to its Convention that has the status of a formal international treaty. Within the ITU, there are two consultative bodies that collate information and formulate policy in relation to telephony and radio communications, respectively. These are the *ITU—Radiocommunication Sector (ITU-R)* and the *ITU—Telecommunication Standardization Sector (ITU-T)*. A third body, the *Telecommunication Development Sector (ITU-D)*, is concerned with the technical requirements of the developing world.

The recommendations of the ITU-R are in turn considered by the World Radio Conference (WRC) [2] which may then be incorporated into the ITU Radio Regulations [3]. These are binding on signatories of the ITU Convention.

Among the important system parameters subject to international regulation are the available frequency bands, the orbital location and the maximum permitted power flux density at the Earth's surface. Table 12.2 shows frequencies assigned to satellite communications as of the start of the 21st Century. Most of the satellite frequencies are shared with other services. Full details are given in Reference [3].

So far, the bulk of telephony and television traffic has been in the six and four GHz bands (C-band), but serious overcrowding in this part of the spectrum has led to the increasing use of 14.5 GHz (Ku-band) and 11 GHz (X-band). Signs of overcrowding are already appearing in some areas at these higher frequencies, and there is a growing interest in 30 and 20 GHz (Ka- and K-bands). So far, however, transmissions at these frequencies have been mainly of a pre-operational rather than a fully commercial nature. The exception to this is the feeder links used by the *IRIDIUM*™ satellite personal communications network (S-PCN). The use of the Ka-band is set to be exploited further shortly with the introduction of broadband Internet access to fixed networks. Mobile satellite services operate in the L- and S-bands but, as with fixed networks, the desire for broadband services is generating interest in the Ka-band.

The position of a satellite in the geostationary orbit is negotiated through the Space Service Department. The main requirement is that there should be sufficient separation between locations to allow a ground-station antenna of reasonable size (and hence beamwidth) to discriminate between adjacent satellites. This implies a separation of about 2° (or more) of longitude between satellites using the same frequency. Some parts of the orbit are relatively empty, but in others, such as the Atlantic region, where traffic density is very high, the choice of location may be very restricted.

In addition to the regulatory bodies set up under the United Nations Organization, there are several multi-national companies and international organizations whose aim is to coordinate and rationalize the commercial exploitation of communications satellites. Of these, the most established is Intelsat Ltd, which owns the satellites listed in Table 12.1 and leases their capacity to a variety of users including, principally, the national post, telegraph and telecommunications organizations. Intelsat began life as an inter-governmental organization in 1964, expanding rapidly from 11 to over 180 member states. In July 2001, Intelsat was privatized to form Intelsat Ltd. The USSR did not initially join INTELSAT but instead set up its own organization, *INTERSPUTNIK* in 1971, serving mainly Eastern Europe. With the collapse of the Soviet Union, the situation is more fluid. INTERSPUTNIK has now expanded from its original 9 member states to 24, operating on a more commercial basis than in the past. In addition to INTELSAT and INTERSPUTNIK, a number of national and regional organizations have been established—notably *EUTELSAT*, which initially provided services mainly to Western Europe and the Mediterranean region, although recent satellite additions to its fleet of 18 satellites has enabled coverage to be extended to the Middle East, Africa, South-West Asia and the Americas. Like Intelsat, Eutelsat began life in 1977 as an inter-governmental organization, changing to a private company in July 2001. Eutelsat provides TV, radio and Internet access as well as mobile, news gathering and fleet management services. The anticipated demand for broadband service delivery to the home, driven by the need for fast Internet access, has resulted

Table 12.2 Frequency allocations for telecommunication satellites[1]

Service	Frequencies (GHz)	Direction	Scope[2]
Fixed satellite	2.500–2.655	Down	R2
	2.500–2.535	Down	R3
	2.655–2.690	Up/down	R2
	2.655–2.690	Up	R3
	3.400–4.200	Down	W
	4.500–4.800	Down	W
	5.150–5.250	Up	W[3]
	5.725–5.850	Up	R1
	5.850–6.700	Up	W
	6.700–7.075	Up/down	W[4]
	7.250–7.770	Down	W
	7.900–8.400	Up	W
	10.70–11.70	Down	W
	10.70–11.70	Up	R1
	11.70–12.20	Down	R2
	12.50–12.75	Up/down	R1
	12.50–12.75	Down	R3
	12.70–12.75	Up	R2
	12.75–13.25	Down	W
	13.75–14.80	Up	W[5]
	15.43–15.63	Up/down	W[4]
	17.30–17.70	Up	W
	17.70–18.4	Up/down	W
	18.40–19.30	Down	W
	19.30–19.70	Up/down	W[6]
	19.70–21.20	Down	W
	24.75–25.25	Up	R2 & R3
	27.00–27.50	Up	R2 & R3
	27.50–31.00	Up	W
	37.50–42.50	Down	W
	42.50–43.50	Up	W
	47.20–50.20	Up	W
	50.40–51.40	Up	W
	71.00–76.00	Down	W
	81.00–86.00	Up	W
	123.0–130.0	Down	W
	158.5–164.0	Down	W
	167.0–174.5	Down	W
	209.0–226.0	Up	W
	232.0–240.0	Down	W
	265.0–275.0	Up	W
Mobile satellite	0.137–0.138	Down	W[7,8]
	0.148–0.15005	Up	W[8,9]
	0.312–0.315	Up	W[10]
	0.387–0.390	Down	W
	0.3999–0.40005	Up	W[8,9]
	0.40015–0.401	Down	W[8]
	0.406–0.4061	Up	W[11]
	0.454–0.456	Up	R2[8]

(continued overleaf)

Table 12.2   (*continued*)

Service	Frequencies (GHz)	Direction	Scope[2]
	0.459–0.460	Up	R2[8]
	0.608–0.614	Up	R2[12]
	1.492–1.525	Down	R2
	1.525–1.559	Down	W[13]
	1.610–1.6605	Up	W[13,14]
	1.6138–1.6265	Down	W[10]
	1.675–1.690	Up	R2
	1.930–1.970	Up	R2[10,15]
	1.980–2.010	Up	W[15]
	2.010–2.025	Up	R2[15]
	2.120–2.160	Down	R2[10,15]
	2.160–2.170	Down	R2
	2.170–2.200	Down	W[15]
	2.4835–2.520	Down	W[13,16,17]
	2.670–2.690	Up	W[17]
	14.00–14.50	Up	W[12]
	19.70–21.20	Down	W[18]
	29.50–31.00	Up	W[18]
	39.50–40.50	Down	W
	43.50–47.00	—	W
	50.40–51.40	Up	W[10]
	66.00–71.00	—	W
	71.00–74.00	Down	W
	81.00–84.00	Up	W
	123.0–130.0	Down	W
	191.8–200.0	—	W
	252.0–265.0	Up	W
Broadcasting satellite	1.452–1.492	Down	W[19]
	2.520–2.670	Down	W[19]
	11.70–12.50	Down	R1
	11.70–12.20	Down	R3
	12.20–12.70	Down	R2
	12.50–12.75	Down	R3
	17.30–17.80	Down	R2
	21.40–22.00	Down	R1 & R3
	40.50–42.50	Down	W
	74.00–76.00	Down	W
Space operation	0.137–0.138	Down	W
	0.272–0.273	Down	W
	0.401–0.402	Down	W
	1.427–1.429	Up	W
	1.525–1.535	Down	W
	2.025–2.110	Up/down	W
	2.200–2.290	Up/down	W
Inter-satellite links	22.55–23.55	—	W
	24.45–24.75	—	W
	25.25–27.50	—	W

Table 12.2 (*continued*)

Service	Frequencies (GHz)	Direction	Scope[2]
	32.00–33.00	—	W
	54.25–58.20	—	W
	59.00–71.00	—	W
	116.0–123.0	—	W[20]
	130.0–134.0	—	W
	167.0–182.0	—	W
	185.0–190.0	—	W
	191.8–200.0	—	W
Amateur satellite	0.028–0.0297	—	W
	0.144–0.146	—	W
	5.830–5.850	Down	W[10]
	10.45–10.50	—	W[10]
	47.00–47.20	—	W
	76.00–77.50	—	W[10]
	77.50–78.00	—	W
	78.00–81.00	—	W[10]
	134.0–136.0	—	W
	136.0–141.0	—	W[10]
	142.0–144.0	—	W
	241.0–248.0	—	W[10]
	248.0–250.0	—	W

[1] Data links for e.g. meteorological, scientific and navigational satellites are not included.

[2] W: Allocated on a world-wide basis.
R: Allocation on a regional basis: R1–Region 1, R2–Region 2, R3–Region 3.

[3] Limited to feeder links of non-geostationary satellite systems in the mobile satellite service.

[4] Downlink limited to feeder links of non-geostationary satellite systems in the mobile satellite service.

[5] The use of the band 14.5–14.8 GHz is limited to feeder links for the broadcasting satellite service.

[6] The use of the band 19.3–19.6 GHz (Up) is limited to feeder links for non-geostationary satellite systems in the mobile satellite service.

[7] Allocated on a secondary basis in bands 137.025–137.175 MHz and 137.825–138 MHz.

[8] The use of the frequency bands 137–138 MHz, 148–150.05 MHz, 399.9–400.05 MHz, 400.15–401 MHz, 454–456 MHz and 459–460 MHz by the mobile satellite service is limited to non-geostationary satellite systems.

[9] The use of the frequency bands 149.9–150.05 MHz and 399.9–400.05 MHz by the mobile satellite service is limited to land mobile satellite services until 1 January 2015.

[10] Allocated on a secondary basis.

[11] Limited to low power satellite emergency position indicating radiobeacons.

[12] Except aeronautical. Secondary allocation.

[13] The frequency bands 1525–1544 MHz, 1545–1559 MHz, 1610–1626.5 MHz, 1626.5–1645.5 MHz, 1646.5–1660.5 MHz and 2483.5–2500 MHz may be used by administrations wishing to implement the satellite component of International Mobile Telecommunications-2000 (IMT-2000).

[14] The band 1610–1626.5 MHz is allocated to the aeronautical mobile satellite service (R) on a primary basis.

[15] Identified for the use of the satellite component of IMT-2000.

[16] May be used by administrations wishing to implement satellite component of IMT-2000.

[17] The bands 2500–2520 MHz and 2670–2690 MHz may be used by administrations wishing to implement satellite component of IMT-2000; in the longer term, may need to be used for terrestrial component, depending on market conditions.

[18] Secondary allocation in R1 and R3 in bands 19.7–20.1 GHz and 29.5–29.9 GHz.

[19] Limited to digital audio broadcasting.

[20] The band 116–122.25 GHz is limited to use by satellites in geostationary orbit.

in a number of initiatives including the SPACEWAY™ constellation of three geostationary satellites, which will operate in the Ka-band, providing high-speed communication links.

It has already been noted that Inmarsat offers global communications to shipping, aeronautical and land mobile users, with over 200 000 subscribers. Inmarsat began life as a maritime inter-governmental organization in 1979, before forming a limited company in 1999. The introduction of INMARSAT-3 satellites with spot beam capability has enabled terminal size to decrease and data rates of up to 64 kbit/s to be provided. The introduction of the 200+ spot beam INMARSAT-4 satellites in 2004, will further increase the data rates to as much as 432 kbit/s.

Additionally, geostationary satellites are also used to provide regional mobile services to Australia, North America (M-SAT), Asia (ACeS) and the Middle East (THURAYA). Presently, the use of non-geostationary satellites is limited to mobile services, through the IRIDIUM™ and GLOBALSTAR™ constellations, although satellite constellations addressing fixed network services are planned for the near future, notably by the TELEDESIC™ and SKYBRIDGE™ systems.

In what follows, we will be concerned primarily with the technical aspects of the system design. Later we will look at some of the units that make up the telecommunications payload but first we must examine some of the principles underlying the operation of the system as a whole.

12.2 TECHNIQUES OF RADIO COMMUNICATIONS

12.2.1 Introduction

Although most of the important elements of a satellite communications link are shown in Figure 12.1, it must be emphasized that this is a very much simplified picture. In particular, only two communications ground stations are shown and the system is a simplex (i.e. one-way) link. Nevertheless, the block diagram will serve to illustrate some of the features that are common to practically all systems. In fact, the extension to two-way (duplex) operation is often trivial. The same satellite transponder may carry both outgoing and return traffic provided that the same uplink and downlink frequencies are available at the two ground stations. However, this is not always the case. For instance, in a system serving mobiles, the satellite–mobile link may operate at L- or S-bands (1.5 to 1.7 GHz), and the satellite-fixed Earth station link at C/Ku/X- or Ka-band. In this case the satellite must carry separate forward and return transponders.

12.2.2 Modulation

Types of modulation

The signals to be transmitted by a communications system normally consist of a band of rather low frequencies, ranging, for instance, from a few tens of Hz to a few KHz in the case of speech or from a few tens of Hz to a few MHz in the case of television. These *baseband* frequencies—those which constitute the original signal—are unsuitable for direct transmission as radio waves.

For transmission purposes, the signal is imposed on a 'carrier' wave of much higher frequency—a process known as *modulation*. This is represented in Figure 12.1 by the *modulator* block in ground station 1. The reverse process—recovery of the baseband signal from the received signal—is known as *demodulation*. If we represent the high-frequency signal by a cosine wave,

$$V = V_c \cos(\omega_c t + \phi_c) = V_c \cos(2\pi f_c t + \phi_c) \tag{12.1}$$

the baseband signal may then be represented by a variation with time of either the carrier amplitude (V_c), its frequency (f_c) or its phase (ϕ_c).

Figure 12.2 illustrates these three basic types of modulation for the simple case where the baseband signal is itself a cosine wave. As shown in the algebraic expressions representing the waveforms in Figure 12.2, the magnitude of the modulation is represented by a *modulation index*, m (also called 'modulation depth') in the case of amplitude modulation (AM) and β in the case of phase modulation (PM) or frequency modulation (FM). PM and FM are different forms of 'angle modulation'. Any FM waveform can be represented as phase modulation (albeit by a slightly different baseband signal) and *vice versa*. For a sinusoidal modulating waveform, β, the peak phase deviation (in radians), is equal to $\Delta f / f_m$ where Δf is the peak frequency deviation and f_m the modulating frequency.

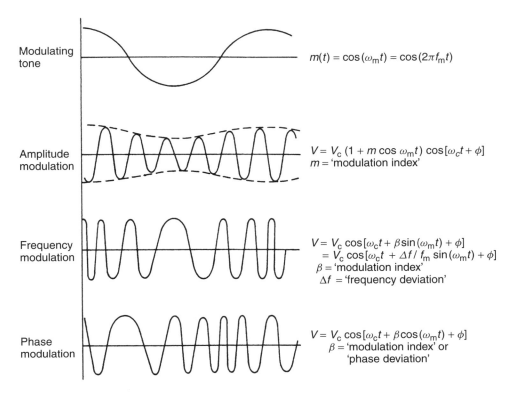

Modulating tone	$m(t) = \cos(\omega_m t) = \cos(2\pi f_m t)$
Amplitude modulation	$V = V_c (1 + m \cos \omega_m t) \cos[\omega_c t + \phi]$ $m = $ 'modulation index'
Frequency modulation	$V = V_c \cos[\omega_c t + \beta \sin(\omega_m t) + \phi]$ $\quad = V_c \cos[\omega_c t + \Delta f / f_m \sin(\omega_m t) + \phi]$ $\beta = $ 'modulation index' $\Delta f = $ 'frequency deviation'
Phase modulation	$V = V_c \cos[\omega_c t + \beta \cos(\omega_m t) + \phi]$ $\beta = $ 'modulation index' or 'phase deviation'

Figure 12.2 Types of analogue modulation. (In the interest of clarity the frequency of the carrier is shown very much reduced compared to that of the modulating signal)

Analogue and digital signals

The modulated carrier waves illustrated in Figure 12.2 are examples of analogue signals. They are characterized by the fact that the instantaneous value of the baseband signal may lie anywhere within a certain range.

An increasing proportion of telecommunications traffic consists of digital signals, that is, signals that can take only a finite number of discrete values—often only two values, corresponding to the binary digits 0 and 1. When the signals are in this form, the three types of modulation described above are known as *amplitude-shift keying (ASK)*, *frequency-shift keying (FSK)* and *phase-shift keying (PSK)*.

Within these general categories there are many variants whose different characteristics make them suitable for differing applications. Some of the more commonly used techniques are illustrated in Figure 12.3 where, for instance, ASK is shown as 'on–off keying', with a binary 1 represented by full amplitude and a 0 by zero amplitude. FSK is illustrated by continuous-phase *fast frequency-shift keying (FFSK)* in which there are no discontinuous changes in phase at the frequency transitions and the two frequencies are separated by half the bit rate (i.e. number of binary digits per second). This choice of frequency separation has certain advantages concerned with the ease of detection of signals in the presence of noise (see Sections 12.2.4 and 12.2.5).

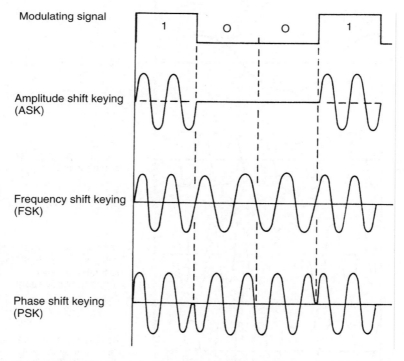

Figure 12.3 Examples of digital modulation. The examples illustrated are: ASK — on–off keying FSK — fast frequency-shift keying PSK — phase reversal keying (In the interest of clarity, the frequency of the carrier is shown to be very much reduced compared to that of the modulating signal)

PSK often uses two phases separated by π radians (as illustrated in Figure 12.3) and is then known as *phase-reversal keying (PRK)* or *Binary Phase-Shift Keying (BPSK)*. However, another important type of PSK uses four phases separated by $\pi/2$ radians. This is *quadrature phase-shift keying (QPSK)*. More complicated multi-phase or multi-phase plus multi-amplitude schemes may also be used, particularly where there is a need to achieve high data rates in a limited bandwidth. (The *bandwidth* is the width of the range of frequencies included in the signal, and transmitted by the system.)

Digital signals occur naturally, for instance, in the transmission of numerical data or in communication between computers. It is also possible to convert analogue signals into digital form. Already most telephone traffic is transmitted by digital techniques, and digitized television and audio broadcasts are now becoming the norm.

The conversion of analogue signals into digital form may be described in terms of three distinct steps — sampling, quantization and encoding.

The sampling process is illustrated in Figure 12.4(a,b). Clearly, if it is known that the signal varies with sufficient smoothness between the sample times, the entire signal can be reconstructed with considerable accuracy from the sample values. It can be shown that this is guaranteed if the signal to be sampled contains no frequencies greater than half the sampling frequency. (This is known as the *Nyquist criterion*.)

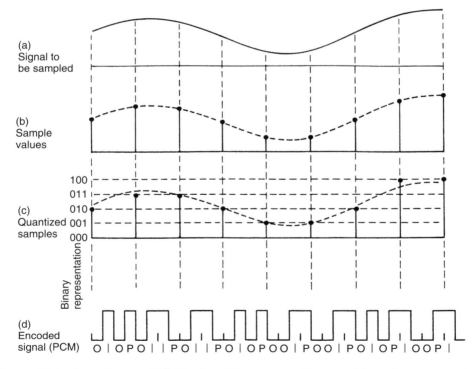

Figure 12.4 Sampling and digitization. The process illustrated here is an example of pulse code modulation (PCM). Each sample value is represented by a binary number with, in this case, three digits and one extra bit for parity checking.(a) Signal to be sampled, (b) sample values, (c) quantized samples and (d) encoded signal (PCM). The symbols marked 'P' are parity bits

The process of quantization involves rounding off each sample to one of a finite number of allowed values. This introduces errors that cannot be compensated in subsequent processing of the signal. In order to reduce this 'quantization noise' to an acceptable value it may be necessary to increase the number of allowable signal levels. However, the greater the number of levels, the greater is the number of digits that are necessary to identify a particular level and hence the greater the required bit rate. This reduces the transmitted energy associated with any one bit and tends to aggravate the signal degradation caused by mechanisms other than quantization (see Section 12.2.5). The system designer must seek a balance that results in best overall signal quality.

The final step, that of encoding, may take many forms. The simplest is to express each sample value directly as a binary number and then combine the numbers (normally with some extra digits for synchronization—that is, to allow the receiver to identify the start of each binary word) to form a long sequence of binary digits. This is the system usually understood by the term *pulse code modulation (PCM)*. However, a number of more subtle techniques may be introduced at this stage with a view, for instance, to making the signals less vulnerable to transmission impairments, providing encryption or modifying the spectrum. Thus digital modulation is in many ways much more flexible than its analogue counterpart.

Spectrum and bandwidth

In the design of a telecommunications system, one of the most important signal parameters is the bandwidth. It is this (among other things) that determines the design of the various filters shown in Figure 12.1. The effect of modulation of a carrier wave is to produce frequencies other than that of the unmodulated carrier and the resultant spectrum depends both on the baseband signal and on the type of modulation in use. Some particular examples are considered below.

In the case of *amplitude modulation* by a co-sinusoidal signal, the spectrum may be inferred from standard trigonometrical identities:

$$V_c(1 + m \cos \omega_m t) \cos \omega_c t = V_c[\cos \omega_c t + (m/2) \cos(\omega_c + \omega_m)t$$
$$+ (m/2) \cos(\omega_c - \omega_m)t] \tag{12.2}$$

The components at frequencies $(\omega_c + \omega_m)$ and $(\omega_c - \omega_m)$ are known as *sidebands* and the term $V_c \cos(\omega_c t)$ is the *carrier component*. Increasing the modulation index m results in increased sideband levels but no change in the range of frequencies present in the signal. This is illustrated in Figure 12.5(a).

A simple development of equation (12.2) shows that if the baseband signal, $m(t)$, contains more than one frequency, each component of the baseband produces a pair of sidebands. In general, if $m(t)$ is not composed of discrete frequencies, its spectrum may be represented by its Fourier transform, $M(\omega) = \int m(t) \exp(-j\omega t) \, dt$. Whatever the baseband spectrum, the spectrum of the modulated carrier has the same form but in bands located symmetrically above and below the carrier frequency. The required channel bandwidth is twice the base bandwidth (irrespective of modulation depth).

The spectrum of a *frequency (or phase)-modulated signal* is considerably more complicated. In the case of co-sinusoidal modulation with a small modulation index, β, the

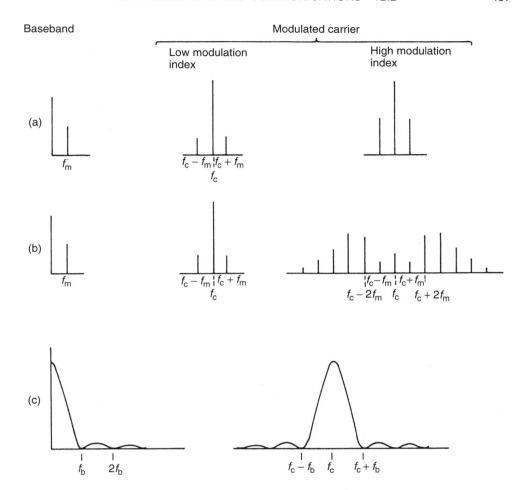

Figure 12.5 Spectra of modulated signals: (a) amplitude modulation (amplitude spectrum); (b) frequency modulation (amplitude spectrum) and (c) phase-reversal keying (power spectrum). The AM and FM spectra show modulation by a single tone of frequency f_m. The PRK spectrum is for a random binary sequence at a bit rate of f_b bits/s

presence of sidebands similar to those for AM may again be demonstrated by standard trigonometrical identities. If, however, the index is not small, the spectrum of FM is in striking contrast to that of AM. Fourier analysis of expressions, such as $\cos(\omega_c t + \beta \sin \omega_m t)$, yields an infinite set of components at frequencies $(f_c \pm n f_m)$ (where n takes all integral values) and with amplitudes that can be evaluated from Bessel functions of the first kind. The number of side frequencies having significant amplitude (and hence the effective bandwidth) increases with increasing β (Figure 12.5b).

When the baseband is not a single sine wave, the side frequencies are no longer related in a simple way to the baseband frequencies. For instance, if the baseband contains two frequencies f_1 and f_2, the modulated signal contains all frequencies of the form $f_c \pm n f_1 \pm m f_2$ (n, m are integers). In strictly mathematical terms, the bandwidth is infinite but there is a useful semi-empirical expression (*Carson's formula*) for the bandwidth containing about 98% of the power:

$$\text{Carson bandwidth} = 2(\Delta f + f_\text{m}) \qquad (12.3)$$

where Δf is the peak frequency deviation and f_m is normally taken to be the highest frequency present in the baseband. Frequency modulation with index sufficiently small so that only the first-order sidebands are significant (β less than about 0.5) is known as *narrow-band FM (NBFM)*, while for larger β the term *wide-band FM (WBFM)* is used.

In the case of *phase-shift keying* the spectrum depends both on the type of modulation (PRK, QPSK etc.) and on the encoding procedure used in producing the bit stream. As an example, we may consider PRK by a random binary signal (i.e. by a random sequence of 0s and 1s occurring with equal probability). A signal of this type does not have a Fourier transform but we may define a power-density spectrum $S(f)$ that is the power per Hz of bandwidth at the frequency f. For the random binary signal:

$$S(f) = V^2 T (\sin \pi f T)^2 / (\pi f T)^2 \qquad (12.4)$$

where V is the signal voltage and T is the length (in time) of one bit.

Since PRK may be regarded as a type of ASK (the amplitude is switched between $+A$ and $-A$), the spectrum of the modulated signal is similar to the baseband spectrum but centred on the carrier frequency. This is illustrated in Figure 12.5(c). Strictly speaking, the overall bandwidth is infinite because, in this case, the baseband has infinite bandwidth. However, the power in the 'tails' of the spectrum is not very significant, the first subsidiary maxima already being some 13 dB below the central peak. In practice, therefore, it is usual to transmit at most, that part of the spectrum that lies between the first zeros. In many cases less will suffice. This has the effect of partially smoothing out the phase transitions, though without destroying the baseband information completely. Since $1/T$ is the number of bits per second, the effective bandwidth for the random binary signal is between one and two times the bit rate.

If the binary signal results from digitizing an analogue signal, the bit rate depends both on the base bandwidth and on the required level of quantizing errors. For instance, for speech (bandwidth 4 kHz) quantized to 256 levels the bit rate is 64 kbit/s since there must be at least 8000 samples per second with 8 binary bits per sample.

Much effort has been concentrated on devising digital techniques that make more efficient use of bandwidth in particular applications. An important example is speech encoding, when used in mobile communications. PCM is an unnecessarily powerful and general technique for this purpose, since it can reproduce all waveforms satisfying the Nyquist criterion—including waveforms that are not produced by the human voice and make no contribution to the intelligibility of speech. Encoding techniques modelled on the mechanisms of voice production or sensitive to the statistics of the speech waveform allow speech transmission of adequate quality at a rate of 2.4 kbit/s. Digital television, by using the Moving Pictures Expert Group (MPEG-2) coding standard, requires bit rates of the order of 1 to 12 Mbit/s, a significant reduction of the 100 Mbit/s that would be required if a simple PCM approach were to be adopted.

12.2.3 Multiple access

It is a requirement of most systems that several users (in some cases very many users) can pass signals through the satellite simultaneously. There are three main techniques for doing this. In *frequency-division multiple access (FDMA)* each user transmits a signal on a different carrier frequency, and at the receiving station the signals are separated by frequency-selective filters. The signals in this case are often analogue in nature.

In the case of digital signals, *time-division multiple access (TDMA)* or *code-division multiple access (CDMA)* may be used. In TDMA, an individual user transmits short bursts of digits in a particular time-slot within a repeating time-frame. Other users occupy different time-slots within the same time-frame. Synchronization signals are included, which allow the receiver to identify the start of the frame, and select those time-slots that contain a particular 'message'. The concentration into short bursts has the result that each signal occupies the whole of the channel bandwidth. However, no two signals occur at the same time.

In CDMA, each signal uses the full channel bandwidth but the signals may well all be present at the same time. In Direct Sequence CDMA (DS-CDMA), the carrier signal is modulated twice, first by the required signal and then by a pseudo-random sequence of binary digits at a much higher bit rate. A receiver using the correct pseudo-random code can undo the effect of the second modulation and so recover the original signal. It does not, however, recover the signals of other users who have employed different codes. An alternative to DS-CDMA is frequency hopping. This involves changing the transmission frequency pseudo-randomly across the available bandwidth. At the receiver, the same pseudo-random sequence is used to synchronize with the transmitted frequency, to allow demodulation of the original signal. Hybrid multiple access schemes, comprising of combinations of the three techniques, are also employed. For example, an FDMA–TDMA scheme involves the division of the available bandwidth into channels (FDMA), each of which contains a TDMA frame.

12.2.4 Noise

Telecommunications would present few problems were it not for the presence of 'noise' in all electrical systems. In radio communications, electrical noise is a result of the random thermal motions of atoms and electrons in matter, which reveal themselves as small randomly varying electromotive forces and currents. Each resistive element in a circuit is a source of thermally generated electrical power of kT watts per Hz of bandwidth in the radio frequency range, where $k = 1.38 \times 10^{-23}$ J/K (Boltzmann's constant), and T is the temperature (in Kelvins) of the resistor. (Throughout this section it should be assumed that temperatures are measured on the absolute scale, that is, at $0°$ Celsius, $T = 273$ K.) Quantum mechanics predicts a power spectral density given by the expression

$$P_0(f) = \frac{hf}{\exp(hf/kT) - 1} + hf/2 \tag{12.5}$$

where $h = 6.625 \times 10^{-34}$ J s is Planck's constant. If $hf \ll kT$ this reduces to $P_0(f) = kT$.

The essential differences between a noise voltage and a typical signal are illustrated in Figure 12.6. The signal has a regular quasi-periodic structure whereas noise is essentially

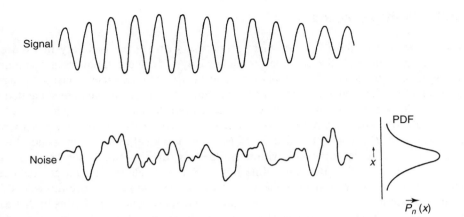

Figure 12.6 Signal and noise voltage waveforms

irregular and unpredictable. The root-mean-square (r.m.s.) value may be used as a measure of noise magnitude but a complete specification must include the probability density function (PDF), that is, the probability $p_n(x)\delta x$ that the instantaneous voltage lies in the range from x to $x + \delta x$. For many sources of noise the PDF is approximately Gaussian, in which case $p_n(x)$ takes the form

$$p_n(x) = \frac{1}{\sigma_n \sqrt{(2\pi)}} \exp\left(\frac{-x^2}{2\sigma_n^2}\right)$$ (12.6)

where σ_n is the r.m.s. value.

When two independent sources of noise are combined, it is their powers (or mean squared values) that must be summed. In particular, the noise powers contained in different frequency ranges are additive, so that for a constant power spectral density, $P_0(f) = kT$ W/Hz, the total power in bandwidth B is given by

$$P = kTB$$ (12.7)

The importance of noise is that it sets a fundamental limit to the sensitivity of a telecommunications receiver. The crucial factor that determines the overall performance of the system is not just the signal power itself, but rather the ratio of signal power to noise power (the *signal-to-noise ratio, SNR*).

In any communications systems there are many possible sources of noise. However, the signal power reaches its lowest level at the end of the ratio transmission path and it is therefore the noise generated in the input circuit of the receiver, which is most critical in determining the system performance. For this reason, noise is shown in Figure 12.1 as an extra input to each of the receivers—a convenient fiction that has much the same effect as the reality.

There are two commonly used ways of expressing the 'noisiness' of an amplifier (or radio receiver). In the first, the expression for thermal noise power available from a resistor, $P_0(f) = kT$, is used to express the noise output as an equivalent temperature. The concept is illustrated in Figure 12.7.

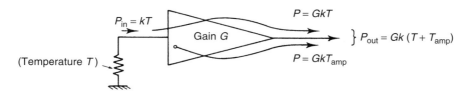

Figure 12.7 Definition of noise temperature

If the amplifier were an ideal noise-free device, the noise power density kT delivered by the source resistance would give an output power density GkT. In practice, the output is greater than this and we can write

$$P_{\text{out}}(f) = Gk(T + T_{\text{amp}}) \qquad (12.8)$$

where T_{amp} is known as the *noise temperature* of the amplifier. The noise temperature is thus the additional noise power originating in the amplifier expressed as an equivalent increase in the source temperature. (Note that noise temperatures, like noise powers, are additive.)

In the second method of expressing the noise performance of the amplifier, we assume that (in the arrangement of Figure 12.7) the source resistance is at a standard temperature T_0. Were the amplifier noise-free, the output power spectral density would then be kT_0G but in reality it is greater than this by a factor F, the *noise factor* (or noise figure):

$$P_{\text{out}}(f) = FGkT_0 \qquad (12.9)$$

From equations (12.8) and (12.9) it can be seen that

$$F = 1 + (T_{\text{amp}}/T_0) \qquad (12.10)$$

The standard choice of T_0 is 290 K—a typical 'room temperature'.

It can be shown that the overall noise temperature T_{tot} of an amplifier chain (Figure 12.8) is

$$T_{\text{tot}} = T_1 + T_2/G_1 + T_3/G_1G_2 + \cdots \qquad (12.11)$$

and, in terms of noise factor,

$$F_{\text{tot}} = F_1 + (F_2 - 1)/G_1 + (F_3 - 1)/G_1G_2 + \cdots \qquad (12.12)$$

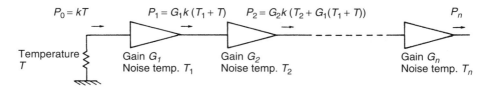

Figure 12.8 Noise temperature of cascaded units. The overall noise temperature of the amplifier chain is $T_1 + T_2/G_1 + T_3/(G_1G_2) + \cdots$

As might be expected, the total noise temperature can never be better than that of the first stage, and moreover for a given first stage noise temperature, the total noise temperature is minimized by making the first stage gain, G_1, as large as possible.

It may happen that one of the units in the cascade is an attenuator rather than an amplifier. This usually occurs at the most critical point in the entire system—at the very input of the receiver, where the antenna feed (the transmission line or waveguide connecting the antenna to the receiver) has a small, but not negligible, attenuation. For this 'stage' the gain, G, is of course less than unity, and its noise temperature can be shown to be $T(1/G - 1)$ referred to its input (where T is the physical temperature of the cable or attenuator).

In addition to the noise originating in successive amplifier stages, the overall system temperature T_{sys} will include background noise received by the antenna (usually represented by an *antenna temperature*) and a (usually small) contribution transmitted with the signal.

Antenna noise results mainly from thermal radiation from various surfaces, which may be either in the 'main beam' of the antenna or in the less sensitive 'sidelobes' (see Section 12.2.8). These include, for instance, the Earth's surface (at about 300 K), the Moon (about 200 K) and the Sun (about 6000 K). Also, the Earth's atmosphere emits thermal radiation when its attenuation becomes significant, as during a rainstorm (see Section 12.2.7). The situation is similar to that in a circuit attenuator except that in this case the temperature must be referred to the output. If the attenuation factor in the rainstorm is α (α is a factor greater than unity), the contribution of the atmosphere to antenna temperature is

$$T_{atmos} = T(1 - 1/\alpha) \qquad (12.13)$$

where the temperature, T, of rain is about 275 K. In the majority of geographical locations, in order to achieve the required limits on outage times, it is necessary to design the system for operation with significant atmospheric attenuation (see Section 12.2.7). Under these conditions the atmosphere may sometimes be one of the main sources of antenna noise (and indeed of system noise) for a ground station, where the receiving antenna is pointing towards the normally 'cold' sky. For an uplink, however, in which the on-board antenna points towards the Earth's surface, the antenna temperature is always in the region of 290 K and is little affected by the atmospheric attenuation (though of course the signal level is affected).

In the case of a satellite downlink, in which a transparent satellite transponder is employed, the transmitted signal is contaminated by noise originating on the uplink. The system is often designed so that the contribution of the uplink to the system noise is about 10 dB or more below that of the downlink. Under these conditions the transmitted noise causes a 0.4 dB degradation of downlink signal-to-noise ratio.

12.2.5 Output signal-to-noise ratio

The preceding section has outlined some of the factors influencing the system noise levels at radio frequencies (RF). The system user, however, is concerned with the SNR at the final (baseband) output of the system $(S/N)_0$. Although $(S/N)_0$ must clearly depend upon the RF signal (or 'carrier')-to-noise ratio (C/N), the relationship can be complicated, depending in particular on the type of modulation in use. It is usually convenient to express C/N in terms of the signal-to-noise-power-density ratio, C/N_0, in which N_0 is the noise power per Hz of bandwidth. Furthermore, in comparing the performance of different types of modulation it is usual to refer to the *input SNR*, $(S/N)_i$, which is the ratio of the RF signal power to the noise power in an RF bandwidth equal to the base bandwidth.

We have

$$C/N = C/(N_0 B) \text{ and } (S/N)_i = C/(N_0 F_m) \tag{12.14}$$

where B is the bandwidth of the receiver and F_m is the maximum baseband frequency.

When referenced to the input of the receiver, N_0 is related to the system temperature by $N_0 = kT_{sys}$. Thus, for a given transmitter power, system temperature and modulating signal, it follows that $(S/N)_i$ is fixed. Differences in $(S/N)_0$ are then because of the modulation system.

For *AM systems* with typical non-sinusoidal baseband signals, $(S/N)_0 \approx 0.05(S/N)_i$. Some improvement in performance may be achieved (at the cost of increased equipment complexity) by suppressing the central carrier component (see Figure 12.5 or equation 12.2), giving *double sideband suppressed carrier (DSBSC)* modulation. It is also possible to reduce the bandwidth by omitting one sideband (*single sideband* modulation—SSB). For both DSBSC and SSB, $(S/N)_0 = (S/N)_i$.

For *FM systems* the effect of noise is equivalent to random fluctuations in carrier phase (see Figure 12.9). A good $(S/N)_0$ can generally be achieved by making the phase variation due to the modulation much larger that the random variations. Typically, $(S/N)_0 \approx 0.3\beta_a^2(S/N)_i$ where β_a is the *deviation ratio*. ($\beta_a = \Delta f/F_m$ where Δf is the peak frequency deviation and F_m the maximum baseband frequency.) Since we may choose $\beta_a \gg 1$, there is a very substantial improvement over AM. However, large β_a implies large receiver bandwidth and hence low C/N (since N is proportional to bandwidth). When C/N falls below a certain threshold (about 11 dB for a conventional demodulator or about 8 dB for the phase-locked loop demodulator used in most Earth stations), $(S/N)_0$ declines rapidly and eventually becomes less than that for DSBSC (see Figure 12.10).

In *digital systems* the mechanism by which system noise affects the output of a communications link differs markedly from the analogue case. The demodulator must now contain one or more threshold detectors, which allocate to each received symbol, one of the permitted values. If the symbol is correctly identified, the noise has no effect whatsoever. Occasionally, however, the noise voltage is large enough to cause the receiver

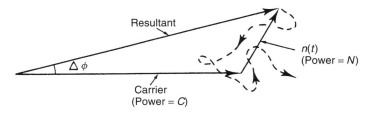

Figure 12.9 Noise in an FM system. The tip of the noise vector follows a random path such as that shown by the dotted line. The consequent random phase variation $\Delta\phi$ has an r.m.s. magnitude $\sqrt{(N/2C)}$ if C/N is large

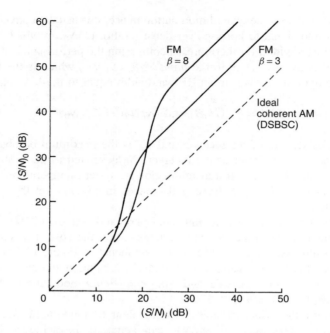

Figure 12.10 Output signal-to-noise ratio for frequency modulation by a single tone

output to lie on the wrong side of the detector threshold and so result in an error in interpretation.

The probability of error in any one bit—known as the *bit error rate (BER)*—depends on the product of the received signal power and the length of the bit interval, that is, on the received energy per bit. For typical transmission systems the BER, P_e, may lie in the range 10^{-3}–10^{-9}. The dependence of BER on the ratio of bit energy to noise power density (E_b/N_0) is shown for two types of binary PSK demodulator in Figure 12.11(a). These are ideal theoretical curves. In practice the value of E_b/N_0 needed for a given error rate may be 1 to 2 dB higher than shown.

The error rate on the final data stream (or the SNR on the reconstituted analogue signal) depends both on P_e and on the system of encoding. For simple PCM (Section 12.2.2) with sinusoidal modulation the output SNR is given by

$$(S/N)_0 = \frac{3(2^{2n-1})}{1 + 4P_e(2^{2n} - 1)} \tag{12.15}$$

where n is the number of binary bits per sample and P_e is the BER. This result is illustrated in Figure 12.11(b) for the error curves of Figure 12.11(a).

The BER P_e can be significantly improved by the use of encoding techniques that permit error detection and correction at the receiver. (This is *forward error correction—FEC*.) The simplest form of error detection is illustrated in Figure 12.4(d). The values of the parity bits are assigned so that the number of 1's in each binary 'word' is (in this case) even. If at the system output a word is received with an odd number of 1's, it is known that an error has occurred. (It is assumed in this case that the probability of two errors

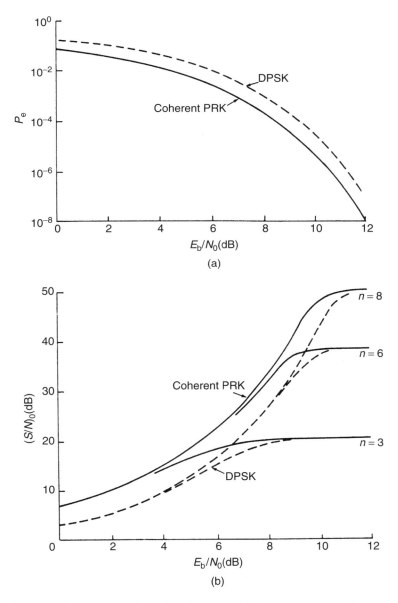

Figure 12.11 The effect of noise on digital systems. (a) Bit error rate *versus* E_b/N_0 for two types of system. In coherent PRK, the phase of each received bit is compared with that of a reference carrier derived separately from a long sample of the signal. In differential PSK (DPSK), the phase of each bit is compared with that of the preceding bit. White Gaussian noise, perfect synchronization and optimum detection are assumed and (b) Output SNR for simple PCM assuming the error rates shown in (a). n is the number of bits per sample

in the same word is negligible.) The same principle—that of transmitting more bits than are strictly required by the data—is used in more sophisticated coding techniques, which not only reveal the presence of errors but also (except in very improbable circumstances) make it possible to identify and so correct the offending bits. The consequent reduction in error rate more than offsets the effect of the required increase in bit rate.

12.2.6 Choice of modulation and access systems

The output SNR is one of the main factors in determining the choice of modulation technique. Other factors of primary importance are the equipment complexity, the payload electrical efficiency, the spectral occupancy and the general system efficiency, flexibility and adaptability.

If simplicity of the equipment and efficient use of spectrum were the only significant requirements, the natural choice would be AM. However, currently, this is rarely, if ever, the final choice.

Wide-band FM allows a specified S/N_0 to be achieved with considerably less power than is required for AM. Moreover, unlike in AM, the transmitter may be operated in a more efficient non-linear mode. Wide-band FM has been used extensively in the past for transcontinental television and *multi-channel per carrier (MCPC)* telephony but today, the approach is to employ digital methods. When many small Earth stations or mobile stations are involved, it is usual to employ PSK with a *single channel per carrier (SCPC)*. As with FM, the use of PSK results in a constant amplitude signal, hence it does not rely on strict linearity to avoid signal distortion, as would be the case in AM. The composite signal formed by adding several individual carriers is not itself of constant amplitude. Distortion of the amplitude waveform by a non-linear circuit results in the generation of *intermodulation (IM) products* whose frequencies are related to (but not the same as) the input frequencies (see Section 12.3.2). It is usually necessary to operate the transmitter at reduced efficiency in order to keep the in-channel carrier-to-intermodulation power ratio (C/I) above a specified value—typically 15 to 20 dB—when the transponder is fully loaded.

The most effective way of avoiding IM products is to use TDMA. In this system no two signals are amplified at the same time, so there is no intermodulation even when the transmitter is operated in a highly non-linear mode. The increased transmitter efficiency is partially offset by the need to transmit extra pulses for synchronization, for channel identification and to give the demodulators time to lock on to each signal in turn.

When the signals originate in analogue form, TDMA suffers from the fact that PCM requires more bandwidth than analogue modulation. However, there is a compensation in the much greater flexibility of TDMA in switching and routing signals and in multiplexing signals of differing bandwidths and differing types in a variety of configurations. For instance, satellites of the Intelsat VI series employ satellite-switched TDMA (SS/TDMA). A high-speed microwave switch matrix rearranges the beam-to-beam connections between TDMA bursts. Thus routing is possible on a channel-by-channel basis.

12.2.7 Radio propagation

Radio waves consist of a system of electric and magnetic fields that travel through free space at a velocity of about 3×10^8 m/s. The transit time for a round trip to a satellite in geostationary orbit and back is thus about 0.24 s—much greater than the propagation delays normally encountered in communication links. For many systems this delay is unimportant, but it cannot always be ignored in packet-switching systems (where different message segments may take different routes), in systems depending on or providing accurate radio location, or in systems involving some form of closed-loop control (for instance, to uplink power). The effect is particularly marked if two links (Earth–satellite or inter-satellite) are connected in tandem.

Polarization

In propagation through uniform isotropic media, the electric and magnetic fields are at right angles both to each other and to the direction of propagation. In a *plane-polarized* wave, the direction of the electric field lies in a fixed plane as shown in Figure 12.12(a). If a second wave travelling in the same direction has its plane of polarization at right angles to the first, then the two waves will propagate independently and indeed may carry quite different signals. A suitable (ideal) receive antenna could absorb all the power from one polarization while completely rejecting the other. The two polarizations are said to be orthogonal.

If the two orthogonal plane-polarized waves are identical, apart from a phase difference of $\pi/2$ radians, then the combination can be regarded as a different type of polarization. In this case the electric field vector at any point in the propagation path rotates, the tip (in a geometrical representation) following a circular path if the two plane-polarized waves are of equal amplitude. This is known as *circular polarization*. It may be either right- or left-handed, depending on whether the vector rotates clockwise or anticlockwise, when viewed by an observer looking in the direction of propagation. Figure 12.12(b) shows the right-handed version. More generally, if the two plane-polarized waves are not of equal amplitude, orthogonal and in quadrature, the result is *elliptical polarization*.

Left- and right-handed circular polarizations are orthogonal and may be received independently by suitable antennas (see Section 12.3.3).

Propagation in the Earth's atmosphere

For most of their journey the signals from a satellite propagate through what is essentially free space, and they travel without change, apart from the steady diminution in intensity in proportion to $1/r^2$ as the distance r from their source increases.

In the last few kilometres, however, as they pass through the Earth's atmosphere, they encounter phenomena of an unpredictable nature that may significantly affect the system performance. These atmospheric effects can be divided into two categories, those that occur in the ionosphere and those that occur in the troposphere. Ionospheric effects are very important at low and medium frequencies but in general their magnitude varies as $1/f^2$ and at the microwave frequencies used for satellite links, they can normally be ignored. Tropospheric propagation phenomena, however, can be of considerable significance. There are three main effects—refraction, attenuation and scintillation—but of these it is usually only attenuation that can cause serious problems to the system designer.

Atmospheric refraction causes a slight shift in the apparent elevation of the satellite. The magnitude of the shift depends on the elevation as well as on the atmospheric pressure

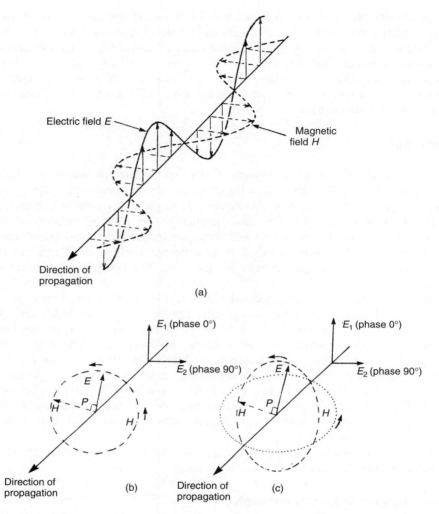

Figure 12.12 Field vectors in electromagnetic waves: (a) a 'snapshot' of a linearly polarized wave at a particular instant in time. The electric and magnetic fields are represented here by geometrical vectors; (b) representation of the E- and H-fields at a particular point, P, in space in a circularly polarized wave and of the E-fields of component linearly polarized waves with a 90° phase difference and (c) the same for elliptical polarization

and water vapour content. For a standard atmosphere, the shift is given approximately by

$$R = \frac{0.02}{\tan[e + 0.14 + 7.32/(e + 4)]} \text{ degrees} \qquad (12.16)$$

where e is the true elevation in degrees. Minor variations caused by differing climatic conditions can readily be accommodated by causing the ground-station antenna to track the satellite.

Attenuation in the troposphere has two causes. The first is molecular absorption by gases, illustrated by Figure 12.13. This shows the attenuation as functions of frequency for oxygen and water vapour, which together account for nearly all of the gaseous absorption in the atmosphere. Both curves display strong absorption bands, and although these occur at higher frequencies than have so far found extensive use in satellite communications, the water vapour band centred on 22.2 GHz will certainly be a significant effect at 20 and 30 GHz. It is also of some interest that the strong absorption by oxygen at 60 GHz makes this frequency very suitable for inter-satellite links, operating above the atmosphere that provides a screen against interference from terrestrial sources. Apart from the microwave absorption bands, the background attenuation shows a steady rise with frequency owing to the tails of absorption bands in the far infrared. This clear-air attenuation is generally negligible at frequencies below 10 GHz but is not entirely insignificant at Ku-band and above.

Much more dramatic attenuation effects are caused by rain. Water droplets both scatter and absorb radiation, the effect being strongest when the drop size is of the order of a wavelength. Thus in the microwave range, for which the wavelength is larger than a raindrop, the attenuation increases with frequency. Semi-empirical curves may be obtained showing attenuation as a function of frequency for various rainfall rates assuming reasonable models of drop-size distribution and of rain-cell size. Since attenuation by rain is very variable, the system designer must seek some way of deciding what is a reasonable performance margin to allow for the occasional deep fade. It is usual for the customer to specify an allowable outage time and of this some will be allocated to loss of signal because of rain. The designer must then attempt to predict the atmospheric attenuation that will not be exceeded for more than this allocated time. Ideally, experimentally determined statistics of attenuation may be available for the site of the ground station. Otherwise, it will be necessary to use an approach such as that provided by the ITU-R, which has adopted a general method for the conversion of rainfall statistics into predictions of attenuation [5]. The method provides attenuation values as a function of probability of occurrence, radio frequency, ground station location and satellite elevation. When local rainfall statistics are not available, use may be made of maps of rainfall rate given in Reference [6], an example of which is shown in Figure 12.14. For temperate climates, typical fade margins are in the region of 2 dB at 11 GHz and 3.5 dB at 14.5 GHz, for a link reliability of about 99.9%.

Scintillation, or rapid fluctuation in signal amplitude analogous to the twinkling of stars, is normally a small effect that for most purposes can be ignored. However, at low elevations, because of the longer atmospheric path, all propagation effects are greatly enhanced and under these conditions scintillation depths of several decibels are not unusual. As a general rule, satellite links are not operated at elevations less than about $10°$ but in some cases this cannot be avoided. The importance of scintillation is that, along with clear-air attenuation, it causes a slight reduction in capacity (and hence, in a commercial system, loss of revenue) at all times. In the long term this may be as serious as the occasional deep signal fade—especially in systems that are able to compensate for short-term fades in some of the signals by, for instance, redistributing the power allocated to individual channels or by changes in coding parameters.

Another consequence of rain is signal *depolarization*. Because raindrops are slightly flattened, they absorb one plane of polarization rather more than the other. Since the axis of the drops is in general inclined to the plane of polarization of the signal, this

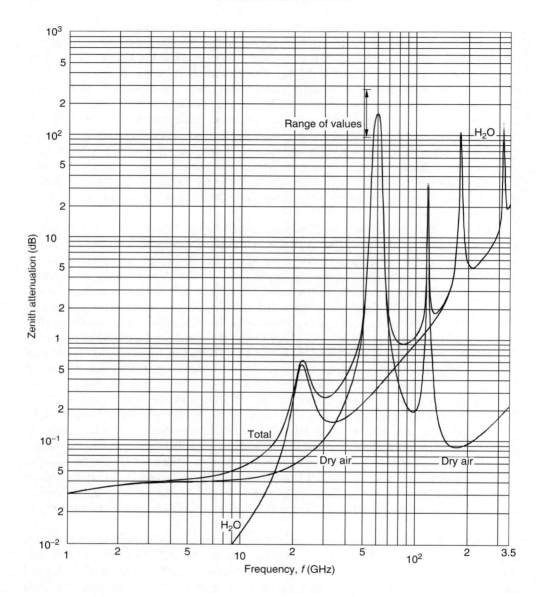

Surface pressure: 1 013 hPa
Surface temperature: 15°C
Surface humidity: 7.5 g/m^3

Figure 12.13 Total dry air and water vapour attenuation at the zenith from sea level [4] (Reproduced by permission of International Telecommunication Union)

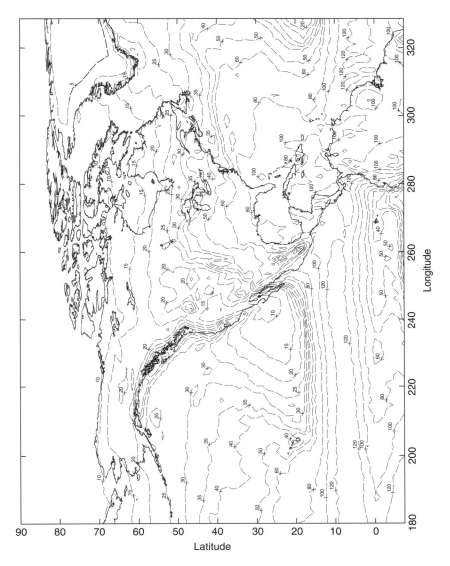

Figure 12.14 Rain intensity (mm/h) exceeded for 0.01% of the average year (Reproduced by permission of International Telecommunication Union)

differential absorption causes a slight rotation of the plane, or in the case of a circularly polarized signal, a slight ellipticity of polarization. The resulting power loss due to mismatch between the signal polarization and that of the receiving antenna is not significant. However, the presence of the orthogonal polarization, albeit at a level well below that of the wanted polarization, can have a serious impact on systems employing frequency reuse by polarization discrimination (see Section 12.2.9).

One further propagation effect that requires mention is multi-path—that is, the simultaneous presence of two or more transmission paths between the satellite and Earth terminal.

This may happen occasionally because of abnormal tropospheric conditions but it is more commonly caused by reflections from the ground or buildings. Signals arriving by the indirect path may interfere either constructively or destructively with the main signal, resulting in signal enhancement or fading. Since the conditions of interference depend on the difference *in wavelengths* between the path lengths, the transmission loss may vary rapidly with frequency across the channel bandwidth, causing signal distortion. Multi-path does not usually present any problems in the fixed-satellite service where the Earth stations are well sited and employ narrow beam antennas (which can select just one of the arrival directions). However, in the mobile services using low gain Earth terminal antennas, possibly in an urban environment, additional link margins are required to allow for multi-path (as well as for shadowing by buildings etc.).

12.2.8 Antennas

This section covers the general properties of antennas as they affect the telecommunications system. Various types of on-board antenna will be discussed later (Section 12.3.3). The description will, in general, be expressed in the terminology of transmission but will apply equally well to reception. The main properties of an antenna—such as gain and beamwidth—are the same for the two functions.

Radiation pattern and beamwidth

No antenna is strictly 'isotropic'. That is, no antenna radiates or receives signals with equal intensity (or sensitivity) in all directions. Indeed, most antennas are designed specifically so that they radiate very strongly in just one direction.

The directional properties can be represented by a polar plot of the radiated field intensity or power as a function of direction. A two-dimensional section through this pattern is known as a polar diagram (see Figure 12.15).

Referring to the highly directional pattern of Figure 12.15(b), it is seen that most of the radiation is emitted within a narrow range of directions known as the *main beam*. There are other directions in which radiation emerges, albeit at considerably lower level, known as *sidelobes*. Their importance lies in the fact that they may contribute significantly to antenna noise and to interference from, and to, other systems. Figure 12.15(b) also illustrates the concept of beamwidth, which is usually measured between −3 dB points (as shown) but

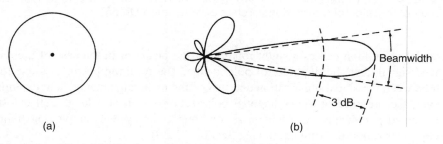

(a)   (b)

Figure 12.15 Antenna radiation patterns: (a) isotropic antenna and (b) directional antenna (large aperture)

is sometimes specified between other limits such as $-1\,\mathrm{dB}$, $-4\,\mathrm{dB}$ or between the first minima in the radiation pattern. Beamwidth is related to antenna size. Many microwave antennas consist of a large physical area, or aperture, illuminated by a single primary feed or by many small radiating elements. The most common arrangement is a paraboloidal reflector illuminated by a horn (see Section 12.3.3). The radiation pattern measured at a large distance from such an antenna is the Fourier transform of the distribution of electromagnetic field in the aperture. It follows directly from the scaling property of the Fourier transform that beamwidth is inversely proportional to antenna size. A useful rule of thumb for calculating the beamwidth is

$$3\,\mathrm{dB\ beamwidth} = 70\lambda/D \text{ (degrees)} \tag{12.17}$$

where λ is the wavelength and D the antenna diameter. This formula applies to a circular aperture with a typical distribution of illumination but it may be used as a rough guideline for most antennas of the 'aperture' type.

It is not usual to design for uniform illumination. By 'tapering' the illumination at the edges of the antenna, it is possible to achieve lower sidelobe levels as well as reduced 'spillover' of radiation from the primary feed. The consequent slight increase in beamwidth and decrease in gain (see below) is often a price worth paying.

Antenna gain

The importance of using highly directional antennas is that they provide signal *power gain* as compared with the same system using (hypothetical) isotropic antennas.

In any direction

$$\text{directive gain} = \frac{\text{actual power flux density}}{\text{power flux density from an isotropic radiator with the same total radiated power}}$$

$$\text{power gain} = \frac{\text{actual power flux density}}{\text{power flux density from a loss-free isotropic radiator with the same input power from the generator}}$$

It is assumed that the flux densities are measured at a fixed large distance from the antenna. The power gain is slightly less than the directive gain because of the power dissipated in losses in the antenna.

For a uniformly illuminated antenna with physical area A, the directive gain at the centre of the main beam is given by

$$G = 4\pi A/\lambda^2 \tag{12.18}$$

We may use the same relation to define an *effective aperture* A_e for any antenna. Thus:

$$A_e = \lambda^2 G/4\pi \tag{12.19}$$

Non-uniform illumination results in the effective aperture being less than the physical area. We may define an *aperture efficiency* by

$$\eta = A_e/A \tag{12.20}$$

For typical microwave antennas η is in the range 0.5 to 0.7. An overall antenna efficiency may be defined in the same way by using the power gain in place of the directive gain.

In a satellite system, where the coverage area is specified, the optimum antenna size depends on a trade-off between antenna gain and beamwidth. In order to optimize performance at the edge of coverage (which is usually the critical requirement in the system design) the gain in this direction should normally be about 4.2 dB less than at beam centre.

Polarization axial ratio

Most antennas are designed either for pure circular or for plane polarization but in practice the radiation is always elliptically polarized. In either case the antenna quality, from this point of view, can be expressed as an *axial ratio*. This is defined as the ratio (dB) of the powers radiated in two orthogonal planes of polarization, in which the planes are chosen to maximize the ratio.

For circular polarization the axial ratio should ideally be 0 dB. In a ground-station antenna, which points directly at the satellite, 1 dB should be achievable, but for an on-board antenna the axial ratio may be as high as 2 to 3 dB at the edge of coverage.

Because the transmitting and receiving antennas may not have their planes of maximum gain aligned, allowance must be made for possible 'polarization loss'.

In the case of a plane-polarized antenna, the axial ratio should be as large as possible in order to discriminate against signals that may be present on the other polarization. An axial ratio (or *cross-polar discrimination, XPD*, as it is usually called in this case) as high as 30 dB is not unreasonable.

12.2.9   Frequency reuse

It has already been pointed out that overcrowding of the spectrum has led to a gradual shift towards higher frequencies. However, since higher frequency usually means less transmitter power and increased atmospheric loss, there are considerable advantages in making maximum use of the lower-frequency bands. Many satellites now use the same frequency for two or more signals. This may be achieved either by spatial discrimination, in which two signals at the same frequency may be transmitted on separate narrow beams to different regions on the Earth's surface, or by polarization discrimination, in which signals at the same frequency may be transmitted over the same path on orthogonal polarizations. In the latter case, antenna imperfections and atmospheric phenomena may give rise to cross-polar interference. Although this can be troublesome, special receivers can use the signal in one channel to 'null out' the unwanted signal in the other, thus making the technique viable. The saving of space in the radio spectrum can be considerable.

12.2.10   The link budget

We are now in a position to discuss the calculation of the transmitter power requirement. We have seen that, for a given type of modulation, the overall performance depends upon the RF carrier-to-noise-power-density ratio, C/N_0, at the receiver. It remains to show how C/N_0 is related to the transmitter power.

The definition of antenna gain implies that a transmitter with output power P_T associated with an antenna of gain G_T can be replaced, for the purpose of this calculation, by an isotropic radiator with output power $P_T G_T$. The quantity $P_T G_T$ is known as the *equivalent isotropic radiated power (EIRP)*. From an isotropic radiator this power would spread out uniformly so that the power flux density (i.e. the power flowing through a unit area) at a distance r from the source is

$$S = P_T G_T / (4\pi r^2) \tag{12.21}$$

If atmospheric attenuation results in power loss by a factor L_A, then the flux density at the receiver is

$$S = P_T G_T / (4\pi r^2 L_A) \tag{12.22}$$

The effective area of a receiving antenna with gain G_R is

$$A_R = \lambda^2 G_R / 4\pi \tag{12.23}$$

Thus the signal power at the input to the receiver is

$$C = P_T G_T G_R (\lambda/4\pi r)^2 (1/L_A) \tag{12.24}$$

Finally for a system temperature T_{sys} the noise power density referred to the receiver input is kT_{sys}, giving a signal-to noise-power-density ratio of

$$C/N_0 = P_T G_T (\lambda/4\pi r)^2 (1/L_A)(G_R/T_{sys})(1/k) \tag{12.25}$$

In this expression, which is known as the *telecommunications equation*, the factor $P_T G_T$ can be regarded as a figure of merit for the transmitter and the term G_R/T_{sys} as a figure of merit for the receiving system. Leaving aside the constant $1/k$, the remaining factors refer to the propagation path. The quantity

$$L_S = (4\pi r/\lambda)^2 \tag{12.26}$$

is known as the *free-space loss*. We may think of it as the attenuation between two isotropic antennas separated by a distance r. It may seem curious that the space loss should depend on frequency but this is an artefact of the (slightly arbitrary) way in which we have separated out the factors related to the transmitter and to the receiver.

For a satellite in geostationary orbit (for which r is 3.6×10^7 m), the space loss from the transmitter to the sub-satellite point is given in decibels by

$$L_S(\text{dB}) = 183.6 + 20\log_{10} f(\text{GHz}) \tag{12.27}$$

If the ground station is not at the sub-satellite point, the increased path length results in additional space loss of up to 1.3 dB, depending on the elevation of the satellite as viewed from the ground station.

In terms of the space loss C/N_0 is given by

$$C/N_0 = P_T G_T (1/L_S)(1/L_A)(G_R/T_{sys})(1/k) \tag{12.28}$$

Table 12.3   Example of a link budget[1]

Transmitter output power (per carrier)		-14.4	dBW
Multiple carrier loss[2,3]	0.2		dB
Transmitting circuit loss[2]	0.9		dB
Transmitted carrier power		-15.5	dBW
Transmitting antenna gain		18.0	dB
EIRP		2.5	dBW
Space loss[2,4]	206.1		dB
Polarization loss[2,5]	0.1		dB
Atmospheric and multi-path losses[2,4]	4.0		dB
Total transmission loss[2,4]	210.2		dB
Ground terminal G/T		35.0	dB/K
Boltzmann's constant		-228.6	dBJ/K
Received C/N_0		55.9	dBHz
Transmitted C/N_0[6]		69.2	dBHz
Resultant C/N_0[7]		55.7	dBHz
Carrier-to-intermod-density ratio, C/I_0[8]		67.0	dBHz
Overall C/N_0 (including intermods)		55.4	dBHz
Required C/N_0[9]		55.3	dBHz
Margin		0.1	dB

[1]On the basis of the shore-to-shore link in a study of a maritime satellite system. The budget given here is for the 11.7 GHz downlink carrying high-quality voice traffic (without FEC) on a SCPC basis. Global coverage is assumed.

[2]Loss factors are shown here as numbers greater than unity (i.e. >0 dB). Thus power is *divided* by the loss factor. In dB, the loss factor is subtracted.

[3]This correction allows for robbing of transmitter power by noise and intermodulation products.

[4]For the minimum elevation of $5°$ at the ground terminal.

[5]Antenna axial ratios: satellite 3 dB; ground terminal 1 dB.

[6]From the uplink budget.

[7]The resultant C/N_0 is calculated by adding the noise powers, for example, $(N_0/C)_{tot} = (N_0/C)_{up} + (N_0/C)_{down}$.

[8]The ratio of carrier power to the power spectral density of IM products (regarded here as a type of 'noise').

[9]From the customer's requirements.

Since the required C/N_0 can be determined from the system specification, this expression allows us to calculate the required transmitter power, P_T. A typical link budget is shown in Table 12.3. (In this table the calculation starts from P_T and derives a value for C/N_0 in order to demonstrate compliance with the customer's requirement.)

It is usually convenient to work in decibels and for this purpose in Table 12.3 all quantities that are proportional to power are expressed in decibels relative to the appropriate unit. For instance, a power of 1 W is 0 dBW (decibels relative to one watt) or 30 dBm (dB relative to one milliwatt), a bandwidth of 1 MHz is 60 dBHz (decibels relative to one Hz) and Boltzmann's constant is -228.6 dBJ/K (decibels relative to one Joule per Kelvin). Note also that C/N_0, being power/(power per Hz), has the dimensions of frequency and may properly be expressed as dBHz.

12.3   THE COMMUNICATIONS PAYLOAD

12.3.1   The transponder system

Figure 12.16 is a simplified block diagram of a typical satellite repeater, which together with its associated antenna subsystem would make up a complete on-board transponder.

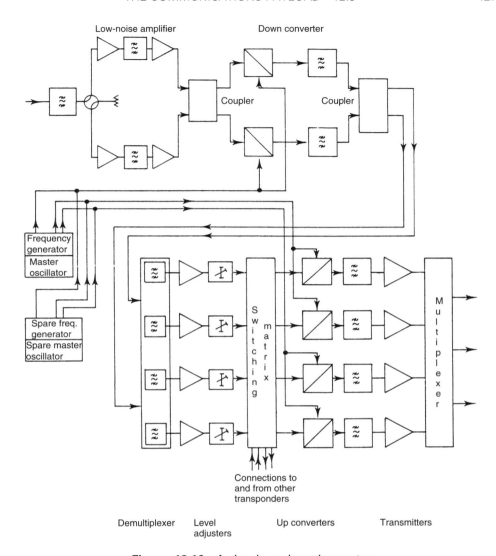

Figure 12.16   A simple on-board repeater

Depending on the purpose of the satellite, there may be just one or perhaps many such repeaters within the same payload.

Before considering any of the details of the units we shall trace the signal's path through the transponder, listing the main subsystems that it encounters and outlining their functions. The scheme adopted is a dual-conversion payload. Some alternative arrangements will be mentioned later.

The signal path

1. The *antenna subsystem's* function is to collect the incident signal power. Clearly the main requirements are that the aperture should be of sufficient size to receive a

signal of adequate strength, and that it should be pointed in the right direction (an obvious need but the source of many non-trivial problems in implementation). In a system relying on frequency reuse, the antenna subsystem must distinguish between signals on the basis of direction of arrival and polarization. Different signals on the same frequency must be presented at different output ports.

2. The *low-noise amplifier (LNA)* must amplify the weak signals arriving at the antenna to a level at which they can be processed without causing degradation of signal quality by noise in the electronic circuits. In addition, the LNA usually incorporates some preliminary filtering with the main purpose of attenuating any strong signals that may be present at adjacent frequencies.

3. The *down-converter* converts the signals to a lower frequency (the *intermediate frequency, IF*) at which much of the amplification takes place. This enables the creation of narrow-band channels that are suitable for many types of communication and may be essential for the implementation of on-board routing and processing functions. Frequency conversion also effectively eliminates the possibility of the amplifiers breaking into oscillation because of coupling between the output and the input of the repeater.

The down-converter includes filters at both its input and its output. In Figure 12.1 the input filter is shown as part of the LNA. Its purpose is to give a more sharply defined RF passband and to reject the 'image channel' (see Section 12.3.5). The output filter rejects the many unwanted frequencies that are generated in the process of down-conversion, in favour of the one wanted at the IF.

4. The *IF processor*. The first part of the processor is normally a demultiplexer or set of filters that divide the broadband output from the down-converter into a number of separate channels. Although the proliferation of equipments in a multi-channel transponder carries a penalty in mass and cost, it also has advantages such as a reduction in IM products (see below) in the high-power stages, greater flexibility in routing signals between different antenna beams and different frequency bands and the provision of 'graceful degradation' of the system (meaning a gradual reduction in performance as equipment failures occur, rather than a sudden and total loss of capacity).

The second part of the processor is a switching matrix to perform routing operations and, in the case of equipment failures, to select channels that are still working. And finally there are the IF amplifiers, one for each channel, that provide most of the transponder power gain.

5. The *up-converter* reverses the function of the down-converter by translating the amplified IF signals to the higher frequency at which they will be transmitted. Like the down-converter it produces many unwanted outputs as well as the one at the right frequency. There is consequently a need for further filtering before the signals pass on to the final amplifiers.

6. The *transmitters* (or *high-power amplifiers, HPAs*) raise the signal power to the required output level. The power amplifiers must achieve this with the maximum efficiency (minimum drain on primary satellite power) consistent with acceptable distortion. In a channelized system the signals must then pass to a multiplexer that re-combines the signal bands to form a single output band. There will then be some filtering to remove the harmonics and other unwanted out-of-band signals that are always generated in the non-linear power amplifier. If the same antenna is used

for both transmitting and receiving, the transmitter output filter and the receiver input filter may be combined in a single unit, the duplexer, which has the added function of ensuring that the incoming and outgoing signals are routed to their correct destinations.

Other payload units

There are several other units that form part of the communications payload but are not directly on the signal path. There is the local oscillator that provides continuous sine wave (CW) signals used by the down- and up-converters in order to provide the required frequency shifts. Typically the unit would provide several (perhaps many) outputs, one for each frequency converter. Possibly some or all of these will need to be harmonically related. Even where it is not a system requirement it is usually good practice to derive all the required frequencies from a single master oscillator.

In order to achieve stable gain and frequencies, highly regulated power supplies are needed. Some power conditioning may be provided in the service module but further regulation is normally provided by units within the payload. In particular, where specialized power conditioning is needed, as for instance, for a travelling wave tube (TWT) or solid-state power amplifier (SSPA), there will be a dedicated unit for each equipment.

Finally the communications payload may include part of the Tracking, Telemetry and Command (TT&C) equipment. In many cases, once the satellite is on station, the telemetry and telecommand signals are transmitted at frequencies within the communications bands. The telecommand signals are extracted at the input demultiplexer or the IF processor and may be further down-converted or partially demodulated for the interface with the TT&C subsystem. Similarly, the telemetry signals are injected into the communications downlink either in the IF processor or in the output multiplexer. During the launch and orbit transfer phases (i.e. prior to switching on the communications equipment) and also as a back-up while on station, the TT&C system uses a dedicated transponder operating at one of the standard space operations frequencies. This is normally part of the satellite bus.

Other types of transponder

A few years ago, the dual-conversion repeater was in most cases the cheapest and most effective way of providing the necessary transponder gain and implementing on-board routing and processing. However, there is a useful reduction in complexity if the up-converter is eliminated, leaving a single conversion from the receive frequency to the transmit frequency. The bulk of the amplification and the other functions of the IF processor must in that case be performed at one of the radio frequencies (usually the downlink frequency, since this is normally lower). With current technology the single-conversion scheme may be viable for downlink frequencies up to 20 GHz (Ka-band).

The system described above is known as a *transparent* repeater. An alternative scheme is on-board demodulation of the signals followed by modulation of a separate downlink carrier. This arrangement is known as a *regenerative* transponder. Its use can lead to an improvement in SNR in digital systems where uplink noise is significant. This might be the case, for instance, if very small ground stations are in use so that uplink power is limited. In a regenerative transponder, digital signals can be 'cleaned up' at baseband so that only those noise peaks that are large enough to cause bit errors will be transferred to the output. Since the BER is normally very sensitive to the SNR, there is an advantage in a system where error rates rather than noise powers are cumulative. A further advantage of a regenerative transponder is that it allows sophisticated signal processing at baseband. In addition to routing on an individual signal basis this may include, for instance, changes in the coding structure and adaptation of error correcting codes in response to changing signal transmission conditions.

The use of redundancy

As is usual in all payload systems, the communications payload includes cold spares of the most critical units. The redundancy inherent in a channelized system has already been mentioned as a source of graceful degradation. Units that are common to all or many of the signal paths, such as (in the payload illustrated in Figure 12.16) the low-noise amplifiers, down-converters and local oscillators, are usually provided with one or more spares, but where there are several parallel signal paths a lower proportion of spares may be appropriate. Wherever possible the spares are connected into the repeater using passive power splitters and combiners or hybrid couplers so that the selection of the operating unit can be effected simply by switching the power supplies on or off. The use of passive components rather than RF switches leads to greater reliability. However, since the use of any of these components involves some loss of signal (in the case of a power splitter or hybrid coupler, at least 3 dB loss) they cannot be used at the input to the receiver where the noise level is critical or at the output of the HPAs where the loss of transmitter power cannot be tolerated. In these positions low-loss switches must be used.

Mass and power budgets

One of the most important aspects of the payload system design, and a determining factor in the choice of payload architecture, is the estimation of the total mass and the power requirements. The way these quantities are apportioned between equipments shows substantial variations from one payload to another. The illustrative budgets shown in Tables 12.4 and 12.5 should not therefore be considered of general application. They are, however, broadly representative of fairly large satellites providing a mix of fixed-satellite and direct broadcast services.

The budgets given in the tables are for a 20-channel single down-conversion payload providing regional coverage. Differences between the numbers of equipments assumed in the mass budgets and those used for calculating power are due to the inclusion of redundant spares. The power budget shows the maximum demand when all 20 channels are driven at their highest power level. Particularly striking is the dominant role of the HPA in the mass of the payload and in the power consumption not only of the payload but of the entire spacecraft.

12.3.2 Payload system performance

It is the task of the payload system designers to specify the performance parameters of individual equipments in such a way that the required performance is achieved for the payload as a whole. The key payload electrical parameters are normally:

Table 12.4   Example of a spacecraft power budget

	Quantity	Consumption (W)		Dissipation (W)
		Equipment	Total	
Receivers (LNAs, down-converters, local oscillators)	2	20	20	20
Channel amplifiers	20	70	70	70
TWTs	20	3310	3310	1250
Power conditioning units (for channel amplifiers and TWTs)	20	3710	330	330
Total payload consumption			**3730**	
Spacecraft bus DC power (equinox)			340	
Losses/battery charge (equinox)			670	
Demand at power source			**4740**	

Table 12.5   Example of a spacecraft mass budget

	Quantity	Total (kg)
Receivers (LNAs, down-converters, local oscillators)	4	8.5
Demultiplexers	4	16.0
Switch matrix	2	11.0
Channel amplifiers	28	8.5
TWTAs (including isolators, output switches etc.)	28	85.0
Multiplexer	2	9.0
RF cables/waveguides	—	10.0
DC harness	—	10.0
Fixings/mounting plates etc.	—	7.0
Total repeater mass		**165.0**
Antenna subsystem		60
Total payload mass		**225**
Structure/propulsion		290
Thermal control		60
Attitude determination and control		50
Telemetry, telecommand, ranging		50
Solar array		125
Batteries		180
Power conditioning and harness		80
Balance		3
Total platform mass		**838**
Spacecraft dry mass		**1063**

Note: TWTA, travelling wave tube amplifier.

- antenna coverage area
- G/T (the ratio of receive antenna gain to system noise temperature)
- EIRP.

Other important requirements relate to, for instance:

- linearity
- group delay variation
- spurious ouputs
- incidental phase modulation
- isolation between channels.

Some of the main elements of payload system performance analysis are outlined below. It should be understood, however, that this account is by no means exhaustive.

Antenna coverage area

It has already been noted (Section 12.2.8) that the antenna coverage area and gain are closely related to the antenna dimensions (which are often determined by the available accommodation). Thus in calculating other key aspects of performance (such as EIRP and G/T) the coverage area and gain are usually regarded as predetermined parameters. Since performance requirements usually apply to the worst case, it is normally the antenna gain at the edge of the service area that is critical.

G/T

For a given antenna gain, improvement in G/T can be achieved only by reducing the system noise temperature, T_{sys}. The main contributions to T_{sys} are noise from the first amplifier stage and antenna noise due to thermal radiation from the Earth. Assuming that a 'state-of-the-art' amplifier device is used, there is little that can be done to reduce these contributions. Other contributions more amenable to the control of the designer come from later stages in the repeater, from noise generated in the transmitter and not fully removed by the output filter and from the receiver input losses.

Input losses include attenuation in the antenna cable or waveguide, loss in the input filter, loss in a test coupler (where fitted) and mismatch losses. In designing the layout of the spacecraft, the LNAs should generally be placed as close as possible to the antennas in order to minimize cable losses. Filter loss depends on the filtering requirements and on the type of construction (see Section 12.3.8). Mismatch losses are caused by reflection of signal power at interfaces between equipments whose electrical impedance differs slightly from the correct design value. Reflections from a single interface can usually be kept as low as 1 or 2% of the power but the combination of several interfaces can result in a small but not negligible loss.

Although if the LNA has sufficient gain the noise from later stages can be made small, the IF processor sometimes makes a significant contribution. This is because the first IF amplifier is often preceded by power splitters, mixers and filters, which introduce attenuation and so undo some of the effect of the high gain LNA.

Detailed calculations of the system noise temperature based on equation (12.11) or (12.12) are used to demonstrate compliance with the G/T requirement and to apportion the required gain between the various equipments.

EIRP

Since the antenna gain is essentially a fixed parameter, the EIRP is determined by the power capability of the HPA. Again, however, there is a need for careful assessment of circuit losses due to the filters (or multiplexer), the cables (or waveguide), the test coupler and impedance mismatch. The output losses vary considerably from one payload to another but typically they amount to 1 or 2 dB. They are thus responsible for roughly a 25 to 45% increase in the HPA power requirement and have a significant impact on the spacecraft primary power demand. It is desirable to minimize the cable lengths but there may be limited scope for this since the HPAs dissipate considerable amounts of heat. Their location may thus be determined by the thermal design.

Linearity

An ideal transparent transponder would employ perfectly linear elements (amplifiers etc.). Perfect linearity cannot be achieved in practice, particularly in circuits designed to handle high signal powers with good electrical efficiency. It has already been noted (Section 12.2.6) that the effect of non-linearity is to generate unwanted outputs or *IM products*. It may be shown that, in response to an input of the form

$$A \cos(2\pi f_1 t) + B \cos(2\pi f_2 t) + C \cos(2\pi f_3 t) + \cdots$$

the output of a non-linear device contains all frequencies of the form $(l f_1 + m f_2 + n f_3 + \cdots)$ where $l, m, n, \ldots$ are integers. The parameter $(|l| + |m| + |n| + \cdots)$ is known as the order of the product. Of particular importance are the third-order products of the form $f_1 + f_2 - f_3$ or $2 f_1 - f_2$. These lie inside or close to the transponder pass band and so form a type of co-channel interference. Where there are many carriers and therefore many IM products their effect is noise-like. It is usually satisfactory to estimate an intermodulation noise power density, I_0, and consider this as an addition to the noise power density N_0. (This is the procedure adopted, for instance, in the link budget, Table 12.3.) However, in estimating the cumulative effect of non-linearities in the transponder it must be remembered that successive non-linear stages produce IM products at precisely the same frequencies. Thus it is usually appropriate to add the IM products, as voltages rather than as powers. The cumulative effect even of quite small non-linearities can be significant. Nevertheless, the principal contributor of IM products is the HPA. The importance of the assessment and apportionment of IM production lies in its impact on the electrical efficiency of the HPA and hence on the primary power demand.

Another important group of IM products are those produced by the transmitter but lying within the receive band. Such products are generally of rather high order, depending on the bandwidth of the transmitter and the separation between the transmit and receive bands. If f is the lowest transmit frequency and $f + \Delta$ the highest, these frequencies combine to give a third-order product at $f + 2\Delta$, a fifth-order product at $f + 3\Delta$ and so on. If the lowest receive frequency is $f + n\Delta$, a product of this type falling within the receive band must be of order greater than $2n - 1$. The relevant orders are usually very high when the separation between transmit and receive bands is large (as in the 20/30 GHz or 11/14 GHz bands) but may be relatively low ($\sim$7th) in the case of the mobile service bands at 1.5/1.6 GHz. In general the magnitudes of IM products decrease with increasing order but in practice the levels are rather unpredictable. Occasionally a very high-order product, presumably caused by some particularly marked non-linearity and possibly enhanced by some unplanned resonance, can be the cause of serious problems.

Group delay variation

Group delay is the time delay experienced by a *modulation waveform* in passing through the equipment. (This is not necessarily the same as the delay experienced by individual RF components.) It may be shown that the group delay is equal to the slope of the phase *versus* frequency characteristic. If the characteristic is non-linear (slope varying with frequency), different components of the modulation experience different delays, resulting in distortion of the baseband signal. Consequently it is usual for the system specification to impose stringent limits on the variation of group delay across the channel bandwidth. In practice these limits apply mainly to the filters and especially to the narrow-band channel filters, which are the main source of group delay variation.

It is particularly important to achieve a good group delay characteristic in the sections preceding the main non-linear elements of the system (i.e. the HPAs). In a perfectly linear system, group delay distortion occurring in one element can be compensated by 'equalizing' circuits at a later stage. However, after the signal has passed through a non-linear device equalization is no longer possible.

Spurious products

The payload specification invariably places limits on the levels of unwanted outputs from the transponder. One aspect of payload design is the prediction and where necessary the elimination (by the inclusion of filters or by other means) of these emissions, which may cause harmful interference to users either of the same system or of other systems or may cause problems in the operation of the payload itself.

Spurious products have a variety of origins including:

- spurious mixing products at the frequency converters,
- spurious local oscillator outputs,
- intermodulation products,
- harmonics of the output frequency,
- power supply ripple components at, for instance, the switching frequency of the voltage regulators.

Although it is often possible to identify by inspection the most likely sources of troublesome outputs, there is ultimately little alternative to a systematic examination of each possible unwanted product to determine where it may get to in the payload system and what harmful effects it may cause.

Incidental phase modulation

Signals at the output of the transponder have, in addition to their original modulation, some unwanted phase modulation that is caused by the transponder itself. The main source of this 'incidental phase modulation' is spurious phase modulation of the local oscillators (see Section 12.3.6). All frequency generator outputs display low-level modulation sidebands, which normally include both continuous and discrete spectral components. The level of the continuous component (the phase noise) is usually proportional to the square of the carrier frequency. Components very close to the carrier (typically within about 1 kHz) generally originate in the master oscillator and are thus common to all outputs from the frequency generator. Their contributions at different frequency conversions add as voltages and their effect at the down-converters is partially cancelled by that at the up-converters. Further from the carrier the sidebands may originate in separate high-frequency oscillators. These sidebands add as powers. The discrete modulation components also may be generated by different mechanisms at different outputs. Thus in order to apportion the allowable incidental phase modulation between the various stages of the transponder, the system designer usually needs to know or to define the architecture of the frequency generator.

Channel frequency characteristic

Variations in equipment gain and phase response across the channel bandwidth can lead to distortion of the signal. Consequently the cumulative values of these variations for the whole transponder must be kept under careful control. The main contributions normally arise in the filters in the narrowband channelized section of the repeater but there may also be contributions from other equipments such as the input and output filters—particularly for channels near the edge of the overall passband. Filter characteristics are discussed further in Section 12.3.8.

12.3.3   The antenna subsystem

The antenna subsystem is often a critical factor in the spacecraft design because of its impact on total mass and stability, the possible need for stowage during launch and erection in orbit and the requirement for Earth pointing, if necessary by the provision of a de-spun platform. From the point of view of communications, the first constraint on antenna design is the required coverage area that determines the beamwidth and hence the antenna size (Section 12.2.8). Table 12.6 gives antenna diameters for three down-link frequencies and for two extremes of service area: Earth coverage, corresponding to beamwidth of 17° from GSO and a (directive) gain of 18.5 dB (at the edge of coverage), and a 'spot' beam covering, say, the British Isles, with a beamwidth of about 1.5° and an edge-of-coverage gain of 40 dB. It is seen that for large coverage areas (comparable with Earth coverage) and especially at the higher frequencies, antennas of quite modest size are sufficient.

There are several possible advantages in the use of spot beams:

- There may be a requirement for a strictly limited coverage area in order, for instance, to avoid interference with other services.

Table 12.6 Spacecraft antenna diameters from geo-stationary orbit

	Diameter (m)	
Frequency	Earth coverage	UK coverage
1.5 GHz (L-band)	1.0	11.0
4.0 GHz (C-band)	0.37	4.1
12.5 GHz (Ku-band)	0.11	1.3
Beamwidth	17°	1.5°
Gain (at edge of coverage)	18.5 dB	39.8 dB

- High antenna gain may be needed in order to serve very small ground terminals. The coverage area may in this case be served by many spot beams.
- More efficient use can be made of the available transmitter power if areas of high traffic density are served by high gain antennas (requiring relatively little transmitter power per carrier).
- Spot beams may enable frequency reuse by spatial discrimination.

The trade-off to decide on the number and size of the spot beams is clearly very complicated and involves many different aspects of the overall system design—the spacecraft structural design, antenna deployment, attitude control, power supplies, thermal control, and so on, as well as telecommunications capacity and market prediction.

Antenna types

1. The *horn antenna* can readily provide the small aperture needed for Earth coverage at 4 GHz or higher frequencies. In its simplest form it is a section of rectangular or circular waveguide spread outwards at the end to give the required aperture dimension (Figure 12.17a). In order to give improved radiation patterns, this simple structure is often modified by, for instance, the inclusion of steps or corrugations in the flared section.

2. *Helical antennas* (Figure 12.17b) are 'end-fire' antennas, preferred to the horn at frequencies below 4 GHz, since even for Earth coverage a horn antenna would be excessively large. For instance, Navstar uses helical antennas at 1.5 GHz to give Earth coverage. Their use is generally limited to rather wide beams and gains less than about 14 dB. They are also restricted to circular polarization.

3. *Reflectors*, such as a paraboloid illuminated by a horn, are usually the most satisfactory solution when a narrow beam is required. The most usual configurations are the front-fed arrangement where a waveguide runs through or round the reflector to the focus of the paraboloid (Figure 12.17c), and the offset feed in which the reflector is a segment of a paraboloid taken from one side of the axis (Figure 12.17d). The front-fed arrangement has the disadvantage that the feed and its support structure cause blockage of the beam, reducing the gain and scattering power into the sidelobes. Moreover, the offset arrangement is often mechanically better, since the feed horn can be mounted rigidly on a face of the spacecraft. On the other hand

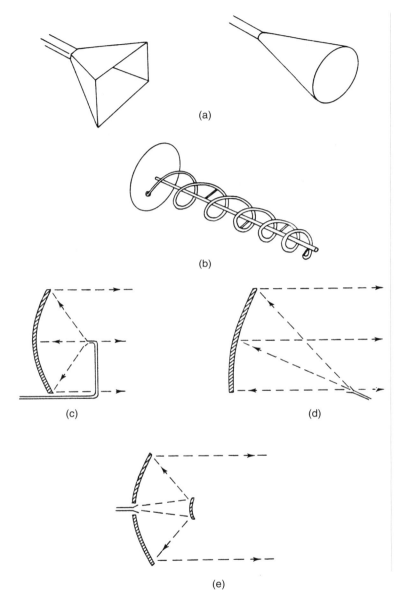

Figure 12.17   Antenna types: (a) horns; (b) helical antenna; (c), (d),
(e) paraboloidal reflector antennas; (c) front-fed; (d) offset feed and
(e) Cassegrain

its lack of symmetry results in poorer XPD when linear polarization is in use, and
in the case of circular polarization causes beams of opposite polarizations to point
in slightly different directions. The Cassegrain system (Figure 12.17e), which is
common on ground stations, is not often used on the satellite because blockage by
the sub-reflector usually makes it unsatisfactory for a small antenna.

In some cases, where the service region is irregular in shape, there may be an advantage in forming an appropriately 'shaped beam' which, in a reflector antenna, is achieved by replacing the single primary feed by a cluster of horns. The reflector must be large enough to form the smallest features of the beam pattern.

4. *Phased arrays* are based upon the principle illustrated in Figure 12.18. The aperture is excited by many separate radiating elements that individually have only very weakly directive properties. Their combination may, however, have a very narrow beam because the radiation in some directions interferes constructively and in others destructively. The figure shows a one-dimensional array but the principle can readily be extended to the two-dimenisonal case.

The advantages of this arrangement are that one array can produce a large number of beams simultaneously and that these can be steered electronically over a rather large angular range without the need for mechanical pointing systems. The distribution of transmitter power between many output paths reduces the demand on any one-power amplifier and facilitates the use of solid-state devices. Finally, the array can be mounted rigidly and can often be made to conform to some convenient surface. The major disadvantage is the complexity of the associated equipment. Each equipment in a signal path to or from an array element must have a precisely controlled phase and amplitude characteristic and this must be maintained over a wide range of temperature, signal level, and so on. Since the number of elements may be large (possibly 100 or more) the programmes for manufacture, alignment and testing assume formidable proportions.

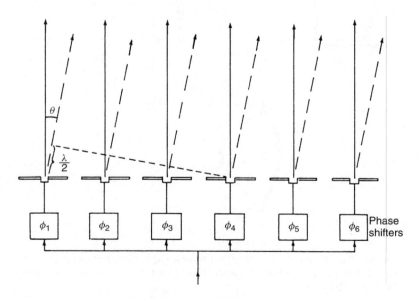

Figure 12.18 Phased array antenna. If all of the phase-shifts, ϕ_1, ϕ_2, ... are equal, there is constructive interference in the broadside direction. At a small angle θ from this there is no emission since radiation from each element interferes destructively with that from an element three spacings away (for the six-element array shown here). By changing the ϕ_i, the beam can be steered electronically without any movement of the antenna

12.3.4   The low-noise amplifiers (LNA)

The primary requirements for the LNA are a low-noise temperature and sufficient gain to ensure that the noise contributions from the succeeding stages are small. A variety of amplifier types have been used in the past, including travelling wave tubes (TWT), tunnel diodes, parametric amplifiers and transistors.

In most current designs a transistor amplifier is chosen for the 'front-end'. In the first stage of amplification, in which the highest possible performance is required, an individual transistor would normally be used. For the later, generally less critical stages, monolithic microwave integrated circuits (MMICs) are now widely used. These have an advantage both in reduced size and mass and in improved reliability and reproducibility.

One further requirement for the LNA is good linearity over a wide dynamic range. Since the amplifier handles a multi-carrier signal (it is placed ahead of the demultiplexer that separates the signals into channels) any departure from linearity can give rise to IM products (see Section 12.3.2), which cannot be removed by subsequent filtering.

12.3.5   Frequency converters

Any non-linear device will, in principle, serve as a frequency converter or mixer since in response to an input containing frequencies f_1 and f_2 the output will contain every frequency of the form $nf_1 \pm mf_2$ (see Section 12.3.2). Usually the required output frequency is $f_1 + f_2$ or $f_1 - f_2$. Most conventional mixers depend on resistive diodes and these are usually arranged in a symmetric configuration, either singly or doubly balanced, which suppresses some of the unwanted outputs and provides better isolation between ports. (In a balanced mixer the signal is applied at one point in the circuit and the local oscillator at another, ideally isolated, port.)

Best performance is achieved when the local oscillator waveform is at a much higher level than the signal, so that the signal is in effect switched on and off at the local oscillator frequency, f_{Lo}. As a general rule the local oscillator level should be at least 10 dB above the maximum signal level in order to minimize the *conversion loss* (the reduction in signal level from input frequency to output frequency), and to keep the level of IM products as low as possible.

A mixer produces many output frequencies in addition to the wanted one. It is also sensitive to frequencies other than the expected input frequency. In particular, in a down-converter, if the input frequency is $(f_{LO} + f_{IF})$, then the mixer also responds to $(f_{LO} - f_{IF})$, which is down-converted to the same IF (and *vice versa*). This is known as the *image response* of the down-converter. Noise and interfering signals in the image channel must be filtered out prior to down-conversion. The levels of the various mixing products for a typical single-balanced mixer are shown in Table 12.7.

The noise temperature of a mixer is closely related to the conversion loss. To a reasonable approximation a down-converter can be regarded as an IF amplifier fed through a resistive attenuator. Semiconductor noise generated by the diodes can usually be made relatively insignificant. Thus the noise figure of the down-converter (in decibels) is given roughly by the sum of the IF amplifier noise figure and the conversion loss. Noise figures in the region of 2 dB are possible for mixers operating at 14.5 GHz

Table 12.7   Spurious outputs of a typical singly balanced mixer

m	$n =$	0	1	2	3	4	5	6	7	8
0			36	40	45	55	55	56	56	55
1		28	0	30	12	35	33	40	35	45
2		60	65	70	65	80	60	75	70	100
3		75	60	65	70	70	55	75	55	75
4		80	80	95	85	95	85	95	85	90
5		85	80	85	70	90	65	90	65	85
6		95	90	95	90	100	95	100	90	100
7		100	95	100	90	95	100	100	85	95

Note: The entry in row m and column n gives the level of the product at frequency $mf_{in} \pm nf_{LO}$. For $m > 0$ the levels are expressed in decibels below the level of the wanted output at $f_{in} \pm f_{LO}$. For $m = 0$ the levels are in decibels below the local oscillator input level. This table is applicable for a reference signal input level $P_{ref} = -40$ dBW and for a local oscillator input -23 dBW. At a different input level P_{in} (dBW) the levels of spurious products are increased, relative to the ($f_{in} \pm f_{LO}$) level, by $(m - 1)(P_{in} - P_{ref})$ dB.

12.3.6   Local oscillators

Apart from the frequencies and signal levels, the two most significant aspects of a local oscillator's performance are the frequency stability and the phase noise. The specification for long-term stability usually limits the frequency drift to less than about 1 p.p.m./year, and this requires a crystal controlled source. Since crystal oscillators operate at less than about 150 MHz, frequency multiplication is necessary. A non-linear device such as a varactor diode or step-recovery diode may be used to generate harmonics, the appropriate microwave frequency being extracted either by filtering or by phase-locking a high-frequency oscillator to the correct harmonic (see Figure 12.19).

A complex communications payload may require many local oscillator frequencies that are not necessarily related by simple ratios. In this case frequency synthesis techniques such as that shown in Figure 12.19(c) are used. The incorporation of a divider in the phase-locked loop allows frequency multiplication by a large (and possibly programmable) number while the mixer allows the loop frequency, f_1, to be inserted as an offset from a second reference frequency, f_2. Both f_1 and f_2 would in turn be derived from the crystal controlled master oscillator frequency. Since in the process of frequency generation many frequencies are produced in addition to the wanted one, careful filtering is required. The unwanted frequencies often appear as discrete low-level phase-modulation sidebands on the final local oscillator output.

Short-term random frequency jitter can be regarded either as a noiselike phase modulation of the oscillator or as low-level continuous sidebands in the spectrum of the oscillator output (see Section 12.2.2). When the oscillator is used in a frequency converter, this phase noise is transferred to the signal as phase modulation with the same index. The oscillator sideband level is highest close to the carrier frequency. Thus phase noise has most effect where low baseband frequencies are important—particularly where low-rate digital signals are to be transmitted.

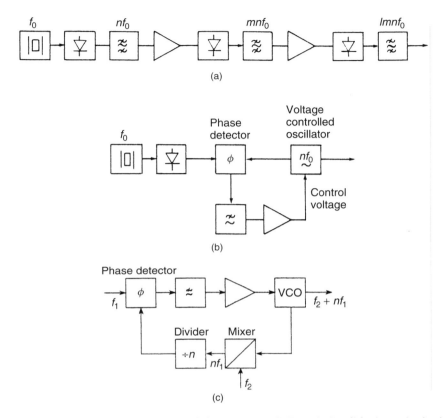

Figure 12.19 Frequency multipliers: (a) direct multiplier chain; (b) phase-locked loop multiplier and (c) synthesizer loop

12.3.7 IF processors

The last decade has witnessed considerable advances in *digital signal processing (DSP)* payload technology. DSP enables a number of payload functions to be performed including signal modulation/demodulation, channelization, channel power control, TWTA linearization, antenna control and autonomous packet switching. This results in a flexible approach to the service provision to be employed, a key requirement in the digital era, in which Internet access and Digital Audio and Video transmissions are likely to vie for satellite resources. Current DSP technology is capable of supporting up to 2000 DSP application specific integrated circuits (ASICs) per payload, with a throughput of 10 Gbit/s.

Where DSP payloads are not employed, the main functions of the IF processor are to provide most of the transponder gain, to define the frequency response of each channel and, where necessary, to perform beam-to-beam routing functions. Thus the principal components are filters, transistor amplifiers and possibly electronic switching circuits. Where the routing requirements call for the division of the spectrum into many narrowband channels, a rather low intermediate frequency would be chosen. At intermediate frequencies less than about 200 MHz, the best choice of filter in terms of bandwidth, frequency stability and weight is almost always some type of crystal or surface acoustic wave (SAW) device. For broader channels and at higher frequencies, helical resonators provide a satisfactory performance while at frequencies above 2 GHz, metal rod (coaxial) resonators or a thin-film technique would probably be used. (This is a method of forming microwave circuits by etching a metal film deposited on a substrate of high dielectric constant.)

The amplifier blocks would normally use bipolar transistors except at the higher frequencies. Printed circuit board or thick-film construction (which depends on the deposition of conducting materials by a silk screen printing technique) could be used at frequencies up to 2 GHz. Above 2 GHz Field Effect Transistors (FETs) may be used in thin-film construction.

12.3.8 Filters

The need for RF filters at various points in the transponder has already been noted. Most of these can be described by one of the labels, 'low-pass', 'high-pass', 'band-pass' or 'band-stop' (see Figure 12.20).

A typical RF filter consists of a series of resonant elements (or reactive elements for low- and high-pass filters) with precisely controlled electrical coupling between them. The most important electrical characteristics are

- the attenuation within the passband,
- the rate of roll-off (increase of attenuation with frequency outside the passband),
- the flatness (lack of amplitude variation) of the in-band response,
- the phase linearity or group delay characteristic.

Some of the more important RF filters in the transponder are listed below.

- The input filter: band-pass; main requirements, low in-band loss with adequate attenuation out of band.
- The channel filters: band-pass; main requirements, rapid roll-off together with moderate loss and a sufficiently flat in-band amplitude and group delay response.

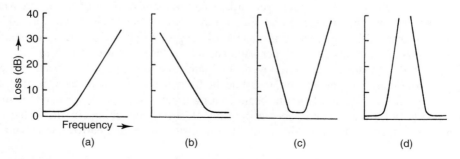

Figure 12.20 Filter types: (a) low-pass; (b) high-pass; (c) band-pass and (d) band-stop

- The output (harmonic) filter: low-pass or band-pass; main requirements, low in-band loss with adequate attenuation at specified higher frequencies.
- The multiplexers and demultiplexers. In general these can be regarded as combinations of band-pass and band-stop filters. There may be a requirement for very rapid roll-off if the channels are closely spaced. The multiplexer, coming after the high-power amplifier, must have low in-band loss but this requirement is usually less stringent for the demultiplexer.

There is much in common in the procedures for the design of the various types of filter. Any of the types shown in Figure 12.20 can be synthesized initially as a low-pass design, followed where necessary by the application of standard rules to transpose the impedances to those required for a high-pass or band-pass characteristic.

Standard response curves and tabulations of component values are available for low-pass prototype filters optimized according to various requirements. Two commonly used criteria are maximally flat in-band response (Butterworth filter), and maximum roll-off for a given in-band amplitude variation (Chebyshev filter). These characteristics can be obtained by including coupling only between adjacent filter sections. If coupling between non-adjacent sections is allowed, other responses are possible of which the most important is the elliptic filter response. This has a very sharp roll-off to attenuation maxima just outside the passband.

The theoretical response curves can be used to determine the number of sections (resonators) needed to achieve the required roll-off while maintaining adequate flatness in-band. It is then possible to predict other filter properties such as group delay response and in-band attenuation.

The attenuation in the passband depends mainly on the number of sections, n, and on the so-called quality factor, Q, of the resonators (or reactive elements for a low-pass or high-pass type). For a band-pass filter with a bandwidth B, the attenuation at the centre frequency f is given approximately by

$$L(\text{dB}) \approx 4.8(Q_l/Q_u)n \qquad (12.29)$$

where $Q_l = f/B$ and Q_u is the 'unloaded' Q of the resonators. Q_u depends on the method of implementation and may be typically of the order of 50 to 100 for microstrip quarter-wave lines, 2000 for quarter-wave resonators consisting of metal rods mounted in a box, or 8000 for rectangular waveguide cavities. Higher Q values can be achieved in large cavities. Thus low in-band loss, which implies high Q, may be bought at the cost of increased size and mass. Conversely, the cavity size may be reduced by incorporating some dielectric material. 'Dielectric resonators' can have Q values comparable with those of waveguide cavities.

In designing a microwave filter for a space application, it is important to allow adequate margins for temperature variations. The main effect is a shift of centre frequency that for aluminium filters may be as much as 1 part in 10^3 over a 50 °C temperature range. For Invar construction the corresponding variation is about 1 in 10^4 but there is a significant mass penalty. Alternatively the filter design may incorporate temperature compensation based on the use of two materials with different coefficients of expansion. Many filters are now made of carbon fibre based materials. These have both low mass and low thermal expansion but their use in some applications is limited by difficulties in manufacture.

12.3.9 Transmitters

There are only two generally accepted means of amplifying the signals to the level required for downlink transmission. These are the TWT and power transistors. The TWT is the most commonly used but as transistor technology improves, solid-state amplifiers are finding an increasing number of applications, mainly because of their higher reliability and lower mass. TWTs, however, are currently the only choice of transmitters at high power levels, particularly at the higher frequency bands.

Travelling wave tube amplifiers (TWTA)

In a TWT, amplification is achieved by interaction between an electron beam and a signal in the form of an electromagnetic field travelling along an elongated guiding structure. For a significant effect, the electrons and the signal must travel at almost the same velocity, which implies that the signal must be slowed down very considerably by the guiding structure. In low- and medium-power tubes the 'slow-wave structure' is a helix of wire held in place by ceramic rods. The signal modulates the density of the beam and the beam in turn transfers energy to the signal, which increases exponentially in amplitude as it travels along the tube to the output port. Helix tubes are capable of up to about 200 W output. For higher powers a more robust slow-wave structure may be required, such as a series of electrically coupled cavities. However, for most space applications, including direct broadcast television, a helix tube is preferred.

A disadvantage of a TWTA is the requirement for a complicated high-voltage power supply (see Figure 12.21). The accelerating voltage is several kilovolts and must be very well regulated. The efficiency of a TWTA can be high—as much as 60 to 65% for the tube itself or about 55 to 60% including the power supplies.

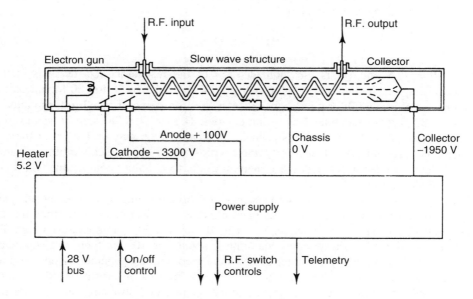

Figure 12.21 Simplified TWTA configuration

TWTs can now be made sufficiently reliable for most missions but they do suffer from a gradual deterioration in performance due to loss of cathode emission during their lifetime. In some cases a facility is provided for a telecommandable increase in heater current in order to restore performance. Alternatively, the correct cathode current may be maintained by a control loop that varies the anode or control grid voltage.

Though attractive from the point of view of gain, efficiency and power output, a TWT is a rather non-linear amplifier. When amplifying a multi-carrier signal, it both generates IM products and converts signal amplitude variations into spurious phase modulation. More linear operation can be achieved by 'backing-off' the tube to a lower power level, but this also results in a loss of efficiency.

It is becoming usual where a multi-carrier operation is required to incorporate a *linearizer* in the TWTA. This is a non-linear driver amplifier that pre-distorts the signal in such a way that the non-linear TWT characteristic is partially compensated. The tube may be operated closer to its maximum power (saturation) level (keeping the efficiency high) whilst IM product levels remain low.

Transistor power amplifiers

Transistors are now available that are capable of providing an output in the region of 2 watts at Ka-band up to about 60 watts at L-band. It is possible to combine the power from two or more transistors in order to extend their range of application.

When compared with the equivalent TWTA, a SSPA has lower mass, higher reliability and lower cost. There is also the significant advantage of eliminating the need for the TWT's complicated high-voltage power supply. The simpler, low voltage supply required for an SSPA affords a further improvement in mass and reliability. Transistors are, however, at a disadvantage with respect to efficiency. The microwave power at the input to a transistor is a significant fraction of the output power, so that in computing the efficiency, the power requirements of the driver stages cannot be ignored. For an L-band (1.5 GHz) transistor amplifier under a multi-carrier operation, the overall efficiency might be around 25 to 30%, appreciably less than for a saturated TWTA. But the efficiency of a TWTA decreases rapidly with back-off. If good linearity is required, the transistor amplifier may in some circumstances prove to be more efficient than a TWTA.

The employment of phased arrays allows the use of distributed power amplifiers, where each individual SSPA has a fairly modest output. As in the case of an LNA, the critical stages (in this case the output stages) are implemented using individual transistors but, in order to reduce size and mass, the pre-amplifier and driver stages and any necessary control functions may be realized in MMIC technology.

Transistors are also gradually replacing TWTs in C-band transmitters for the trunk telephone and television service. The output per transponder is typically a few watts to a few tens of watts. Powers up to about 30 W are within the present capabilities of SSPAs at 4 GHz and X-band SSPAs are planned for use in the fairly near future.

Passive intermodulation and multipaction

Having achieved sufficient power from the final amplifier, the designer faces two further problems associated with the output circuits and antenna.

Although these assemblies contain no active devices such as transistors or vacuum tubes, they may display a slight electrical non-linearity caused by magnetic materials or imperfect contact between metal surfaces contaminated with dirt or a thin layer of oxide. The resulting passive IM products (PIMs), though very weak compared with the transmitted signal, may be comparable in power with the received signal. If IM products within the receive band are generated in a unit (e.g. antenna, cable or duplexer), which is common to the transmit and receive paths, they cannot be removed by filtering. They must be controlled by extreme care in manufacturing—that is by extreme cleanliness, suitable choice of materials, very precise machining and high contact pressures between mating surfaces.

The second problem is a type of electrical breakdown that occurs only in high vacuum and at high frequencies (UHF and microwaves). *Multipaction* is caused by acceleration of electrons by an RF field in a gap between metal surfaces. An electron arriving at one of the surfaces may cause secondary emission and for certain combinations of frequency, voltage and gap size an electron avalanche can be produced. The resulting leakage current causes IM products, broadband noise, and in extreme cases, physical damage. Figure 12.22 shows an example of the conditions under which theory predicts the possibility of multipaction between parallel metal plates. In practice the multipaction zones are found to extend somewhat beyond these theoretical limits and the onset of the phenomenon is sensitive to surface finish and contamination. Thus a generous design margin must be allowed. Where units cannot be configured to avoid operation in the multipaction region special

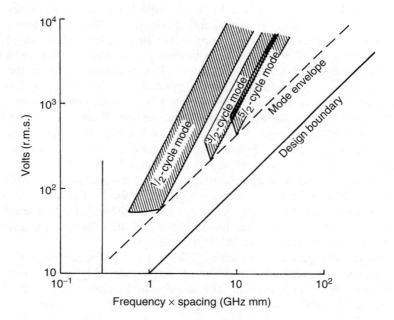

Figure 12.22 Multipaction between metal surfaces. The shaded areas show a typical theoretical arrangement of multipactor regions for parallel plate geometry. Regions are shown corresponding to modes in which there are one, three and five reversals of field during the transit time of an electron (Reproduced by permission of European Space Agency)

precautions must be taken, such as filling with a dielectric foam or pressurizing with enough inert gas to prevent the formation of an electron cascade.

12.4 CONCLUSION

A modern communications satellite is a far cry from the simple transponders of Telstar and Early Bird. What seemed in the 1960s, a bold and imaginative step is now a mature technology. Nevertheless, it is a technology that continues to develop both under the stimulus of advances in component technology and in response to new markets brought about by the digital era. Geostationary satellites continue to increase in size and power, providing, in addition to traditional telephony and TV transmissions, mobile and Internet services. Technology advances have not been limited to the geostationary satellite. The introduction of Satellite Personal Communication Network services in the latter part of the last decade has demonstrated the viability of using the LEO, although limited market take-up may have dampened the enthusiasm for this type of solution. The less sophisticated 'little' LEO satellites have also been introduced to provide services such as remote monitoring and tracking. However, in terms of non-geostationary satellites, it is the navigation services that are, at present, the most successful.

As the service requirements of the 21st century citizen continue to evolve, the unique opportunities offered by satellites will ensure that this form of delivery will carry on playing an important role in the provision of telecommunication services for the foreseeable future.

REFERENCES

[1]   Clarke, A. C. (1945) Extra-terrestrial relays, *Wireless World*, **51**, 10.
[2]   FINAL ACTS WRC-2000, World Radiocommunication Conference, ITU, Geneva.
[3]   Radio Regulations (1998) ITU, Geneva.
[4]   Recommendation ITU-R P.676-4 Attenuation by Atmospheric Gases (2001) ITU.
[5]   Recommendation ITU-R P.618-7, Propagation Data and Prediction Methods Required for the Design of Earth-space Telecommunication Systems (2001) ITU.
[6]   Recommendation ITU-R P.837-3, Characteristics of Precipitation for Propagation Modelling (2001) ITU.

The reproduction of Figures 12.13 and 12.14 is made with the prior authorization of the International Telecommunication Union as copyright holder. The complete publication of the CCIR from which the figures reproduced in the present publication are extracted can be obtained from:

ITU General Secretariat
Sales and Marketing Division
Place Des Nations CH-1211 Geneva 20
Switzerland

TELEMETRY, COMMAND, DATA HANDLING AND PROCESSING

13

Nigel P. Fillery[1] and David Stanton[2]

[1]*Astrium Ltd, Portsmouth, UK*
[2]*Keltik Ltd, Hampton Hill, UK*

13.1 INTRODUCTION

The *telemetry, command, data handling and processing* functions provide for the two-way flow of information between a spacecraft and its ground control station(s). There are transmission (downlink) and reception (uplink) functions to perform, as well as the tasks of gathering and processing data ready for transmission and the processing and routing of command data from the ground receiver. In addition, there will be a transponder for ranging, and support services for the payload.

The mission, orbit, type of payload and the selected ground control station(s) play a considerable part in determining the nature of the design. This is illustrated by studying *Intelsat* and *Inmarsat*, in Geostationary Earth Orbit (GEO) with communication payloads; *Earth Resources Satellite* (*ERS*), *Envisat*, *METOP* and *TerraSAR*, in a near-polar Low Earth Orbit (LEO) with passive and active sensors to monitor the Earth; and the *International Space Station* (ISS), in a non-sun-synchronous LEO.

The *spacecraft/ground station link* from GEO can be a continuous direct one, with the spacecraft visible at all times from an appropriate ground station. This allows for a relaxed transmission link compared with that from a non-geosynchronous orbit. Envisat, for example, is visible to its ground station for only a few minutes at each ground pass. It therefore needs on-board data storage and a rapid data-transfer link. The ISS requires a continuous link, and this can be provided by a two-way link with any ground station

Spacecraft Systems Engineering (Third Edition). Edited by P. W. Fortescue, J. P. W. Stark and G. G. Swinerd
© 2003 John Wiley & Sons Ltd

in sight, and/or by a relay system [Tracking and Data Relay Satellite System (TDRSS)], which makes the ground link via special-purpose spacecraft in GEO.

Spacecraft destined for GEO will be non-synchronous in their early phases, such as during launch and intermediate orbits, and will require special ground support before they are handed over to their dedicated ground controllers. The Intelsat spacecraft use a worldwide Launch Support Network operating on C-band, both during launch and subsequently. Other spacecraft may use the ESA network operating in S-band for initial orbits, with control being transferred to a dedicated station operating in Ku-band when it is on-station.

The *telemetry downlink* must provide the ground control team with information about the functioning of the subsystems in the craft, so that they may detect whether it has its correct orientation, or whether any fault has developed, and if so, they must be able to diagnose its cause. It may also be the channel for passing information from the payload to the ground when the mission is scientific or Earth-observation. When the payload comprises communications equipment, that equipment will serve as an alternative route for the telemetry data, once it is deployed.

The *command uplink* must enable the ground controller to change the role of the spacecraft, such as to reorientate it, to correct a fault, to operate a mechanism, or for other reasons. It must do so in a highly reliable way, giving confirmation that the instructions have been carried out.

The *ranging transponder* forms part of the system by which the ground controller tracks the spacecraft and determines its orbit.

The *payload* may require significant control, data handling, data storage and processing functions. As spacecraft designs evolve towards autonomous operation, the bus itself may also require extensive data processing and storage functions.

13.2   SYSTEM ARCHITECTURE

All spacecraft need most of the functions described in this chapter. However, the physical distribution and implementation vary considerably.

13.2.1   Relationship with other subsystems

The context of the telemetry, command and data-handling subsystem with respect to the other main subsystems is illustrated in Figure 13.1. Whilst it is usual to consider any subsystem under discussion as the centre of the system, in this case it is true, for it is the digital system that spacecraft operators and users 'see' and interact with.

Commands are received via, typically, an S-band link, decoded and placed in a queue for either internal distribution or distribution to the other subsystems. The verification of each command is achieved by the feedback of telemetry, usually from each stage in the command distribution and after the operation of the command. This telemetry is checked to verify that all is well at each stage. The operation of all of the other subsystems is achieved in this way. Some subsystems are not demanding in terms of processing requirements but are still critical to the mission, for example, deployment. Others are very demanding but not so critical, for example, instruments. Separate downlinks, at X- or Ka-band, may be

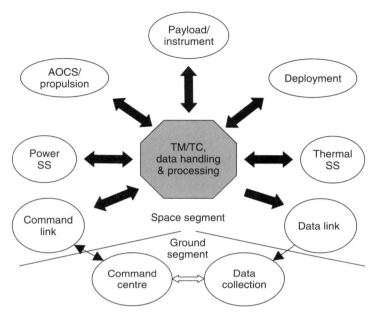

Figure 13.1   System context

provided for the payload data, particularly when the data rate is too great for the normal telemetry link. The system protocol interacts with an on-ground command centre, and sometimes a separate data-collection centre. The system usually operates in a number of different modes—ranging from a basic 'safe' mode, in which the spacecraft is kept alive from a command, thermal and power point of view, up to a fully performing platform and payload.

13.2.2   Avionics

The term *Avionics* is often loosely used. It can apply just to the electronic part of the Attitude and Orbit Control System (AOCS) subsystem but sometimes it is used to cover all of the platform electronics hardware and software, that is, elements for the AOCS, spacecraft-level telemetry/telecommand (TM/TC), data communications, data processing, data storage and the transponder. In a modest spacecraft, all of the processing for these functions may be in a single on-board computer. In large spacecraft, a number of distributed and sometimes different processors and devices are used. Avionics does not usually include the payload electronics such as that required by instruments and experiments.

13.2.3   On-board data-handling functions

The emphasis on the data-handling functions required varies according to the mission type. A typical communications satellite only has requirements for basic telemetry and telecommand functions in order to maintain the health and position of the platform on which the payloads are situated. Communications satellites with signal processing capability may need a little more interaction with the ground.

For all other mission types, the primary product of the mission is in the data, which is sent to the ground via telemetry. These missions also tend to operate in a number of modes according to mission phase or data product required, and thus involve more interaction with ground operations. The data-handling systems are therefore more directly involved with the mission product generation and tend to be more complex than for communications satellites.

Typical functions that can be required of the on-board data-handling (OBDH) system include the following:

- Enabling the flow of housekeeping and science data.
- Receiving and distributing commands.
- Performing telemetry and telecommand protocols.
- Time distribution around the spacecraft—required for synchronization, and the time stamping of information (datation). This is usually derived from an ultra-stable oscillator and a mission-specific epoch.
- Providing data storage.
- Executing commands and schedules.
- Controlling payloads and subsystems.
- Monitoring spacecraft health.
- Making autonomous decisions.
- Performing data compression.

13.3 TELEMETRY DATA FORMATTING

13.3.1 Classification of data

Telemetry data fall into three basic categories: housekeeping, attitude and payload.

Housekeeping data, sometimes known as engineering parameter data, needs to be monitored to keep a check on the health and operating status of the on-board equipment. Such data can arise in many forms and some typical examples are as follows:

- *Temperatures* of equipment boxes, solar arrays, attitude-control thrusters and plenum chambers, parts of the structure and so forth. Thermistors are used to convert temperature into an analogue voltage. For high temperatures, thermocouples are used and the output of a few mV d.c. is amplified to a level suitable for the telemetry encoder.
- *Pressure* in fuel tanks, plenum chambers and so forth. Various forms of pressure transducers are used.
- *Voltages and currents* of equipment power supplies. The rail voltages are scaled to a common full-scale range, which is often 0 to +5.12 V. Current monitoring may involve a variety of circuit techniques.
- Individual digital bits representing the *operational status* of equipment, each indicating that a particular functional mode is either selected or deselected. For proportional status information, such as amplifier gain settings, a number of bits are grouped together into words of appropriate length.

- *Redundancy status* that furnishes each relevant relay or switch with a set of contacts to provide a status bit provides information on whether the 'main side' or the 'cold redundant side' of equipment is in use.
- *Deployment of mechanisms*, separation from the launcher and so forth. A microswitch is fitted in order to provide an appropriate status bit.

On a modern, large communications satellite there will be several hundred to about 1000 engineering parameters to be monitored, and the result of every command is usually checked via the telemetry. The great majority of these will only need sampling at infrequent intervals of typically 30 s to 2 min, so the bandwidth required is quite small. A bit rate of a few hundred bits per second is sufficient to transmit the total information.

Attitude data arises from a variety of sensors such as Sun, Earth and star sensors, gyroscopes and accelerometers (see Chapter 9). The data can be analogue, digital or a mixture of both.

During transfer and intermediate orbit phase, the attitude and velocity will change rapidly, and frequent sampling is needed, typically from once to four times per second. For GEO operations, a reduced rate may be provided, selected by command.

Although only a few channels of attitude information are required, the high sampling rate needed during some mission phases may lead to a bandwidth that exceeds that which can be provided conveniently by a standard pulse code modulation (PCM) data system, and a separate wideband system may have to be provided.

Payload data is very variable and each case needs to be considered individually. Science and Earth-observation missions may generate very large data volumes, for example, imagery is typically very data-intensive. Often only a few channels of data are required, but their rates may be as high as many Mbits or Gbits per second. An entirely separate high-rate system may then have to be provided, such as on Envisat (see Chapter 19), and data compression may also be used to reduce the rate. Whilst high-rate downlinks can be made available for LEO missions, deep space images may take hours or even days to transmit to Earth. The capacity to acquire such data may be limited by the amount of on-board storage as well as the bandwidth of the communication links.

A communications payload gives rise to many channels of engineering parameter data in addition to that generated in the service module. Because of its complexity and the large amount of redundancy employed, it is likely to demand considerably more channels than the service module.

Typical monitoring requirements of a communications payload are as follows:

- Temperatures of travelling wave tube amplifiers and other repeater equipment.
- Power supply voltages and currents for each of the main equipment.
- Operating and redundancy status monitors for the many waveguide switches.
- Analogue power monitors for the signal levels at each main RF equipment interface.
- Telemetry monitoring of the digital gain settings telecommanded to the various channel amplifiers.

13.3.2 Influence of mission phase

The phases of a mission run from ground testing through the launch phase to the orbit phases. In each of these, the telemetry plays a part.

During *ground testing*, there is access to inter-unit connections, and test and diagnostic connectors, in addition to the normal telemetry data. On satellites with an avionics data bus, direct access to the bus may be possible. This allows higher data rates for telemetry and command, with the benefit of faster ground testing. However, RF links are still used for overall performance testing.

During the *launch phase*, satellite data is usually minimal and is restricted to a few housekeeping parameters such as battery condition, some key temperatures, reaction control equipment (RCE) pressures and deployed item status. Payloads are not usually switched on until in orbit and until out-gassing has been completed as necessary, but some telemetry activities may occur at separation. This data is sent down via the launch vehicle.

The *in-orbit* phases include the transfer and intermediate orbit phases of geostationary missions. In these cases, telemetry contact is not continuous unless relay satellites are used. Spacecraft operation must be autonomous as far as possible in order to avoid the need for intervention by ground control. For example, majority-voting techniques may be used instead of cold redundancy.

13.3.3 Telemetry data encoding

All the data considered so far arises in three basic forms: analogue, digital bi-level, and digital serial. The data is time-division multiplexed (TDM) into a PCM bit stream, which modulates the downlink RF carrier (see Chapter 12).

The first step in conditioning *analogue data* is to scale the data to a common full-scale range, usually 0 to +5.12 V. This is done at the source of the data. Frequency components greater than half the sampling frequency need to be removed by a low-pass filter to prevent aliasing errors. It is good practice to include a simple low-pass filter in each analogue line.

After filtering, the channels are sampled in turn by analogue switches and each sample is converted to a digital word, which is mixed in with the digital data as described later. For most data, an overall accuracy of about 1% is sufficient, and an 8-bit analogue-to-digital converter is used to achieve this.

Analogue commutation is invariably carried out in two stages: by a mainframe multiplexer sampling at a relatively fast rate and a slower sub-multiplexer sampling at a binary sub-multiple of the mainframe rate. The larger number of housekeeping channels that need sampling only once or twice per minute use the sub-multiplexer, and the few channels requiring fast sampling go straight into the main multiplexer.

This arrangement is very convenient because some of the sub-multiplexers can be remotely located in subsystems or payloads, reducing the complexity and mass of the harness.

Digital bi-level data arising from relay contact closures and so forth, is first conditioned to appropriate logic levels in which the 'off' state is represented by nominally zero voltage and the 'on' state by a positive voltage suitable for the integrated circuit (IC) logic family used. Individual bits are then grouped together into 8- or 16-bit words and sampled by logic gates whose outputs are serialized in a parallel-to-serial converter and mixed in with the main PCM data stream.

Digital data is usually acquired in serial form, thereby simplifying the cable harness. Such data is initially stored as an 8- or 16-bit word in a shift register located in the equipment generating the data. In modern equipment using processors, blocks of words in random access memory are also used.

In normal PCM mode, the 8-bit parallel words from the analogue-to-digital converter and the serial digital multiplexer are loaded, together with bi-level status data, into a shift register in the parallel-to-serial converter and clocked out at the telemetry bit rate as a continuous PCM bit stream. Unique synchronization and identification codes are inserted in the bit stream so that all data can be identified when it reaches the ground. Finally, the bit stream is bi-phase modulated on to a coherent sub-carrier at an integral multiple of the bit rate before routing it to the two TM transmitters.

In many encoders, an alternative to the above allows sampling to be concentrated on a particular word, to the exclusion of the other data. A 'dwell' mode is achieved by loading the address of the desired channel into the control and timing block, which then sets up the input multiplexers permanently to that channel instead of cycling round all the channels.

Wideband analogue signals bypass the PCM section of the encoder altogether, and frequency modulates a voltage-controlled oscillator (VCO) instead, using one of the US Inter-Range Instrumentation Group (IRIG) standard centre frequencies. Since the phase-shift keying (PSK) sub-carrier and VCO frequencies are different, the PCM and FM channels can operate simultaneously.

13.3.4   Telemetry list and data format

The first step in designing the telemetry system is to draw up a telemetry list for the spacecraft. For each item, this will contain the signal identification, the type of data (analogue, digital bi-level, or digital serial), the required accuracy and the sampling rate required. This list is first established during the initial 'Phase A' feasibility study and it evolves with the project. It is important to allow enough spare channels at the outset to cater for natural growth as the project matures.

The next stage is to lay out the format of the PCM message, and a simple format consisting of eight frames, each containing 64 8-bit words, may be used. The first two words of each frame contain a fixed synchronization code, 16 bit long, which the ground station equipment will recognize as the start of a frame. The next word is a frame-identification channel. The ground station de-commutation process is then able to identify every channel in the format and display the parameter contained in it on the 'quick look' facility at the station. Further processing by computer provides outputs in engineering/scientific form for users.

The rest of the format consists of data channels. A channel is typically sampled once in every frame, and such a channel is known as a mainframe or prime channel, and could be used, for example, for a nutation sensor or other data source needing fast sampling. When a still higher rate is required, the same data can be put into more than one mainframe word—such a channel is said to be super-commutated.

Housekeeping parameters requiring infrequent sampling can be sub-commutated by sampling only once per format.

By extending the principles of sub- and super-commutation over a longer sequence of frames, a wide variety of sampling rate needs can be accommodated, and once this is done, the bit rate needed to provide every parameter with at least its minimum sampling rate can be worked out. The sub-commutation depth of eight frames used in the above example would not be sufficient for a typical communications satellite.

These PCM systems are used with spacecraft, such as Intelsat X, in which there may be existing infrastructure and/or equipment, or where there is no need for the additional utility of Consultative Committee for Space Data Systems (CCSDS).

13.3.5 Packet telemetry

The *CCSDS* has produced a series of recommendations for data systems such as 'Packet Telemetry' [1], which have been adopted by ESA through the issue of its own 'Packet Telemetry Standard' [2]. These supersede the PCM Telemetry Standard. They represent the evolution from the time-division multiplex methods used in the past and come into their own on multiagency projects such as the ISS, where increased interoperability and standardization are essential. The data flow in a packet telemetry system is illustrated in Figure 13.2.

Application data is first encapsulated by the source into a 'source packet' by prefacing the data with a standard label known as the 'packet header'. This is used to route the data through the system and must therefore contain identification of the source and its particular applications process, the number of the packet in the sequence of packets produced by the source (so that the packets can be delivered in the right order at the data sink end of the link) and the length of the data field attached to the header. The provision to enable source packets exceeding a prescribed length to be segmented into several shorter 'telemetry packets' that can then be interleaved with packets from other sources has been eliminated from the latest CCSDS standards.

The next stage is to assign each group of sources to a so-called 'virtual channel'. The packets are inserted into fixed-length frames known as virtual channel data units (VCDU), which are then multiplexed into a single transfer frame for transmission on the downlink. If necessary, each virtual channel can be sampled at a different rate, appropriate to the bandwidth of that channel.

After transmission, the transfer frame can be de-multiplexed in the normal way, and the telemetry and source packets can be reconstructed and routed to their destinations using the headers. Labelling the source packet with all the necessary information about its routing and interpretation is the key to the flexibility of the packet system.

A number of extensions to the packet system have been developed, providing even more utility and higher levels of protocol. The need is for reliable file transfer and internet-type operation. The CCSDS 'Advanced Orbiting Systems (AOS)' [3], and 'Space Communications Protocol Specification' [4] are two examples (see 13.5).

The advantage of using packet telemetry is its inherent adaptability to changing data-transmission requirements. This is due to the variable-length nature of the packets and the fact that the packet headers include fields that can be used to identify and interpret the contents of the packet. The data length and periodicity are not constrained by having to occupy a particular position within a TDM frame. Bandwidth is not wasted in sending periodic data to the ground, which is then discarded. Data packets can be issued asynchronously by any data source, and the packets can be of whatever length to accommodate the data requirements at that instant. The service provided by packet telemetry is also directly compatible with higher layer application services such as networking and Internet applications.

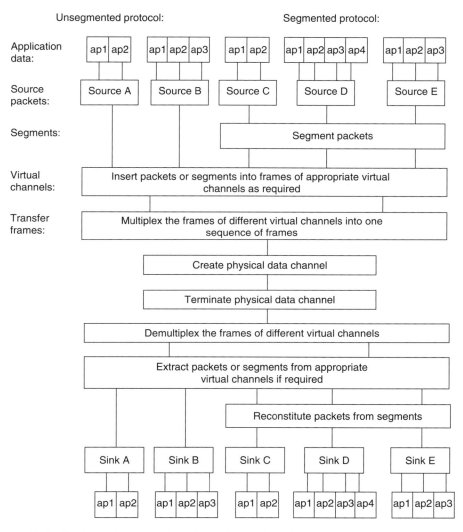

Figure 13.2 Packet telemetry data flow (Reproduced by permission of the European Space Agency)

CCSDS packet telemetry is now the dominant technology in use for missions all around the world. Space-qualified chip sets, software libraries, and global networks of earth stations are available, and there is a great body of expertise within the space industry.

13.3.6 Error-control coding

On communications spacecraft, there is usually enough transmitter power for a bit error rate (BER) of more than 1 in 10^5 on the downlink. The loss of even a complete frame is not catastrophic, so there is no great need for correction of such errors—nevertheless an error-checking code is sometimes included in the frame.

In a typical scheme, this is the 16-bit remainder that results from dividing the data portion of the telemetry frame by a specified polynomial such as $x^{16} + x^{12} + x^5 + 1$. The remainder is placed in the last two words of the telemetry frame and transmitted to the ground with the rest of the data. On the ground, the complete frame including the check bits is again divided by the same polynomial, and if a non-zero result is given, then the frame is flagged as being in error. This scheme is not able to correct errors and offers no saving in transmitter power since the actual error rate on the link is unaffected.

When *forward error correction* is needed, a scheme based on convolutional coding provides good correction capability in a Gaussian noise channel and is simple to implement in hardware. Significant further improvement (particularly with bursts of errors) can be obtained by concatenating a Reed–Solomon (RS) block code with the convolutional code. The RS code is used as the outer code while the convolutional code is the inner one.

One application of forward error correction is in packet telemetry. This needs nearly error-free channels for the successful routing and control of the data packets. The coding techniques are used either singly or together and are an integral part of the standards.

Missions involving very long transmission path lengths, such as the *Huygens Probe* destined for Saturn's moon Titan, are good candidates for the application of coding. The telemetry transmitter may then dominate the power budget, and the coding gain of several dB can be exploited to reduce the RF power needed for a given BER.

Joint NASA/ESA telemetry coding guidelines have been issued to standardize on particular convolutional and block codes to facilitate network cross-support on joint missions. The convolutional code recommended is the constraint length $k = 7$, rate $r = \frac{1}{2}$ code that is in common use by ESA, and within NASA by the Goddard Space Flight Centre and the Jet Propulsion Laboratory. CCSDS recommendations have been adopted by ESA through the issue of a 'Telemetry Channel Coding Standard' [5] that supersedes the 'Telemetry Coding Standard'.

The selected RS code is a (255, 223) code, which can be used either non-interleaved ($I = 1$) or interleaved to a depth of $I = 5$. This means that 223 I bytes of uncoded information are fed into the RS encoder and emerge unaltered with 32 I check symbol bytes appended. The whole code block of 255 I bytes is then transmitted, followed by a further 223 I information bytes, and the cycle is repeated indefinitely. The RS code is capable of correcting bursts of errors in up to 16 I consecutive bytes, and it can be seen that the interleaved form is considerably more effective in combating long burst errors.

The RS code generator is a more complex proposition than the convolutional coder as it involves binary multiplication and the intermediate storage and processing of up to 1275 bits in a shift register or equivalent RAM. However, custom chips, designed by both NASA and ESA, reduce the hardware to manageable proportions. Software solutions are also used where the data rate is relatively modest and may be used increasingly as the performance of space-qualified processors increase.

Using soft-decision *Viterbi decoding*, convolutional coding typically gives a 5.1 dB increase in link margin (reducing the power requirement for a given rate to nearly a quarter). RS encoding can give a further 2.5 dB increase.

13.3.7 Downlink frequencies and modulation

A spacecraft uses one or more of the worldwide ground station networks for command, telemetry and ranging, at least in the early mission phases. The frequency bands are constrained to be those that are supported by the chosen ground stations.

The first step in choosing the band(s) to use is to contact the Frequency Management Office of the networks concerned, with a request for the favoured frequency band. The

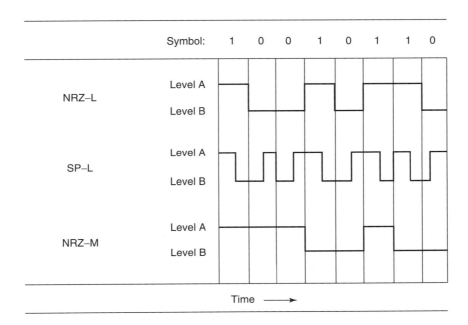

NRZ–L   ● Level A signifies symbol '1'

        ● Level B signifies symbol '0'

SP–L   ● Level A during the first half-symbol followed by level B during the second signifies symbol '1'

        ● Level B during the first half-symbol followed by level A during the second signifies symbol '0'

NRZ–M   ● Level change from A to B or B to A signifies symbol '1'

        ● No change in level signifies symbol '0'

Figure 13.3 PCM waveforms

second step is the selection of discrete frequencies within the allocated bands. This choice is a very complex and lengthy process because of the need for coordination with other space agencies and national authorities.

Many networks provide telemetry support in the ultra high-frequency (UHF) band, an exception being the Earth stations covering Intelsat missions.

Phase modulation of the RF downlink is universally used, and there is a degree of compatibility between the ESA and NASA standards, which allows cross-support, provided certain limitations are observed.

At low bit rates, a PSK-modulated sub-carrier is used to prevent modulation sidebands occurring too near to the RF carrier and upsetting the phase-lock loops in the ground station receivers. At higher bit rates a preferred alternative, achieving the same result, is to modulate the RF carrier directly with PCM-SPL data as defined in Figure 13.3.

Pre-modulation filtering of the square-wave PCM bit stream is usually provided to attenuate high-order modulation sidebands.

13.4   TELECOMMAND

13.4.1   Telecommand user interface

For relatively simple missions, there are three basic types of commands, as follows:

- *Low-level on–off commands.* These are logic-level pulses used to set or reset logic flip-flops.
- *High-level on–off commands.* These are higher-powered pulses, capable of operating a latching relay or RF waveguide switch directly. Typically these may be 12 to 28 V pulses lasting several tens of milliseconds and capable of supplying 90 mA for a relay drive, or up to 600 mA to drive an electromechanical RF switch. Separate 'on' and 'off' pulses are supplied on two different lines to drive the two coils of the latching device.
- *Proportional commands.* These are complete digital words, which may be used for purposes such as the reprogramming of memory locations in an on-board computer, or for setting up registers in the attitude control subsystem.

Enough spare channels of each type need to be provided to allow for natural growth, otherwise a redesign may be necessary at a later stage. This is controlled through the generation and maintenance of command lists.

More advanced missions require more sophisticated services. Most ESA missions currently use the ESA Packet Utilization Standard (PUS). This provides standard facilities for the above functions as well as features for loading timelines, loading, checking and dumping memory, software patching and so forth. The PUS also includes standard telemetry services.

13.4.2   PCM telecommand standards — Intelsat and the US Air Force Satellite Control Facility (SCF)

The agencies responsible for the ground station networks publish standards covering space–ground interface requirements and procedures to be adhered to for the transmission of commands. Their purpose is to ensure compatibility between the spacecraft and the equipment at the ground station. Conformance to these standards is mandatory and any deviations for a particular mission have to be justified and approved by the agency before being included in the design.

In GEO, with continuous visibility from several Earth stations, it is very seldom that a sequence of commands has to be sent within a short period of time. The Intelsat command standards are therefore based upon a command-verify-execute strategy in which each command is held in the satellite decoder and verified through telemetry before it is executed.

Figure 13.4 shows a simplified block diagram of a typical decoder for an Intelsat spacecraft. The explanation relates primarily to the Intelsat V system, but more recent spacecraft in the series have used fundamentally similar principles.

The command uplink signal consists of a sequence of tones that represent data 0, data 1 and execute information 'bits'. These are frequency-modulated on the 6-GHz carrier and demodulated by the two command receivers. The ground operator is able to choose which receiver is to be used by selecting either of two alternative data 0 tone frequencies.

Each digital command message includes an introductory series of zeros to select the receiver and to synchronize the bit detector clock, and this is followed by a decoder address word, a command vector, and finally an on–off command word or a proportional data command word. The 0 and 1 bits comprising the message are stored in a shift register.

Each command unit has a unique address, which is hardwired, and only messages containing this address will be accepted. The data held in the command message store are telemetered to the ground for verification, and assuming this is satisfactory, an execute tone burst is sent. This turns on the execute power switch, which then outputs the command to the specified user channel. The combination of power switching and the use of diode isolation and redundant components in the output circuits allows the outputs of the two command decoders to be combined in a fail-safe manner without loss of reliability due

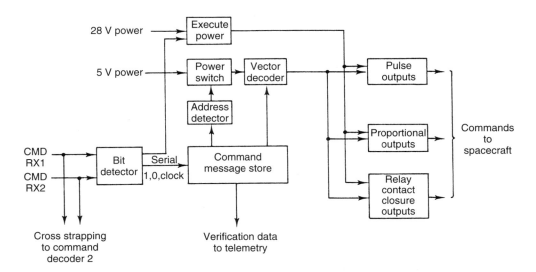

Figure 13.4 Telecommand decoder

to single-point failure modes. The US Air Force SCF tracking network used a basically similar ternary (three-tone) modulation scheme using frequency-shift keyed tones.

13.4.3   PCM telecommand standards — ESA and NASA

Agencies supporting scientific and LEO missions often need to send rapid sequences of commands in a short time period, such as during a brief ground station pass. Although the ESA and NASA systems do make use of ground verification, the structure of the telecommand message allows the checking and correction to take place inside the satellite's decoder itself.

The ESA telecommand message starts with a fixed number of acquisition bits to initialize the message, usually fifteen 'zeros' followed by a single 'one'. The message sequence proper then consists of an unbroken sequence of one or more 96-bit frames similar to that shown in Figure 13.5. A single 16-bit spacecraft address and synchronization word (ASW) is used to terminate the message. The initial 16-bit ASW is used to identify individual spacecraft and to synchronize the decoder.

A 4-bit mode selection word using 2-out-of-4 redundant coding is included in each frame after the ASW. This word is repeated once and provides means for selecting on–off pulse commands, proportional commands or time-tagged commands for delayed execution.

Since a single-bit error will give an invalid mode word, the acceptance criterion in Figure 13.5 ensures the acceptance of a command even if a single-bit error has occurred, and its rejection if more than one error has taken place. In addition, a command receiver squelch cuts off the input to the bit detector if the signal-to-noise ratio is too low, thus ensuring a low bit error probability (normally better than 10^{-5}) and a correspondingly high probability of frame acceptance.

The three data words following the mode word are each of eight bits and represent respectively the address of the 16-bit memory to which the data are to be sent, the first eight bits of data and the final eight bits of data. Each data word is repeated once to increase the probability of acceptance, and four Hamming-code check bits are appended to each word to permit error detection and correction in the spacecraft. By comparing the received check bits with a set generated locally from the data word, it is possible to detect all single- and double-bit errors and, in the case of single-bit errors only, to determine which data bit is in error and correct it.

The end-to-end probability of command rejection can be reduced to less than 1 in 10^6, and the probability of an erroneous command to less than 1 in 10^8.

Spacecraft that use the ESA standards are compatible with the NASA ground network and procedures, whose modulation and coding techniques are to a large extent compatible.

13.4.4   Packet telecommand

The PCM telecommand standards are being superseded by a packet-based system, which has some similarity to the packet telemetry system. CCSDS recommendations have been produced and are adopted by ESA through its 'Packet Telecommand Standard' [6].

The packet telecommand system is arranged in a number of layers, which provide the mechanism for the flow of data from the ground (source) to the spacecraft subsystem (sink). These layers are as follows:

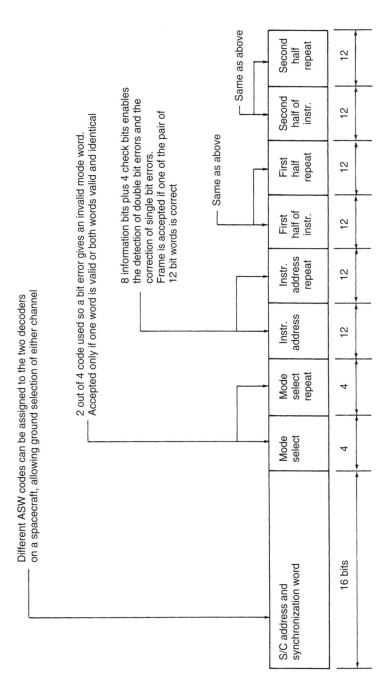

Figure 13.5 Memory load command frame structure

- An *Application Process Layer* allows in-orbit payloads or subsystems to be controlled remotely by a user, providing the interface from humans and their computers. It can also be used to support advanced higher-level protocols such as file management.
- A *System Management Layer* provides the translation between the commanding language used in the Application Process and the detailed communications and control language used in the Packetization layer.
- A *Packetization Layer* allows the user to optimize his application data with minimum constraints and independently from other users. The application data is formed into TC packets consisting of a fixed-length packet header of 48 bits and a variable-length packet data field. The header provides identification, sequence control and packet length. Data field headers and packet error control may be added to the data field as options.
- A *Segmentation Layer* provides the mechanism for a number of variable-length TC packets to be formed into one virtual channel. It also enables the multiplexing of TC packets from a number of different sources and control of the data flow. The standard structure of a TC segment consists of an 8-bit segment header, a variable-length segment data field and an optional variable-length trailer.
- The *Transfer Layer* uses TC transfer frames to provide error-free transfer of data over the RF link. Two types of transfer frame provide for sequence control, and one for an expedited service typically used in recovery situations. The standard structure of a TC transfer frame consists of a 40-bit frame header, a variable-length data field (containing one TC segment) and a 16-bit frame error-control field.
- A *Coding Layer* provides the error correction and the synchronization required by the receiver. Each transfer frame is placed into a single command link transfer unit (CLTU). The CLTU consists of a 16-bit start sequence, n code blocks of up to 64 bits and a tail sequence the same length as a code block. Here n is an integer, sufficient to encompass one encoded transfer frame, with fill data being used when required. The code applied is a (63,56) using the polynomial $x^7 + x^6 + x^2 + x^0$ to produce a 7-bit parity covering up to 56 bits of data.
- The *Physical Layer* is the radio link itself and operates in one of four Carrier Modulation Modes. These are

1. unmodulated RF carrier,
2. modulated with Acquisition sequence,
3. modulated with one protocol data unit (CLTU),
4. modulated with Idle sequence.

The receiver, demodulator/command decoder and OBDH operate on the same layers but in reverse order. At each stage, telemetry reporting is fed back to ground, providing confirmation of good receipt and successful command operation when applicable.

As is the case for packet telemetry, packet telecommand protocols provide the user with greater flexibility than the earlier standards. It accommodates the dynamic commanding needed for advanced satellite operation. It makes possible complex interactions between ground and spacecraft such as software uploads, on-board file system manipulation and even access to web sites on the satellite from anywhere on the terrestrial internet.

13.4.5   Error-control coding

In general, error-control encoding is a less intensive processing task than decoding. For telemetry links, this places the more complex error-control tasks in the ground segment. It is difficult to implement powerful telecommand error-control decoding in the space segment because of the constraints on on-board electronic systems. For this reason, the error coding in telecommand standards is rather basic. The CCSDS telecommand recommendations choose a Bose–Chaudhuri–Hocquenchem (BCH) algorithm whose major strength lies in error detection. Indeed, the failure of this code is used to denote the end of a telecommand frame.

In any case, error control via coding still leaves a small but significant possibility of error, which may not be important for telemetry but could be disastrous in a mission critical command. Command links in general, therefore, use an automatic retransmission mechanism (the CCSDS protocol is known as COP-1) to guarantee command delivery by detecting errors and retransmitting failed frames. The protocol also ensures that telecommands are always received and executed in the correct sequence.

13.4.6   Uplink frequencies and modulation

Command uplink frequencies are typically in the S-band (2025–2120 MHz) or X-band (7145–7235 MHz) and support relatively modest data rates. The early ESA RF and Modulation scheme is based on a sub-carrier and a maximum data rate of 4 ksymbol/s. However, newer applications require higher rates and thus the CCSDS developed a modulation standard that supports rates of up to 256 ksymbol/s using Split Phase-Level/Phase Modulation (SP-L/PM) directly on the carrier. This scheme is incorporated in the ESA RF and modulation standard [7].

Still higher rates are possible by using direct bi-phase-shift keying (BPSK) carrier modulation, although, unlike the lower rate systems, this will not support simultaneous ranging.

13.5   COMMUNICATION TECHNIQUES AND PROTOCOLS

13.5.1   Ranging

Ranging is achieved by means of a transponder, which is integrated into the telemetry-command subsystem. This retransmits ranging tones through the telemetry channels in response to tones received via the command route.

On Intelsat X, the output of a command receiver on a 6-GHz carrier can be connected to a beacon transmitter that phase-modulates the ranging tones on to the 4-GHz downlink carrier. On Earth-observation satellites such as Envisat these operations are carried out using the S-band links. The signals are tones as specified in the ESA ranging standards. These may be phase-modulated on to the uplink carrier in place of command signals and on to the downlink simultaneously with telemetry signals. Interference on the downlink is avoided by the choice of sub-carrier frequencies. Also, integrated Doppler measurements are made in order to provide spacecraft tracking—see Reference [8].

13.5.2   Advanced orbiting systems (AOS)

In the late 1980s, the requirements of the ISS, and space transportation systems such as Hermes and Earth-observation systems outgrew the capabilities of the CCSDS packet telemetry scheme.

The deficiencies were in the areas of

- support for isochronous (e.g. voice/video) services,
- support for high-volume downlinks (e.g. Earth-observation imagery, including ssynthetic aperture radar (SAR), video, etc.),
- support for interactive services (e.g. two-way fast file transfer).

The CCSDS recommendations addressed these issues and also took the opportunity to adopt a service-driven structure to the recommendation rather than the protocol/data format-oriented system of the packet TM recommendations. AOS is used in the downlinks of the ISS and Envisat. AOS also introduced networking to the CCSDS standards, with the incorporation of the efficient connection-oriented path protocol and support for the ISO 8473 connectionless network protocol—subsequently amended to support the Internet Protocol (IP).

13.5.3   Proximity links

CCSDS Proximity-1 is applicable to power-constrained links and is specifically designed for use in inter-satellite links, constellation missions and planetary orbiter/lander/rover configurations.

Proximity-1 operates symmetrically—that is, there is no distinction between telemetry and telecommand or forward and return links. Proximity-1 includes facilities for the following:

- Automatic detection of link availability and the negotiation of link parameters. A medium access control (MAC) sub-layer provides for the establishment of a link before data is transferred, its maintenance and termination. It also controls link characteristics, such as data rates, and the state of the data link and physical parameters.
- Reliable, in-sequence, complete delivery in both directions through the use of automatic retransmission. This uses a process known as 'persistence' to verify correct transfer.
- Convolutional and RS coding in both directions (suitable space-qualified decoding devices are becoming available at the time of writing).
- Use of fixed-length or variable-length frames.

Exploration missions to Mars are typical examples in which Proximity-1 may be used. It uses features adapted from both the CCSDS packet telemetry and telecommand standards.

13.5.4 Space communications protocol standards (SCPS)

NASA initiated the SCPS development in collaboration with the US DoD. It was driven by perceived requirements for military and civil satellite constellations and sensor networks. SCPS takes the existing protocols used in the networking, transport and application layers of the terrestrial Internet and extends them to cater for the more onerous requirements of space networking. The components of the SCPS protocol suite are the SCPS network protocol (SCPS-NP), security protocol (SCPS-SP), transport protocol (SCPS-TP) and file protocol (SCPS-FP)

The application of SCPS to space missions has been slow, although this may change as more complex space constellations are implemented. However, there has been much interest in its use in terrestrial applications that have constraints similar to the space environment.

13.5.5 CCSDS file delivery protocol (CFDP)

CFDP provides a reliable data-transfer service. It is an application process (see Section 13.4.4) that operates over CCSDS or other link and networking protocols. The major features of CFDP are as follows:

- *Store and forward networking* differs from real-time networking in that an end-to-end path does not have to be present at the same time to allow networked communications. The protocol takes account of a dynamic network configuration with disjointed links.
- *Continuous file delivery* supports the delivery of unbounded TM streams. It can deliver file segments as they arrive, as well as deliver what can be recovered from a failed file transfer.
- *Negative automatic report queuing (ARQ) scheme* guarantees data delivery and is optimized for large bandwidth/delay products and for a low overhead in the return path. The retransmission requests can be optimized for operations; for example, they can be issued as errors occur or can be bundled together until a suitable opportunity for retransmission occurs.
- *Proxy transfer facilities* provide for third-party control of data transfer.
- *Graceful suspend/resume* for predictable link outages.
- *Garbage clearance facilities* protocol recognizes that software may have faults or may be disrupted by single-event upsets (SEU) and ensures that systematic errors do not accumulate and require operator intervention.
- *File Manipulation* allows for the remote naming, deletion, copying and management of on-board file systems.

The CFDP protocol has emerged from requirements to support spacecraft with on-board file systems and to provide a reliable service for bulk data transfer. This is particularly the case for spacecraft telemetry in which advances in on-board autonomy and data compression, which reduced the data redundancy in the downlink, making uncorrected bit errors less intolerable.

13.5.6 The interplanetary internet (IPN)

The IPN is an initiative of the Internet Research Task Force (IRTF). The IRTF is an Internet-based group largely responsible for the future direction of Internet development. The IPN Research Group (IPNRG) carries out this development. Despite its name, the initiative is intended to reach out into all space networking applications and incorporates both near-Earth and deep space applications.

The IPN consists of the following:

● *In situ Internets* are deployed Internets such as the Earth's Internet, a Mars Internet or an Internet local to a satellite constellation. These are composed, in general, of dedicated sub-network protocols supporting common transmission control protocol/Internet protocol (TCP/IP) for the network and transport.
● *The Interplanetary backbone* provides a long-haul transport service between remotely deployed Internets. The long-haul transport protocol (LTP) is currently under development and uses some of the techniques developed for the CFDP.
● *Interplanetary gateways* connect deployed Internets to each other or to the interplanetary backbone.
● *Inter-Internet dialogues* provide the store and forward networking necessary to operate with either long delays or disjointed communications resulting from orbital or planetary motion. It provides the store and forward networking required to interconnect deployed Internets.

The inter-Internet dialogues are based on the concept of bundling. Just as the IP networking layer provides a common means of data transfer across heterogeneous subnets in real time, bundling provides a common means of data transfer across heterogeneous Internets in a store and forward environment.

It is expected that, in the interplanetary environment, special bundle applications will need to be developed to avoid an excess of short discrete transfers, as would be the case if some common email exchange protocols, such as SMTP, were used.

13.6 ON-BOARD DATA HANDLING (OBDH) AND PROCESSING

13.6.1 Platform and payloads

The OBDH functions reside both on the spacecraft platform and within the payloads. They provide both the command and data management associated with the telemetry and telecommand operations discussed earlier. Communications spacecraft such as Inmarsat have separate and very complex payloads for the receipt and transmission of the user data. Remote sensing and scientific spacecraft often have separate provision for the handling and transmission of instrument data. The latter may require special measures to cater for very high data rates. For example, radar instruments, such as the Advanced Synthetic Aperture Radar (ASAR) and TerraSAR, can generate data at rates from around 100 Mbps to over 1 Gbps.

Classical OBDH architectures are based upon a central processor, typically connected via a video or digital path to the RF communications subsystem. This central processor will communicate with the platform subsystems such as the AOCS and the payloads using a serial bus with high data integrity. ESA spacecraft typically use an ESA standard OBDH bus operating at 250 kbps with separate Interrogation and Response lines, transformer-coupled and Litton-encoded. A variety of terminal devices are used depending on the particular need.

The Ariane launch vehicle uses a bus based on the Military Standard 1553B. The 1553 bus is a serial bus capable of operating at 765 kbps, using transformer coupling and Manchester encoding. The speed of such buses is gradually being increased. It is interesting to note that certain spacecraft have been designed with mixed bus architectures to allow flexibility in equipment procurement.

The *central processor* has two primary functions. These are to provide for telecommand decoding and to operate as the central terminal for the OBDH bus. Depending upon the spacecraft needs, these functions may be implemented using from one to five separate units. The precise arrangement will depend upon the mission. It may be necessary to be able to extract, prior to any processing, certain high-priority commands for config-uration switching operations. The subsystem may be involved in the arming and firing of pyrotechnic devices. Also, expanded macrocommands and time-tagged commands are often required. Ultrastable oscillators may be incorporated for accurate timing of control functions and for datation purposes. Certain missions where ground contact is not contin-uous will require a degree of autonomous operation or at least a fail-safe survival mode. An example of a remote sensing platform is given in Figure 13.6, in which the central processor has become a data-handling subsystem.

Terminal devices vary considerably and can be either stand-alone units or embedded within an instrument or subsystem. Stand-alone units may be unintelligent, as in the case of a remote terminal unit (RTU), or sophisticated communications processors. The RTU

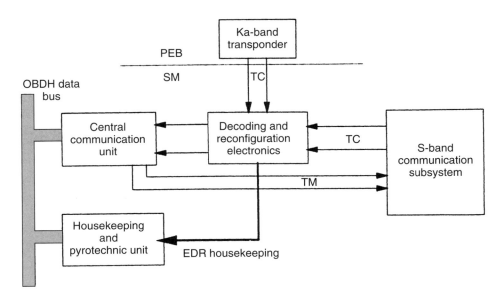

Figure 13.6 Data-handling subsystem (central processor)

caters for bus coupling and the Litton code. It provides the distribution of simple 16-bit serial memory load and single-bit discrete commands, and high-level commands with the capability to drive relays directly. It provides for the collection of 8/16-bit serial and discrete bi-level telemetry. The RTU may also include the conditioning and analogue-to-digital conversion of analogue telemetry. An RTU will typically use a switched matrix to provide for a high number of interfaces whilst minimizing the amount of internal circuits. It will include multiplexing and de-multiplexing functions.

Communications processors operate as Intelligent Terminal Units and can provide data-packetization services for the payloads. The interfaces between a Communications Processor and the payload may be either through dedicated ports or a bus.

Embedded terminals can operate as either mini-RTUs or more sophisticated remote bus interfaces (RBIs), which provide for direct memory access within the payload. In both these cases, separate units handle bus coupling and Litton-coded functions.

The more complex instruments may include their own processor-based intelligent control units (ICU) utilizing an embedded RBI. The processing power of some payloads sometimes exceeds the capability of the platform computers.

Payload processors can provide the functions of a communications processor, ICU and science data processor.

A *typical on-board computer* will use a 16- or 32-bit microprocessor, a microcontroller or digital signal processor (DSP)—built in radiation hard or tolerant technology. The embedded software will be produced to high standards controlling the software life cycle by addressing aspects such as documentation, review, structure, margins and verification. Secure methods and tools are used including object-orientated techniques and high-level languages such as ADA. Optimized operating systems are an option. Normally, the system would be designed to allow the flight software to be patched and dumped, allowing modification and investigations to be carried out in orbit.

The production of on-board code is evolving from the situation where every line of on-board code was written for that particular mission, or at the very most, inherited from a very similar mission. The processing power and storage available on-board are following terrestrial industrial processors, and the mission requirements are also evolving towards those prevalent in terrestrial control applications. These trends manifest themselves in requirements for complex data organization, autonomous decision-making, intensive signal processing and multitasking, and the coordination of large distributed development teams.

The production of such software systems for highly reliable application has been addressed in terrestrial systems and has resulted in the availability of high integrity, real-time operating systems. These are now finding their way onto on-board processing systems and bring the advantages of

- faster, cheaper development,
- ease of maintenance and developer team coordination,
- off-the-shelf development systems,
- off-the-shelf processing and communications libraries,
- built-in file handling, multitasking and inter-task communication,
- large investment in software reliability.

The instrument data is often collected by a separate data-handling system, one example being known as the *payload data handling and transmission* (*PDHT*) system. These systems allow for the direct and independent collection of data at rates much higher than what can be handled by standard buses. They have their own dedicated RF ground link. Such systems will embrace high-speed multiplexing of data packets and provide data storage when direct ground contact is not available. Magnetic tape recorders were used for a long time with capacities of 10 Gbits and more. However, the availability of very high-density semiconductor memories has enabled modern spacecraft to use solid-state data stores. They offer the advantage of random access, plus the option of applying architectures capable of operating with peak rates of over 1 Gbit per second and capacities of over 500 Gbits.

13.6.2   Instruments, experiments and sensors

Figure 13.7 illustrates the main elements that might exist in a spacecraft generalized instrument, experiment or sensor. A payload may consist of one or more of these, some typical examples being

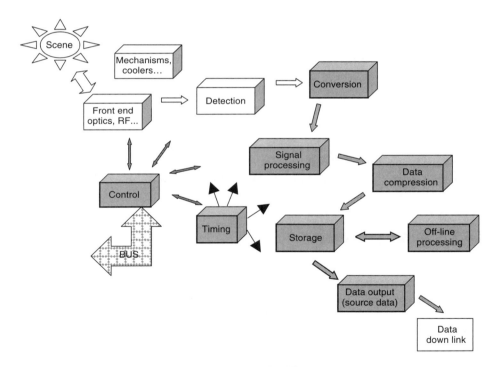

Figure 13.7   Generalized instrument

- active radars, such as Altimeters or the Advanced Synthetic Aperture Radar (ASAR) on the Envisat spacecraft;
- passive microwave sounders, such as the Micro Humidity Sounder (MHS) on the NOAA and METOP spacecraft;
- IR and visible spectrum radiometers, such as on the METEOSAT series of spacecraft;

- multi- or hyper-spectral instruments, such as the Compact High-Resolution Imaging Spectrometer (CHRIS) on the PROBA spacecraft;
- laser-based instruments such as the ALADIN Lidar for the Aeolus mission;
- star sensors used for attitude control;
- science instruments, such as on the Rosetta Spacecraft.

Except for the simplest case, most systems will have a controller that provides the functional interface with the platform and provides local and typically real-time operation. In addition, there may be digital signal processing and data-handling functions to support. The data output from these systems will typically be in *source packet* format. Many instruments are multimode.

The *controller* will typically be a standard microprocessor, such as the 486 or the ERC32. Alternatively, a DSP device, such as the TSC21020F operating at 20 MIPS and 60 MFLOPs, may be used where signal processing is required in addition to the instrument control function. It is anticipated that in some applications standard processor designs such as the LEON, operating at 100 MIPS, will be embedded within application-specific integrated circuits (ASICs), with appropriate intellectual property arrangements. These standard designs are known as 'IP cores'.

Digital communication within an instrument may be based on a multi-drop bus or dedicated parallel and point-to-point serial links.

Multi-drop buses, such as the ESA OBDH, the Mil Std 1553B and the CAN bus, are used where one device (master) needs to communicate with more than one other device (terminal). They are used for commanding and gathering telemetry from intelligent devices. They are also used for gathering science or instrument data at modest rates of up to around 1 Mbps.

Point-to-Point Links, such as the Mil Std 1355, ESA SpaceWire [9], HotLinks and GigaLink, are used for direct communication between two devices. They are particularly suited for the transfer of blocks of data and can support much higher rates than the multi-drop bus. Rates can vary from 10 Mbps to 1 Gbps plus. A number of parallel point-to-point links may be used to support higher rates and data-routing functions.

13.6.3 Signal processing

Signal processing in this context is applicable to platform sensors and operation, and to payload experiments and instruments, in both analogue and digital form. The nature of the processing and the speed at which it is done vary considerably.

Platform operation and sensors—control and navigation subsystems rely on a number of sensors providing information on the Sun, Earth, stars, spacecraft inertia, and so forth as inputs to the control algorithms (see Chapter 9). Sensing elements require signal conditioning, conversion to digital form, detection, and the manipulation of the data acquired to yield the information. Charge-coupled device (CCD) and Active Pixel detector arrays, as found in conventional digital cameras, require timing, read-out circuits, and analogue-to-digital conversion. Such sensors produce data at rates of typically 20 Mbps.

Payload experiments and instruments often require the processing of a baseband signal in either the time or frequency domain. Some examples are digital filtering, compression, feature extraction, data formatting, error control, and encryption. Operation with a number of single channels, each at up to 200 Mbps, may be required.

13.6.4   Compression and storage

Data compression can be achieved by eliminating the redundant or duplicate content of the data. Image compression can be achieved by eliminating, for example, unwanted information or by reducing the resolution.

On-line Data Compression such as block-adaptive quantization may be used to reduce the data rate to more manageable proportions and would typically reduce the rate by factors from 2 to 8. They operate in real time on the data stream, perhaps taking advantage of first-in-first-out (FIFO) buffering.

Off-line Data Compression usually works in conjunction with a large data store and would run in dedicated processors or ASICs. There are many algorithms available that can give lossless compression with a reduction factor of perhaps 2, or lossy compression with reduction factors of up to and beyond 100. The selection of the algorithm really depends on the information required from the data and the data reduction needs. Compression algorithms are available to established standards such as Java Picture Experts Group (JPEG) and CCSDS [10].

Data storage is an important function for missions in which the data volume produced is greater than the link capacity and/or availability. Where tape recorders were once employed, solid-state memories using dynamic RAM are used, providing very large capacities of several hundred Gbits and more. Typically, they do not have external addressing to each RAM location but operate on a block or file basis. The block may be of the order of 1000 bytes or sometimes it will correspond to one source packet. It is important to maintain the integrity of the data and so data coding and error correction is employed to protect against errors such as those caused by SEUs induced by cosmic radiation. The detailed arrangements of the input and output busses have to take into account the data rates and volumes produced by the instrument and the corresponding capability of the following downlink. To provide for flexible operation and degradation in the memory array, a data-routing function is usually included on both the input and output.

The large capacity of on-board mass memory has led to the same challenges in organizing stored data as is found on disk storage in terrestrial computers. The data is best arranged hierarchically, with data files identified by name rather than exact locations. This is leading to two trends—firstly, the adoption of industrial real-time operating systems, as mentioned earlier, and secondly, the requirement for remote file manipulation facilities as provided by the CFDP protocol.

13.6.5   Data downlinks

A typical *System Architecture* will consist of a data-handling function followed by a RF transmit chain or chains. The data handling needs to provide for data routing, the buffering of different input and output data rates (using a simple FIFO or a large data store), data formatting, carrier modulation, amplification and the generation of an RF beam.

Modulation techniques are normally quadrature phase-shift keying (QPSK), or of a higher order. In the case of QPSK, a 100-Mbps data stream would require an RF bandwidth of 55 MHz. The bandwidth available for LEO missions in X-band is 8025 to 8400 GHz, which with guard bands can support around 500 Mbps. However, permission has to be obtained from the International Telecommunications Union (ITU) and it is unlikely that all of the frequency range would be granted to one mission.

Amplification to achieve an equivalent isotropic radiated power (EIRP) of typically 15 dBW would normally require travelling wave tube or solid-state amplifiers and an antenna system with some gain. There are many *antenna* solutions ranging from fixed pseudo-omni-reflectors to very focused electronically steered beams. The beam has to be well controlled to achieve a reliable link over the required area of the earth and to limit the power flux density to satisfy World Administrative Radio Conference (WARC) regulations.

Link Layer data processing is needed to convert source packets into physical channel access protocol data units (PCA-PDU). This includes the formatting to establish VCDU, as required by the selected CCSDS standard; pseudo-randomization to smooth out the spectral content of the data and the addition of synchronization bits to enable the ground processing to recover the randomized data, forming channel access data units (CADU); differential encoding to resolve phase ambiguities; and the splitting of the data into in-phase and quadrature (IQ) channels prior to the modulator. Forward error correction may optionally be included to improve the link budget.

The *Link Availability* is an important system design consideration for LEO missions and it depends upon the spacecraft orbit and the location of the ground station. Typically, the link to a reasonably high latitude station will be available from 3 to 10 minutes during one orbit of 100 minutes. Low latitude stations will have a significant number of orbits with no availability.

13.7 TECHNOLOGY

Spacecraft data-handling systems are implemented using electronics technology. The space environment places major constraints on the use of this technology in a number of ways. The spacecraft resources available to electronic systems are limited. The three major parameters that are at a premium are mass, power and volume. The physical environment is very stressful for electronics components and subsystems. These must be chosen and designed to withstand the rigours of launch vibration and in-orbit temperature extremes. They must also be compatible with the vacuum environment in not having pressurized voids in the packaging (common in plastic packages) and having no outgassing materials that may contaminate other spacecraft equipment, for example, optical surfaces.

A major consideration when choosing appropriate components is the radiation environment. There are three effects to be considered (see also Chapter 2):

● *Single Event Upsets* (*SEU*) are temporary effects due to ionizing radiation changing the state of an electronics node on a semiconductor device. The effects are unpredictable. SEUs can cause noise in data, random switching events, changes in memory locations, random jumps in software code or illegal state-machine configurations. Data stored over a long period in on-board memory is subject to randomization by cumulative SEUs. The effects can be mitigated using error-correction codes, which are checked on a regular basis and the data is corrected if necessary. This is known as memory scrubbing. Other circuits may be protected using voting logic (e.g. triple module redundancy) and by incorporating software watchdog timers.

● *Total Dose* damage is due to the cumulative effect of ionizing radiation over time. The result is a catastrophic device failure. A device's tolerance to total dose is measured in krads, and the immunity required varies from 1 krad to around 100 krad depending on the spacecraft trajectory and lifetime. Devices can be shielded from total dose effects by the spacecraft structure, by the equipment housing, by specific device packaging or spot shielding.

- *Latch up* is another catastrophic condition and is caused by a single energetic ion initiating a runaway current flow in the device leading to failure. Again, shielding can help (although the emission of secondary ions can exacerbate the effect). An alternative strategy is to protect the device with current sensing and limiting circuitry that blocks the runaway effects and returns it to normal operation when the event has passed.

Note that the above catastrophic effects are significantly reduced when devices are unpowered. Another mitigation strategy is therefore to power-down electronic subsystems when not required.

Microprocessors that are specifically designed for use in the space environment emulate those available for terrestrial uses. However, they tend to lag terrestrial technology by a number of years. This is not necessarily a problem. Routine spacecraft processing tasks such as attitude control, housekeeping reporting, systems management, command execution and timeline sequencing require very little processing power. In this case, simple 8-bit microcontrollers are adequate for the task.

It is worth remembering that most support devices such as RAM, ROM, buffers, and so forth come in 8-bit wide packages. Choosing a microprocessor with a wide data bus causes a proportionate increase in the number of support devices, and a minimum bus width should always be selected.

For more demanding applications, 16- or 32-bit processors and signal processors are available in space-qualified technology. These are limited in processing power compared to their terrestrial counterparts, but find application in hosting complex spacecraft or payload software, or in performing intensive signal processing tasks.

Although some processors have been produced in space-qualified technology, the same cannot be said for peripheral devices. Universal asynchronous receiver and transmitter's (UART), dynamic RAM controllers, I/O controllers and communications devices usually need to be implemented on a custom basis using field programmable technology. This may be difficult to find in space-qualified form.

The electronics design cycle for spacecraft electronics used to be very costly. Past solutions used discrete digital devices with typically less than 50 gates per device. This resulted in large assemblies and a number of generations of costly and time-consuming printed circuit board (PCB) layout, before a satisfactory flight design was reached. The size of the electronics systems was greatly reduced by the advent of ASICs. The ASIC design and modification process is, however, costly and time consuming.

Radiation-hard field programmable gate arrays (FPGAs) are now in common usage. These can host a huge number of functions (at this time around 100 000 gates per device) and can be designed and programmed using a desktop computer. This significantly decreases the time required to complete a design. Some devices can be continuously modified; others can only be programmed once. For critical or mass production applications, FPGAs are often used to prototype ASIC designs. The use of FPGAs can result in a single-board layout cycle in which the philosophy is to route all signals on the board into the FPGA and finalize the FPGA functionality later. In conjunction with desktop

PCB design, FPGA technology is increasing the available functionality and shortening development timescales for on-board electronic systems.

13.8   TOOLS AND CONTROLLING DOCUMENTS

Because of the relatively high cost of equipment development and production, the criticality of the application, and the need to manage risk, it is essential to keep good control of the requirements and design. It is also essential to ensure that the solution is properly verified at each stage in its life cycle.

In the data handling and related fields, this is increasingly being achieved using computer tools such as very high-speed integrated circuit hardware description language (VHDL), mathematical modelling of digital signal processing, simulation of analogue and digital circuits, and auto layout and routing. There are a number of good commercial software packages available, running on workstations and personal computers.

The definition and control process is also achieved through the use of formal documentation such as national and international standards, specification and user manuals, telecommand and telemetry directories, hardware/software interface control documents, and electrical and mechanical interface control documents. These are just some examples of the documents produced and maintained throughout a product's life cycle.

Some of the ESA PSS specifications mentioned are being gradually replaced by the European Cooperation for Space Standardization (ECSS) documents or later versions of the CCSDS documents.

REFERENCES

[1]   *CCSDS Packet Telemetry*, 102.0-B-5 Issue 5 (November 2000).
[2]   *Packet Telemetry Standard*, ESA PSS-04-106 Issue 1 (January 1988).
[3]   *CCSDS Advanced Orbiting Systems*, 704.0-B-1 Issue 2 (November 1992, reconfirmed June 1998).
[4]   *CCSDS Space Communications Protocol Specification*, 713.0-B-1 Issue 1 (May 1999).
[5]   *Telemetry Channel Coding Standard*, ESA PSS-04-103 Issue 1 (September 1989).
[6]   *Packet Telecommand Standard*, ESA PS-04-107 Issue 2 (April 1992).
[7]   *RF & Modulation Standard*, ESA PSS-04-105 Issue 2.4 (November 1996).
[8]   *Ranging Standard*, ESA PSS-04-104 Issue 2 (March 1991).
[9]   *SpaceWire*, ESA ECSS-E-50-12 draft 2 (December 2001).
[10] *CCSDS Lossless Data Compression*, 121.0-B-1 Issue 1 (May 1997).

Links:

http://www.ccsds.org/blue_books.html
http://www.estec.esa.nl/ecss/menu1.html
http://esapub.esrin.esa.it/pss/pss-ct03.htm
http://www.ipnsig.org
http://www.irtf.org

14 *GROUND STATIONS*

Richard Holdaway

Rutherford Appleton Laboratory, Chilton, United Kingdom

14.1 INTRODUCTION

This chapter describes the main functions performed at a ground station in support of an operational spacecraft. These are highly complex functions, usually involving the following tasks:

1. Tracking to determine the position of the satellite in orbit;
2. Telemetry operations to acquire and record satellite data and status;
3. Commanding operations to interrogate and control the various functions of the satellite;
4. Controlling operations to determine orbital parameters, to schedule all satellite passes and to monitor and load the on-board computer;
5. Data processing operations to present all the engineering and scientific data in the formats required for the successful progress of the mission;
6. Voice and data links to other worldwide ground stations and processing centres.

Within the above list there are four main components of a ground station, namely, hardware, software, people and operations. This chapter deals with each of these in turn, discusses recent advances in autonomous operations and concludes with a look at the future challenges for Ground Control.

14.2 HARDWARE

The main hardware components of a ground station are an antenna, a receive–transmit system, data recorder(s), computer(s) and their peripherals and control consoles. This is a basic list that does not significantly change with the type of spacecraft being controlled, and therefore applies equally to Space Science, Earth Observation/Remote Sensing and

---

Spacecraft Systems Engineering (Third Edition). Edited by P. W. Fortescue, J. P. W. Stark and G. G. Swinerd
© 2003 John Wiley & Sons Ltd

to Telecommunications. It is also as applicable to low earth orbits as it is to geostationary orbits or interplanetary orbits.

14.2.1   Antennas

The main hardware component of a ground station is the antenna, whose support functions may include tracking, telemetry, command, space-ground voice and television capabilities. Figure 14.1 shows a typical configuration, in this case the 12 m antenna at the Rutherford Appleton Laboratory (RAL), UK. As another example, the Spaceflight Tracking and Data Network (STDN) of National Aeronautics and Space Administration (NASA) has antenna systems operating at a number of different frequencies. The most frequently used is the S-band system, of which STDN has many ground stations, but an increasing number of systems now employ K, L and X-band transmission.

S-band systems employ monopulse autotrack principles to maintain the antenna pointing towards the spacecraft's transmitted signal. To aid initial acquisition, computer-controlled programme modes are usually used. This uses orbital prediction data to generate the required pointing angles. Initial acquisition of the spacecraft's RF signal may be facilitated by means of a small, wider beamwidth acquisition parabolic antenna, often mounted at the edge of the large one. Other operating modes may include manual position and

Figure 14.1   RAL antenna at night

velocity, slave and manual programme. $X-Y$ mounts are capable of tracking through zenith but have a restriction on their azimuth rotation rate approaching zenith. Coverage patterns are further restricted at most stations by the surrounding terrain, as shown in Figure 14.2.

The *antenna diameter*, D, required for any particular mission is primarily a function of maximum satellite range, d (km), carrier frequency f_c (Hz), data rate b (bits/s), and satellite transmitter power t_p (watts). A useful approximation for D can be found from

$$D = \frac{6 \times 10^3 \, d}{f_c} \sqrt{\frac{b}{t_p}}$$

Thus, for a polar orbiting satellite, for example, transmitting 1 Mbit/s data at S-band with a 1 watt on-board transmitter, D is calculated to be of the order of 9 m (see also Sections 12.2.8 and 12.2.10 of Chapter 12).

STDN ranging equipment operating in conjunction with multifunction receivers and S-band exciters provide precision range and Doppler measurements for a variety of spacecraft. For spacecraft carrying an S-band phase-locked transponder, this will provide unambiguous range data to distances greater than 500 000 km. Systems are designed for spacecraft dynamics of over 15 000 m/s and 150 m/s^2. They employ sinusoidal modulation and extremely narrow-band processing techniques to provide high-accuracy data with low received signal strength. Figure 14.3 shows a typical antenna configuration, and Table 14.1 gives the location of the STDN S-band antenna stations.

Of increasing significance as an aid to high-accuracy orbit determination, is the worldwide network of *laser tracking stations* (Table 14.2). These provide highly accurate range measurements to satellites equipped with optical retroreflectors, based upon the propagation time of a laser pulse from the tracker to the spacecraft and back. Corrections for

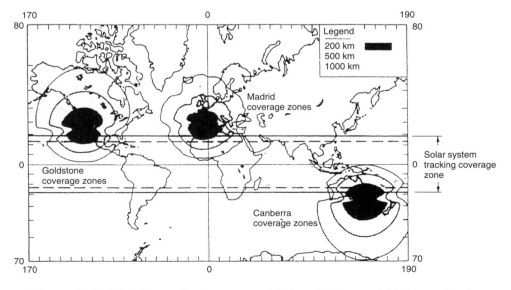

Figure 14.2   DSN 26 m subnet coverage, 200 km, 500 km, and 1000 km altitude

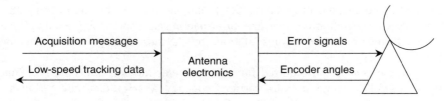

Figure 14.3 Antenna configuration

Table 14.1 Location of main STDN stations

Station	System	Latitude[1]	Longitude (E)	Height above ellipsoid (m)
Ascension Island (ACN)	9 m USB	−7°57′17.37″	345°40′22.57″	528
Santiago, Chile (AGO)	9 m USB	−33°09′03.58″	289°20′01.08″	706
	VHF RARR	−33°09′06.06″	289°20′01.07″	706
	Interferometer	−33°08′58.10″	289°19′54.20″	694
Bermuda (BDA)	9 m USB	32°21′05.00″	295°20′31.94″	−33
	FPQ-6 radar	32°20′53.05″	295°20′47.90″	−35
Grand Canary Island (CYI)	9 m USB	27°45′51.61″	344°21′57.88″	167
Engineering training	9 m USB	38°59′54.84″	283°59′54.84″	−1
Center, Maryland (ETC)	9 m USB (ERTS)	38°59′54.08″	283°09′29.21″	4
	Interferometer	38°59′57.25″	283°09′38.71″	−5
Goldstone, California	26 m USB	35°20′29.66″	243°07′35.06″	919
(GCS)	9 m USB (ERTS)	35°20′29.64″	243°07′37.45″	913
Guam (GWM)	9 m USB	13°18′38.25″	144°44′12.53″	116
Hawaii (HAW)	9 m USB	22°07′34.46″	200°20′05.43″	1139
	FPS 16 radar	22°07′24.37″	200°20′04.02″	1143
Madrid, Spain (MAD)	26 m USB	40°27′19.67″	355°49′53.59″	808
Merritt Island, Florida	9 m USB No 1	28°30′29.79″	279°18′23.85″	−55
(MIL)	9 m USB No 2	28°30′27.91″	279°18′23.85″	−55
Orroral Valley, Australia	Interferometer	−35°07′32.19″	148°57′15.15″	926
(ORR)				
Quito, Ecuador (QUI)	Interferometer	−00°37′22.04″	281°25′16.10″	3546
Rosman, North Carolina	4.3 USB	35°11′45.99″	277°07′26.96″	810
(ROS)	VHF RARR	35°11′42.02″	277°07′26.97″	810
Tananarive, Malagasy	4.3 USB	−19°01′13.87″	47°18′11.87″	1368
Republic (TAN)	VHF RARR	−19°01′16.34″	47°18′11.83″	1368
	Interferometer	−19°00′31.66″	47°17′59.75″	1347
	FPS 16 radar	−19°00′05.52″	47°18′53.46″	1307
Fairbanks, Alaska (ULA)	9 m USB	64°58′19.20″	212°29′13.39″	339
	VHF RARR	64°58′17.50″	212°29′19.12″	339
	Interferometer	64°58′36.91″	212°28′31.89″	282

[1] A minus sign indicates South latitude.
Note: USB, Upper side band VHF, Very high frequency.

internal system delays and refraction are made on-station. Ranging can be performed during night and day up to distances of several thousand kilometres, provided atmospheric conditions are favourable. Nominal ranging accuracy is of the order of 2 cm, and angle data is also provided. The laser telescopes normally have no autotrack, pointing being computer driven according to *a priori* orbit information; thus the angle measurement is

Table 14.2 Location of main SLR (Satellite Laser Range) stations

Station			E. longitude	Latitude	Height (m)	Single-shot precision (cm)	Agency
1181	Potsdam	Germany	13.0652	52.3803	148	20	ZIPE
7086	Fort Davis	USA (TX)	255.9841	30.6770	1964	7	NASA
7090	Yaragadee	Australia	115.3467	−29.0465	245	2	NASA
7105	Greenbelt	USA (MD)	283.1723	39.0206	22	4	NASA
7109	Quincy	USA (CA)	239.0553	39.9750	1110	3	NASA
7110	Monument Peak	USA (CA)	243.5773	32.8917	1842	4	NASA
7112	Platteville	USA (CO)	255.2740	40.1828	1505	11	NASA
7121	Huahine	Tahiti	208.9588	−16.7335	47	10	NASA
7122	Mazatlan	Mexico	253.5409	23.3429	34	12	NASA
7210	Haleakela	Hawaii	203.7440	20.7072	3069	4	NASA
7833	Kootwijk	Netherlands	5.8098	52.1784	94	17	THD
7834	Wettzell	Germany	12.8780	49.1449	661	7	IFAG
7835	Grasse	France	6.9	43.7	1320	7	CERGA
7838	Simosato	Japan	135.9370	33.5777	102	11	SHO
7839	Graz	Austria	15.4933	47.0671	540	4	Obs. Lust buehel
7840	Herstmonceux	UK	0.3361	50.8674	76	4	NERC
7843	Orroral	Australia	148.9	−35.6	950		DNM/NASA
7907	Arequipa	Peru	288.5068	−16.4657	2492	15	NASA/SAO
7935	Dodair	Japan	139.2	36.0	850		SHO
7939	Matera	Italy	16.7046	40.6488	536	15	SAO

Note: SAO, Smithsonian Astrophysics Observatory.

only as accurate as the laser beamwidth. The laser transmitter and the sighting telescope are mounted on each side of the receiving telescope.

The accuracy of laser ranging is critically dependent upon precise timing. Typically, the sequence of events begins with a 1 pulse/s output of the timing subsystem, which is generally referenced to Greenwich Mean Time (GMT) or Universal Time Coordinated (UTC). When the laser is fired, the 1 pulse/s is gated to the laser power supply, producing the high-voltage pulse required to energize the laser. A time delay occurs between the timing system 1 pulse/s output and the actual firing of the laser. This time delay is measured and recorded for accurate time-tagging of the tracking data. A sampling of the laser output pulse is used to start the time interval unit, which counts the time in tenths of nanoseconds until the received pulse from the target terminates the count, providing accurate ranging information. Additional (system) delays are removed by calibrating the laser with a surveyed target of known distance.

14.2.2 Tracking and data relay satellite system (TDRSS)

In the early 1980s, NASA set up space-based systems for Tracking, Telemetry and Command (TT&C) between ground and space, with the tracking and data relay satellite system (TDRSS). TDRSS comprises six operational satellites and one in-orbit spare. The operational ones are separated by 130° of longitude, two at 41°W and two at 171°W, and are

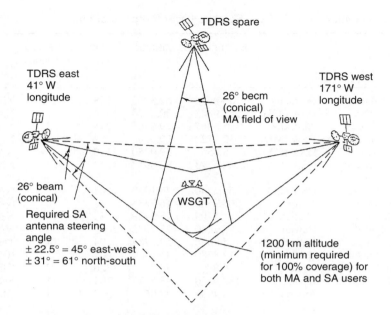

Figure 14.4 TDRSS configuration and coverage limits

centred about the White Sands Ground Terminal (WSGT) in New Mexico (Figure 14.4). Whilst the STDN ground network only provides about 15% visibility coverage, TDRSS can provide 85 to 100% for many spacecraft, including the Shuttle.

Each Tracking and Data Relay Satellite (TDRS) is equipped with two user service antenna systems. The high-gain system comprises two steerable, 5 m dual S/K-band antennas, known as the Single-Access System (S-band Steerable Antenna (SSA) and K-band Steerable Antenna (KSA)). The low-gain system consists of a 30-element S-band phased array, which can provide one forward link and multiple, simultaneous Pseudo-random Noise (PN) code division multiplexed return links; this is known as the Multiple Access (MA) system.

Space-to-ground communications (Figure 14.5) for command, telemetry, and user signals, are routed through a 2 m K-band space-to-ground link (SGL) antenna. During periods of maintenance and K-band outages, TT&C can be supported via an S-band omni TT&C antenna.

The WSGT is configured with three 18 m K-band elevation-over-azimuth (az-el) antennas, a 6 m S-band az-el TT&C antenna and roof-mounted S- and K-band simulation/verification antennas. The communications equipment in the WSGT can simultaneously support two-way communications for six SSA, six KSA, and three MA services, as well as a total of 20 MA return links.

User-tracking equipment can provide 9 ranging and 19 Doppler services, simultaneously. The Doppler observation is a continuous count of a bias plus the Doppler frequency, resolved to 0.001 Hz at S-band and 0.01 Hz at K-band. The range observation is the four-leg round-trip light time resolved to one nanosecond(ns). The range measurement is ambiguous in multiples of the ranging code period, about 0.086 s (13 000 km one way).

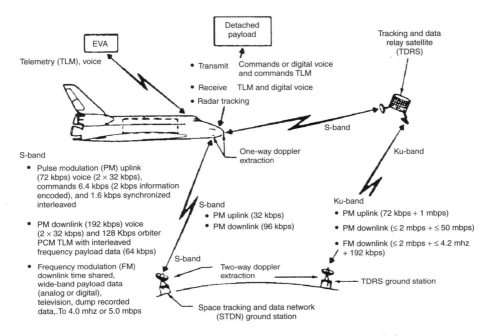

Figure 14.5 Orbital communications and tracking links

The observations are strobed on the whole second, formatted and transmitted to NASA at selectable sample rates of 1, 5, 10, 60 and 300 s.

The first TDRS was launched in June 1983, and since then many missions have used the TDRS system, including the Space Shuttle. The geometric configuration provides the ability to determine the ephemerides of all three TDRSs to an accuracy of at least 100–200 m. Unmanned user spacecraft, currently supported operationally by the ground S-band network, provide an excellent opportunity to evaluate the end-to-end navigation accuracy of user spacecraft.

14.2.3 Typical ground station configuration

Although the antenna is the main item of hardware at a ground station, the other items identified in Sections 14.2.1 and 14.2.2 are an equally integral part of its facilities.

To support spacecraft and their payloads, the ground system must command and control them, monitor their health, track them to determine orbital position, and determine spacecraft attitude from sensor information, although some of these functions are currently carried out on-board the spacecraft. The ground system controls the spacecraft and its instruments or payloads by transmitting command data to the spacecraft and uses spacecraft housekeeping telemetry and mission data to carry out these functions. It acquires mission data from a spacecraft and its instruments and transfers it to the data users. The ground system also supplies any telemetry and tracking information that they may need. Most space missions allow the user's evolving requirements to influence changes in the ground system's data relay and control functions.

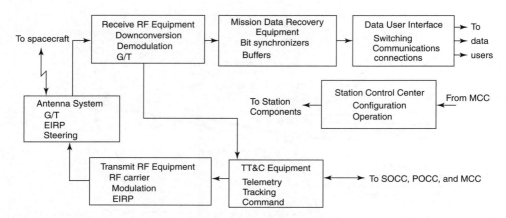

Figure 14.6 The basic ground station has only the minimum components needed to control a spacecraft and to relay mission data to users

A complete ground system (Figures 14.6 and 14.7) consists of a ground station and control centre working together to support the spacecraft and the data user. The prime components [1] are the RF communications link, provision of spacecraft/payload control, issuing commands, determination of orbital parameters, monitoring and processing of telemetry, determination of spacecraft attitude, transmission of spacecraft and payload data and mission operations support. Generally, the ground system commands and controls the spacecraft based on requests from the data user to the control centres. Except for communications satellites, users do not usually send commands independently to the spacecraft.

A single antenna system includes the antenna and mount, its associated electromechanical actuators, the consoles and servo circuitry that control the antenna and the feeds and transmission lines that carry RF signals to and from the RF equipment. The antenna, along with the receive RF equipment, satisfies the required receive gain-to-noise temperature ratio at the frequency of the downlink carrier. It also works with the transmit RF equipment to provide the required effective isotropic radiated power at the uplink carrier frequency. The antenna steering must also provide the look angles required by the mission. For low-Earth orbit missions, these can cover essentially all the visible hemisphere. It must also provide the required steering modes, such as programmed computer steering and autotracking. *Autotracking* refers to the use of the received spacecraft signal itself to steer the antenna. In this case, the antenna system usually provides continuous pointing coordinates to the tracking component at the ground station.

The receive RF equipment is generally in suites of racks, located to minimize transmission-line losses to the antenna. This equipment accepts the downlink carrier frequency from the antenna system, down-converts it to intermediate frequencies and demodulates it to baseband signals for the equipment devoted to mission data recovery and TT&C.

Also in racks near the antenna system, the transmit RF equipment accepts tracking and command signals from the ground system's TT&C component and modulates them onto the RF uplink, which it also generates. In the case of communication satellites, it also modulates user data onto an uplink carrier.

After the RF receive equipment demodulates the signals, the mission data recovery equipment conditions the mission data before relaying it to data users and ground system

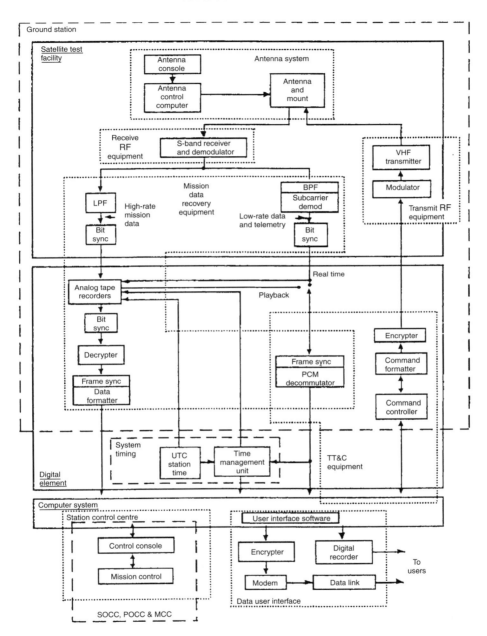

Figure 14.7 Block diagram for a ground system

components. It typically has its own location in the system, but it may be intermingled with the receive RF suite for simple data streams.

The data user interface connects the mission data recovery equipment and the data user. If all parts of the ground system and the data user are co-located, this interface generally

consists of no more than manual or electronic patching of the data lines between the
ground station and the user facilities.

The TT&C equipment conditions and distributes received telemetry and tracking signals. It also electrically formats, authenticates and times transmitted command and tracking
signals. It usually processes these tracking signals and data on the antenna-pointing angle
to inform users about range, range-rate and spacecraft position. TT&C functions are
usually highly automated because of the need for speed, timeliness and accuracy.

The station control centre controls the configuration of and the interconnects between the
ground station components. Operating under instructions from the ground system's mission
control centre, it keeps the ground station configured to support mission operations.

Ground systems operations require time coordination, so one system element maintains
a clock precise enough to meet mission requirements; it distributes clock time and reference frequencies to the other system elements, moving through the co-located elements
of a ground system by wire or cable. It is accurate to within microseconds or better. Its
usual one-per-second timing pulses are synchronized to within a few microseconds or
less to a world time scale, such as UTC. Satellites that are able to transfer time even
more precisely, such as Global Positioning System (GPS), will permit synchronization
well below a microsecond.

14.3   SOFTWARE

There are four major areas of software that either run at or are controlled from a ground
station. These are the pre-pass, real-time, post-pass and on-board software. Except for
the most basic spacecraft systems, some aspects of each of these will be required in the
ground station. In addition, there are a number of very important configuration control
procedures that need to be adopted to safeguard the integrity of the software part of the
ground system.

14.3.1   Pre-pass software

As its name implies, pre-pass software is required to run in advance of the pass of
an active spacecraft over a ground station. There are nominally four types of software
required, namely, Orbit Determination and Prediction, Observation Planning and Scheduling, Command List Generation and Simulation.

Orbit determination and prediction

In determining and predicting the orbit of a near-Earth satellite, there are usually three
factors [2] that determine the accuracy required. These are the ability

1. to track the satellite correctly during a station pass,
2. to reconstruct the orbit and hence the position of the satellite at any time during the
   mission and
3. to predict ahead many weeks for the purposes of mission planning.

It is usually item (2) which has the least tolerance on allowable errors. In order to track from a ground station, the maximum allowable error in azimuth and/or elevation can typically be ± 30 arc min for a large antenna. This is interpreted in the worst case as a time error of about one second in the position of a satellite at 1000 km altitude. Clearly, the allowable error decreases slightly with decreasing altitude.

For orbit reconstruction, the requirement is based on observations taken at times before and after a reconstructed epoch. For instance, in a project that requires mapping features of the Earth, position reconstruction has to be exceptionally accurate (within metres), and this is definitely not possible using a single ground station, even with the aid of advanced techniques such as a laser ranging. However, for many satellites an accuracy of ± 1 km in altitude and ± 5 km in along-track position is sufficient. For orbit prediction, the required accuracy depends not only on the accuracy of orbit determination but also on the accuracy of the orbit propagator. Long-term predictions are usually required for advance planning of pass times, eclipses, attitude manoeuvres or experiment observations. Typical accuracy requirements are a few seconds per week, accumulative.

For highly accurate orbit determinations to less than a 1 m error in position, it is necessary to use laser or microwave tracking data together with highly sophisticated numerical and analytic algorithms such as those incorporated in the NASA GSFC GEODYN [2] programme or the industry-standard STK. For most purposes, however, a simpler orbit determination and prediction process is sufficient.

Observation planning and scheduling

In planning the observation schedule for a typical scientific satellite, it is necessary to know where the satellite will be in the future. The antenna control software needs to know where to find it during passes, and the data processing programmes need to know where it was when the data was accumulated. The data exchange between these software packages is summarized in Figure 14.8.

The observation planning and scheduling software will take in all requests from the spacecraft 'user'. It checks that observation modes, attitude control or spacecraft constraints are not violated. It will then schedule all observations into an optimized order, taking particular care not to violate further spacecraft constraints during attitude manoeuvres. An important part of the optimization process is to minimize the slew time of these manoeuvres, as this is generally non-productive for data gathering.

Command list generation

A pass schedule is constructed as far ahead of the pass as possible. It contains a set of instructions or commands that define the actions to be taken during the pass, the tracking data, the new satellite plan and any other data to be loaded into the computer memory. Normally, the pass schedule will be created automatically using a set of default options, but if necessary it can be constructed interactively, overriding some or all these options.

In ideal conditions, the pass schedule will control the entire pass with minimal intervention from the satellite controller. Each table entry will be interpreted by the commanding system software, which will invoke other modules as necessary.

A typical pass schedule might include instructions such as requested dumps of stored data, confirm that the previous satellite plan is unchanged or load or enable a new satellite

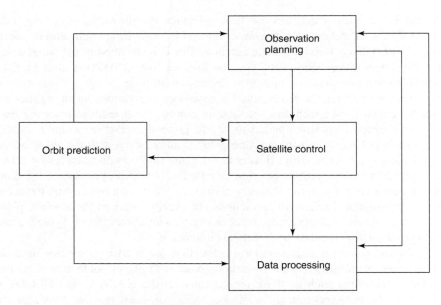

Figure 14.8 Observation planning interfaces

plan. The data to be dumped during the pass may have been stored on an on-board tape recorder or in a part of the computer memory.

Simulation

No operator of spacecraft would ever dream of building and launching a spacecraft without having available on the ground a spacecraft simulator (often the Engineering Model). This is used post-launch to aid problem-solving in the event of partial spacecraft or subsystem failure. For exactly the same reasons, it is prudent to provide well in advance of launch a simulator for the complete end-to-end ground system. This includes the capability to simulate both the spacecraft housekeeping data as well as the instrument data. It is also useful for training operation staff (see Section 14.5) and checking the satisfactory handling of corrupted data.

14.3.2 Real-time software

Real-time software operates during the whole of the period when the spacecraft is visible from the associated ground station. It includes computer control of the antenna tracking, command uplink and verification, data reception and status checking of all critical system parameters.

Tracking

At the start of a pass of a satellite over the ground station, the antenna is commanded by the computer to point towards the horizon in the direction in which acquisition of signal (AOS) is expected. Two possible courses of action now exist. The first is called 'programme control', in which the antenna drive is programmed to follow exactly the predicted path of the satellite across the sky right through to loss of signal (LOS) near the opposite horizon. The second is called 'automatic control', in which the error between the satellite position and antenna-pointing direction is automatically and continuously fed back into the antenna drive encoders, thus enabling the satellite to be tracked in closed-loop automatic mode. Antenna-pointing direction is logged throughout the pass to be used as input to future orbit determination.

Command uplink and verification

The process of command uplink and verification is covered in full detail in Chapter 13. Essentially, the tasks are to uplink individual and/or strings of commands from the previously generated pass schedule, and to verify that those commands have been correctly received and implemented (or stored) by the satellite on-board the data handling system. Although command strings always contain parity codes for verification, it is often standard practice for commands to be retransmitted by the satellite to ground in order to make a comparison between transmitted and received commands. This process then helps to detect uplink errors.

Data reception

During a satellite's ground-contact, data from the tape recorders is transmitted to the ground over a high-speed telemetry (HST) link—anything from tens of kbps to many Mbps—and stored on analogue tape for digitization after the pass (Figure 14.9), or directly

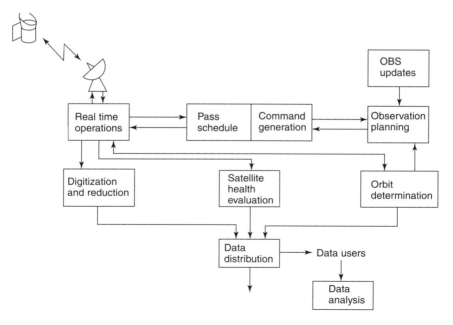

Figure 14.9 OCC software

on to digital tape or solid-state memory. The contents of the computer memory form part of the data that is transmitted continuously over a low-speed telemetry (LST) channel, usually at the rate of a few kbps. The rest of the frame contains command verification data and satellite housekeeping data. All the LST data transmitted during the pass is received and processed in real time.

Status checking

Both before and after the uplink of commands to the satellite, real-time data from the housekeeping telemetry frames are received at the ground station. These data contain parameters that are critical to both the health and operation of the satellite. In general, these data are a mixture of status flags (e.g. subsystem on/off) and engineering parameter values, and must be checked and analysed immediately. The status checking software trips this critical data out of the main telemetry stream, converts it (where appropriate) to engineering values, compares it with in-limits criteria and displays it on control consoles for immediate notification and if necessary, action, by subsystem engineers.

As each low-speed frame is received, its satellite clock time is extracted and saved, together with the current value of the Universal Time (UT). This set of timing points will be used to calibrate the satellite clock during post-pass processing. The command verification data from each frame are used to check that the commands sent to the satellite were received correctly. The housekeeping data contains a set of satellite status indicators and sensor read-outs, and the values of a selected group of these parameters can be extracted from each frame and displayed on control consoles at the ground station. For most analogue parameters, a set of engineering and working limits have been defined, and any limit violations that are detected are also displayed on the consoles.

During the pass, a copy of the contents of the whole of the on-board computer memory will be built up, and new data will be loaded into some parts of it. Normally the whole memory is dumped cyclically, but during the pass commands may be sent to the satellite to request that particular areas be dumped immediately. Parts of the memory, such as the programme area, should remain unchanged and will be dumped and compared with the expected contents. Parts of the memory that have been reloaded will be dumped and verified. Error-free copies of the memory areas used for the storage of engineering data will be built up.

If necessary, the satellite controller can interrupt the pass schedule and enter commands from the console. Control may be returned later to the pass schedule, either at the interrupted entry or at another. If an error is detected during processing, the satellite controller will always be informed so that appropriate remedial action can be taken.

14.3.3 Post-pass software

The immediate post-pass tasks include extraction of housekeeping and science/technology data for quality control and health assessment, data processing, orbit determination and data analysis.

Health assessment

After a pass, the ground software extracts and processes the engineering data in order to assess the success of the observations' or experiments' schedule on the satellite, and it supplies the instrument and spacecraft support teams with the information they need to monitor the spacecraft's health and performance.

The data derived from the LST link includes a set of timing points and a copy of the satellite event data, together with stored housekeeping data from the computer memory. The timing data is used to derive a single timing point that correlates the satellite clock with UT. All the data from the satellite is labelled with the satellite clock value, so an accurate correlation with UT is essential.

The event data include a record of all the attitude calibrations and changes, and this is combined with some of the stored housekeeping data to give a compendium of engineering data, which are needed during analysis of the science data.

Data processing

The task of data processing (other than that contained within the Status Checking and Health Assessment described above) is two-fold: the first is further processing of house-keeping data to provide subsystem engineers with all their products (such as the full range of temperatures, pressures, voltage/current levels). The second is processing of science/technology data from the raw telemetry (Level 0). In both cases telemetry data are demultiplexed, telemetry is quality checked and flagged, time-tagged and converted to Level 1 (calibrated and annotated raw) data. The data are then passed on to the specialist engineers and scientists for further detailed analysis.

Orbit determination

Whereas the orbit prediction process described in Section 14.3.1 is used for planning the mission and its operations, orbit determination is the process whereby the previous best estimate of the satellite's orbit (and its position within the orbit) is updated as a result of measurements taken during the ground station pass, and from other measurements taken during the processes described in Section 14.2.1. For most general purposes, it is sufficient to determine the satellite along-track position within its orbit to an accuracy of a few kilometres. However, for remote sensing types of satellite (e.g. ERS-2, TOPEX, ENVISAT), the satellite radial position is required to be of extremely high accuracy, in some cases better than 10 cm.

Data analysis

Data analysis software often accounts for the largest share of the budget for software development. It encompasses the software for analysis of trends in the engineering data as well as the complete analysis of all the science/technical data through Levels 2 and 3 (e.g. fully processed data providing maps of the sky, celestial bodies, Earth surface, resources). It includes all the analysis and graphics facilities, as well as the often worldwide distribution of data to the users.

14.3.4   On-board software

This software resides in the spacecraft's own on-board computer; there are two basic types. Read-Only Memory (ROM) contains the basic instructions and safeguard modes during operations, and is built into its computer before launch; once launched, it cannot be changed. Random Access Memory (RAM) software contains more subtle instructions for the spacecraft, giving it a greater deal of sophistication and flexibility. It can be programmed before launch, but has the added advantage that it can be modified or built up after launch by uplinking from the ground station. Data handling is covered in greater depth in Chapter 13.

14.3.5 Configuration control

Controlling the design, development, integration, testing and modification of a large software system is a complicated task in any business. Software should be developed according to a standard software life cycle (i.e. requirements, design, development, test, integration and documentation) [3]. For a large space project, the problems are compounded for a number of reasons:

1. There are interfaces not just between local system boundaries, but between systems designed and developed in different locations.
2. The requirements of the software system are regularly changing to match increasing capabilities of the hardware designers.
3. There is often a need, post-launch, to be able to modify any of the software extremely quickly. A satellite cannot be turned off and put on ice until the problem is solved some days or weeks later.

Throughout the design, build, integration and test period it is necessary to keep a strict control on changes to the design of software. A number of procedures are involved and are now described (see also Figure 14.10).

- *Configuration control board (CCB)*. A CCB needs to be formed, usually consisting of the Quality Assurance (QA) manager, software manager and subsystem managers. Their function is to approve/disapprove of all software change requests, to approve the completion of all test reports and to approve the final delivery of all software products.
- *Local reviews*. In-house there are regular internal reviews of software design and build. These are accompanied by 'walk-throughs' of each programme.
- *Test plans*. The plans are written for all software items, and test reports produced on the results. All errors, or omissions from the requirements specification, are documented on non-conformance report forms, and these are also reviewed regularly to ensure progress in countering the errors.
- *Simulation data*. Data simulators are always required, often one to simulate the real-time system, one to simulate the housekeeping data and the spacecraft orbit and environment and one to simulate the science data, if present. The three sets of simulated data are merged and a series of tests on the complete software system is run by passing the data right through the system. Initially the data is error-free; then the simulation data is variously and deliberately corrupted to be more like the expected operational situation, and to ensure that the software system can cope adequately.
- *Operator control*. A number of training exercises are held, to train the Operations Control Centre (OCC) operators in running the different software packages. Similar

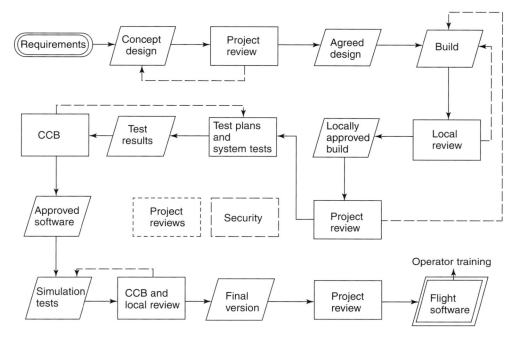

Figure 14.10 Software control mechanisms

exercises are held to train the data analysts in the art of identifying software and spacecraft problems. In this respect, several of the training exercises consist of 12-, 24- and 48-h mission simulations.

● *Security.* A sophisticated system of security is required to ensure (1) that no programmes and/or files are inadvertently erased, (2) that copies of programmes and files are available in case of system crashes, and (3) that no individual user can inadvertently corrupt part of their, or another individual's software. At the same time, it has to be possible to make changes quickly when required. Consequently, a number of *flight directories* are set up containing the mission software. No software can be modified without prior approval of the software manager, and all such directories are *password* protected. All software modifications are made and tested in external directories before approval is given for the modified software to be incorporated in the flight directories. Security tape dumps are made at regular intervals, both before and after launch, for all software programmes and files that are updated, so protecting the system against computer failure.

● *Post-launch control.* In a perfect world there would be no need for post-launch control of the software system, but life is never perfect and space projects are certainly no exception. It is recognized that errors will appear after launch that have not shown up during pre-launch testing. It is also envisaged (correctly) that satellites do not behave in a totally predictable way, and so design changes are needed in the software. Therefore, it is desirable from the outset to retain the CCB post-launch and adhere to all procedures adopted pre-launch.

14.4 PEOPLE

As with any large operational set-up, there is a requirement for people to run the facility, ably abetted by well-defined operational procedures. This includes the following:

- *Site and project management*—required for the supervision of day-to-day operations both on the ground and in orbit.
- *Operations shift staff*—to staff the ground station, often on a 24-hours per day, 7-days per week basis. This task is particularly onerous given that operations and processing are often continuous over the period indicated, and that spacecraft problems are just as likely to occur in the middle of the night as they are in the more sociable middle of the day. Even a small ground station (single-spacecraft operations) is likely to need about five staff in attendance at any one time to cover this function. In order to cover every hour of the week, normally five shifts of staff will be required. Thus a total of 25 staff is required for round-the-clock operations. It is partly because of this high level of staffing required for each ground station that the concept of TDRSS is so attractive.
- *Hardware staff*—for maintenance of ground station equipment (e.g. the antenna electrical and mechanical systems, the RF equipment, the computing hardware).
- *Software staff*—for the maintenance, modification and possible correction of ground system and on-board software, and the full life cycle requirements of new software development.
- *Data and engineering support staff*—to plan operations, discuss results, and to be on hand at times of crisis.
- *Administration*—secretarial and administrative support covering, for example, typing, production and distribution of documentation, photocopying, archiving, personnel advice, travel arrangements and meeting logistics.
- *Specialist engineers*—for detailed monitoring and analysis of data from the spacecraft and instrument housekeeping data streams.
- *Specialist scientists/technologists*—for detailed monitoring and analysis of data from the instrument science/technology data streams.

At the time of launch of a mission, and at least for the period of in-orbit-checkout (typically 3 to 9 months) the above staff can easily number 250 people even for a small spacecraft. Once the mission is into routine operations, the number can decline to less than 100. Excluded in these totals are end users of data. These are usually located remote from the ground station, and can often number over 1000 people.

14.5 OPERATIONS

The preceding sections have outlined in varying degrees of detail the three main 'physical' components of a ground station. The fourth component, namely operations, is the part that brings the hardware, software and people together into a unified programme. The operations team is the fundamental human unit that integrates the mission software and hardware into as effective a routine process as is possible. It is becoming more commonplace these days for ground stations to operate several satellites at once, albeit with the increasing assistance of automated procedures. However, the success of a mission will very much depend on the standard of training and experience of the operations team.

In order to ease the process of operating a spacecraft, a number of well-documented procedures are developed in parallel with the main hardware and software development. These documents include:

- *Interface documents*—describing how each instrument interfaces its data to the main spacecraft bus, how the spacecraft command and data handling facility interfaces with the ground and laying out the detailed (bit-level) description of data formats.
- *User manuals*—describing how the spacecraft and each instrument works (i.e. operation options), what the normal operating modes are, what the non-routine/emergency modes are, and how the ground hardware modes operate (e.g. antenna, data processing systems).
- *Operational procedures*—describing the orbit-by-orbit and daily work tasks, and the emergency procedures. (It is of enormous benefit to have considered all or at least most of the possible failure mechanisms of the spacecraft, payload and ground station, before launch. Corrective procedure plans can then be produced and used for reference during actual failure in the mission).
- *Operations simulator*—available per spacecraft, per instrument, per day, per pass, used for routine and emergency purposes, for end-to-end data systems testing and for staff training. The operations simulator should be available at least one year before launch.

14.6   COST-EFFECTIVE AND AUTONOMOUS GROUND SYSTEMS

Spacecraft hardware, instruments and launchers have undergone a trend towards lower costs per unit output in recent years. This is similar to the trend experienced in the computer industry. As the capability of software systems has risen, so has the percentage mission cost of the Ground System and Operations. There is therefore a clear need to focus cost reduction techniques on the Ground System and Operations.

Well before Mr Daniel Goldin's famous dictum 'Faster, Better, Cheaper' there was growing pressure from users to reduce the cost of access to space. At the same time, the search began (and is still continuing) for lower launch costs.

14.6.1   User needs

Before we consider the methods available for cost reduction, let us consider our requirements.

The starting point in defining the requirements on the Ground Station is to consider what the User actually needs (as well as what he/she wants, which may not necessarily be the same!). Overall, a rough guide to the main requirements may be considered as

- lowest possible cost, *but*
- reliable operations (not missing passes or losing data)

● fast return of critical data
● regular return of bulk data
● rapid response for critical commanding
● ease of access to data.

In order to achieve the low-cost goal, it is not, however, unreasonable to expect some compromises to be made. These may include

● acceptance of occasional (1 in 1000?) lost passes
● acceptance of some (<1%) lost data
● non-rapid return of non-urgent data

14.6.2   Cost reduction scenarios

The cost of mission operations represents a significant portion of the total programme costs, typically 30 to 40%. Thus the ground segment configuration (i.e. hardware and software) and the operational modes (i.e. their complexity) have a significant influence on total costs and must be given serious consideration in overall system design.

Reliability versus cost

For large missions, it has always been normal practice to maximize the reliability of the ground system despite the associated increase in cost. This is not unreasonable for man-rated missions, but is often an unnecessary expense for most other missions. There is a very sizeable potential reduction in cost to be obtained by making just a very small reduction in system reliability. It is proposed here that it should be agreed 'up-front', that a small percentage (perhaps 0.1%) of satellite passes can be lost through ground system outage. This may (though not necessarily) lead to some data loss, but even so a data loss of a few per cent is not usually significant. By agreeing to this potential reduction in reliability, the level of hardware redundancy (and perhaps software complexity) required in the ground system can be significantly reduced, and hence, the cost is reduced. Likewise, if the number of passes required per day to support the mission operation can be reduced through a slightly less than optimal coverage programme, the cost of operations can also be reduced.

Data availability

There is no doubt that for all missions it is essential to be able to process some subsets of the data in real time and/or near real time. However, the less data that has to be processed in this manner, the simpler the immediate ground system complexity becomes. For the majority of small satellite missions, it should only be necessary to process instrument/bus health data as a matter of urgency, thus decoupling the task of satellite operations from that of off-line data processing.

Data transfer

There are basically two different methods of transferring data from the operations part of the ground system to the user or data processing centres. The first (and most expensive) is via one of the many space or terrestrial data links. This is the common route for most satellite data and ensures that the data gets to the end-user very quickly. However, it is more often the case that although the end-user likes to have this data 'as quickly as possible' it is not often an absolute necessity. In this case, the alternative route via mailed magnetic tapes/optical disks can be just as satisfactory. Possibly some (small) percentage of the data can still be transmitted via a low bandwidth (and lower cost) data link such as the World Wide Web; it is important to try to avoid the exclusive dedicated use of these links as this too adds to the cost.

Data access

There are as many different philosophies regarding methods of data access as there are concerning designs of satellite. Generally, however, the most cost efficient and practical method is the concept of a Centralized Data Handling Facility that is accessible by users over local data networks. This concentrates the pipeline data processing in one place, whilst allowing the individual users both the ability to develop their own specialized software and to make full use of centrally developed software.

The following are some general considerations concerning the ground segment configuration and operation.

System modularity

In exactly the same way as satellite costs can be significantly reduced by greater use of common modularized subsystems, ground system configurations can also be modularized. Instead of developing individual Electrical Ground Support Equipment (EGSE) and Ground Segment equipment for every instrument and/or satellite, there are standardized off-the-shelf equipment now being developed that can subsequently be customized to individual needs, at much lower cost. Within the ground system itself, computing power is currently sufficient to combine the tasks of TT&C into a single low-cost workstation. Of even more potential benefit, is the reuse of previous mission software for many of the data analysis functions. As an example of this, the data analysis software for the JET-X instrument, which will fly in 2005 a part of the Spectrum-X mission, is almost entirely based on software developed for the ROSAT mission launched in 1990. This scenario alone has cut the software development cost for this mission by a factor of three.

National facilities

Probably the greatest potential for cost reduction of the ground system is by making greater use of national facilities. Agency facilities are clearly required for large (manned and unmanned) missions, but are often too cumbersome, inflexible and expensive for small missions. It has usually proven far more cost-effective to employ national facilities—ideally utilizing just a single ground station. For instance, the two European AMPTE spacecraft were controlled from single stations in Germany and England, respectively.

The UK station was developed at very low cost by updating the original Infra-Red Astronomical Satellite (IRAS) control centre to the requirements of the AMPTE mission. Although new software and operational procedures were necessary, very little new hardware was required. Similarly, it is now possible to receive data using rooftop antennas and command/receive using desktop PCs.

Autonomous operations

The main cost element of operations at a ground station is that of the personnel. For most space missions of the past, operations have entailed staff coverage 24-hours per day, 7- days per week. To undertake this safely requires five shifts of anything from 5 to 50 people/shift, which is a very large human resource. However, with the very much higher standards of reliability and power for modern computing systems in relative terms, particularly for PCs, it is feasible to automate many of the 'routine' operations at a ground station. Furthermore, with the science of artificial intelligence and expert systems now having reached maturity, many more of the non-routine operations can be considered as candidates for automation and indeed for autonomy. (In this context automation can be regarded as an open-loop execution, and autonomy as a closed-loop execution.)

The following subsystems are regarded as candidates for either automation or autonomy:

1. *Tracking*. In the first instance, open-loop auto-tracking of high altitude satellites is possible by setting up predicted tracking data in advance. A second stage is closed-loop feedback of error-signals into the orbit determination process in order to update the orbit parameters autonomously for future prediction.
2. *Automatic retuning of the frequency set-up* of the ground station RF equipment in order to handle downlink signals at different frequencies from different satellites, allowing further processing and analysis at convenient times. With modern processor-controlled synthesizers, this is quite practical. It is also possible to estimate Doppler shift for each pass, with either real-time estimates being fed to the controller or a look-up table prepared before each pass.
3. *Automatic reception and storage of downlink telemetry*. Modern desktop computers are capable of handling telemetry in real time at a rate that extends into the Megabyte per second range. Creating suitable data files for storage also presents little difficulty, either on tape or disk.
4. *Automatic conversion of critical raw data to engineering units*. Given either algorithms or look-up tables, there is no difficulty in making the necessary conversions. Real-time indication of trends or other statistics is also straightforward.
5. *Automatic and eventual autonomous checking of critical data* during real-time passes to identify parameters that are out-of-limits. There is also the potential for automatic switching to on-board backup systems.
6. *Automatic computer dial-out to on-call engineers*. These expert subsystem engineers will have modem-link accessibility to the operations computer whenever 'called' by the computer for human fault-finding. All elements of a dial-out system are now commonplace. A telephone 'bleeper' may be an integral part of the system.
7. *Automatic pipeline data processing*. This is already commonplace.
8. *Automatic distribution of data* (engineering and science) *to end users*. With the widespread use of the Internet, distribution of data to anyone with suitable equipment

is straightforward for moderate amounts of data. Larger quantities may require other methods.

9. *Automatic production of summary telemetry quality*, spacecraft health and science data outputs.
10. *Automatic uplink of pre-prepared commands* with the possibility of modifications or selection of alternative sequences depending on automated checkout. Commands for a given pass will clearly be generated and stored in a command file, which will then be uplinked at the appropriate time during the pass.
11. *Automatic close-down at end-of-pass*, and automatic set-up for following passes. Some degree of intelligence needs to be built-in here. For example, if a pass is closely followed by another, it may not be necessary to stow the antenna, which could instead be pointed to the start position of the following pass. Setting up for subsequent passes will probably involve no more than opening the correct files for configuring the TT&C equipment and accessing the relevant look-up tables.

14.6.3   So who is doing what?

The main focus of cost reduction techniques is as follows:

- *Commercial off-the-shelf (COTS)*. Several commercial organizations (including Loral Aerosys, Storm Integrations, AlliedSignal, Hughes) are developing COTS products for general ground station use.
- *Autonomy*. Perhaps the most significantly widespread methodology is through ground station autonomy. Although not yet generally applicable for large (complex) missions, and certainly not yet for manned missions, many small missions are now being operated with a significant level of autonomy. Good examples of this are the EUVE mission operations at University of Berkeley, and the ACE downlink operations at RAL. Jet Propulsion Laboratory (JPL), the US Air Force Phillips Laboratory, and Surrey Satellite Technology are also developing autonomous systems, as are several other well known organizations.
- *Data standards*. The pioneering work of the Consultative Committee for Space Data Systems (CCSDS) has been highly influential in the way telemetry data is handled, processed, transferred and archived more effectively and efficiently.
- *Use of national facilities*. The growing trend of making greater use of local/national facilities has occurred primarily because costs are lower. Co-sharing operations staff with data processing staff is both effective and lower cost, as evidenced at the above-mentioned organizations.

14.7   LOOKING TO THE FUTURE

Until the early 1980s, typically 90% of the cost of a mission was the cost of the space segment (i.e. the spacecraft plus launch), with only 10% going into the ground system. During the 1990s, more effort was put into ground systems as the power of ground-based computers and related software increased. For the new generation of high data rate space systems, (such as the NASA Earth Observing System (EOS)), as much as 50% of the cost of a mission is planned for the ground system. To give an idea of the size of task at hand, EOS, together with parallel programmes in ESA (ENVISAT/METOP) and Japan (ADEOS II), will provide 15 years of operation at a continuous data collection rate of about 50 Mbits/s. This gives a total of around one terabyte/day of processed data!

There are a number of key technologies that need innovative development in order to handle the vast quantities of data that will be forthcoming at the end of the decade:

- *On-board data storage*—Although many spacecraft will operate continuously through systems like TDRSS, there is still a great need for large on-board mass storage, for example, solid state memory, bubble memory, high capacity tape recorders.
- *On-board data compression*—Even with the availability of mass storage on-board the spacecraft, the data still has to be transmitted to ground. To overcome partly the very high data rates required, the concept of on-board data compression and subsequent ground data decompression/reconstruction is attractive.
- *Ground data storage*—Having got the raw data on the ground, it has to be stored in a readily accessible form, such as high density optical disks and the associated 'jukebox' type carousels.
- *Fast processing*—the on-going march towards ever faster (parallel) processing of data.
- *Automated data analysis*—with the ever growing use of knowledge-based systems including artificial intelligence, smart (expert) systems, autonomous TT&C, and even the automated research assistant!

REFERENCES

[1] Larsen, W. J. and Wertz, J. R. (Editors) (1999) *Space Mission Analysis and Design*, 3rd Edition, Kluwer Academic Press.
[2] Martin, T. V. (1987) GEODYN Descriptive Summary. NSA 5-11734-MOD 65.
[3] *ESA Software Engineering Standards, Issue 2*. (1991) ESA PSS-05-0, ESA Publications Division, Estec, Noordwijk, The Netherlands.

15 SPACECRAFT MECHANISMS

H. Mervyn Briscoe[1] and Guglielmo S. Aglietti[2]

[1] European Space Research and Technology Centre (ESTEC), European Space Agency
[2] Aeronautics & Astronautics, School of Engineering Sciences, University of Southampton, United Kingdom

15.1 INTRODUCTION

Spacecraft usually contain mechanisms that perform functions that are essential to their operation. These mechanisms often have a single point failure. The effect of this upon the mission is never less than serious and is frequently catastrophic when it occurs. In this sense, mechanisms are critical. Usually, they will be part of one of the major subsystems such as the Attitude and Orbit Control System (AOCS) or the power supply system. This, at once makes reliability the primary goal for every mechanism designer.

A convenient approach to the study of spacecraft mechanisms is to divide them into two basic categories:

- *One-shot devices*, which, as the name suggests, are mechanisms required to function only once during the spacecraft mission, e.g. satellite release mechanisms, deployment mechanisms for whip antennas, most of the deployment mechanisms for solar arrays, and so on.
- *Continuously (or intermittently) operating devices*, which include all those mechanisms that are required to run continuously (e.g. momentum and reaction wheels) or intermittently (e.g. antenna pointing mechanisms (APMs), solar array drives (SADs) etc.) throughout the life of the spacecraft.

The *general requirements* [1,2], are determined by the launch and space environments that have been described in Chapters 2, 7, 8 and 11, and the need to achieve near-absolute reliability. The launch conditions provide the worst mechanical environment, and some ball-bearing systems may need to be specially protected.

The space environment is generally not very hostile to mechanisms, with the two important exceptions of tribology and temperature, especially thermal gradients. Nearly

---

Spacecraft Systems Engineering (Third Edition). Edited by P. W. Fortescue, J. P. W. Stark and G. G. Swinerd
© 2003 John Wiley & Sons Ltd

every mechanism failure in space is due to either poor choice of materials, an inadequate understanding of space tribology or poor estimation of thermal gradients, which can lead to high loads and high torques. The requirement for very long life in rotating machinery puts an even stronger emphasis on these three areas. Many failures of space mechanisms are caused by poor engineering practice and an inability to assess risk. Every decision taken by the designer clearly involves an element of risk, and an ability to appreciate and assess that risk is, in the author's opinion, *condicio sine qua non* for success.

15.2 ONE-SHOT DEVICES

Typically, the function of one-shot devices is to change the structural configuration of the spacecraft. These changes can be dramatic events, such as stage separations of a launch vehicle or major changes in a satellite configuration such as deployment of large solar arrays or they may be relatively minor events such as the deployment of a small whip antenna. In all cases, the change in configuration is produced by allowing and then forcing the relative motion of some of the parts of the mechanism that were previously constrained. Consequently, in most systems it is possible to distinguish at least two separate functions carried out by two (or more) different types of devices.

The first function is to restrain firmly one or more parts of the mechanism for all the time up to the moment when the system is finally 'fired'. The actuation is generally carried out by pyrotechnic devices (e.g. explosive bolts or pyro cutters). When fired, the moving part of the mechanism is released, directly or by activating/opening a more complex latching mechanism. Currently, there is a trend towards the use of non-explosive devices for applications on-board satellites for various reasons that will be discussed later in this chapter.

The second function is to enforce a predetermined movement (i.e. deployment) of particular parts of the mechanism. In order to perform this function, most mechanisms store some form of potential energy, which, during the release/deployment, powers the movement. Since these devices are designed to work only once, one of the most commonly used methods is to store the potential energy in springs. The springs are loaded when the mechanism is 'charged', and they store the energy until the moment when mechanism is activated, which can be months or years later (as in the case of the release mechanisms for the *Rosetta* lander or planetary probes such as *Huygens*). The release of the elastic energy stored in the spring is a relatively sudden phenomenon and the moving part(s) might acquire a relatively large kinetic energy. Consequently, damping devices are often included, in particular, to control the arrest of the mechanism when fully deployed. Devices for separation, for example launch vehicle stage separation or satellite release mechanisms, do not generally require the addition of dampers, since the kinetic energy stays with the parts that separate, causing them to drift apart. Other mechanisms are powered by the expansion of gas (e.g. nitrogen) previously compressed and stored in cylinders. In this case, the release of the gas is controlled by valves and the expanding gas can be used to move linear pistonlike devices, or to deploy inflatable structures. For spinning spacecraft, the deployment of radial appendices can by driven by 'centrifugal' force (e.g. Geos), but this is the exception rather than the rule. Some mechanisms, for example those to deploy booms, are sometimes required to be retractable, and in this case they are generally actuated by electric motors. However, these mechanisms cannot be classified as truly one-shot devices.

15.2.1   Separation systems

In every mission, various separation systems are used, and classic examples are the mechanisms used to release satellites from their launch vehicle. Most of these mechanisms are based on the use of a *Marmon clampband*, which has been in use since the early days of space exploration. It remains one of the most reliable and commonly used mechanism [3].

The basic version of this device, whose cross-section is shown in Figure 15.1, is a circular belt with shoes having a V-shape horizontal groove, secured to its internal surface (see also Figure 8.4 of Chapter 8). When the clampband is taut, the V-groove on the shoes engages two circular mating flanges (one flange is part of the launch vehicle structure and the other one is part of the satellite structure) holding them together. The mechanism is activated by a pyrocutter or a similar device that cuts a retaining bolt. Once this bolt is cut, the belt opens up (usually helped by torsional springs) pulling out the shoes, and thus freeing the flanges. At this moment, the satellite is pushed away by springs (separation springs) secured to the structure on the launch vehicle side. For flanges with a diameter larger than 1 m, the clampband is usually divided into two or more segments of equal length, joined by the pyrotechnically actuated bolts. Figure 15.2 shows a view of all the mechanical components of a large separation mechanism. In this case, these devices have to be perfectly synchronized to make sure that the belt opens up at all the junctions simultaneously. The main advantage of this type of mechanism, apart from its simplicity, is that the two flanges are held together along the whole perimeter, thus transferring the loads in a relatively uniform manner. This is particularly advantageous when the two structures to be constrained are cylindrical, which for launch vehicles is always the case. Other types of release mechanisms for spacecraft have been designed, tested and used successfully, but the Marmon clampband, in its various versions, is still the most widely used.

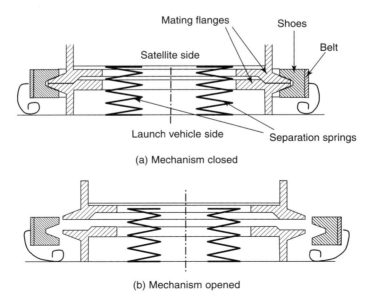

(a) Mechanism closed

(b) Mechanism opened

Figure 15.1   Cross-section of Marmon clampband type release mechanism

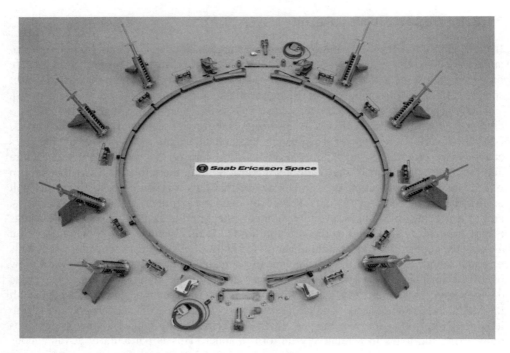

Figure 15.2 Components of the Saab Ericsson Space separation mechanism (Reproduced by permission of Saab Ericsson Space.)

15.2.2 Mechanisms for deployable structures

Various types of mechanisms can be used in order to deploy elements of the spacecraft, which are generally referred to as *appendages*.

The simplest method to deploy an appendage is to use a knuckle joint located at the root of the appendage. This joint can be a simple hinge coupled with a torsional spring. An example of this is shown in Figure 15.3, which is used to deploy a whip antenna. The torsional spring forces the rotation of the hinge, driving the deployment of the antenna and keeping the structure fully deployed once the hinge reaches its end-stop.

Sometimes deployment mechanisms must be required to be self-locking, in order to guarantee that the deployed part of the structure does not fold back when loaded. For example, this is the case for the joints used to deploy the legs of the Rosetta lander's landing gear (Figure 15.4). Here it is crucial that the legs, once deployed are kept in position, and therefore a device called a 'mechanical diode' is used to guarantee that the deployment movement of the legs cannot be reversed by the forces applied during touchdown.

A very simple type of self-locking joint that has found many applications in space is the *tape spring hinge*. This flexible linkage is made by two tape springs parallel to each other, which connect two structural segments as is shown in Figure 15.5. The joint is folded by allowing the tape spring internal to the hinge to buckle and then bending the deployable part of the mechanism. Once folded, strain energy is stored in the tape springs and, as soon as the constraint to the deployable part of the mechanism is removed, the tapes spring straight, thus opening the joint. Once straight, this joint has a high rotational/bending

Figure 15.3 Hinge of whip antenna (Reproduced by permission of AEC-ABLE Engineering Inc.)

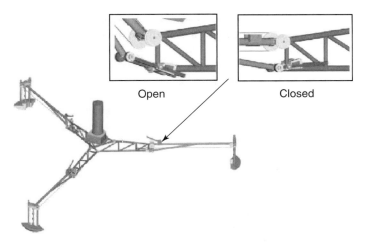

Figure 15.4 Rosetta lander landing gear (Reproduced by permission of Dr H. Rosenbauer)

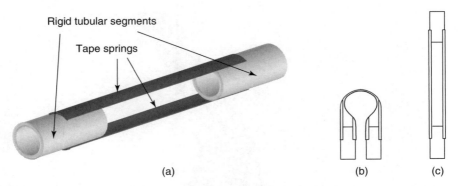

Figure 15.5   Tape spring joint (a) three-dimensional view of the joint, (b) side view of the joint (folded) and (c) side view of the joint (deployed)

stiffness (produced by the high buckling load of the tape springs) but relatively low shear stiffness. This mechanism does not control and restrain the parts as precisely as a hinge, but it is much simpler and cheaper. Also, this mechanism can be exploited to deploy relatively long booms made up of shorter segments connected by tape spring hinges. This approach has been used for the design of a 2.5 m long boom used on the micro-satellite *FedSat*. The boom, made of six rigid tubular segments connected by tape spring hinges, carries a magnetometer on its tip, and is used to stabilize the satellite exploiting the gravity gradient torque.

A common problem in the design of these types of mechanisms is that the structure to be deployed is made up of one or more relatively long structural members, which have to be constrained against the external surface of the spacecraft prior to deployment. Furthermore, another mechanism has to be used in order to hold down and then free the deployable structure. This problem can be overcome using telescopic booms, and in fact these mechanisms have been commonly employed on spacecraft. The most commonly used device to extend the boom is an axially stiff strip, rolled on a drum at the base of the tube assembly, and connected to the interior tube of the telescopic assembly (Figure 15.6).

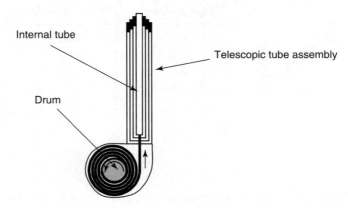

Figure 15.6   Device for the extension/deployment of telescopic assemblies

When the drum rotates, the strip is fed progressively inside the assembly and pushes out the interior tube. This tube progressively drags out all the other tubes, thus extending the telescopic mast.

Another device that can be used to extend the boom is a lead screw that rotates inside the interior tube of the telescopic assembly, thus pushing out the tubes one after the other. However, telescopic booms require a relatively large stowage volume, having its longest dimension in the direction of the boom.

A solution to this problem is to use devices similar to a carpenter's measuring tape, which is stowed rolled on a reel, and deployed by unrolling the reel. The concept is illustrated in Figure 15.7. The boom is elastically flattened into a strip (tape) and stowed rolled-up onto a drum. When the tape is unrolled to extend the boom, it reacquires its original shape, thus gaining the bending stiffness necessary to keep the boom straight.

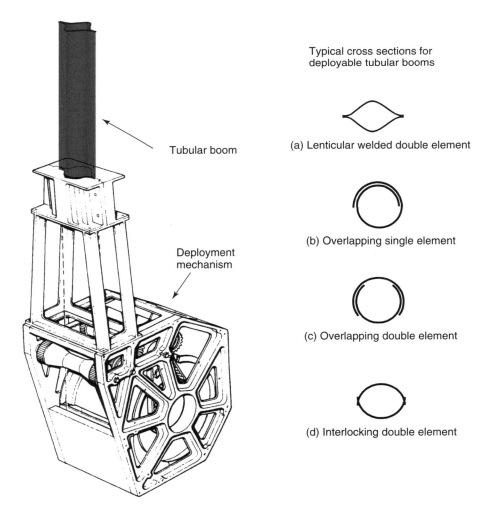

Typical cross sections for
deployable tubular booms

Tubular boom

(a) Lenticular welded double element

Deployment
mechanism

(b) Overlapping single element

(c) Overlapping double element

(d) Interlocking double element

Figure 15.7 Collapsible tube mast deployment mechanism (Reproduced by permission of Sener Ingenieria y Sistemas, S.A.)

This type of boom has good structural properties when extended. However, high shear stresses arise in the welding between the two halves when the boom is rolled onto a small reel. Simpler devices can be built using just one element (Figure 15.7b). However, this type of boom has inferior structural properties, and in particular, it has a very low torsional stiffness. To improve the structural properties of the deployed booms there are several possibilities, as shown in Figure 15.7(c) and (d). However, these types of devices require two reels, and therefore, add complexity to the mechanism.

Another class of deployment mechanisms is the deployable lattice mast. A particularly successful example of this is the AEC-ABLE CoilABLE™ mast, shown in Figure 15.8. This mast is an integral part of the Imager for Mars Pathfinder (IMP) system, holding aloft a stereoscopic camera to provide panoramic views of the Red Planet. When fully deployed, the open lattice mast is about 0.7 m long. However, when stowed, it retracts to 10% of its extended length, coiling up inside a 76-mm long canister. The stored strain energy in the structure allows the CoilABLE™ mast to self-deploy without expensive motors that add weight and complexity. The nimble steel and fibreglass construction provides a steady, rigid base for the IMP camera, with a mass of less than 0.7 kg.

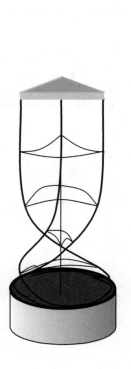
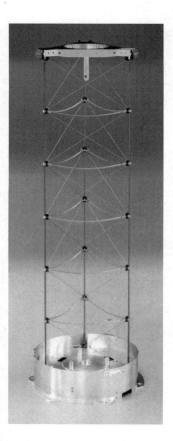

Figure 15.8 Mars lander deployable camera mast (Reproduced by permission of AEC-ABLE Engineering Inc.)

15.2.3   Surface deployment devices

Solar arrays

A solar array may be a simple assembly of two, three or more panels hinged together, folded and secured for launch and then unfolded in space. This simple concept can be full of hidden failure mechanisms and there are too many cases of satellites flying with partially unfolded solar arrays to encourage complacency.

The hold-down and release mechanism must be stiff enough to prevent movement during launch, and the possibility of friction welding at the load points is very real. The hinge system will need to accommodate out-of-plane movement, so spherical rod end-bearings are commonly used. There have been cases of distortion due to thermal gradients producing torques high enough to stop deployment. The latch-up mechanism must stiffen the array and eliminate backlash.

In all spring-driven deployment systems, a problem needing careful resolution is the control of the stored energy. The torque to be provided should never be less than four times the estimated resisting torque, and must ensure final latch-up. To absorb the surplus energy a wide variety of devices have been used, including clutter mechanisms, friction brakes, crushable honeycomb plugs, eddy current and fluid dampers. The balance between stored and absorbed energy is always one that requires very careful judgment.

The Hubble space telescope (HST) has a more complex solar array, each of which is 6 m long by 2.9 m wide. In the original deployment mechanism, the blankets are rolled onto a single drum supported by an arm that enables the whole assembly to be folded into a stowed position for launch. The primary deployment releases the arm, rotates it outward and locks it. The secondary deployment unrolls the blanket on each side of the arm to its final configuration. The system is complicated by the requirement for both automatic and astronaut-assisted re-stowage. The primary deployment system can be seen in Figure 15.9. The arm, supported on its bearing, is linked to the drive mechanism by a connecting rod that also provides a measure of overcentre locking in the fully deployed position, although it is augmented by a spring detent lock. The drive unit is powered by two 1200 steps per revolution stepper motors to provide redundancy and the stators are aligned to enable them to be operated in tandem should maximum torque be required in an emergency. This technique is now commonly applied in spacecraft drive systems. The principle of

Figure 15.9   The primary deployment system for the Hubble space telescope (HST) solar array (Reproduced by permission of Oerlikon-Contraves.)

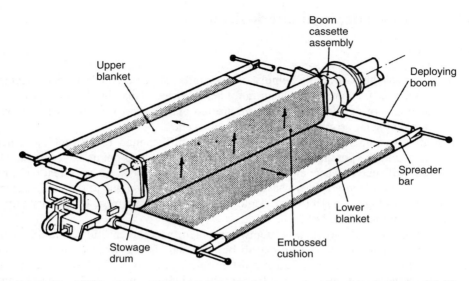

Figure 15.10 HST solar array secondary deployment mechanism (schematic) (Reproduced by permission of Astrium Limited)

the secondary deployment system can be seen in Figure 15.10. The two blankets unroll from either side of a single drum drawn by four extending booms driven in unison. An embossed Kapton cushion sheet separates the blankets on the stowage drum and during deployment is rolled onto its own storage drum. Clearly the mechanism to operate this system is not simple. A central drive unit for all four booms is required, constant torque spring motors act against drum rotation to provide tension in each blanket, the electrical harness must unwind as the drum rotates and redundant drives must be provided. The whole system must be capable of reversal at the end of five to seven years in orbit to enable recovery for refurbishment.

Antennas

Antennas (see Section 12.3.3 of Chapter 12) vary from the elementary single rod whip antenna to the complexity of the Tracking and Data Relay Satellite System (TDRSS) and Galileo unfolding umbrellas (Figure 15.11). In this design, the antenna surface is a wire mesh in which the tension of every wire is predetermined and set. The antenna is then folded as an umbrella and when unfolded in space the surface shape must be precisely recovered. Since the diameter of such an antenna is about 4 m, the mechanical problems are formidable. The design, manufacture and assembly are driven by the need for extreme precision in the final shape of the antenna, which, in turn impacts on every mechanical item in the system. Once in orbit, unfortunately the Galileo antenna did not deploy completely, to the disappointment of scientists and engineers. The reason for this malfunction was most probably the excessive vibrations undergone by the mechanism during terrestrial transportation. In fact, the launch of the Galileo spacecraft was delayed a few years, during which time it was transported back to the Jet Propulsion Laboratory (JPL), and then again to its launch site. This episode emphasizes how the life of a mechanism has to be followed by its designers, and how the impact of changes in the mission planning

Figure 15.11 Umbrella-type antenna, showing an 18-rib wire mesh high-gain antenna, deployed (Jet Propulsion Laboratory, California Institute of Technology, Pasadena, California.)

has to be assessed in order to avoid undesirable results. With the benefit of hindsight, the Galileo mission to Jupiter will be seen as one of the major achievements in the history of astronautics, despite this significant mechanism failure.

15.3 CONTINUOUSLY AND INTERMITTENTLY OPERATING DEVICES

In general, this covers a large variety of devices, but in limiting ourselves to spacecraft mechanisms it is useful to divide the category into

- *mechanisms that operate throughout the life of the satellite* such as momentum wheels, SADs, de-spin mechanisms and horizon scanners;
- *mechanisms that operate intermittently* or on demand, such as APMs and retractable appendages—for example, the Hubble solar arrays.

The problems common to all electro-mechanical systems apply equally to space mechanisms, but here their significance increases by an order of magnitude. Failure modes are so numerous that to eliminate them all is impossible, but each one must be recognized and evaluated for the possibility of occurrence and the criticality to the system. Redundancy must always be considered, but with the proviso that it can decrease reliability rather than improve it.

To illustrate these points we will look at some examples.

15.3.1 Continuously rotating mechanisms

Solar array drive (SAD)

These are needed to decouple the solar array from the satellite to maintain a Sun-pointing direction. The speed for a geostationary orbit will clearly be one revolution per day whilst in Low Earth Orbit (LEO) it may be about 15 revolutions per day.

Figure 15.12 reveals the construction of a typical SAD mechanism. The shaft is mounted on two angular contact bearings lubricated by a film of lead and preloaded by a titanium diaphragm, which also accommodates thermal expansion and the movement required to off-load the bearings for launch. Allowing the bearing to slide on the shaft is discouraged in all spacecraft systems because of the risk of seizure. Both the shaft and the housing are made from beryllium, to take advantage of its high stiffness, good thermal conductivity and a coefficient of expansion matching that of steel. For off-loading, the bearings are displaced by a lever system onto a conical seating. Figure 15.12 shows the system in this state. The lever is retained by a pyrotechnically operated pin-puller. Two brushed DC motors are provided for redundancy, and are acceptable because the number of rotations required for a seven-year life is under 5000. The slip-ring system is conventional with two dry-lubricated brushes running on each ring. The drive system is totally redundant with control signals being generated by Sun sensors mounted on the solar array. Although an old design, it has a very good operational history and shows the main features of a SAD.

Giotto de-spin mechanism

With the *Intelsat* series of dual spinners, there is significant experience of the design and operation of de-spin mechanisms in the USA. However, in Europe such experience is

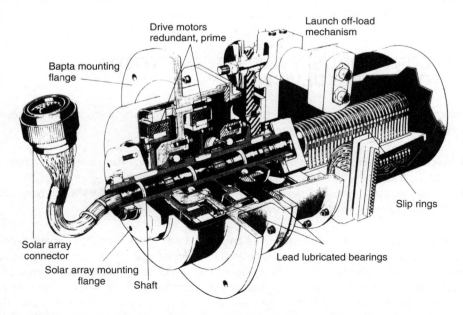

Figure 15.12 Schematic of a typical solar array drive mechanism (Reproduced by permission of Astrium Limited)

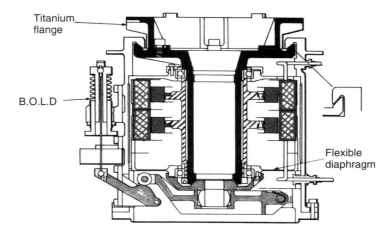

Figure 15.13 Giotto de-spin mechanism (Reproduced by permission of Astrium Limited)

relatively limited—a good example of a European design is the *Giotto* de-spin mechanism. As a spin-stabilized spacecraft, Giotto needed such a mechanism to de-spin the high-gain antenna from 14 rpm, with a further requirement to nullify the speed with no jitter. A cross-section of the mechanism is shown in Figure 15.13. Normally, a very stiff servo loop needing some 180 updates per revolution would be needed. But to simplify the system, the designer decided on one update per revolution, thus throwing the problem back to the bearing designer to achieve a low and consistent torque throughout the life of Giotto. Tests at the European Space Tribology Laboratory (ESTL) demonstrated that lead-lubricated bearings could achieve the desired performance and they were adopted. The off-load mechanism, released by a pyrotechnic, can be seen together with the two drive motors. These were the same 1200 steps-per-revolution motors as used in the Hubble primary deployment system (Section 15.2.3), but here they run as synchronous motors to achieve absolute speed control. The success of the Giotto mission is now space history.

Momentum and reaction wheels

The different roles for these two types of wheels in the AOCS are made clear in Section 9.4.7 of Chapter 9. The momentum wheel has a large momentum of 50 to 200 Nm s and a maximum speed up to 10 000 rpm. The reaction wheel will have a capacity of about 2 Nm s and a speed range of about 3500 rpm in both directions. Electromechanically they are similar and use the same technology, but there are some very important differences. The speeds of both types of wheel are limited by the bearing system and by power consumption. From a mass point of view, a very small wheel at 30 000 rpm would appear attractive but the power needed to produce the same torque from it would be very large since the energy in the wheel is a function of the speed squared. The bearing system is critical to success and almost all ball-bearing mounted wheels follow a similar approach. Figure 15.14 shows the Olympus reaction wheel. The angular contact bearings are preloaded by carefully designed springs to avoid variation of torque with temperature. Since this is a reaction wheel and will run from -3500 to $+3500$ rpm, the

Figure 15.14 Olympus reaction wheel (Reproduced by permission of Astrium Limited)

lubrication system in the bearing will be either boundary layer or elasto-hydrodynamic, according to speed. At zero speed, there will be a torque spike since the static coefficient of friction is higher than the dynamic. The quantity of oil in the bearing must be small and constant to avoid torque variation, but must never dry out during the life of the satellite. A porous nylon oil store is filled with a fine cut hydrocarbon oil to achieve long life. To prevent loss of oil and to maintain extreme cleanliness, the wheel is encased in a hermetic canister filled with helium at 1 torr to reduce windage loss and improve heat transfer. It also allows the wheel to be tested under ground conditions [4,5]. However, ball-bearing lubrication remains the principal life-limiting factor for momentum and reaction wheels.

To avoid some of these problems, wheels supported on magnetic bearings have been the subject of intense development for more than 25 years in France, Germany and the USA. To support a wheel magnetically in five degrees of freedom (DOF), a choice must be made between a passive permanent magnet and an actively controlled electromagnet. In the nineteenth century, it was shown to be impossible to support all five DOF with permanent magnets alone; the system is fundamentally unstable and one degree at least *must* be actively controlled. Permanent magnets are attractive in their simplicity and almost-zero power consumption—they do in fact have a very small drag, but low stiffness.

It has become common practice to identify space wheels by the number of actively controlled DOF. The first to fly in Europe was a French 1 DOF wheel, in SPOT 1, 2 and 3 and in ERS 1 and 2 (Figure 15.15).

In this type of wheel, the actively controlled DOF was the axial displacement of the rotor, which leads to a relatively large axial dimension of the wheel. Also, this configuration makes the passive damping of radial oscillations difficult.

Subsequently, actively controlled 2 DOF wheels were developed, where the wheel is controlled along two orthogonal radial directions. This design results in a flatter geometry and higher momentum-to-mass ratios. 2 DOF wheels have flown successfully on SPOT 4, 5 and Helios and their use is planned for METOP. These types of reaction wheels,

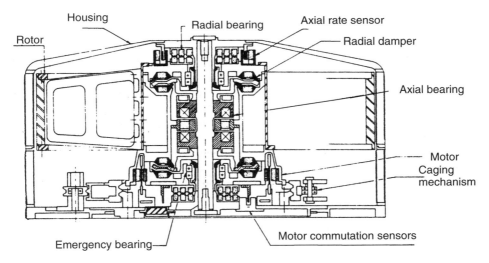

Figure 15.15 SPOT magnetically suspended momentum wheel (prototype) (Reproduced from Centre National d'Études Spatiales, France)

developed by Aerospatiale, have a maximum angular speed of 2500 rpm. However, the new generations of 2 DOF wheels can reach up to 10 000 rpm.

The 5 DOF wheel, in an advanced state of development in Germany, can be regarded as future technology and will embody all the capabilities of magnetic suspension. All magnetic bearing wheels include ball-bearings as safety touch-down elements in case of failure.

Each type of wheel has advantages and drawbacks. The ball-bearing wheels are relatively simple and therefore cheap, and give reliable lives up to 10 years or more.

They are, however, mechanically noisy because of the bearings, there is difficulty in achieving perfect balance, and they cannot act effectively as nutation dampers. Magnetically suspended wheels eliminate some of these problems. The 1 DOF wheel can be made to run beyond its first critical speed to avoid out-of-balance effects—a facility not available to the reaction wheel, of course. By removing bearing wear, the speed limitation is now governed by power consumption and the root stress in the wheel. The need for a hermetically sealed case disappears.

The 5 DOF wheel adds further sophistication. With the availability of full control of each DOF, the wheel can be fully balanced, and the damping capability in each loop can be used to control nutation of the spacecraft. Yet more significant is the ability to tilt the wheel axis through a small angle, currently $\pm 1.7°$, allowing either three-axis control or cross-momentum storage with only one wheel. The resulting gyroscopic torque about the orthogonal axis is controlled by a special tilt control law. The torque impulse is 1.8 Nm s and the gimballing torque is 3.9 Nm. Further extension of the system by adding mechanical gimbals for a larger angle of tilt is feasible or, alternatively, two skewed contra-rotating wheels can be used to achieve full attitude control. This is future technology and will yield highly sophisticated attitude control systems for future projects [6].

15.3.2 Intermittently operated mechanisms

As examples of this type of mechanism, we look at *antenna pointing mechanisms (APMs)*, although in some modes of operation the APM may be continuously trimming the antenna direction.

All APMs will be designed to operate in one or more of three modes as follows:

- *steady-state pointing*, maintaining alignment with any predefined angle on both axes to an accuracy of $\pm 0.01°$;
- *re-pointing*, when it shall be capable of re-pointing the antenna through any required angle to an accuracy of $\pm 0.01°$;
- *tracking*, when it shall be capable of maintaining the antenna on a station, following a ground-based RF beam in closed loop mode to $\pm 0.015°$.

In addition, in the event of complete failure of a drive, it must be possible to return the antenna to a nominal zero position and to lock it.

APMs are required to rotate the antenna in the direction of a specific 'target'. In general, to perform this function, it is necessary to control independently two rotational DOF. These are controlled by gimbal-drives and, at least in theory, the axes of these drives could have any orientation. In practice, it is desirable for the gimbals to have axes along mutually orthogonal directions, one of which is either 'horizontal' or 'vertical'. This leads to two possible configurations [7].

In the first configuration called *az-el*, the elevation of the antenna is controlled by a horizontal-axis gimbal-drive, which, in turn is supported by a vertical-axis gimbal-drive. This configuration, schematically shown in Figure 15.16 is very common in the driving mechanisms for ground station antennas. The main drawback of this configuration is that to maintain pointing on a target moving in the region close to the zenith of the mechanism, large (and rapid) rotations of the azimuth drive are required. The advantage is that a full coverage can be achieved with the elevation movement limited between 0 and 90 degrees.

The second type of configuration, called *x-y*, is shown in Figure 15.17. This mechanism allows the deployment of large antennas situated on the side of the body of the spacecraft

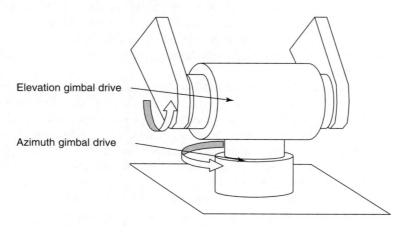

Elevation gimbal drive

Azimuth gimbal drive

Figure 15.16 Schematic view of an azimuth-elevation pointing mechanism

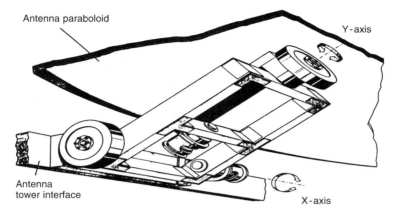

Figure 15.17 Dornier antenna pointing mechanism: television satellite version (schematic) (Reproduced by permission of Dornier GmbH, now Astrium Limited)

and closed in a clamshell fashion for launch. By displacing the two axes of the APM, as shown in Figure 15.17, the outer becomes also the deployment axis whilst the inner remains a pointing axis of limited amplitude. This development required enlargement of the outer axis bearings, a reconsideration of the lubrication system and stiffening of the bearing mountings, none of which changed the fundamental features of the concept.

An example of an alternative design is the Olympus APM. This mechanism utilizes the swash-plate or rotating-wedge principle. The structure, shown in Figure 15.18(a), consists of four cylindrical and co-axial elements. The two end sections act as the spacecraft and payload interfaces and are prevented from mutual rotation by a bellows. The two centre sections are wedge-shaped and are able to rotate independently about the common axis by means of a bearing system. The two swash-plates may therefore be considered to have pointing vectors, equal to their wedge angles, which may be rotated independently. Their vector sum is the resultant pointing angle of the mechanism. By driving these wedges, any pointing angle may be achieved within a conical pointing range with a semi-cone angle equal to twice the swash angle; for Olympus the range is 8.6°.

The realization of this principle into an APM is shown in Figure 15.18(b). The four elements are separated by single four-point-contact Kaydon thin ring-bearings, which are lubricated with Braycote 601 grease. The choice of a liquid lubricant gives a much higher thermal conductivity through the bearing than with dry lubrication, which, for this particular machine, is very important. The drive is achieved by 20 degree-per-step stepper motors with redundant windings, mounted on the stationary plates. The drive to the wedge-plates is made through gear-rings of nitralloy steel. The body is made from beryllium to take advantage of its high thermal conductivity, high stiffness and low density. That the coefficient of expansion almost matches that of steel is a further advantage. In the event of complete failure of one motor, the mechanism may be driven back to the zero-pointing position by rotating the still operating wedge. In the extreme case of both motors failing, the mechanism retains its position and its integrity without the need for a separate locking device.

The excellent mechanical strength and stiffness of the mechanism removes all need for an off-load device during launch. Its relatively small size, its substantial load capacity

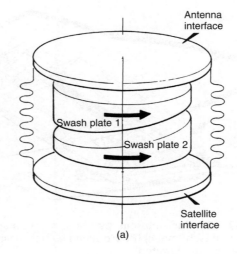

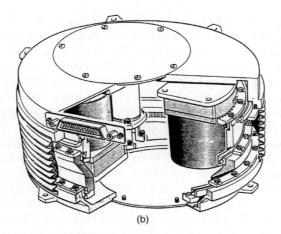

Figure 15.18 Olympus antenna pointing mechanism (a) principle of operation and (b) the mechanical arrangement (Reproduced by permission of Astrium Limited)

of 50 kg, and its advantage of being a closed and self-contained unit makes for easy integration in the spacecraft and flexibility of use for other pointing applications [8].

Pointing mechanisms for optical systems

Pointing mechanisms for optical systems generally require a pointing accuracy far greater than antenna pointing mechanisms. In particular, the narrower the optical beam that has to be 'pointed', the greater the accuracy required from the mechanism.

During the last two decades, various systems have been designed for optical communication between satellites using a laser beam. Since the divergence of a laser beam is of the order of micro radians, the accuracy required from the pointing mechanisms is of the order of fractions of a micro radian. This pointing accuracy has to be maintained during the transmission of data, and given that both satellites move with respect to each other, the pointing mechanisms have to be able to maintain this accuracy whilst tracking its target. The steering of the laser beam is usually performed by a fine-steering mirror, driven by actuators. To achieve the required performance, these pointing mechanisms usually implement closed-loop active control systems with a bandwidth up to a few hundred Hertz. This is required in order to be able to compensate for the mechanical disturbances (e.g. vibration) of the mounting platform.

In addition to fine pointing and tracking, the system is usually required to be able to acquire the target. Target acquisition is generally performed by scanning a relatively large angular sector until the target is detected. This, in turn, requires the capability of relatively large rotations, which would be problematic for a fine-steering mirror. Fortunately, the precision required during the scanning is smaller than that required during the tracking. It follows that most systems are composed of two devices assembled in series, a coarse pointing mechanism (CPM) (capable of large rotations for the acquisition of the target), which supports a fine pointing mechanism (FPM), for accurate pointing and the tracking of the target.

This type of design has been implemented in the *Silex* [9] experiment conducted by ESA. The CPM, shown in Figure 15.19, is an L shape bracket (moved by a 'vertical axis' rotational drive) with a 'horizontal axis' rotational drive (mounted on the vertical part of the bracket), which supports the whole optical bench. Its functioning is similar to the *az-el* APM seen in the previous section, with the difference that this mechanism rotates an optical bench. The FPM is included in the optical bench. This is just one example of a pointing mechanism for an optical communication system, but many others have been designed and tested.

Figure 15.19 Silex coarse pointing assembly (Reproduced by permission of Astrium-Space Ltd.)

15.4 COMPONENTS

15.4.1 Electric motors

Since these are the only source of continuous mechanical power on spacecraft, they are perhaps the most important mechanism component. Reliability is paramount and in critical systems they will be either double-wound or fully duplicated.

In space applications, the most commonly used electric motors are *DC Motors* (brushed or brushless), or *Stepper Motors* [10].

DC motors used in space are usually permanent magnet machines but may be either brushed or brushless. As they give maximum torque at zero speed and, in the brushed configuration, can be used with minimum electronics, they are attractive for many space applications. Dissipation of heat is a major problem and space motors may have to be derated by as much as 70%. Brush wear is of course, the life-limiting parameter (see Section 15.4.2). Brushless motors have the advantage of being very controllable at the expense of considerable electronic complexity. They are now widely used both in space and on the ground to take full advantage of the DC motor's performance linked to excellent controllability.

In general, the design of brushless DC motors tends to achieve a particular type of performance profile. Where high torque is required (up to and beyond 1000 Nm), the motor tends to have a relatively large diameter and short axial length. These types of motors are generally called torque motors, or 'pancake motors' but the drawback of this type of design is that the angular velocity is relatively limited, usually below 200 rpm. If higher speeds are required (generally up to 15 000 rpm, where the maximum speed is generally limited by the rotor bearings), the ratio between axial length and diameter of the motor will be higher. Also, if specifically smooth speed/torque characteristics are required, so-called 'toothless' brushless DC motors are employed. However, it should be remembered that mass, size and power consumption for the electronics necessary in a brushless motor are larger than that of a brush assembly. The use of brushes in some applications in space may also have unacceptable drawbacks in terms of the brush's behaviour in vacuum (wear, maintainability etc.), therefore making a brushless motor more attractive.

In the SAD (Figure 15.12), the motors are brushed only because at one revolution per day the lifetime total will not exceed about 5000 revolutions.

Stepper motors have simple driving electronics, which, together with the incremental stepping motion of the shaft, makes them ideal for several applications, in particular, for open loop positioning/pointing devices. These types of motors can be divided into three main groups:

- Hybrid stepper motors
- Reluctance stepper motor
- Permanent magnet stepper motors.

However, in space applications, hybrid stepper motors are by far the most utilized.

This type of motor has a high power/torque ratio (especially at low speed), maximum torque in the region of 5 to 10 Nm, number of steps per turn between 50 and 1200 (for motors with an external diameter between 15 mm and 150 mm) and a speed between 10 and a few hundred steps per second. Reluctance stepper motors have a poor torque/power ratio. They are, however, much simpler devices, which make them suitable for applications in cryogenic conditions.

Stepper motors are very versatile as both drive and servo prime movers and find frequent application in spacecraft systems. For the Giotto de-spin application, the motors are 1200 step-per-revolution steppers but run in the synchronous mode at 14 rpm.

All the motors considered so far are rotational devices. However, spacecraft mechanisms often require linear motors. These often use the same principles exploited in rotational motors, and they are usually grouped into two categories:

- Short stroke linear motors, with a stroke length usually shorter than 50 mm, generally having a single phase winding that makes the electronic control relatively easy.
- Long stroke linear motors, with a stroke longer than approximately 50 mm, which is achieved by exploiting multi-phase windings.

Motors for space are often designed specifically for each application since performance, power consumption, heat dissipation and materials must all be closely controlled. Adapting commercial motors for space, whilst not impossible, is always hazardous and is acceptable only in conjunction with a very stringent evaluation program [11,12].

15.4.2   Slip rings

There are a number of space mechanisms in which an electric current must be transmitted across a rotating joint, the SAD being an obvious example. In the case of the HST, the rotation of the solar array was limited to 340° to avoid sunlight entering the telescope, so a wire harness could be used. For complete rotation, slip rings are needed. These may be axial, as in Figure 15.12, or radial in the form of a disc (see Figure 15.20), but the material

Figure 15.20   Pancake slip ring (Reproduced by permission of John Wiley & Sons, Inc.)

requirements are the same. The brush material must have good electrical conductivity, a low wear-rate and a self-contained lubrication system. Graphite cannot be used, as it becomes an abrasive in vacuum. Several compact materials based on copper, silver and molybdenum disulphide with a wear-rate between 10^{-14} and 10^{-15} m^3/Nm have been widely used, but a lower wear-rate would be preferable for long-life mechanisms since the control of 'wear debris' in space is still a problem. The ring material is normally copper, plated with gold or silver, although the latter suffers too readily from sulphide contamination in air.

Oil can be used as a lubricant but must be only vestigial. In one case of dry-lubricated slip-rings in a 60 rpm de-spin mechanism, the wear-rate of the brushes was markedly reduced by oil vapour from the liquid-lubricated bearings. The mechanism ran on test for over six years.

Honeywell have developed their Roll-Ring design to the stage at which it can be used with confidence in low-speed applications. The concept is based upon a flexible hoop rolling between the stationary and rotating elements to carry the current.

15.4.3 Gears and bearings

These two machine elements may be discussed together. They are both essential components of a variety of space mechanisms and their technologies are both critically dependent on tribology and materials selection.

Space gearboxes differ from industrial units of similar size in their much reduced permissible tooth-loading and their limited choice of both lubricant and material pairing. The tooth module, defined as the pitch circle diameter divided by the number of teeth, will define the tooth size. As a general rule, the minimum module is 0.08 but the choice is a balance between tooth load, accuracy of transmission and gear ratio.

The two types of tooth stress governing performance are the contact, or Hertzian, stress and the bending root stress. In practice, the former has been identified as the more important in failures because it also controls the sub-surface shear stress and, by implication, the fatigue failure. A useful rule of thumb is to limit the tooth load of metal gears to a maximum of 10 N per mm tooth width, and to reduce this figure by a factor of 10 for plastic wheels. Metallic materials for gears include carbon, stainless and maraging steels, cast iron, bronze, aluminium and titanium, with suitable surface treatments. Favoured plastics are polyimide and polyacetal, often loaded with molybdenum disulphide or carbon fibre. Plastic gears should always run against metal gears, usually stainless steel, although aluminium and titanium are sometimes used.

The lubrication of lightly loaded plastic gears is usually achieved by the incorporation of a solid lubricant such as polytetrafluoroethylene (PTFE) or molybdenum disulphide in the plastic. For heavily loaded metal gears, the choice is between thin films of soft metal, for example, gold or lead, or molybdenum disulphide. Grease and oil may also be used but control against migratory loss will be needed.

Sometimes it is possible to achieve redundancy in a gear drive by incorporating a differential gear, one half of which is locked and released only if the other half should seize. Such complexity can only be justified in very critical systems. It is well to remember that a gearbox is nearly always reversible, and achieving true irreversibility, even with a worm gear, can be very difficult.

Harmonic Drives are often used in space. These are characterized by very high transmission ratios with a high torque capability, together with elevated positioning accuracy (virtually zero backlash) and repeatability [13]. The first major space application of Harmonic Drives was in 1971, on the Lunar Roving Vehicle of the Apollo 15 mission, and since then they have been used quite frequently where compact and powerful positioning drives are required.

The principle of operation of these devices, illustrated in Figure 15.21, is relatively simple and it is based upon three mechanical components: the wave generator, the flexspline and the circular spline, which are shown in Figure 15.22. The flexspline is basically a flexible gear that is deformed into an elliptical shape by the wave generator. The wave generator rotates inside the flexspline forcing the teeth of the latter to mate with those of the circular spine, which is fixed (thus acting like the case of a gear box). The manner in which this functions is illustrated in Figure 15.21, and highlighted by the movement of the black dot that is fixed to the flexspline. In this case, after half a turn of the wave generator the flexspline has rotated only 1/48 of a circumference, which corresponds to a transmission ratio of 24 : 1. In practice, the number of teeth in real devices is much higher, and it is possible to reach transmission ratios from 30 up to 320 : 1.

For *bearings*, the application of the apparently simple ball-bearing to space mechanisms might be thought to be a straightforward adaptation, but it has taken more than thirty years of research and more than two million hours of testing in vacuum to reach the present state of knowledge, and there is still more to be learned. Speed, load, reverse rotation, type of cage, material and both wet and dry lubrication exert major influences on the performance of this machine element. The duty cycle and life required of the bearing will be important in the choice of these factors. As an example, a lead-lubricated bearing

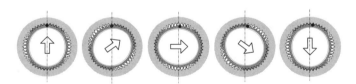

Figure 15.21 Harmonic drive principle of operation

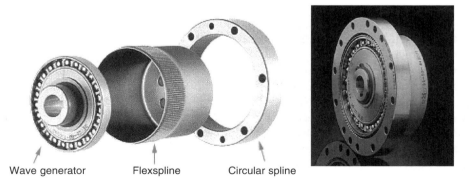

Wave generator   Flexspline   Circular spline

Figure 15.22 Harmonic drive basic components and final assembly (Reproduced by permission of Harmonic Drive AG.)

with a lead bronze cage will be excellent in a SAD and useless in a momentum wheel. Temperatures, particularly thermal gradients, are most important; the thermal conductivity of a liquid-lubricated bearing can be ten times that of one with dry lubrication. The warning must be never to use a ball-bearing in a space mechanism without the guidance of a space tribology expert (see Section 15.6). Plain bearings are also used in space but almost exclusively for one-shot systems. They are always dry-lubricated and polymer-metal pair based. Spherical rod end-bearings are widely used in deployable elements to avoid high friction from thermal distortion. They must always be validated for space since some commercial materials are unacceptable. Reference [14] is a good general survey and Reference [15] is important.

15.4.4 Sensors

Unless operating completely open loop, which is rare, an electro-mechanical system will embody a sensor, and the choice of this device will be crucial to the success of the system. The angle over which it must operate will limit the choice. Some sensors, such as the pick-offs used in gyros, are extremely sensitive but operate over very small angles. Others such as synchros, resolvers, stepper motors, and encoders can be used over 360°. Optical encoders are commonly used in space, their development commencing in the 1950s. In its simplest form, an optical encoder consists of a collimated light source (often a light emitting diode LED), a rotating code disk (with transparent and opaque radial segments), a grating element (or slit plate) and one or more light detectors. The code disk is attached to the element whose rotation has to be determined, whilst all the other elements are fixed. The light beam propagates from the LED through the code disk, to the grating device and finally to the detector. The rotation of the code disk produces interruptions of the light beam, and from the pattern of these interruptions it is possible to determine the position of the disk. The optical encoders used on the HST have an accuracy better than 1 arcsec.

There are many factors to be considered in the choice of a sensor since many of them are very sensitive to mechanical stability and thermal distortion, particularly over the longer lives now being demanded. Mechanically robust devices such as resolvers can be very attractive for this reason. Wear can ruin a mechanical contact encoder, and optical units are dependent on the reliability of the optical elements.

15.4.5 Release devices

Pyrotechnics have always been, and still are, the principal elements for releasing space mechanisms [16,17]. They are very energy efficient (better than 10 kJ/g), operate very rapidly (less than 10 ms) and when well engineered, are very reliable. Space pyrotechnics always have two initiators for each charge and fully redundant firing circuits. The pressure generated by combustion is used to operate a simple mechanical device such as a pin-puller or pin-pusher, cable-cutter or valve-actuator. They can of course only be operated once and are then discarded, which seriously limits ground and pre-flight testing. They are also difficult to seal hermetically, although there have been a number of successful designs using deformable capsules or bellows. Safety is a very important aspect and in the early days there were incidents in which stray currents and electrostatic discharge,

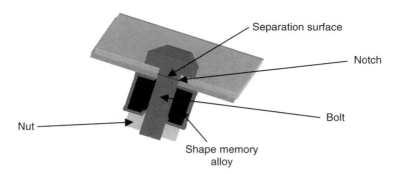

Figure 15.23 Schematic representation of the three-dimensional cross-section of a Frangibolt

even from the human operator, caused untimely ignition. This led to the need for elaborate protection systems to absorb electrostatic discharge, which are now built into the initiators of every space pyrotechnic.

One of the drawbacks of pyrotechnics is the mechanical shock produced by the explosion of the charge. Pyrotechnics are, in fact, forbidden in some spacecraft such as the HST, in which *memory metal actuators* are used. The metal, an alloy of nickel and titanium, can be deformed into a new shape below a critical transformation temperature. On heating above this temperature, less than $100°$, the shape changes back to the original. If this change of shape on re-heating is restrained, then the alloy can be used to generate a significant force capable of doing work, which, in turn, can be used to operate a mechanism. On HST it was used to release the off-loading system of the SAD.

Practical examples of how this phenomenon is exploited in real mechanisms are the Frangibolts produced by the TiNi Alloy Company, which are commonly used. The mechanism, schematically represented in Figure 15.23, exploits the expansion of Shape Memory Alloy to produce an extra tension in the already loaded bolt, which consequently breaks the bolt across the area identified by the circular notch around its shank.

High Energy Paraffin actuators are another possible release initiator. They rely, for their action, on the large expansion of paraffin wax when changing from the solid to the liquid phase. They can generate high forces and the action is indefinitely repeatable when thermally cycled—an important advantage in their favour. In another device, a heated knife is used to cut a tensioned Kevlar cable to release a solar array.

All these devices take time to operate—of the order of one minute—so that simultaneous release of multiple elements cannot be guaranteed. If this is required, then a pyrotechnic will be needed.

15.5 MATERIALS

The selection of the materials for a space mechanism is possibly the single most important decision that the designer will have to make, only equalled by the selection of the lubricant. The requirements follow very closely those specified for space structures and the reader is referred to Chapter 8. Only materials that have been validated for use in space should be selected. If a new material is chosen, then the cost of validation must be accounted. See [18], which is an essential guide.

Of particular importance to mechanisms, is stress-corrosion cracking (SCC) where it is sometimes necessary to use materials that are not generally recommended. For example, the bearing steel 440C, although more susceptible to SCC, is preferred over the widely used 52 100 because of its better resistance to corrosion.

Most mechanism parts are machined from the solid, which is very expensive. The small number required for a project, perhaps only four or six, usually precludes other production methods. However, some space firms have economically used the lost wax casting process for short runs.

Polymers are as widely used in mechanisms as in structures. Polyacetal and poly-imide are used for small components and with appropriate fillers are self-lubricating. But all polymers have high thermal coefficients of expansion and low thermal conductivity, which may be unacceptable, particularly when used with metals. Other materials such as ceramics are now finding application in space. Hot-pressed silicon nitride has been used for bearing balls and indeed for complete bearings. Titanium carbide and titanium nitride are frequently applied to surfaces as hard coatings.

All *plating processes* must be fully evaluated before being applied to a space mechanism. Cadmium, for example, is widely used for earthbound mechanisms, but is totally forbidden in space because of sublimation in vacuum.

In LEO, attack by *atomic oxygen* is an environmental hazard. It is particularly damaging to exposed polymers and can also attack the surfaces of metals that are sensitive to oxidation, such as silver (see also Chapter 2).

15.6 TRIBOLOGY

Tribology encompasses lubrication, wear, surface physics and the physics and chemistry of lubricants, and is thus an essential study for anyone pursuing the design of space mechanisms.

The European Space Agency recognized the importance of the subject to the reliability and integrity of space mechanisms and set up a special facility, the ESTL, to develop space tribology and at the same time provide meticulously controlled test facilities for space mechanisms. The wisdom of this action has been vindicated in the avoidance of many failures in space mechanisms.

The following five precepts summarize the principles to be applied for successful lubrication in space:

1. The optimum lubrication system is an integral part of the mechanism design and not a process to be added when the design is complete.
2. Avoid making the operation of the mechanism dependent upon close control of the coefficient of friction.
3. The lubrication process must be fully codified and documented to ensure consistent repeatability of performance.
4. The lubricant must be approved and validated for space to a recognized specification, and must be source traceable.
5. The test programme to determine the performance of the lubrication system in the mechanism must reproduce all the operational conditions of duty cycle, life, and both space and ground environments that it will experience in application.

The range of *space lubricants* available today is not extensive but is certainly wide enough to permit 'fatal' mistakes by the uninitiated. The apparently excessive requirements of the fifth precept represent the only effective means of avoiding such mistakes. The causes of failure in a tribological system can be very subtle, and unpleasant surprises are still too frequent.

Lubricants can be broadly classified as follows:

Dry:
1. Dichalcogenides, of which molybdenum disulphide is the most useful; it performs better in vacuum than in air because it is degraded by the presence of water.
2. Solid lubricant composites based on polymers loaded with molybdenum disulphide or PTFE.
3. Soft metals of which lead and gold are the most effective.

Wet:
1. Hydrocarbon oils, now limited to Ball Bros. Vackote and Penzane SHFX 2000.
2. Synthetic oils (other than hydrocarbons) that are now limited to the polyfluoralkyl-ethers such as Fomblin Z25, Krytox and others. Even these have limited lives in ball-bearings because of the formation of polymers; special materials such as silicon nitride or titanium carbide must be used to prevent it.

This is another area where expert advice is the only basis for success. Special testing and evaluation for the specific application may be necessary and must not be ignored. (Reference [19] is a good survey of the subject and Reference [20] is still important.)

15.7 TESTING AND VERIFICATION

For a mechanism, meticulous testing is the only method of demonstrating the integrity of both the design and the build of the machine and verifying its ability to operate over its design life. Many faults can be revealed only by testing in an environment that simulates the conditions that it will experience in operation. There are no short cuts. Taking an APM or a SAD as examples, vacuum chambers with the ability to create thermal cycles and thermal gradients in a clean room environment must be provided. Perhaps one of the most difficult problems during testing is to recreate the microgravity environment in which the mechanism will operate. For deployable structures, a relatively standard technique is to suspend each segment of the structure from its centre-of-mass (using light-weight wires), which compensates for the 'one-g' environment. Life testing, qualification testing and testing of individual builds are also essential and will add significant cost to the programme; they could absorb as much as 25% of the budget.

For one-shot systems of the single hinge variety, it is all too easy to accept a few deployments in the laboratory as verification and then be mightily surprised when it fails to deploy in space because of temperature differentials. The remarks above apply equally to such systems. Failures of this character are still much too frequent.

Reference [21] sets out the rationale to be applied to formal testing of mechanisms, and covers also the mechanical environment tests. Thermal vacuum testing is the subject of References [22] and [23].

15.8 CONCLUSION

Mechanisms are fundamental spacecraft subsystems, necessary to perform operations that are crucial to the success of the mission. Currently, there is a relatively large number of mechanisms (e.g. spacecraft release mechanisms, reaction and momentum wheels, SADs etc.) that have reached a certain maturity in their development, and thus have relatively standard designs. However, progress in engineering that produces continuous improvements and upgrades in the mechanisms design is always necessary. Furthermore, there will always be missions that require mechanisms with completely new requirements, and their design is a challenging and exciting endeavour. This very brief look at mechanisms shows the range of skills needed for their design. Specialist knowledge in such diverse areas as motors, sensors, pyrotechnics, tribology and structures will be needed together with a wide knowledge of materials. Testing is yet another area in which specialist expertise will be required.

The application of product assurance (PA) to mechanisms has not been addressed in this chapter and the reader is directed to Chapter 17. The reliability and avoidance of failure in space mechanisms depends, in no small measure, upon the strict application of the principles and practice of PA throughout every stage of the design process. Meticulous attention to detail, the refusal to take anything for granted, and a carefully planned and executed test programme are fundamental ingredients for success. References [24,25 and 26] are important.

For the reader interested in expanding his or her knowledge of Spacecraft Mechanisms, References [27,28 and 29] are excellent sources.

REFERENCES

[1]   General Specification for Moving Mechanical Assemblies for Space and Launch Vehicles (1988), MIL-A-83577B (USAF).
[2]   Labruyere, G. and Urmston, P. (1996) ESA mechanisms requirements, *Proceedings of the 6th European Conference on Spacecraft Structures, Materials and Mechanical Testing*, Paris.
[3]   Anckarman, B. and Thuswaldner, M. (1998) The development and testing of the ACU2624 separation system, *Proceedings of the European Conference on Spacecraft Structures, Materials and Mechanical Testing*, Braunschweig, Germany.
[4]   Bosgra, J. A. and Prins, J. J. M. (1982) Testing and investigation of reaction wheels, *Automatic Control in Space Symposium*, ESA and IFAC.
[5]   Standing, J. M. and Sheppard, J. M. (1977) Design, development and qualification of a 2 Nm s reaction wheel, *Attitude and Orbit Control Systems Symposium*, October 1977, ESA SP 128.
[6]   Sindlinger, R. S. (1977) Magnetic bearing momentum wheels with magnetic gimballing capability, *Attitude and Orbit Control Systems Symposium*, October 1977, ESA SP 128.
[7]   Pritchard, W., Suyderhoud, H. and Nelson, R. (1993) *Satellite Communications Systems Engineering*, ISBN 0137914687, Prentice Hall, New York.
[8]   Brunnen, A. J. D. and Bentall, R. H. (1982) Development of a high stability pointing mechanism for wide application, *16th Aerospace Mechanism Symposium*, NASA CP2221.
[9]   Tolker-Nielsen, T. and Guillen, J. C. (1998) Silex: The First European Optical Communication Terminal in Orbit, ESA Bulletin 96.
[10] Favre, E., Brunner, C. and Piaget, D. (1999) European electric space rated motors handbook, *8th European Space Mechanisms and Tribology Symposium*, Toulouse.
[11] Slemon, G. R. and Straughan, A. (1980) *Electric Machines*, Addison Wesley, New York.

[12]   *Second International Conference on Small and Special Electrical Machines*, September 1981, Institution of Electrical Engineers.

[13]   Uera, K. and Slatter, R. (1999) Development of the harmonic drive gear for space application, *8th European Space Mechanisms and Tribology Symposium*, Toulouse.

[14]   Todd, M. J. (1987) Models for steady and non-steady Coulomb torque in ball-bearings, *3rd Space Mechanisms and Tribology Symposium*, ESA SP279.

[15]   Gill, S., Forster, D. J. and Rowntree, R. A. (1992) Thermal vacuum performance of cycloid and harmonic gearboxes with molybdenum disulphide and liquid lubrication, *5th European Space Mechanisms and Tribology Symposium*, ESA SP334.

[16]   Catalogue of European Pyrotechnic Devices (2nd edn), (1982), ESA SP1021.

[17]   Explosifs et Pyrotechnie, *Proceedings of Symposium* (1979), ESA SP144.

[18]   Data for Selection of Space Materials, ESA PSS-01-70.

[19]   Tribology International-Special Issue on Space Tribology, Vol. 23, No. 2, April 1990.

[20]   First European Space Tribology Symposium, April 1975, ESA SP111.

[21]   Briscoe, H. M. (1983) A rationale for the testing of space mechanisms, *First European Space Mechanisms and Tribology Symposium*, Act. 1983, ESA SP196.

[22]   Robbins, E. J. (1975) Tribology tests for satellite application, simulation of a space environment, *First European Space Tribology Symposium*, April 1975, ESA SP111.

[23]   Parker, K. and Duvall, J. A. (1983) Achievement of reliability by thermal vacuum testing of spacecraft mechanisms, *First European Space Mechanisms and Tribology Symposium*, Act. 1983, ESA SP196.

[24]   Basic Requirements for Product Assurance of ESA Spacecraft and Associated Equipment, ESA PSS-01-0.

[25]   Failure Rates, ESA PSS-01-302.

[26]   Failure Mode Effects and Criticality Analysis, ESA PSS-01-303.

[27]   Conley, P. L. (1998) *Space Vehicle Mechanisms—Elements of Successful Design*, John Wiley & Sons, New York.

[28]   Sarafin, T. P. (ed.) (1995) *Spacecraft Structures and Mechanisms: From Concept to Launch*, (Space Technology Library) Wiley J. Larson.

[29]   Fusaro, R. L. (1995) Space Mechanisms Lesson Learned Study, Vol. 1&2, NASA-TM-107046 and 107047.

16 SPACECRAFT ELECTROMAGNETIC COMPATIBILITY ENGINEERING

Ken Redford

British Aerospace, Bristol, United Kingdom

16.1 INTRODUCTION

Electromagnetic Compatibility (EMC) for a system or equipment requires that

1. It does not cause interference with other systems or equipment.
2. It is not susceptible to emissions from other systems, equipment or electrical environments.
3. It does not cause interference within itself that can cause the system or equipment to malfunction or behave in an undesirable manner.

EMC problems were first noticed on domestic radio receivers in the early part of the twentieth century. These early vacuum-tube receivers were sensitive and able to pick up interference clicks and buzzes generated by electric motors, overhead wires and connections used in electric transport vehicles.

EMC has become much more important as technology progresses. More complex electronics performing many new functions are packed into smaller and smaller enclosures.

This trend is particularly true for space vehicles. The wide variety of communications, weather, robotic, remote sensing and scientific spacecraft, and more recently, small satellites launched as constellations, all pose complex EMC problems.

The trend has caused spacecraft contractors to impose stringent EMC requirements on both spacecraft and ground support equipment.

Spacecraft Systems Engineering (Third Edition). Edited by P. W. Fortescue, J. P. W. Stark and G. G. Swinerd
© 2003 John Wiley & Sons Ltd

EMC provisions for all electrical and electronic equipment have now become law in all EEC member countries. It is illegal to sell any electrical or electronic product without a 'CE' mark certifying that the equipment meets the required European EMC standards.

Further reading on EMC can be found in References [1] and [2].

16.2   EXAMPLES OF EMC PROBLEMS

In domestic situations, most people have experienced interference from car ignition systems and on TV and radio reception. The recent discussions on the safety of operating mobile phones close to the brain and whether close proximity to overhead power lines affects health are all examples of EMC problems where one system could cause another to malfunction or behave in an undesirable manner.

EMC is fundamental to the military sector since interference can directly affect the operation of military equipment and is indeed used as a weapon in Electronic Warfare.

Fortunately, EMC problems in spacecraft are relatively rare since major contractors are careful to make early EMC provisions on all projects. Most EMC problems encountered on spacecraft result in a slightly degraded performance but some have had more serious effects. Electrostatic Discharge (ESD) problems between thermal blankets and the spacecraft structure on early maritime communications satellites were solved by careful grounding. The interference from these tiny spark discharges was sufficient to cause telemetry status latches monitoring the power subsystem to flip over. This initiated shedding of the payload communications power until reset by ground.

These examples illustrate the importance of EMC provisions in all electronic applications.

16.3   EMC SPECIFICATIONS

EMC Requirements Specifications are derived and written for each spacecraft depending upon its mission and they generally contain

- Spacecraft level requirements and derived subsystem/unit requirements broken down into the EMC categories described later.
- Details of how particular requirements are to be verified are also specified. This verification is done either by Inspection/Analysis or Test, and an indication is given of which tests are to be carried out at spacecraft/subsystem or individual electronic unit levels.
- Design guidelines are sometimes included to address particular EMC problems. For instance, if a scientific spacecraft is attempting to measure the magnetic field environment of a planet with a sensitive magnetometer, then there may be limitations on the use of magnets or magnetically permeable material in the spacecraft build.

EMC test methods are generally based on the American Military Standard MIL—STD-461 [3]. This is widely used in both Europe and the USA as a military standard for both ground-based military and spacecraft electronic hardware. The standard covers EMC requirements and test limits for electrical, electronic and electromechanical equipment, subsystems and systems and establishes techniques to be used for measurements.

EMC requirements for ground support equipment used for testing at spacecraft/ subsystem and unit level are sometimes covered in separate requirements documents.

16.4 ELECTROMAGNETIC COMPATIBILITY — TERMS AND DEFINITIONS

16.4.1 Common terms used

Commonly used terminology in the EMC area is now introduced.

EMC — Electromagnetic Compatibility

This is a general term addressing all categories of electromagnetic emissions and susceptibility. It is generally split into four areas:

- *Radiated emissions* measures all electric or magnetic fields emitted from the equipment.
- *Conducted emissions* measures all power, analogue and digital signals that are conducted *via* cables, harnesses and structural parts of the spacecraft.
- *Radiated susceptibility* measures the ability of the spacecraft to operate satisfactorily when placed in an environment that contains externally generated electric or magnetic fields.
- *Conducted susceptibility* measures the ability of the spacecraft to operate satisfactorily when externally generated conducted interference signals are directly injected into the spacecraft's harness cables or structure.

EMI — Electromagnetic Interference

This is a general term referring to conducted and radiated EMI produced by a system.

RFI — Radio Frequency Interference

A more specific version of EMI referring to emitted radio frequency electric fields—generally in the Radio frequency (RF) bands from 3 KHz up to 3000 GHz.

ESD — Electrostatic discharge

An ESD will occur if two equipments or systems that are electrostatically charged at different potentials are brought together.

A lightning strike on a spacecraft launch vehicle or a spark ignition on a car engine are both examples of ESDs. ESDs to the human body can sometimes be felt when touching electrostatically charged car bodies.

Two main effects have to be considered when an ESD occurs:

1. The effect of the spark discharge that generates *radiated* electric and magnetic fields.

2. The effect of the resultant *conducted* discharge current through the equipment or system. This discharge can rise to hundreds or sometimes thousands of amperes flowing at the ESD point in a very short period.

EMP — Electromagnetic Pulse

This is the intense electromagnetic wave produced when a nuclear detonation occurs. It is characterized by extremely high electric and magnetic fields occurring in an extremely short period.

It was first noticed about 1943 when electronic equipment used to monitor the first atomic bomb explosions were destroyed by the EMP.

Nuclear 'hardening' has since become a significant requirement for all strategic military equipment, government communications equipment and critical electronic systems.

16.5   EMC FUNDAMENTALS

16.5.1   The basic EMC problem and its solution

All EMC problems have these aspects in common:

1. a source or transmitter that produces emissions,
2. a receiver that receives the emissions,
3. a transfer or coupling path between transmitter and receiver.

Interference occurs if the received signal causes the receiver to misbehave in some way.

There are three possible means to reduce the interaction between the transmitter and receiver.

1. Reduce the transmitted emissions.
2. Alter the coupling path between the transmitter of interference and the receiver by physical separation.
3. Make the receiver less susceptible to the interfering signal.

Clearly, reducing the transmitted emission is the most desirable since this is the cause of the problem. However, this is not always possible.

For example, the prime function of a telemetry transmitter on a spacecraft is to generate an RF signal and to send it to Earth *via* an antenna on the spacecraft. The transmitted power is governed by the link budget and the performance of the spacecraft and ground systems. Primary RF emissions cannot, therefore, be reduced for EMC reasons.

In contrast, a Switch Mode Power Converter also generates RF emissions. However, its primary purpose is usually to convert direct current (DC) power to different regulated voltages. It is, therefore, very important, and possible, to try to reduce the secondary RF emissions from the converter, even if this makes the converter slightly less efficient.

Altering the coupling path between transmitter and receiver by physical separation can be effective in some cases. An example of this can be found on the Ulysses Spacecraft. One of its functions is to measure the magnetic environment around the polar regions of the Sun. The Magnetometer sensor is mounted on a 5.6 m radial boom to minimize magnetic interference from the spacecraft body.

Finally, the third option of making the receiver less susceptible to the transmitted interference is almost always possible. Examples of this are filters on spacecraft electronic units' interfaces to eliminate conducted interference from pulses on power and signal lines on the spacecraft.

16.6 THE SYSTEMS APPROACH TO EMC

One of the most important aspects of EMC is that it must be considered as an integral part of the specification, design, manufacturing and testing phases of any spacecraft.

A 'top down' systems approach to EMC must be adopted at the same time that the basic requirements of the spacecraft are defined and specified. This involves examining each aspect of the spacecraft mission requirements, considering how each EMC category is influenced and deriving detailed EMC requirements and specifications.

System margins and budgets are attached to EMC performance in the same way as power budgets are applied to Power Systems or pointing budgets are allocated to Attitude and Orbit Control Systems.

16.6.1 Safety margins for EMC

These margins are defined as the difference between system susceptibility levels and the level of system emissions.

For instance, the electronics units mounted on a spacecraft platform will be required to perform their required function in the presence of RF emissions back scattered from the telemetry transmitter antenna sending signals to Earth. The systems designer may have calculated the worst case field strength *emission* from the transmitter at any location on the spacecraft as, say, 1 V/m. However, in the *susceptibility* specification for other electronic units the designer may impose a requirement that the 'units shall not be susceptible to a field strength of 10 V/m'. The designer has then assigned a 20 dB ($\times$10) safety margin between the *emission* and *susceptibility* specifications.

The size of this margin may reflect

1. the uncertainty in the designer's calculations,
2. that there are perhaps several 'sources' of interference from several equipments that can 'add up' at system level,
3. the criticality of the subsystem or equipment.

Generally, the size of the margin between susceptibility and emission is at least 6 dB ($\times$2) but it could be as high as 20 dB for safety critical systems such as pyrotechnic release mechanisms or military systems.

16.7 EMC CATEGORIES

As introduced in Section 16.4.1, EMC categories can be split into *radiated* and *conducted* main areas, each of which can be further subdivided into *emissions* and *susceptibility* as shown in Figure 16.1.

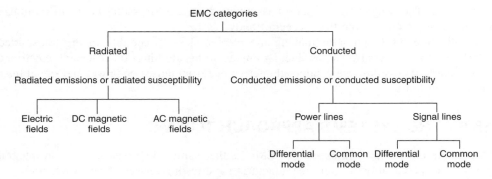

Figure 16.1 EMC categories

16.7.1 Radiated emissions

Radiated emissions are electric fields and AC or DC magnetic fields, which are emitted from either the subsystem units, or from the harnesses and cables that interconnect the subsystems.

Electric fields

These are radio waves generated by RF transmitters or signals radiated from the wires in electronic units and associated interconnecting harnesses acting as small antennae. They are formally defined as sinusoidal RF radiations in volts per metre over a frequency range 3 KHz to 3000 GHz. On most spacecraft, they are usually of concern only in the 15 KHz to 30 GHz region.

RF emissions are measured on a Spectrum Analyser using a variety of different antennae to pick up emissions. The equipment under test and the antennae are enclosed in a screened room to avoid external EMC influences.

Electric field emissions can be reduced as follows:

- Adopting a good grounding and screening philosophy for the equipment or subsystem under test.
- Choosing the slowest digital and analogue technologies consistent with the mission objectives—particularly on interface circuits between subsystems and electronic units.
- Effective shielding of all cables, harnesses and connectors both inside units but, more importantly between units. This is done by the use of screened wires and connector shrouds, coaxial cable, twisted wire, overall shielded braids on wires, wire bundles and so on. Almost all Electric Field emission problems can be minimized by adequate shielding and grounding of harnesses, cables and connectors between units and subsystems.
- Effective shielding and grounding of all electronic units by encasing all units in metal screened boxes or enclosures. Holes or apertures in units should be minimized and lid-fixing screws should be close together to avoid apertures. In extreme cases, RF

gasketing material can be used between adjacent metallic parts (e.g. lids/access parts) although this is only effective in the higher frequency ranges above a few MHz.

● Metal parts/panels should be electrically bonded together—giving typically less than 10 mohms resistance between adjacent parts.

DC magnetic fields

These are fixed DC magnetic fields, usually measured in units of *PicoTeslas* at 1 m distance. They do not vary with time and are produced by permanent magnets or DC currents flowing in any circuit. Their magnitude is important, and should be minimized on scientific spacecraft that use magnetometers to measure magnetic field environments. It is impossible to construct a spacecraft that does not have some level of DC magnetic field, and so if sensitive measurements are required, the magnetometer sensors are generally mounted on booms several metres in length, away from the spacecraft body.

DC Magnetic Fields can be reduced by

● minimizing the use of ferromagnetic or permeable materials,
● reducing DC currents and minimizing the loop area around which they flow.

In practice, this cannot be perfectly achieved. For instance, transformers, inductors and relays rely on ferromagnetic material. Flow control valves and latch valves controlling gas jets for attitude control purposes contain magnetic material. Even the mounting cans and connecting wires on integrated circuits and transistors are made of magnetically permeable nickel alloy.

Compensating magnets, or magnetic screening using 'Mu-metal' alloy material, can be used with some success in critical applications. DC magnetic fields can be further reduced by 'de-perming' the electronic units. This generally involves placing the unit in a coil that is energized to produce an AC magnetic flux density and slowly reducing this field to zero.

AC magnetic fields or B fields

These are alternating magnetic fields that vary with time (AC) and are produced by inductive circuits such as transformers, inductors, and, indeed, any wire loops or circuits in which alternating current flows. They are also measured in *PicoTeslas* at 1 m distance. AC magnetic fields are mostly of concern in the 50 Hz to 100 KHz frequency region. Above 100 KHz, electric fields tend to become more important than magnetic fields.

AC magnetic fields can be reduced by adopting the following methods:

● Reducing the magnitude of AC currents in the equipment and harnesses.
● Reducing the loop area around which these AC currents flow. If this is not possible, then cancellation or screening techniques can be used—although usually only with limited success. Some reduction can also occur if cancellation techniques or 'shorted turns' are used. A copper strap around the outside of a transformer body is an example of 'shorted turns' that do not affect basic transformer action. However, any stray external magnetic field present will induce currents in the strap that will, in turn, produce a compensating magnetic field.

Magnetic Screening with Mu-metal can sometimes be effective but only at frequencies less than about 1 KHz. Magnetic Fields tend to be very directional and some EMC problems can be minimized by simply physically moving or rotating the offending source of interference.

16.7.2 Radiated susceptibility

Units or subsystems can be caused to malfunction or to be *susceptible* to the presence of radiated AC or DC magnetic or electric fields.

These externally applied fields can be picked up on the harnesses and cables between subsystems or in internal wiring of the units themselves. These induced signals can produce conducted signals in the units and cables that can cause circuit functions to fail momentarily. In extreme cases, electrical interfaces can be permanently damaged.

Susceptibility testing is essentially the reverse of testing for emissions. RF power, from generators that can be swept over the required frequency range, is fed to a variety of antennae to radiate the equipment or spacecraft under test. Field strengths at the equipment under test are generally measured first. Then the equipment is switched on in its most sensitive mode to detect any susceptibility.

16.7.3 Conducted emissions

These are generally split into emissions from power lines and signal lines and are further categorized into *Differential Mode* or *Common Mode* signals as shown in Figure 16.1.

Noisy circuits and components inside a subsystem can cause conducted emissions to be present on both signal and power lines between units on the spacecraft. These emissions can also be conducted to ground *via* the chassis, case or screens in cables between units.

Measurements are made directly using voltage or current probes monitoring conducted signals, and these can be displayed on an oscilloscope or spectrum analyser.

The difference between a *differential mode* and a *common mode* signal is now explained. Figure 16.2 shows a typical signal interface between subsystems in which, for example, digital data is being sent from Subsystem 1 to the data-handling subsystem (Subsystem 2) for subsequent transmission to ground.

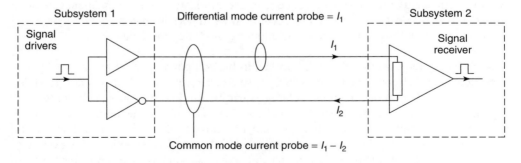

Figure 16.2 Differential mode and common mode signals

Most power, analogue signals and digital signals are transmitted between subsystems on 'two wire' interfaces since any common interference present is equally induced on both wires. This is termed *'common mode noise'*. The signal sent on one wire is designed to be complementary to the other, that is, if a logic '1' digital signal level is sent on one line then a logic '0' is transmitted on the second line. The interface circuit receiving the signal or power is designed only to look at the relatively interference-free signal *between* the two wires to detect the logic state or signal transmitted. The *common mode current* between the two subsystems is the algebraic sum of the currents flowing in the two wires. Two-wire interfaces are also commonly used for analogue signals and for distributing power around the spacecraft, using the same principles.

In the diagram, the *differential current* can be measured using a current probe around either wire and measures either I_1 or I_2. *Common mode current* is measured with a current probe around both wires so that the net forward current measured is $(I_1 - I_2)$. Similar definitions apply to *differential and common mode voltage* signals between interfaces.

16.7.4   Conducted susceptibility

Units or subsystems can be caused to malfunction or to be susceptible to the presence of conducted signals on the spacecraft harness or structure.

These externally applied voltages and currents are injected directly into the harnesses and cables between subsystems or directly on the unit's connector pins. These signals can cause a temporary malfunction, commonly called a 'glitch', or in extreme cases, electrical interfaces can be permanently damaged.

Susceptibility testing is essentially the reverse of testing for conducted emissions.

Sinusoidal and pulse signal generators, which can be swept over the required frequency range, are used as signal sources. These signals are power amplified and injected *via* current probes, transformers or voltage probes directly into the equipment or spacecraft under test. Applied signals are generally measured with an oscilloscope before the equipment is switched on, in its most sensitive mode, to detect any susceptibility.

16.8   ELECTROSTATIC DISCHARGE

This is similarly categorized into two areas, radiated and conducted.

● When an ESD occurs, there is always a 'spark', for example, lightning. The spark discharge occurs between two bodies or parts of a spacecraft—for example, a human body and a car door handle. The occurrence of the spark can generate a very large but extremely short duration burst of electric and magnetic *radiated emissions* over a wide frequency band. It is easy to confirm this by listening to any radio set tuned to the Medium or Long Wave bands during a thunderstorm. If any equipment, harnesses, wires or cables are close to this spark discharge, these fields will induce currents and voltages in the wires and equipment. This can cause either a temporary malfunction or, in some cases, permanent damage to equipment interfaces.

Testing for ESD radiated susceptibility involves generating a spark discharge of known energy usually 30 cm from the equipment and its connecting harnesses, and monitoring the equipment for any temporary or permanent malfunction.

- The secondary effect of an ESD is the *conducted current* and path of the ESD. The magnitude of this current can be up to 50 000 A with rise times in the order of nanoseconds to microseconds. If this conducted current is applied at equipment interfaces, or even discharged through an electronic unit's case, then this can cause either a temporary malfunction or permanent damage to the equipment.

Testing for ESD conducted susceptibility generally involves directly injecting current pulses from spark discharges, or directly from amplified pulse generators, into the equipment structure, harnesses and connector bodies and monitoring for any susceptibility.

The proximity of charged particles in the environment around any spacecraft can cause charge to build up on any isolated conductive surface. An account of this on-orbit charging mechanism can be found in Reference [4]. ESD problems can be eliminated by avoiding this charge build-up by grounding and bonding all parts of the spacecraft to the structure.

The importance of ESDs has increased as technology progresses. Vacuum tubes had an extremely high immunity to ESDs. Semiconductors used in spacecraft, particularly the high impedance metal oxide silicon (MOS) devices, can be very sensitive to even the smallest ESDs caused simply by handling the device without the appropriate precautions. Discharges of about 3.5 kV can occur from the fingers of an engineer and they are of such a short duration that they cannot be seen or felt. However, these discharges can easily destroy sensitive semiconductor devices, some of which are susceptible to voltages as low as 50 V.

For this reason, spacecraft clean rooms and manufacturing areas take special precautions and use grounded conducting floors, benches and chairs to minimize any possible charge build-up. Wrist straps are also used to connect personnel to ground during handling operations.

16.9   SPACECRAFT GROUNDING SCHEMES

Several fundamental questions need to be asked during the initial design phases of any spacecraft:

- What types of digital and analogue signals pass between the subsystems of the spacecraft, between subsystems and payloads, and what are the data transmission rates and bandwidth requirements of each of these signals?
- How much power is supplied by the spacecraft power subsystem to different parts of the spacecraft and payload and in what form? For example, are there direct supplies at main bus voltage levels in either regulated or unregulated form, or converted and regulated supplies distributed at lower or higher voltage levels?

Answers to these questions may perhaps already be known, since they may be customer-specified, or perhaps earlier studies may have answered them. For instance, data rates between the scientific payload of an (earth resources satellite) and the data-handling subsystem telemetering data to ground may be hundreds of Megabits per second. In contrast, communications satellites may only be telemetering a small amount of background housekeeping data.

Similarly, spacecraft with active microwave payloads such as a synthetic aperture radar (SAR) may consume kilowatts of power and would therefore be *directly* supplied by main bus power at voltages generally in the range of 28 to 120 V. Small satellites may supply payloads and subsystems with common *converted* and regulated supplies at, perhaps 5 V and 12 or 15 V levels.

A Spacecraft grounding scheme defines how all analogue and digital data signals and power are distributed and referenced on a spacecraft. All systems have a reference point or points to which all electrical signals and power is referred. This is usually called '0 V', 'earth', 'common', 'ground' or 'chassis'.

On any spacecraft, all these common points would be at the same potential, in an ideal world, regardless of whether signals in the system are 'Direct Current' (DC), (e.g. power buses), or Alternating in nature (AC) such as digital data signals between subsystems. In practice this is not true. Two points 'grounded together' with a direct wire connection, or perhaps *via* a conducting spacecraft structure, will not be at the same '0 V' potential. It depends on the currents that pass along the wire or structure, and their frequencies.

For example, if a DC current of 1 A is flowing in a 24 gauge ground wire of 10 cm length, there will be a very small potential difference between the two wire ends of, perhaps, 10 mV—the wire resistance would be 10 mohm from Ohm's Law. However, if AC signals are flowing in the ground wire at a frequency of 10 MHz, the wire will have a significant impedance—perhaps 10 ohms—due to the *inductance of the wire*. A much larger potential difference will exist if 1 amp flows in this wire at 10 MHz, in this case 10 V. Therefore, the two points apparently 'grounded together' with this wire for DC currents are not at the same potential at this frequency due to the wire inductance.

This example shows how essential a grounding scheme is to define all interfacing and grounding, and to pass only defined signals and power along defined paths or interfaces. This will minimize cross talk between signals (or power supply voltages) by minimizing common paths for the signals and power.

Computer calculations or estimations can also be made of the potential (DC) or noise (AC) at any point in the system with respect to a spacecraft ground point. A grounding scheme diagram is an essential aid to achieving these objectives.

There are three common ground schemes used on Spacecraft:

- Single-Point Grounding (SPG) Scheme
- Multipoint Ground (MPG) Scheme
- A Hybrid SPG/MPG Scheme.

16.9.1 Single-point ground (SPG) scheme

The example of an SPG scheme showing power and signal connections between two subsystems on a spacecraft is shown in Figure 16.3.

An SPG scheme has the following features:

- There is only one 0 V reference point to which all power and signals on the spacecraft are referred.
- Main Bus Power is fed to each subsystem by a 'twisted pair' set of wires.
- All signal and return wire currents pass down wires and therefore, no currents appear in the spacecraft structure. This is important particularly on some scientific satellites (e.g. the European GEOS and Ulysses spacecraft), which contained plasma wave detecting instruments sensitive to interference from spacecraft structure noise.

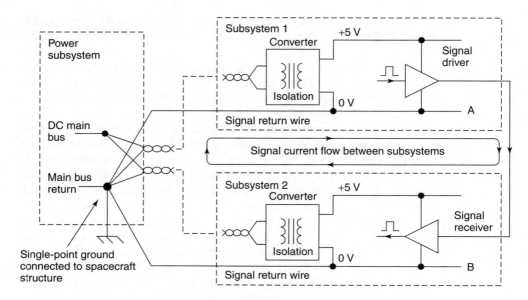

Figure 16.3   Single-point ground (SPG) scheme

- Note that the 0 V lines are grounded to the 0 V reference point at only one point on the spacecraft. This connection is usually made close to the Power Subsystem but can sometimes be located near the Data-Handling Subsystem.
- The single-ended digital signal transmitted between subsystems will be on slightly different logic '1' and logic '0' levels since points A and B will be at slightly different potentials. The signal current path shown indicates that both points A and B, in Figure 16.3, will have 'AC' signal frequency noise present because of the signal current flowing, and the 'AC' impedance between the 0 V point and A, and between the 0 V point and B. In practice, this is not a serious problem since logic chip integrated circuits have built-in 'noise immunity'. However, if many signals at very high frequencies are passing between the two subsystems, then the noise at these points with respect to the reference 0 V level could be sufficient to cause problems. This noise can cause distortions in the shape of the rise and fall time edges of digital signals. This gives rise to 'glitches' or interference pulses on signals between subsystems as the noisy signals cross and re-cross the logic '0' and '1' threshold levels. On modern spacecraft, these problems are invariably overcome by the use of differential drivers and receivers as shown in Figure 16.2 above, or by the increasing use of opto-coupled interface circuits. These pass information by switching infrared signals on and off. Opto-couplers, therefore, eliminate the flow of *electrical* signal currents between subsystems and are immune to noise on ground points between subsystems. However, they are usually not as fast for data-transfer purposes as conventional interface circuits.

A similar situation exists when analogue signals are passed between subsystems—for example, analogue telemetry signals monitoring currents and voltages, or those monitoring temperatures using thermistor sensors. Two wire interfaces are commonly used although they are often grouped with common return lines to minimize wires between subsystems and reduce harness mass.

In general, screened/twisted pair cables are used in the harness between units for both analogue and digital signals to minimize radiated emissions and susceptibility problems.

The main disadvantage of the SPG scheme is that, for large systems, the ground leads can be numerous and long, making the harness quite heavy. Although ideal for low frequency systems or distributing DC power, this can sometimes create problems since at higher frequencies, signal currents flowing in long wires can cause significant radiated emissions. These problems can be solved by using a MPG Scheme.

16.9.2 Multipoint grounding (MPG) scheme

An MPG scheme grounds all signal and power wire returns locally to a common ground plane. This ground plane has very low inductance between any two points on its surface and so ground currents flowing, even at very high frequencies, will not produce significant noise above ground potential on any part of this plane.

The ground plane is a flat, thin and wide conductor preferably made of plated copper. However, it could be any large, flat, and thin conductive surface—like the aluminium honeycomb structure used for many spacecraft platforms. This geometry gives a very low inductance between any part of the ground plane.

Separate flat, wide and thin conductive strips made from plated copper are used as a ground plane for those spacecraft that have carbon fibre structures, since carbon fibre has a comparatively high electrical resistance compared to aluminium.

If many signal interfaces are grounded locally to this ground plane, then many signal currents will flow in the ground plane but the effect of these is negligibly small since the inductance of the plane is so low—that is, 'Almost ideal' ground points are created throughout the spacecraft.

Figure 16.4 shows an example of an MPG scheme.

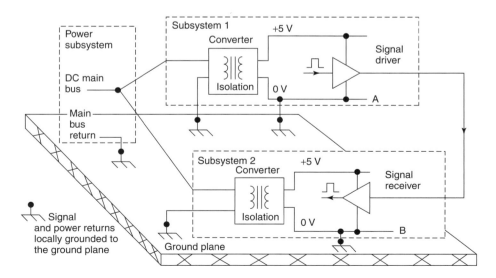

Figure 16.4 Multipoint ground (MPG) scheme

The grounds of logic chips on Subsystem 1 are closely coupled to those of Subsystem 2 by direct connection locally to the ground plane. This will allow interface circuits to operate at much higher frequencies (>30 MHz) using faster logic families. Note that it is still good practice to use differential drivers, receivers and screened cables and, if possible, opto-coupled interfaces between subsystems.

There may be several ground planes in a spacecraft, some inside the individual electronic units in each subsystem. For example, there may be separate planes for logic signals, sensitive analogue signals and high current digital drives for motors and solenoids. These precautions are taken to minimize cross coupling between radically different systems handling different signals. For example, the logic family ground plane may be handling 0 to 5 V logic signals with several milliamps of current in the ground plane. In contrast, the separate ground plane for motor drive or switch mode power converter circuits may be handling 50 V signals with 1 amp spikes induced in the ground plane.

16.9.3   Hybrid SPG/MPG scheme

It is common on spacecraft to adopt a hybrid-grounding scheme to take advantage of the best of both systems. For example, a typical strategy may be an MPG scheme for all digital signal transfers with separate ground planes for, say, high current digital drives, and an SPG scheme for the distribution of power and analogue signals. The SPG point and the many MPG points are shown in Figure 16.5.

This gives the advantage of the good high frequency performance of the MPG scheme together with the excellent isolation and elimination of ground loops for low frequency/analogue or power supply lines of the SPG system.

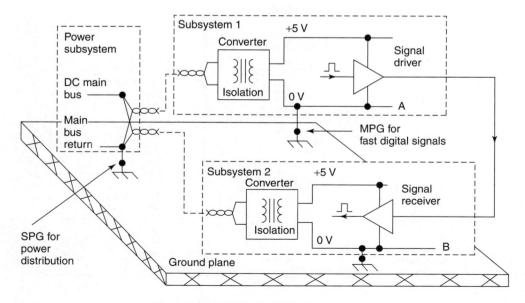

Figure 16.5 Hybrid SPG/MPG scheme

16.10 MAJOR CAUSES OF SPACECRAFT EMC PROBLEMS

16.10.1 Power supplies

Power supplies, particularly Switch Mode Power Converters, are usually major causes of EMC problems on any spacecraft. These generally convert main bus DC supplies down to other regulated voltages for use in subsystems, payloads and individual electronic units. They operate by first converting the DC input voltage into a rectangular AC signal using transistor power switches operating at frequencies up to several MHz. The AC signal produced is then applied to a transformer to produce several different AC voltages, before converting it back to new DC levels. A typical converter in a Data-Handling Subsystem unit would generate one 5 V and two 12 V supplies from one 28 V DC main bus input.

The speed of operation of the switching transistors and the use of magnetic components, combined with high voltages and circulating currents, gives rise to electric and magnetic field emissions at the fundamental switching frequency, and most of its harmonics. RF radiation can extend into the tens or even hundreds of MHz regions if adequate precautions are not taken.

The most effective way of reducing radiated interference from these converters is to reduce the radiations at source by slowing down transistor switching speeds. Reduction of transient voltages and currents, which are inevitably associated with the conversion process, is also necessary. In addition, close attention must be paid to the printed circuit board, unit layout and screening provisions as described in Section 16.7. Early testing on prototype models will identify any problems, since circuit layout problems are very expensive to fix after flight models of equipment have been manufactured and delivered.

Conducted interference currents, either to chassis or to other wires on the spacecraft, are also significant in these converters. Fast transistor switching times and large voltage handling can cause currents to be coupled, *via* stray capacitance effects, into the spacecraft structure or other wires. Since capacitance is inversely proportional to proximity, these problems are exacerbated by the trend towards smaller and smaller power supplies. Close coupling of heat generating components to chassis, to minimize thermal problems, can also cause these problems. In addition, stray capacitance between transformer windings can cause interference to pass back along main bus power leads or out to the converter loads.

Copper foil shields between windings can reduce these problems. Also, a careful choice of inductive and transformer components can reduce magnetic field leakage problems.

Diodes are used to convert the high frequency AC signals back to DC. The fast on-off switching of these rectifier diodes can also cause some problems by producing radiated emissions at harmonics of the converter switching frequency. Special circuits called 'snubbers' (usually a capacitor and resistor across each diode) can slow down the switching times of diodes to reduce this problem.

16.10.2 DC motors and actuators

The inductive nature of motors and actuators, the pulse width modulated nature and fast rise/fall times of the drive signals mean that these components have very similar EMC problems to switch mode power supplies. The resultant voltage and current waveforms and their harmonics can cause both radiated and conducted interference. Similar avoidance techniques are required.

16.10.3 Harnesses and cables

A significant percentage of all radiated and conducted emissions and susceptibility problems appear to be caused by the spacecraft harness. However, the harness is completely passive and cannot, by itself, either generate or be susceptible to conducted or radiated signals. It can radiate emissions and conduct electrical signals that are placed on the harness by electrical transmitters and receivers located at either end of the cable.

Electric and magnetic field emissions from harnesses and also, susceptibility to these signals can be minimized by

- Adopting a suitable grounding and bonding scheme for the systems.
- Partitioning and physically separating harnesses into power, pyrotechnic, noisy (digital) and quiet (analogue) signals.
- Paying careful attention to interface circuit designs, and where necessary, using matched drivers, transmitters and cables such that cables look like transmission lines.
- Matching the required electrical characteristics of each transmitted signal with the capabilities of the associated driver circuit, the harness wire characteristics, and receiver interface circuit properties. For example, for an interface transmitting digital signals between subsystems, this means slowing down logic signal rise and fall times as logic levels change from '0' to '1' and *vice versa*. It also means designing the slowest driver and receiver interface circuits, which, together with the harness wire characteristics, will successfully transmit these signals. Slowing down these signals, and also reducing the voltage and current levels used, prevents unwanted noise and associated harmonics appearing on the harness wires for subsequent electric and magnetic field radiation. Special filter connectors can also be used to slow down or filter signal and power lines.
- Using individual screened or twisted cables between signal interfaces, and making electrical connection to any screen ends *inside* metal backshells on the connectors at either end of the interface. In critical cases, overall shielding or braid can be fitted and electrically bonded with conductive tape to conductive backshells on connectors.

Conducted emission on harnesses depends critically on interface circuit designs and the overall grounding scheme adopted for the system or equipment.

16.11 ANALYSIS METHODS FOR SPACECRAFT EMC ENGINEERING

Computer analysis of radiated electric and magnetic fields is notoriously difficult since it relies heavily on accurate physical models of the environment. It is difficult also to predict with precision the nature of the frequency and time-domain interference that occurs. This is the reason most spacecraft contractors rely on preventative measures during the design phases, and early EMC testing and analysis of test results to characterize and identify problems early in the spacecraft development.

Analysis of conducted interference, using well-tried analysis software such as PSpice^R [5], is much more successful and is widely used. The input and output response of filters on power lines, interface circuits on both digital and analogue lines, and grounding diagrams are easily and accurately modelled and predicted using these computer programs. In addition, it is easy to generate computer simulations of a wide variety of interfering sinusoids and pulse signals to test the conducted susceptibility of all conducted power and digital/analogue interfaces. Differential responses to interfering signals can also be accurately predicted.

However, care must be taken when trying to predict the common mode responses of interfaces. Correct simulation relies heavily on correct models of stray capacitances and couplings of harnesses to each other and to the chassis.

Care must also be taken at frequencies above the 10 to 100 MHz region, since computer models of particular components may not be representative at these frequencies and beyond. A good example of this is the response of low frequency analogue circuits to RF radiation in the GHz region. Unless adequately screened, these RF fields can make small changes to the DC characteristics and offsets of the integrated circuits used. This produces unexpected changes in analogue circuit performance.

REFERENCES

[1] Clayton R. Paul (1992) *Introduction to Electromagnetic Compatibility*, John Wiley & Sons.
[2] Morgan, D. and Peregrinus, P. (1995) *A Handbook of EMC Testing and Measurement*, IEEE Press.
[3] Requirements for the Control of Electromagnetic Interference Characteristics of Subsystems and Equipment, MIL-STD-461E, August 1999, USA Department of Defense Interface Standard.
[4] Hasting, D. and Garrett, H. (1996) *Spacecraft-Environment Interactions*, Cambridge University Press, New York.
[5] PSpice^R, Cadence Design Systems, Inc. 2000, San Jose, CA, USA, Web address: *www.orcad.com*.

17 *PRODUCT ASSURANCE*

Thomas A. Meaker

Applied Technologies Business Unit, Verhaert, Belgium

17.1 INTRODUCTION

The aim of the product assurance (PA) system is to ensure that failure, hazard and degradation aspects of a designed and manufactured item are identified and controlled during the total spacecraft engineering process. This requires that the manufacturing processes and technology are 'stable' and that the design, production and verification processes are 'fully understood and controlled'. These stabilities and controls are essential to establish homogeneity of the overall spacecraft engineering process, hence validating the statistical methods that are the basis for PA from design analysis to parts and material sample testing. Homogeneity is also essential for predicting the occurrence and effects of failures and risk.

The establishment of the above homogeneity is very expensive, and requires detailed knowledge and history in all related areas. This requirement immediately excludes the latest 'Hi-technology' and advanced techniques. In fact, everything must be proven to be adequate to the intended function. The Qualification process (see Section 17.1.2) was conceived to ensure that space vehicles were produced from stable technologies and processes, and hence avoid the proliferation of historical failures. This process was largely successful, but the commercialization of space with its constant requirement that the next spacecraft must be cheaper, with a shorter production time and a higher performance, and the reduction in governmental support (financially as well as for R & D resources), internationally, for commercial and military space systems has resulted in the following situation today:

1. Advanced 'commercial' technologies increasingly being used to meet the 'faster, better, cheaper' commercial and military requirements.
2. Unqualified technologies increasingly being used to meet cost and schedule competition requirements (shorter parts delivery times).
3. The above technologies are referred to as *Civil, Aircraft, Military* (*CAM*) for hardware and *Commercial Off The Shelf* (*COTS*) for software (COTS is also sometimes used for hardware).

Spacecraft Systems Engineering (Third Edition). Edited by P. W. Fortescue, J. P. W. Stark and G. G. Swinerd
© 2003 John Wiley & Sons Ltd

4. The traditional Qualification process being largely superseded by a '*Qualification by Similarity*' process for CAM & COTS items. Hence the justification for use of advanced technologies is now based on data from all areas wherein the technology has actually been used with limited failure information. This data set is then compared with the 'nearest' qualified item and a judgment made, often qualitatively, concerning its acceptability for space. This is known as the *Request For Approval (RFA)* approach. Some details of the RFA approach are given in Section 17.1.2.
5. In many cases failure mechanisms (see Section 17.2) are not completely understood, so risk is increased.
6. The customer or 'next highest contractor' often needs expert resources to judge the RFAs.
7. Technology validation becoming more of an integrated design and PA process is consuming more resources at the commencement of the design process.
8. Small satellites using mainly Commercial and Military quality parts with complex design strategies to combat the increased degradation due to radiation.

The time between technology evolutionary steps is decreasing, and hence the risk sensitivity of the RFA process will increase. There is always a conflict between the desire to use the very latest technologies, for performance and revenue advantages, and the desire to use established technologies with their lower but more predictable performance, and hence lower risk.

Consequently, in order to achieve high system 'effectiveness' or 'availability', it is necessary to understand the dependency of design and manufacturing on PA and *vice versa*. The most elegant design in the world is useless if it cannot be (1) analysed for its weakness, (2) verified that it meets the specified requirements, (3) manufactured such that the design performance is achieved in the completed hardware and software and (4) maintained.

Fundamentally, the objective of PA is to ensure that the consequences of hazards and failures do not 'unacceptably affect' life, the space vehicle mission or the space vehicle itself. The real added value of PA is the *flexibility* with which this can be done in order to preserve the 'essentially needed' performance and Quality. PA engineers should *not* just be 'inspectors or checkers' but must add value by positively contributing to the design and manufacturing processes.

An essential element in all PA disciplines is the feedback cycle, in which the results of design, analysis, manufacturing, development, test and operation are immediately fed back to all 'involved' areas. There needs to be a continuous interchange and updating of all data, and a measure of its significance and status. This presents a significant organizational and management challenge.

The disciplines that collectively constitute PA are reliability, quality, safety, configuration control, parts (electronic, mechanical etc.) and materials and processes evaluation. These disciplines apply to hardware and software.

The particular aims of PA are as follows:

● to protect human life, the investment, the environment, public and private property and the mission;
● to establish and implement parametric derating criteria to ensure that no electrical, mechanical or chemical overstressing occurs during the mission;

- to define the probability that the system will perform successfully according to the specifications;
- to identify and control critical elements within the system such that the success of the mission is not jeopardized;
- to verify that all elements of design, hardware, manufacturing and test are of an adequate standard and are consistent and correlatable during all phases of the programme;
- to ensure that all failures, hazards and non-conformances are identified, their total effects understood and adequate rectification and retest validation carried out.

A general policy of prevention rather than cure is applied.

It will be apparent that PA covers all aspects of spacecraft design, build, test and operation at all contractual levels, and technically, from transistor junctions to the overall space segment. PA is thus an integrated and necessary part of spacecraft engineering. It cannot be emphasized too strongly that an essential ingredient is the application of efficient management techniques to ensure proper and timely identification of status and problems, and resource allocation for their rectification.

17.1.1 History

The extensive implementation of quality control requirements in the space business can be traced back to the failures of the American *Redstone, Jupiter and Vanguard* rockets in the 1950s. Their problems emphasized the need to take a scientific–engineering approach to reliability; a launcher typically contains over 100 000 component parts! Hence, reliability engineering was introduced seriously into the application arena.

In January 1967, a fire in the oxygen environment of the *Apollo* spacecraft caused the tragic death of three astronauts. This resulted in a complete reassessment of the role of PA and of safety in particular. NASA subsequently formulated a policy that resulted in the centralization of most PA functions and responsibilities. This was a major landmark since it resulted in a coordinated approach, detailed requirements and the recognition of PA as an accountable function in spacecraft engineering. The accountability of engineering and design disciplines was also established.

In November 1980, three fuses in the attitude control system of the $235 million *Solar Max* satellite blew, resulting in a lack of fine-pointing capability for four of the six onboard telescopes, and effectively mission failure after only nine months in orbit. The fuses failed because they were incorrectly sized following a circuit modification. The importance of Configuration and Change Control was demonstrated and its vital role in spacecraft engineering underlined.

Now that the space segment utilization is increasing with the introduction of internationally owned and operated manned orbiting laboratories for scientific and commercial purposes, involving enormous public & private investments, the quality and reliability of the equipment, and the safety and sustenance of 'man' have become top priority requirements. With the future habitation, industrialization and further exploration of space, PA-related aspects such as human engineering, maintainability, replenishment, and availability/reliability simulation are already being specified.

Over the past decades, tremendous increases in the complexity of technology, mission performance and lifetimes have taken place, and are continuing. Hence, because of the financial commitments and the impact of the loss of personnel and data, a need has developed to provide an '*a priori*' confidence that the spacecraft and platforms will successfully complete their missions. This is also a prerequisite of the insurance companies. Against this background, which will continue into the future, PA has developed from fundamentally an 'inspection' function to a sophisticated and respected total engineering science.

17.1.2 Qualification

Prior to launch, all elements of a spacecraft, including the launch vehicle, must have an acceptable qualification status according to the contractor's in-house 'risk' requirements and the customer's contractual requirements.

Qualification means that, by test and analysis, the design has been demonstrated to contain adequate margins and factors of safety, such that performance, manufacturing and interactive variabilities will not cause excursions outside the flight operational envelope of any individual spacecraft made to that design. In fact, it is then considered that only random failures will occur. Qualification is applicable at part, equipment and system levels.

Qualification tests involve the application of stresses to demonstrate that indigenous failure mechanisms will not degrade the mission. The 'types' of stress relate to those that the spacecraft would experience from lift-off to end-of-life. The stress levels are higher than those that are expected during operation life in order to demonstrate the margins mentioned above. Generally, they are 1.5 times the expected vibration levels (see Chapter 8), and $\pm 10\,°C$ on the predicted orbit temperatures for thermal testing (see Chapter 11). Test durations are either directly related to the operational environment or representative of it. For example, solar simulation testing includes one or two equinox, solstice and eclipse test cases. Equipments are usually only energized if they would normally be energized during the corresponding part of the mission. For example, an equipment that is switched on for the first time in geostationary orbit would not be energized during vibration (launch simulation) testing. In the author's view, this is a shortsighted policy since intermittent faults, possibly initiated only during vibration testing and the actual launch phase, would not be detected during testing but might cause serious problems during the actual mission.

Historically, two main methods of achieving spacecraft qualification have been used, namely, the prototype (PT) and protoflight (PF) approaches. Parts qualification is addressed in Section 17.3.4. However, for largely commercial reasons (see Section 17.1) the RFA approach is fast becoming the main qualification route.

In the *prototype approach*, which is applicable to entirely new designs and missions, dedicated fully instrumented qualification hardware is manufactured and exposed to the full qualification test programme at both the equipment and integrated system levels. After such testing the hardware is considered unfit for flight, and in many cases some of it is subjected to destructive analysis to evaluate fully the effects of the failure mechanisms.

For this approach, the cost and schedule impact of one complete additional spacecraft and test programme must be considered. For a series of satellites, for example, Intelsats or Eutelsats, the costs can be amortized over the series, and the risk *versus* cost trade-off is more attractive.

The *protoflight approach* was introduced after some experience had been gained from a number of space projects. It was argued that many pieces of hardware were being reused with relatively insignificant modifications from previously qualified missions, and that the application of the prototype philosophy was both inappropriate and unnecessarily expensive. The requirement for the application of the PF approach is that a qualification 'heritage' from similar space missions is demonstrated for each equipment proposed for the new project. This process often forms a major point of altercation between a customer and potential contractors due to different claims being made for 'qualification similarity'. With this approach the qualification test programme is applied to equipments and the fully integrated spacecraft, but half the full test durations are used at each level. After this the spacecraft is considered fit for a full operational mission, requiring no refurbishment. Instrumentation in the spacecraft is much less than in the PT approach and no additional internal hardware investigation is normally carried out. It should be noted that the PF approach requires that the spacecraft should be *designed* for the mission and the PF testing *with margins.*

The *RFA 'qualification'*, for CAM and COTS items, has been introduced to enable the utilization of commercial and advanced technologies provided they meet certain requirements. The failure mechanism and traceability requirements in the PF approach have been relaxed. However, the use of CAM/COTS technology must be justified *via* a trade-off analysis that estimates the risk and the criticality of the application and provides a minimum data set from the supplier and other users. Operational testing must be carried out at subsystem and system levels. As an example, the following general requirements apply to all CAM items for the *International Space Station* (*ISS*):

- Thermal cycling: orbit temperature predictions $\pm10\,°C$.
- Parts, Materials and Processes (PMP) verification using reverse engineering if necessary (stripping items for materials identification etc.).
- Failure Propagation Avoidance *via* Failure Modes Effects and Criticality Analysis (FMECA), Mean Time Between Failures(MTBF)/lifetime analyses and so on.
- Radiation effects assessment at equipment and subsystem levels with problem avoidance *via* customized screening tests, manufacturing and field history analysis.
- Flammability, 'offgassing' (in an atmosphere) and 'outgassing' (in vacuum), stress corrosion and fracture mechanics required at subsystem and system levels.
- All system-level Qualification (PF) tests must meet requirements.

For COTS SW the following main steps have to be followed:

- Audit of supplier (history, users, quality, future stability).
- Quality and relevance of *Beta Testing* (early field testing by selected experts).
- Verify supplier configuration control system and standards used.
- Carry out design walk-through.
- Run COTS SW with as many functions as possible.

It should be noted that COTS items cannot be used for safety critical functions and more stringent controls apply to the CAM equipments when used in the safety domain.

17.1.3 Acceptance

Once the design has been qualified, then hardware will be manufactured and integrated to form flight spacecraft. At the equipment and integrated spacecraft levels, *acceptance testing* is applied to ensure that no manufacturing or workmanship errors have been introduced that could deleteriously affect the mission. In general, random vibration and thermal vacuum testing are applied. Vibration stress levels equal to predicted orbit levels, and ±5 °C on the orbit temperature predictions are used. Significant repair/rework during testing can require repeat testing, and hence the importance of rigorous qualifications testing and controlled manufacturing is emphasized, to avoid such large additional expenses and schedule impacts.

17.2 FAILURES

A spacecraft must be designed to fulfil its particular mission; it is hoped that it will not be under-designed or over-designed. Under-design will lead to failure in completing the mission whilst over-design will result in a reduction in the payload due to non-optimal utilization of mass, power and so on, for redundancy in unrealistic situations. The focus must be on the fact that the 'payload' is the reason for the space vehicle, even if it is a 'lifeboat' for example, *Crew Return Vehicle*.

Designers in general tend to consider results based on the *successful* functioning of technology. The PA engineer on the other hand, evaluates the consequences of the presence and reaction of *failures, hazards and degradations*. The quantitative 'PA oriented' requirements of a design are stipulated in terms of reliability, mission life, permissible single point failures, availability and maintainability. These are 'system'-level requirements and are manifest at the circuit design level as derating rules, failure mode effects, criticality analyses, *Fault Detection, Inspection and Recovery* (*FDIR*) **and so on**. The overall objective is that during testing prior to launch and operation in orbit, the spacecraft should not experience any 'infant-mortality or wear-out' effects and that all other 'failure effects' would be controlled by the design such that the system performance is not affected by them.

The consideration of the failure potential, or mission risk, of a design and the subsequent hardware and software is an essential part of spacecraft engineering. In order to ensure that the correct design has been established, it is necessary to understand all the possible failure mechanisms that may degrade or terminate the performance or life.

Failure mechanisms are the fundamental events, electrochemical or otherwise, which represent an unacceptable change from a previously defined and stable condition. Failure mechanisms are always present and usually become more significant as time increases; often from interactive effects. It is therefore imperative that their 'end-of-orbit-life' effects are understood and contained within the design envelope. For example, particle radiation can seriously affect technology and man (see Chapter 2). Its effects can be instantaneous or time-dependent.

In order to facilitate the design process, and enable the resolution of otherwise impossibly complex reliability equations, failure mechanisms are divided into two classes: those that occur randomly in time and those that are time-dependent. In practice, the former group contains all failure mechanisms for which time dependency has not been discovered. As missions become more demanding, it sometimes happens that a failure mechanism changes its classification due to a higher stress environment. An example of this is the banning of wire filament fuses on long-life (7 to 10 year) satellite missions due to a gradual glass seal leakage causing the fuse to operate in an unpredictable fashion.

Failure mechanisms for electronic parts, and their associated activation energies are determined during the parts qualification programmes. Life testing is carried out to verify the effects of the time-dependent failure mechanisms, and hence the long duration operational missions will not be deleteriously affected. Life testing is typically of the 'accelerated' variety whereby the component operating temperature, for example, is increased and each test hour is then equivalent to a number of orbit operating hours, according to the Arrhenius relationship (see Appendix 1). It is important to realize that each failure mechanism has its own activation energy and test programmes must be designed accordingly.

Failure modes refer to the way in which failures are manifest, as opposed to the mechanisms that cause them. In electronics, the main failure modes are short circuit, open circuit and parameter drift. Mechanical failure nodes are usually associated with wear, crack propagation, migration of lubricants, spalling and so on, and are therefore time-dependent in nature. Electrochemical failure mechanisms, involving interaction and breakdown effects, also tend to be time-dependent. An example is, metal-to-metal contact occurring after the non-replacement of oxide surfaces or lubricant films, because of the lack of gravitational forces and oxygen, resulting in the occurrence of cold welds and subsequent failure.

Table 17.1 contains a listing of some well-known failure mechanisms and modes for selected technologies.

Table 17.1 Failure modes and mechanism examples

Type	Failure mode	Failure mechanism
Resistor—wirewound	Drift Open	Faulty thermal treatment. Wire corrosion due to contamination
Thermistor	Open	Fracture of glass bead types during vibration and temp. cycling. Bond failure of element to mount
	Drift (calibration change)	Lead fracture. Contamination through poorly manufactured or damaged coating
Inductors	Drift	Cracked ferrite core during vibration temperature cycling (cause: usually incorrect assembly)
	Open	Movement of core clamps
	Short	Poor locking; wire breakage; shorted turns (workmanship). Dielectric breakdown in high voltage and RF transformers (workmanship)
Capacitor—tantalum, solid	Short + leakage	Dielectric breakdown
Diode	Open	Cracked glass
General purpose (glass case)	Drift	Die surface contamination
Transistor—microwave	Drift + open	Unequal current heating in emitter fingers (local heating and metal transportation)
Integrated circuits	Open	Contamination of plating solutions
	Seal faults	Corrosion at lead base; poor integrity of die/header bond

17.2.1 Failure mode effects and criticality analysis (FMECA)

The overall objective of **FMECA** is to identify all failure modes that could occur on all components at all levels, and determine their effects on performance at the next higher functional level, and thence up to system level, for all operational phases. By analysing the effect of all failure modes on equipment, subsystem and system, it is possible to identify, quantitatively, the probability of single-point failures and thus assess the criticality of the design. The consequences of the effects of a failure are clearly critical concerning system level impact, and hence a component redesign, or a system redesign to avoid using the component, may be required. Criticality categories, also involving safety criteria to address the man-rated systems, have been defined to standardize this aspect of the analysis and are as follows:

Cat. 1a	Catastrophic (safety hazard Cat. 1)	Causes death or permanent injury
Cat. 1b	Critical (safety hazard Cat. 2)	Causes severe injury or major property damage
Cat. 2	Major	The system cannot operate
Cat. 3	Significant	The system is partially operable

The FMECA is a 'bottom-up' analysis and procedures exist for carrying it out. It is complemented by *Fault Tree Analysis* (*FTA*), which concentrates on failures at system level and shows their possible causes by a 'top-down' approach to the equipment or electronic parts level.

The importance of proper FMECA cannot be overemphasized. It should be commenced at the conceptual design stage, updated for every design and hardware review and used for validation of all ground testing and orbit operation. Its role in testing is frequently omitted, often with serious consequences. An example of this was the testing of a particular infra-red Earth sensor. The FMECA indicated the existence of a failure mechanism at the upper operating temperature which would only be detectable if the infra-red sensor was stimulated by an infra-red source at that temperature. Unfortunately, the FMECA was not used during the preparation of the test procedure, the sensor was not stimulated during thermal testing and a fault occurred—in orbit! A $100 million project was thus unnecessarily jeopardized.

Figure 17.1 shows an example of a *fault tree*.

17.2.2 Failure rates and confidence levels

The failure rate (the instantaneous failure rate is often referred to as the hazard rate) is a measure of the rate at which failures occur. Whether it is constant, increasing or decreasing with respect to mission time is clearly of critical importance. The two general curves in Figure 17.3 are typical of the failure rate *versus* time characteristics for 'standard' electronic and mechanical parts, such as transistors, ball-bearings, battery cells and so on.

Failure rates are derived from test and operational data, and since these represent a sample of the total population it is necessary to quote a confidence level to indicate how closely the sample data statistically correlate to the population. In practice, failure rates are usually quoted at a confidence level of 60%. If the sample relates to an exponential distribution, then the chi-square statistic can be used to derive confidence levels. They

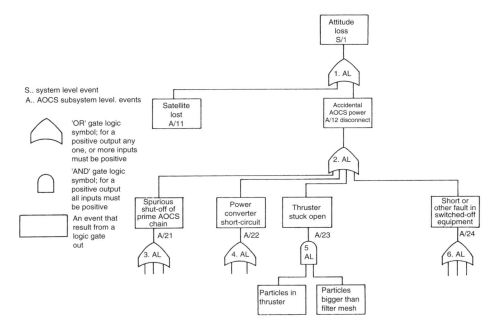

Figure 17.1 Fault tree analysis (contingency analysis)

can also be derived from Weibull plots but special statistical tables must be used, which give the probability that a defined number of weak parts remain in the batch (sample).

Failure rates are used in reliability calculations in order to evaluate the probability that an electronic circuit, for example, will successfully function according to specification for the required time. The failure rate is usually subdivided according to the relative likelihood of occurrence of the applicable failure modes. Hence, for example, in the design process, the effect of the short-circuit failure mode of a transistor on circuit, subsystem and spacecraft performance can be established. Reliability models generally utilize Quality Factors (QF) that relate to the manufacturing quality of the components to be used. Examples of the relationship between QF and the Component Quality are as follows:

Component procurement system	Quality level	Quality factor
SCCG	Level B	1
SCCG	Level C	2
ER-Mil	FR level S	1
ER-Mil	FR level R	2
Mil-S-19500	Jan S	1
Mil-S-19500	Jan TXV	2
Mil-M-38510	Class S	1
Mil-M-38510	Class B	2

Note: ER-MIL, Established Reliability-MIL

Table 17.2 Typical 'failure rate and % occurrence of related failure mode' data for transistors (unit: FITS = failures per 10^9 hours)

Type	Failure rate (FITS)	Failure mode % of occurrence			Notes	
		Short	Open	Drift	Power (W)	Frequency (GHz)
Linear Si	0.8	20	15	65	<1	—
Logic switching Si	0.4	60	30	10	<1	—
Power Si	1.2	20	15	65	1–5	—
Power Si	1.6	20	15	65	5–20	—
Power Si	2.0	20	15	65	20–50	—
RF Si	8.0	40	10	50	<1	<0.2
Microwave bipolar	15	40	10	50	<5	0.2–1.0
Microwave bipolar	100	40	10	50	5–10	0.2–1.0
Microwave bipolar	600	40	10	50	10–30	0.2–1.0
Microwave bipolar	15	40	10	50	<0.5	1–4
Microwave bipolar	200	40	10	50	0.5–5	1–4
Microwave bipolar	1000	40	10	50	5–10	1–4
Microwave bipolar	7000	40	10	50	10–30	1–4
FET linear	1.5	20	15	65		
FET switching	0.7	20	15	65		
FET RF	5.0	40	10	50		
FET GaAs low noise	50.0	20	15	65		
FET GaAs driver	360.0	20	15	65		

It will be seen that component quality not only has a large effect on reliability but also on price and delivery times. Many authorities now do not accept the quality factors as being representative of the quality of commercial and industrial components, and caution must be exercised in this area. Many satellite authorities collect their own failure statistics.

A selection of failure rates is shown in Table 17.2.

17.3   RELIABILITY

The *reliability* of a spacecraft is defined as the probability with which it will successfully complete the specified mission performance for the required mission time. Reliability is not a 'stand-alone' discipline but is dependent on all others, for example, correct selection of components, correct derating, correct definition of the environmental stresses, restriction of vibration and thermal transfer effects from other subsystems, representative testing, proper manufacturing and so on.

For a typical telecommunications type spacecraft, such as *Hotbird*, with a mission life of twelve years, a reliability of about 0.8 is currently achievable. This means that there is about an 80% chance that the full specification performance will be achieved for the mission time, or conversely a 20% chance that it will not. The latter represents the *risk* that is taken, by the customer and will be manifest in the operational status as down time.

For a 'service' type spacecraft such as a telephony/television communications satellite, down time, or 'unavailability' constitutes loss of revenue, and hence the cost benefits of design improvements to increase reliability can be optimized against their impact on revenue return. This is not a trivial process in practice and can involve a complex Monte Carlo simulation using the spacecraft reliability model to determine the expected availability from failures occurring randomly in time.

Reliability calculations have two main applications:

1. trade-off analysis, usually with other criteria such as cost and mass, to compare, for example, one subsystem with another;
2. assessment of risk, availability, maintenance strategies and so on, using comparative and absolute reliability figures.

Reliability is calculated using failure rates, and hence the accuracy of the calculations depends on the accuracy and realism of our knowledge of failure mechanisms and modes. For most established electronic parts, failure rates are well known, but the same cannot be said for mechanical, electromechanical, and electrochemical parts or man. In modern applications in which computers and their 'embedded' software, are often integrated into the system, the reliability of the software must also be considered. Once again, presently, failure data for software are very sparse and although a number of 'reliability' models exist it will be some years before there is an established industrial standard. Hence, the accuracy of the overall system reliability analysis is dominated by the lack of precise knowledge in the 'non-electric', 'software' and 'man–machine interface' areas. Notwithstanding these limitations, which must be carefully considered by the spacecraft engineer, reliability analyses cover significant portions of the spacecraft design and represent, together with safety hazard analyses, practically the only method of assessing technical risk. Also, it should be noted that as more technology utilization data becomes available, the failure rates usually decrease. Hence initial reliability assessments will tend to be unrealistically high.

Software reliability is generally defined in terms of the indigenous faults within the software, use of the software and inputs to the software. As far as the user is concerned, software faults usually occur randomly in time; the causes of failure, however, are clearly systematic. Typical errors relate to program interactions, syntax errors, logic errors and performance (storage, accuracy and so on). Figure 17.2 shows a software lifecycle management scheme, which indicates the early introduction of PA and the various test phases. The ESA SW standards contain an abbreviated SW lifecycle set of requirements referred to as *SWLITE*. This approach is to facilitate the development and verification of 'small' and COTS SW packages in which the application of the full SW PA requirements would be disproportionate in terms of cost and resources. The *verification* that the SW will function correctly is a major PA element and can be very complex, time consuming (schedule) and expensive. The SW must be designed such that it can be tested down to the smallest module level.

17.3.1 Reliability analysis

As mentioned previously, it is the objective of the 'qualification' analysis/test programmes to verify that the spacecraft design limits and controls the effects of the 'identified' failure mechanisms such that the mission will be successful—with a certain probability. Unfortunately, all variables cannot be analysed or tested. Occasionally manufacturing and materials exhibit high variability and so failures still occur. The latter are assumed

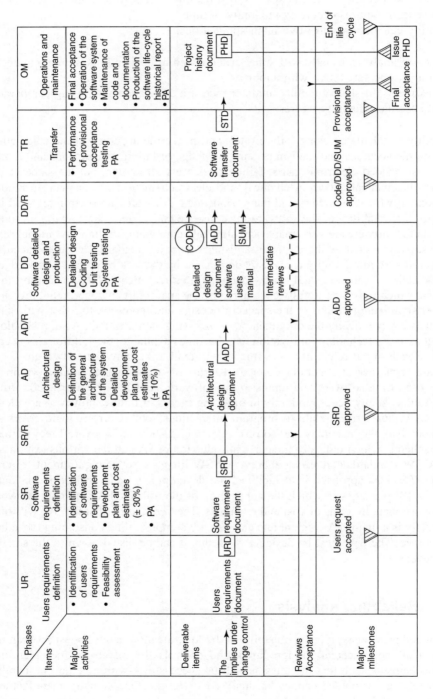

Figure 17.2 Software lifecycle management scheme

to occur randomly in time—by 'chance'. Thus, since we are considering only random variables, probability theory can be applied, and relatively simple mathematics can be used to evaluate reliability.

It is very important that the spacecraft engineer realizes that, in general, reliability calculations involve only random failures and constant failure rates. Hence, the following functions are essential to ensure that all non-random failure mechanisms are removed from, or controlled by, the flight design and hardware:

1. a rigorous qualification programme;
2. expert and detailed investigation of all problems to determine their cause, and rectify accordingly;
3. rigid configuration control to ensure proper design, and hardware modification and analysis.

17.3.2 Probability distributions

Reliability is defined as the probability that a product will operate successfully for a given time. The relative frequency with which particular events occur is indicated by a probability distribution; this is used to define the relative frequency of 'the arrival times at an unsatisfactory state' of the product. For example, the failure rate may be constant, increasing, decreasing or a combination, depending on the failure mechanisms in operation, that is, the distribution of arrival times at the failure condition.

Four probability distributions will now be briefly discussed. Binomial and Weibull distributions are applicable in Figure 17.3(a) (Regions I and II) and in Figure 17.3(b) (Regions I, II and III).

The *binomial distribution* relates to a sequence of mutually independent events and takes the form $(p + q)^n$ where p is the probability of a random event, $q = 1 - p$, and n is the number of trials. It is applicable when an event has only two possible outcomes, and is thus used for single-shot evaluations such as solid apogee boost motor firings. It is also used in the computation of the reliability of complex systems, and the sensitivity of the overall reliability to various degrees of redundancy.

The *Gaussian (Normal) distribution* can be used when the failure rate is increasing as a function of time. The main applications in reliability are as follows:

1. the relationship of a variable operating characteristic to a set of limits;
2. wear-out life of parts;
3. failures due to extended periods of operation.

The distribution is described by two parameters: mean and standard deviation.

The *Weibull distribution* is described by three parameters, relating to scale, shape and location. It is extensively used in the analysis of experimental data—a process that is generally facilitated because of the ready availability of Weibull graph paper that enables the Weibull parameters to be deduced from the plots.

This distribution is used in the evaluation of parts 'burn-in' data, fatigue and other wear-out data.

The *(negative) exponential distribution*, which is a special case of the Weibull distribution, is used extensively when the failure rate is constant, and failures occur randomly

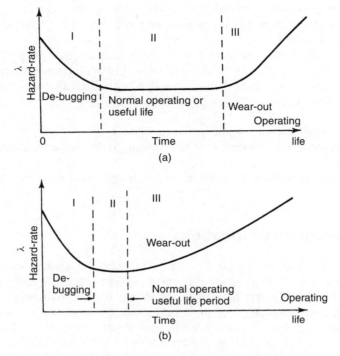

Figure 17.3 Failure rates *versus* time: (a) electronic parts; (b) mechanical parts

in time, for example, in electronics. It is described by one parameter, failure rate, and is applicable to Region II in Figure 17.3(a).

The reliability function takes the form

$$R(t) = e^{-\lambda t}$$

and the failure density function is

$$f(t) = -dR(t)/dt = \lambda e^{-\lambda t}$$

The reliability of elements within a system can thus be computed, and the system reliability calculated by considering the functional way(s) by which the elements are connected. Reliability block diagrams or models can be constructed on a functional basis that incorporates redundant elements and how they are to be brought into operation.

17.3.3   Reliability enhancement techniques

Redundancy of functional elements is used to counteract the effects of failures on the mission: the two main types are standby and active.

In *standby redundancy*, the redundant element is switched into operation when a failure occurs in the prime unit; it is often not energized prior to operation, and in this 'dormant' state it is assumed to have a failure rate equal to one tenth of its energized rate. A special case used increasingly in telecommunication-type satellites is the *ring redundancy*. This requires a large number of three- or four-port switches, which enable any transmitter to be switched to any channel.

Active redundancy involves all elements being operational for the entire mission in such a way that the total load or stress is shared, and thus reduced in proportion to the number of redundant elements. *Voting redundancy* is a special case in which N elements are required to ensure the required performance, but M elements are available, where $M > N$.

Depending on the mission requirements, any mixture of types of redundancy may be used. The impact of different redundancy configurations can be established by carrying out a *sensitivity analysis*. This usually involves the use of the binomial distribution to evaluate the individual probabilities of partially failed configurations; the total number of states is simply the sum of the binomial coefficients. A comparison of the above 'individual' probabilities with the criteria for mission success permits selection of the optimal redundancy design.

With the prospects of multi-spacecraft operation to achieve a particular mission, such as constellation satellites (e.g. Iridium) and intersatellite links, the concept of redundancy being provided by physically detached hardware must be considered. In this scenario, maintenance aspects, that is, repairability and replacement, must also be modelled. The assessment of risk is thus becoming more difficult, and more necessary, and resort must be made to simulation to overcome the complexities and assumptions of analytical techniques.

A summary of *redundancy models* is given in Figure 17.4.

Metallic Shielding and circuit & memory protection techniques are used to enhance reliability against orbit radiation effects.

The possible catastrophic effects of *solar flare* activity on man are beyond the scope of this book. However, its effect on electronics, solar cells and external thermal control paints and materials is considered. The 'shielding effect' of the spacecraft structure, equipment walls, printed circuit boards and so on, is evaluated by carrying out a *sector analysis* of the entire spacecraft. In this process, the reduction of the incident *total dose* (electrons and protons) radiation by the on-board materials is assessed at the location of the sensitive parts. This reduction in damaging radiation energy can be of three orders of magnitude. Hence careful layout may preclude the need for extra shielding and provide more mass for payload.

Depending on the orbit, the spacecraft will be exposed to varying doses of *electron, proton and* high-energy *cosmic radiation*. The major effects are bulk atomic displacement, which can be permanently destructive, and surface ionization, which is usually transient. The actual failure mechanisms concerned are many but include decreased carrier lifetime, mobility and concentration. *Cosmic rays* are problematic in that they can cause changes of state in small-geometry technology such as microprocessors and large memories. These changes, termed *single-event upsets* (*SEU*), are manifest as temporary bit errors or permanent *latch-up*, and are the result of high local ionization and displacement caused by individual high-energy heavy ions. The effects can be extremely serious since, with autonomous on-board control, incorrect commands can be given. Shielding is ineffective with cosmic radiation and the only method of protection is careful system design (e.g.

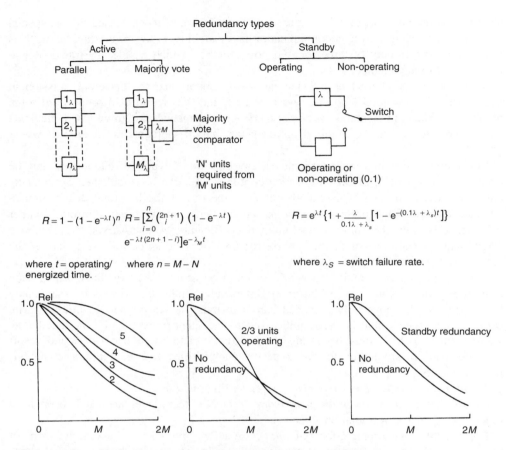

Figure 17.4 Redundancy models. M = MTBF. All units have the same failure rates

anti-latch-up circuits, memory washing, frequent *post-mortem checks*) and part selection. On manned space vehicles certain technologies are prohibited for safety critical functions.

The two main methods of avoiding electron and proton radiation problems are the use of *radiation hard parts* and metal shields. *Dose-depth curves* exist, which indicate the 'stopping power' of various thickness of, for example, aluminium, for various strengths of incident radiation. This information, together with the knowledge of the radiation sensitivity of the part measured during the qualification programme, enables a safe and reliable design to be made. A short note of warning is necessary. When charged particles scatter off each other, as in the case of electrons impacting shielding, for instance, then a secondary emission of X-rays is produced which also has a damaging effect. This phenomenon is termed *Bremsstrahlung* (German for 'braking radiation') and represents the lowest level of radiation that can be achieved by shielding. In general, Bremsstrahlung is more significant for light, high-energy particles and high-density shield materials.

Appendix 2 contains radiation sensitivities for some part types and a note on radiation units.

Derating, in order to reduce the stress levels in essentially electronic and electrical parts, is used as a means of improving reliability.

Parts manufacturers issue data sheets that specify the performance of the particular part. This is termed as the rated performance and represents the maximum stresses to which the part should be subjected in order to provide a nominal operational life. The life and reliability aspects are usually not specified for commercial and industrial quality components, but they are essential for unmanned space vehicles, and mission critical and safety applications. Hence the *Hi-Rel* part specifications were established with firm requirements for material selection, *process stability* and the elimination of all adverse failure mechanism effects by design, inspection and testing.

To ensure, to the maximum extent possible, that parts will not fail when used in safety critical functions or during long, non-maintainable space missions they are utilized in a derated condition. The actual amount of derating necessary depends on the mission requirements and economic considerations. However, many procurement authorities specify derating requirements and stipulate that their specified failure rates are valid only if the derating rules, or better, are applied. The failure rate can be reduced if the stress level is below the required derating level, but in most cases the cost and time to demonstrate such marginal gains is prohibitive. Typical derating data are shown in Appendix 3.

17.3.4   Reliability testing/demonstration

Reliability testing/demonstration is rarely carried out nowadays in space programmes due to the high cost and long durations involved. It is, however, effectively carried out at 'part' level by considering the cumulative test hours accrued by each part type during life-test programmes. Thus in the American *ER-MIL* (Established Reliability) system, passive parts, such as resistors, can be procured against a specified failure rate. The usual objective of a reliability test is to measure the failure rate, or *Mean Time Between Failures* (*MTBF*), of a piece of hardware such that the test results are statistically significant. Such a test has to be extremely carefully designed and implemented and all failures have to be investigated for primary cause, that is, failure mechanism. The test should be continued until at least one failure occurs, and 'truncated' test methods have been developed, such as the *AGREE* and MIL test methods, to minimize test times. Replacement and/or repair of failed elements must be decided, and all interfaces and assumptions carefully considered in the statistical calculations to decide the test result. It must be remembered that a success or failure may determine whether a contractor is paid, and hence all test criteria and all possible test outcomes must be agreed before the test commences.

As an example of the duration of, and investment necessary for, a reliability test the following points are made. Consider a spacecraft equipment that is claimed to have an MTBF of one million hours (equivalent to a failure rate of 1000 FITS) with a confidence level of 60%. In order to demonstrate this claim, the contractor would have to test the equipment for 916 000 hours with no failures or just over 2 million hours with one failure. This is based on MTBF $= 2n/\chi^2(a : 2r + 2)$ where n is the number of test hours, r is the number of failures and a is the confidence level of the χ^2 distribution ($\chi^2 =$ chi-square statistic).

The cost of retaining manpower and test facilities to carry out such a test programme can well be imagined but if the impact of the equipment failure is sufficiently large, or relates to manned safety, then the test may be necessary. The spacecraft engineer should be aware that such testing is not trivial, and hence other methods of risk evaluation and containment, such as the use of Hi-Rel parts and analytical attempts to keep the design point above the stress/strength envelope, are usually used.

17.3.5   Reliability of mechanisms

One of the most difficult areas for forecasting reliability is that of *mechanisms*. These cover a range of devices as discussed in Chapter 15. It is particularly difficult for those mechanisms that have to remain dormant for long periods before actuation, such as may occur on planetary probes.

The fundamental reason for these difficulties is that practically every mechanism needs to be designed, manufactured and tested (verified), specifically for each application. This is necessary because of the sensitivity of mechanisms to slight changes in, for example, temperature ranges, vibration inputs and operating requirements. Commensurate with this is the fact that mechanisms are almost exclusively based on interfacing materials and their interactions, and this itself constitutes a difficult area (see Section 17.5). These difficulties are compounded by the limited extent of the knowledge-bases and databases.

At the time of writing there was no accepted method for predicting the reliability of mechanisms. In the modelling process, it is usually considered to be unity.

Currently, the reliability aspects of mechanisms are covered by:

- ensuring that the experience and expertise of the involved persons are adequate;
- using conservative design margins;
- rigorous testing and qualification. Unfortunately testing to failure, thereby establishing which failure mechanism occurs first, is usually not carried out. Additionally, simulation of zero gravity is normally not possible;
- applying FMECA (see Section 17.2.1);
- rigorous optical inspection and non-destructive analysis such as using radiographic techniques for example, X-ray for high-density materials, and neutron for low density ones.

Of great concern to mechanism designers is that an adequate mass allowance is available to permit proper stiffness, the absence of damaging resonant frequencies, adequate accommodation of inertial loads, and so on. Unfortunately, in a typical space project, mass reduction exercises become the order of the month as developing designs and manufacture see the mass increase. Mechanisms do not escape this predatorial process, sometimes with dire consequences in orbit.

In order to improve the above situation, a method for analysing the reliability of non-electronic systems has been developed at ESA, for example. It is supported by knowledge-bases and databases and software tools. This methodology utilizes failure distributions of basic mechanical building blocks, such as ball-bearings, springs, pyrotechnic cutters, and so on, which have been constructed from quantitative data from tests, and subjective assessments of experts.

The design, and the implementation of manufacturing and quality operations to produce it in hardware form, (often referred to as 'achievement of the design'), are considered separately. Human reliability aspects relating to manufacture and operational activities are included.

The method is based on two typical (macro) failure causes and has thus resulted in a bifurcated approach that requires the assessment of (1) 'theoretical reliability' and (2) 'practical reliability'. These two terms cover the following:

1. the event that the design margins are insufficient to take account of the natural scatter of the performance
2. the event that a failure occurs during the 'achievement of the design' and is not detected by the PA controls

The theoretical reliability assessment considers system margins, safety factors, assumptions, qualification and acceptance/test aspects, and utilizes a functional FMECA, based on an analysis of the effect of failures on the system's functions, such as navigation, guidance and so on. See Section 17.2.1.

The practical reliability assessment covers:

● causes of failure, and solutions, by using the functional FMECA and
● human reliability aspects, utilizing subjective data such as statements by experts, and objective data such as test information that enables the probabilities of the causes of failure to be calculated.

This method uses the data collected and analysed from failure events as they occur.

17.4   THE USE OF QUALIFIED PARTS

When a Qualification Authority exists, such as ESA and the Goddard Space Flight Centre, then it usually issues and maintains a Qualified Parts List (QPL). In general, the qualification criteria are similar for the various authorities.

The spacecraft engineer is constrained to select parts from a Preferred Parts List (PPL), which is compiled from QPLs. The list of parts that are actually needed to design the particular spacecraft is termed the Declared Parts List (DPL).

The problems really begin when parts are required that are not on the PPL. It is then necessary to evaluate or qualify, the 'performance' of the particular application and the 'technology' for its failure mechanisms. Outputs from the qualification programme enable the screening test criteria ('*infant mortality*') and the technology wear-out (life) point to be defined (see Figure 17.3). The reliability (risk) calculations are thus dependent on using properly qualified and screened parts. Currently, approximately 95% of the total production of electronic components is for commercial and industrial applications. Hence designers and PA engineers are increasingly having to devise methods to use and evaluate the risk of commercial components. The RFA approach is oriented to this problem.

A typical evaluation/qualification programme is given in Figure 17.5. Such a programme often proceeds in parallel with the main project, thus introducing additional risk, cost and schedule constraints.

After (qualified) parts are manufactured, they are all subjected to '*screening*' or '*burn-in*' tests to verify that they contain no indigenous or 'infant-mortality' defects (see Figure 17.3). It is possible to integrate the screening tests within the evaluation/qualification programme, and thus use some of the parts directly for flight equipment manufacture.

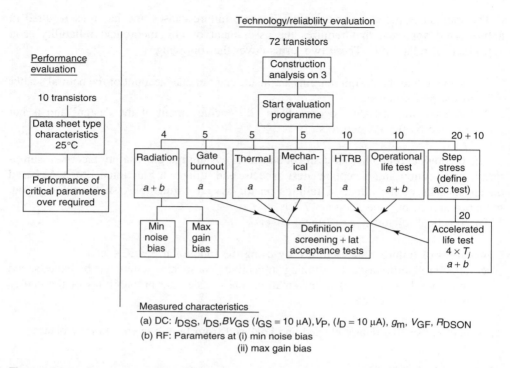

Figure 17.5 Example of a detailed evaluation programme for GaAs FETs (82 transistors)

17.5 MATERIALS AND PROCESSES

Probably the most underrated aspect of spacecraft engineering is that related to the performance of *materials* and *processes*.

The prime properties under consideration are, outgassing, compatibility with other materials, *toxicity, odour, flammability* and *offgassing*. The latter four properties are specific to manned missions and relate to safety aspects.

Outgassing is one of the major problems associated with using materials in a vacuum. The criterion for accepting a material is that its outgassing should not exceed '1% [of its] total mass loss (TML) and 0.1% collected volatile condensable materials (CVCM)'; the latter refers to the material emitted from a sample at $+125\,^{\circ}$C and collected on a surface at $+25\,^{\circ}$C over a 24-hour period. This criterion is only an 'indicator' of acceptability.

A large data store exists, which contains a wide spectrum of information on materials and processes. The job of the materials engineer is to 'fit' materials and processes to the required task within the environment concerned (see Chapter 2). A clear understanding of failure mechanisms and modes is required and 'evaluation' programmes have to be designed and implemented. Also, standards have been developed, such as for soldering, welding, crimping, material plating, outgassing tests and surface property (absorbance/reflectance/emittance) measurements.

Tables 17.3 and 17.4 contain a listing of *non-preferred materials* and some information on materials 'stress/performance' characteristics.

Table 17.3 List of prohibited and non-preferred materials

1. All adhesives must be 100% solid
2. Polyvinyl chloride backing tapes
3. Cellulose, paper, fabric, etc.
4. Varnishes and coatings which rely on solvent evaporation for hardening
5. Canada balsam; organic glasses in high-precision equipment
6. Direct space exposure of most oils and greases
7. Graphite—is an abrasive in vacuum
8. Cadmium, zinc (whisker growth)
9. Paints should be avoided where possible
10. Polyvinyl chloride and acetate; cellulose and acetate, plastic films
11. Potting should be avoided where possible
12. Polyester laminates
13. Polysulphide rubbers; rubbers containing plasticizers; chlorinated rubbers
14. Polyvinyl chloride (PVC) thermoplastic, polyvinyl acetate butyrate, many polyamides

17.6 SAFETY

The overall objective of the safety programme is to ensure that accidents are prevented and all 'hazards', or threats, to people, the system and the mission are identified and controlled. Safety requirements apply to all programme phases and embrace ground and flight hardware, software and documentation. They also endeavour to protect people from 'man-induced' hazards.

Standard listings of typical hazards exist, such as 'flammable materials', and the Hazard analysis has the following general form:

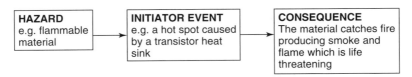

For unmanned spacecraft, safety is largely a matter of industrial protection and involves such aspects as handling toxic fluids, protection against high-voltage and RF supplies, transportation and storage of flight equipment and so on.

In the case of manned spacecraft, safety is a severe design requirement and compliance must be demonstrated prior to launch. Hazards are classified as 'catastrophic, critical or marginal' depending on their consequences (loss of life, spacecraft loss, injury, damage, etc.). For example, functions that could result in critical or catastrophic hazards must be controlled by a single or double *Fault Tolerance* design, which means two or three independent inhibits (barriers), respectively. Premature firing of a liquid propellant propulsion system would probably constitute a catastrophic hazard. The designer therefore has to incorporate three independent propellant flow-control devices, in series, in the propulsion system. This means that after failure of the first barrier the system is still operational. After failure of the second barrier, it is still safe, but cannot be operated, since it would

Table 17.4 Stress/performance characteristics of materials

Parameter → / Material ↓	Main categories	Vacuum	Particulate radiation	Ultraviolet radiation	Temperature High	Temperature Low	Thermal cycling
Adhesives	Epoxies, phenolic, polyurethanes, silicones, cyano-acrylates	Outgassing	Outgassing increased	Optical adhesives darken. Outgassing increased.	Causes degradation 300 °C max. for polyimide, epoxy 170 °C max.	Hardening embrittlement	Unmatched expansion coefficients cause failures
Glasses	Silicates, sapphires, fluorides, polystyrene, acrylic, etc.	Contamination danger only	Most harmful (10³ rad)	Harmful	Thermal shock main problem		
Lubricants	Hydrocarbons, silicones, esters, MoS₂, WSe₂, Pb	Evaporation, dry-off creep contamination	Metal screening usually effective		Accelerated evaporation etc.	No major problem	
Plastic films	Polyolefins, polyesters, fluorinated plastics, etc.	Film stiffeners; contamination	Deformation embrittlement discoloration		Degradation	Embrittlement	Damaging to metallized films, metal film detachment
Potting compounds	Epoxies, silicones, polyurethanes	Contamination corona	Minor problem	Minor problem	Chemical degradation	Shrink rigid increase; internal stresses increase	Cracking debonding
Reinforced and thermosetting resins	Epoxies, phenolics, melamine, polyesters	Outgassing corona contamination	No problem	No problem			Can be a problem due to anisotropy of reinforced plastics
Rubbers	Polybutadiene, polychloroprene, acrylics, nitrile, etc.	Outgassing from additives and de-polymerization of base polymer. Contamination	Will harden or soften	Will harden or soften	Decomposition	Hardening, stiffening, crazing, crushing	See high/low temperature
Thermoplastics	Polyamides, acetal, polyolefins, acrylics, etc.	Degradation due to outgassing of stabilizing additives. Contamination	Discoloration, outgassing, hardening	Discoloration, outgassing, hardening	Softening	Hardening, embrittlement	
Paints	Epoxy, silicone, etc. binders, ZnO, TiO₂, Al, C for white/black pigment	High outgassing	Absorptance severely affected; embrittlement	Absorptance severely affected; embrittlement	Degradation, paint flaking	Minor problem	Degradation of paint flexibility

explode if another failure should occur. This is called a '*Fail-operational/Fail-safe*' system. The result is an increase of mass, instrumentation, cost and testing complexity; and a consequent decrease in mass available for carrying revenue-earning or science-providing payload. This is a major disadvantage of launching spacecraft on manned orbiters, since the spacecraft, although themselves not manned, must meet the manned safety requirements for the launch and injection (and possibly recovery) phases. It may, however, be offset by other considerations. For structures, pressure vessels, composites and joints and so on, damage tolerance (*Fracture Control*) design policy is applicable to manned vehicles and the concepts of 'Safe Life' (SL) and 'Fail Safe' (FS) relate. Basically, the following failure critical conditions must be avoided:

- release of fragments of $>113.5\,g$
- release of tension-preloaded masses of $>13\,g$
- release of hazardous substances
- prevention of safe descent from orbit

Safe Life requires that the largest 'undetectable' flaw will not grow to failure dimensions in four complete service lifetimes. *Fail Safe* is related to the period of time during which the maximum growth of the largest defect will not invalidate the specified performance. For example, a structure can be classified as Fail Safe if the structure is still safe even after losing a redundant load-bearing member. Both SL and FS establish a safety barrier.

In general, the design rationale is that hazards must be controlled, minimized or eliminated using fault-tolerant, derating, safety devices such as power removal interlocks and warning indicators.

It should be noted that, for the ISS, software is prohibited for the control of critical and catastrophic hazards. Also, hazard control software must utilize recognized languages, for example, Columbus ADA. Safety and reliability are complementary. Safety relates to the protection of personnel and the investment; reliability relates to success of the mission. The spacecraft engineer should appreciate that spacecraft must be designed to provide adequate safety and reliability, with strict controls on the build and testing. It is almost impossible to modify a spacecraft, once it is built, to meet the safety requirements. As an example, an Intelsat satellite was refused permission to fly on the Space Shuttle because it did not comply with the safety requirements; design rectification would have been too expensive, so an (unmanned) expendable launcher was finally used.

17.6.1   Sneak path analysis

In the previous paragraphs techniques have been presented, such as FMECA and FTA, which aim to determine the consequences of component failures. However, system malfunctions can also be caused by design errors. The manifestation of these in systems with a complex architecture, such as electrical, electronics and software, is called a *sneak circuit*. It is an unexpected path for a flow of mass, energy or logical sequence that under certain conditions can initiate an undesired function or can inhibit a desired one. Sneak circuits are not the result of a failure, but are latent conditions inadvertently designed into the system.

Sneak Path Analysis is an identification technique. It is a search for unwanted causal links between 'sources' and targets. The former are defined as components that supply the resources such as power, data, mass, and so on, and the latter as components that

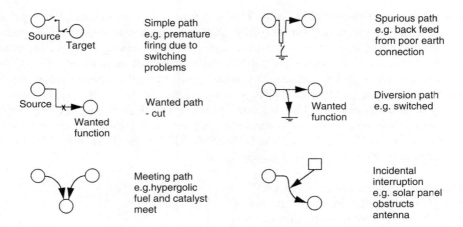

Figure 17.6 Sneak circuit path types

perform or inhibit a function. Sneak Path Analysis makes use of path-tracing (Figure 17.6) between the sources and the targets, and of checklists called clues. These are applied to the whole path (path clues) or to the components along it (component + path clues) in order to help identify the sneak circuits.

The basic steps in Sneak Path Analysis of a system are

- To define the boundaries and the mission phases of the system. This is based on the results of preliminary Hazard Analysis and/or Functional Failure Analysis;
- gather the relevant design data such as drawings and specifications and so on;
- document the basic dynamic characteristics (i.e. the state of the functional input or output of the system during the planned operational modes) in an 'input/output switching matrix';
- identify targets, such as a pyrotechnic firing circuit in a launcher;
- identify sources, such as a battery;
- trace the paths between sources and targets, which are allowed by the system structure and dynamic characteristics;
- apply path clues to the paths, and component and path clues to the items contained in the path;
- fill out sneak circuit reports, giving data about sneak circuits together with recommendations to eliminate them.

Sneak Path Analysis can be applied when a detailed design becomes available, between Preliminary and Critical Design Reviews (PDR and CDR), and should be performed with close cooperation between the *Reliability, Availability, Maintainability and Safety* (*RAMS*) personnel and Systems engineering personnel.

17.6.2 Risk analysis

The risk that the equipment (hardware and software) and people involved in a space system will not function as planned is due to the following:

1. the hazardous nature of the equipments and of man and
2. the effects of failures of any functional element including man.

Risk analysis basically involves the *likelihood* (probability) that a hazardous event will occur and the *severity* of the effect of the hazard on the system being considered.

This risk is identified using safety analysis that comprises deterministic hazard analysis and probabilistic risk assessment.

Hazard analysis is the systematic identification, evaluation and classification of all hazards and associated scenarios in the system design, operation and environment.

Risk analysis addresses the *cumulative* effects of all hazards and their associated scenarios.

An example of a hazard and a failure, as contributors to overall risk is a cylinder of high-pressure oxygen for life support, placed under a crew member's seat. The hazardous characteristics of the contained high energy are apparent because of the siting of the cylinder in such a position. The effects of an explosion due to a failure, are self-evident; the probability of such an event occurring can be calculated from test data.

From the preceding explanations of reliability and safety in this chapter it will be realized that risk assessment requires inputs from, and the integration of, both reliability and safety analyses. The fundamental aim of *safety analysis* is the elimination or reduction of hazards. In order to ensure that an analysis of a complete system is comprehensive, it is assessed from the 'top down' and from the 'bottom up'. The former approach, which is often carried out using fault trees (see Section 17.2.1), considers risk at, for example, the space vehicle level, and then tracks down through the system to identify all hazards that could cause or contribute to that system level risk. In the 'bottom-up' approach, hazards at the lowest level are assessed for their system level impact by tracking them up through the system. FMECA is often used for this activity. An example could be the system level impact of a semiconductor short-circuit failure in a guidance control system.

After the safety analysis has been completed, and hazard elimination and reduction has been implemented to the maximum extent, there may still be some hazards that constitute significant risk. This is termed residual risk, and a formal review process involving top management must be implemented before it can be accepted. If it cannot be accepted, then expensive design or hardware changes usually result.

The process of safety analysis and risk management is summarized in Figure 17.7. Some of its objectives are

- to support cost savings by integrated safety and reliability optimization;
- to enhance the application of probabilistic risk analysis with emphasis on identification and modelling of uncertainties;
- to integrate analyses such as FMECA, human and software dependability, reliability prediction and so on;
- to emphasize a total system approach that integrates hardware, software, man and the environment;
- to increase the dialogue between safety and engineering;
- to support the establishment of lines of communication between engineering and management.

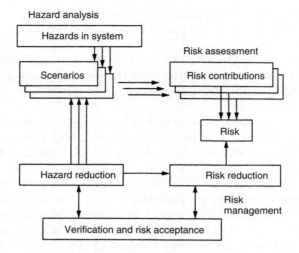

Figure 17.7 Synthesis of safety analysis and risk management

17.7 CONFIGURATION CONTROL

Everything related to a space project, from specifications and batteries to change requests, is represented by a number. These numbers are written on pieces of paper, which are also allocated numbers, and at any particular time the design or build can be defined by a listing of the appropriate numbers. This listing is termed a *baseline*, and in theory anybody could build a spacecraft from the paperwork identified by the Final Design Baseline.

A typical communications-type satellite that contains over 300 equipments, is designed, manufactured and tested by 30 or more different companies, and is subjected to over 500 different designs, hardware, test and delivery reviews. The specification tree probably contains over 100 specifications, and 'other' documentation includes four to five times that number. During the project's life, a total of over 2000 changes would have been raised and processed, each affecting some or all of the above documents.

The problem is one of 'defining, synchronizing and controlling' all the documents and their updates such that all baselines represent a design or build standard that is up to date and consistent, at both its internal and external interfaces. The solution is provided by the configuration control system that requires 'accountability' between all documents and 'reconciliation' between all baselines. Implementation is *via* 'change-control' and 'interface-management' systems.

As an example of the necessity of such a system to the spacecraft engineer, it would be very serious if a long and expensive analysis of, for instance, the functioning of the spacecraft's on-board processor was carried out using drawings and performance documents that were out of date!

17.8 BUILD AND VERIFICATION

This section refers to the manufacturing, assembly and testing of the spacecraft at all levels.

17.8.1 Build

Typically, the flight-standard '*Design Baseline*' is approved ('frozen') at the *Critical Design Review* and permission is given to commence the manufacture of flight hardware. The '*Build Baseline*' is thus initiated. As the build progresses, problems are encountered that require expert judgment by satellite engineers to ensure that the fully built spacecraft will function according to the original design specification.

Many of the design analyses can be invalidated during the manufacturing phase unless proper quality controls and standards are used. For example, failure rates could become incorrect owing to contamination or bad soldering, introducing rogue degradation failure mechanisms. RF losses in a transponder could be unacceptably increased because of incorrect assembly of connectors and cables. In order to represent the design exactly in terms of hardware, every facet of the manufacturing and integration activity is planned, monitored and approved after satisfactory completion by the quality engineers.

17.8.2 Verification

At the CDR a *verification matrix* is approved, which identifies how every design and mission requirement, as stated in the specifications, will be verified. The verification is normally by analysis and/or test.

Prior to the implementation of any verification aspect, the build baseline is compared to the design baseline. All differences must be satisfactorily explained (or 'reconciled' to use the jargon) before the test can proceed. The test programme is a verification activity, and when all verification activities have been successfully completed the spacecraft is qualified for flight.

Thus the long and exacting spacecraft engineering process is finalized with the assurance of a fully verified spacecraft (product) on the launch pad.

17.9 ACKNOWLEDGEMENTS

The author wishes to acknowledge the following persons for permission to use material from their various publications: Mr B. Dore (Sneak Path Analysis), Dr C. Preyssl (Risk Analysis), and Dr A. Holmes-Siedle (Radiation).

APPENDIX 1: THE ARRHENIUS EQUATION

The *Arrhenius equation* was originally formulated by J. J. Hood and demonstrated by Svante Arrhenius as being applicable to most kinds of reactions. It describes the effect of reaction temperatures on the rate of a chemical reaction.

When applied in 'parts' engineering, the reaction rate formula used for the chemical and physical processes causing strength degradation is

$$R = A\exp(-E_a/kT) \tag{A17.1}$$

The temperature-lifetime relationship is given by

$$t = C\exp(E_a/kT) \tag{A17.2}$$

where: $A =$ a constant
        $C =$ a temperature-independent constant
      $E_a =$ activation energy (see Table A17.5)
       $k =$ Boltzmann's constant
      $R =$ reaction rate
       $t =$ time
      $T =$ absolute temperature

Table A17.5 Some typical activation energies

Part type	E_a (eV)
1. Silicon semiconductors	
• surface charge build-up (bipolar)	1.0
• surface charge build-up (MOS)	1.2
• electro-migration	0.6/1.2
• corrosion	0.3/0.6
2. Integrated circuits: (MOSFETS)	
• threshold voltage shift	1.2
3. GaAs microwave transistors	
• contact metal migration	2.3/2.7
4. Carbon comp. resistors	0.6

For a particular activation energy (failure mechanism), equation (A17.2) reduces to

$$\log(t_1/t_2) = T_2/T_1 \qquad (A17.3)$$

This enables accelerated test times to be calculated. For example, a screening test of 240 h at a junction temperature of 200 °C is equivalent to 4368 h at 100 °C, the temperature normally used during orbit operation. Hence the screening test is roughly equivalent to six months in orbit.

APPENDIX 2: PARTICLE RADIATION DATA

The data given are approximate. The changing of one process step, or even part of it, can have a significant effect on the radiation tolerance of the technology. Caution-radiation characteristics can change from one manufactured lot to another.

A1 Radiation-sensitivity characteristics of some component types

Integrated circuits

1. Threshold voltage shifts in the negative direction resulting in an absolute increase for P-type metal oxide semiconductor (P-MOS) and a decrease for N-type metal oxide semiconductor (N-MOS).

2. Supply current: rapid increase for Complementary metal oxide semiconductor (CMOS) as soon as the threshold voltage of the N-MOS reaches zero volts.
3. Propagation delay: increases because of increased channel resistance of the P-MOS.

The following ionization dose sensitivities apply, the range referring to the values at which the 'most' and 'least' sensitive type shows significant degradation:

$$\text{NMOS } 800 - 10^4 \text{ rad (Si)}$$
$$\text{CMOS } 5000 - 10^7 \text{ rad (Si)}$$

Microwave parts: PIN diodes

Increase in surface leakage current. Fundamentally bulk damage sensitivity (mesa construction), and hence some effect (decrease) on lifetime. No significant problem for seven-year geostationary orbit—mainly ionization radiation.

Microwave transistors

DC current gain and collector-base breakdown voltage decrease; collector-emitter breakdown voltage increases. Transistor gain (S21e) is affected.

Degradation can be expected at 5×10^5 rad(Si) but each part should be individually considered.

Operational amplifiers

Main degradations are loss of gain and alteration of offsets.

A/D and D/A converters

Calibration drifts as a function of irradiation.

A2 Radiation units

Ionizing dose

This is the energy per unit mass element deposited by a beam of particles in an array of atoms of a given material. The receptor material must be specified. Silicon is used as the reference material for spacecraft and the dose is thus quoted as 'rad(Si)'.

$$1 \text{ rad} = 10^2 \text{ erg/gram}$$
$$1 \text{ gray} = 10^4 \text{ erg/gram} = 1 \text{ joule/kg}$$

Conversion

In order to convert radiation flux (number of particles passing through a unit area in unit time) to ionizing dose, a rule of thumb is:

1. $1 \text{ rad} = 3 \times 10^7 \text{ electrons/cm}^2$
2. $1 \text{ rad} = 4 \times 10^6 \text{ protons/cm}^2$

These apply to energy levels of 2 MeV for electrons and 30 MeV for protons.

A3   Radiation equivalence

Qualitatively, the effects of all types of ionizing radiation are the same. Provided that the same energy arrives at a particular location, the degradations that develop as the result of increasing flux will be the same for all radiation. Hence, in the laboratory or test house, we can, for example, use X-rays to simulate electrons.

APPENDIX 3: DE-RATING

De-rating is the reduction of electrical and thermal stress of a component in order to increase its useful lifetime. The de-rating factor is usually given as a percentage of the maximum value of the parameter considered, these maxima being defined in the procurement specifications.

Linear integrated circuits ($T_{amax} < 85\,°C$)

Supply voltage	of maximum specified rated value	80%
Input voltage	of maximum specified rated value	70%
Output current	of maximum specified rated value	80%

Fuses

Fuses shall be avoided wherever possible, but when needed cermet fuses are preferred. The largest fuse rating compatible with the source current shall be used.

Absolute maximum allowed ratings:
    65% of the nominal specified current,
    100% of the maximum specified voltage after blow.

Wire link fuses should only be envisaged for short-time application and on special request.

Transistors

De-rating factor of the minimum specified reverse voltage:	65%
De-rating factor of the maximum specified current:	75%
Power de-rating factor of the maximum specified value:	60%
Maximum junction temperature:	115 °C

Worst-case analysis:	Drift to be considered

Leakage currents:	5 times the specified maximum
Breakdown voltages:	5% decrease of the specified minimum
Forward voltages:	10% increase of the specified maximum
Gain:	25% decrease of the specified minimum
	25% decrease of the specified maximum
Saturation voltages (V_{cesat}, V_{besat}):	10% increase of the specified maximum

Currents and voltages shall be maintained at levels compatible with the safe operating area.

For transistor operating *without* heat sink:

$$P_{\text{derated}} = 0.6[(115 - T_a)/90]P_{\max} \qquad T_a > 25\,°C$$

$$P_{\text{derated}} = 0.6P_{\max} \qquad T_a \leq 25\,°C$$

For transistor operating *with* heat sink:

$$P_{\text{derated}} = 0.6P_{\max} \qquad \text{if } T_c \leq 115\,°C - 0.6(175 - T_m)°C$$

$$P_{\text{derated}} = [(115 - T_c)/(175 - T_m)]P_{\max} \qquad \text{if } T_c > 115\,°C - 0.6(175 - T_m)°C,$$

Where $P_{\max}$ = maximum rated power at 25 °C
   T_a = ambient temperature
   T_c = case temperature
   T_m = maximum specified case temperature for maximum rated power

Note that the maximum junction temperature is fixed to 175 °C for the calculation of the de-rated power. For lower specified maximum junction temperatures, additional de-rating may be needed.

Relays

Contact voltage	No de-rating
Coil voltage	No de-rating
Number of operations[1]	50% of qualification number
Contact current[2]	Lower than 75% of specified max. rated number (resistive load)
Contact current transients[3]	Limited such that $I^2t \leq 6(I_{\max})^2 10^{-5}\,(A^2\,s)$
Degradation for worst-case analysis	Four times the contact resistance specified limit for >1000 operations[4]

[1]If the number of operations is below 10 cycles, the specified max. rated current can be used.
[2]It is not recommended to use a relay with a contact current below 20% of the max. rated current except for low level application(<1 mA).
[3]In the case in which the pulse duration is low or equal to 10 μs, the transient current shall be limited to $4I_{\max}$.
[4]For a number of switching operations of less than 1000, an increase of twice the contact resistance can be used.

18 SMALL-SATELLITE ENGINEERING AND APPLICATIONS

Martin N. Sweeting and Craig I. Underwood

Surrey Space Centre, University of Surrey, Guildford, Surrey, United Kingdom

18.1 INTRODUCTION

Early western satellites were necessarily small. However, the need for ever-larger, more capable and more complex satellites led to a natural growth in satellite mass. This trend was first limited by the capabilities of available launchers and then later by finance and technological infrastructure, as space-faring nations required a highly developed technological base with huge investment. This resulted in space becoming limited to a few nations—an exclusive club of space 'haves' with enormous military, economic and cultural advantages over the space 'have-nots'.

Changing world politics and military emphasis in the last decade has brought about a quiet revolution in space. Pressure on both civil and military space agency budgets has meant increasingly that fewer (and bigger) satellites have been commissioned and new ideas, technologies and scientific experiments have found it difficult to gain timely access to space.

In the 1970s, advances in *very large scale integration (VLSI)* led to the possibility of sophisticated functions being built into very small volumes, with low mass and requiring minimal electrical power. This led to the emergence of the modern small satellite—or *microsatellite*—which, in turn, led to the potential for a dramatic cost reduction in satellite programmes. This was initially demonstrated by the pioneering Orbiting Satellite Carrying Amateur Radio (OSCAR) and Radio-Sputnik (RS) series of satellites produced by the international amateur radio satellite (AMSAT) community and their Russian counterparts.

Starting in 1961 with OSCAR-1—a simple radio transmitter in a box—these 'amateur' microsatellites, built by enthusiasts in the USA, USSR, Germany, Australia and Japan, grew in sophistication, until by the end of the 1970s, with OSCAR-8, they had reached a significant level of sophistication in terms of supporting amateur radio communications [1] but still without an on-board computing capability.

Spacecraft Systems Engineering (Third Edition). Edited by P. W. Fortescue, J. P. W. Stark and G. G. Swinerd
© 2003 John Wiley & Sons Ltd

It was at this time that staff and students (all amateur radio enthusiasts) at the University of Surrey embarked on a programme to develop a microsatellite for the UK—which resulted in the design and construction of UoSAT-1 (UoSAT-OSCAR-9), launched by NASA in October 1981 [2].

The 1980s thus ushered in a new era of small sophisticated satellites, characterized by the extensive use of microprocessors and other VLSI technologies. Indeed the success of UoSAT-1 lead directly to a second spacecraft (UoSAT-2—UoSAT-OSCAR-11), being designed and built in a record time of just 6 months for launch in March 1984 [3].

UoSAT-1 continued to operate successfully for 8 years until its re-entry in 1989. UoSAT-2 was still operational in 2002 after more than 18 years in orbit! These two spacecraft established the modern concept of a microsatellite, and Surrey has since become a world leader in providing small-satellite platforms and the associated technological 'know-how' to both existing and emerging space nations through its spin-off company—Surrey Satellite Technology Ltd (SSTL).

In many ways the relationship between a 'microsatellite' and a 'conventional' satellite is similar to that between a microcomputer [i.e. a personal computer (PC)] and a conventional computer. That is, a microsatellite is sufficiently low in cost and yet has sufficient utility to enable it to serve a much wider user community than would otherwise be the case.

The mass market for PCs is sustained and exists (in part) because of the astonishing rate of progress of microelectronic technologies over the last 30 years. The staggering developments in microelectronics, stimulated increasingly by the consumer market rather than military requirements, and the dramatic pace of consumer product development, have caused conventional space technology to often lag considerably behind that taken for granted on the ground. However, these advances have had a direct impact on the power and sophistication of microsatellites through their adoption of *commercial-off-the-shelf (COTS)* microelectronic technologies developed for terrestrial use and adapted to the space environment. The combination of reducing budgets for space and increasing capability of low-power microelectronics have enabled a new breed of highly capable *smaller, faster, cheaper* satellites to realize many space missions (see also Chapter 17)—complementing the conventional large satellite systems still necessary for large-scale space science and communications services to small terminals.

Like their PC counterparts, microsatellites are generally small and relatively inexpensive—typically 1/10th or 1/100th the mass and cost of conventional satellites. This has lead to the following widely accepted definitions, given in Table 18.1, and shown schematically in Figure 18.1.

Table 18.1   Classification of spacecraft by mass and cost

Class	Mass (kg)	Cost (£M)
Conventional large satellite	>1000	>100
Conventional small satellite	500–1000	25–100
Minisatellite	100–500	7–25
Microsatellite	10–100	1–7
Nanosatellite	1–10	0.1–1
Picosatellite	<1	<0.1

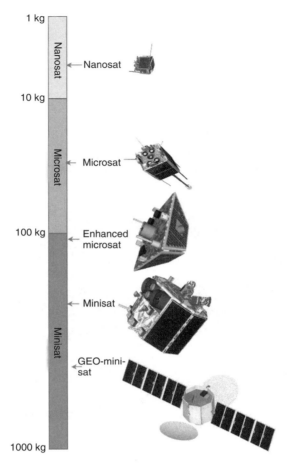

Figure 18.1   Schematic of small-satellite classification

Although microsatellites are physically small, they are nevertheless complex and exhibit virtually all the characteristics of a large satellite—but in microcosm. The typical size of such microsatellites is illustrated in Figure 18.2.

In addition to fulfilling communications, remote sensing and space science applications, these miniature satellites are particularly suitable as a focus for the education and training of young scientists and engineers by providing a means for direct, hands-on experience in all stages and aspects (both technical and managerial) of a real satellite mission—from design, construction, test and launch through to orbital operation.

Of course, such small satellites cannot cater for all mission objectives—especially where large transmitter powers, antenna arrays or optical apertures are required for applications such as direct broadcast TV and mobile voice communications from geostationary orbit, or for very high-resolution Earth or astronomical imaging. Nevertheless, as on-board processing capabilities have advanced, many missions, hitherto only feasible with large satellites, have become possible with microsatellites at a fraction of the conventional mission cost. In particular, microsatellites make affordable constellations possible (see

Figure 18.2 *PoSAT-1* and *HealthSat-2* (1993) — typical examples of the microsatellites developed at the University of Surrey (Surrey Satellite Technology Ltd)

Section 5.5 of Chapter 5) and can thus, for example, provide rapid revisit and high temporal resolution for Earth-observation missions at a realistic cost.

Small, low-cost satellites alone do not result in low-cost missions, and each element of the qualitative equation below needs to be appropriately apportioned, with typical values given below:

$$\text{Total mission cost} = \text{satellite cost} + \text{launch cost} + \begin{array}{c} \text{orbital operations} \\ \text{costs over lifetime} \end{array}$$

$$100\% \qquad\qquad 70\% \qquad\qquad 20\% \qquad\qquad 10\%$$

Similarly, a small satellite is not necessarily a low-cost satellite. Low cost is achieved through the application of a certain *design philosophy* to the entire mission. Indeed, for many, it is not the *size* of the spacecraft as such, but the application of this design philosophy, which distinguishes 'small' satellites from conventional ones.

18.2   SMALL-SATELLITE DESIGN PHILOSOPHY

Modern small satellites are constructed in rapid timescales (typically 12–18 months), at relatively low cost, and make maximum use of state-of-the-art COTS technologies to achieve complex functionality, while at the same time minimizing dependence on complex mechanisms, deployable structures and so on.

The satellites are *engineered to cost* specifically to meet their mission objectives during their design lifetime—and no more. The mission objectives are carefully traded against cost to achieve the minimum necessary to achieve the required outcome—typically aiming to achieve 80% of the (conventional) mission capability at 20% of the (conventional) mission cost. Careful management of risk allows more spacecraft missions to be afforded more regularly. The consequences of this design philosophy are contrasted with those of a conventional mission in Figure 18.3.

The rapid development of modern electronic devices means that small satellites often use state-of-the-art technologies with little or no flight heritage. This poses an immediate problem—how do you know if these technologies will work in space if they have never flown before? This is especially true of COTS technologies, which, by their nature, are unlikely to have been designed to operate in the relatively harsh environment of space.

In conventional space missions, there is usually sufficient time and resources to undertake a thorough programme of ground-based testing prior to flight (see Chapter 17), but this is problematical if cost reduction and short design-to-orbit times are major mission drivers—as they usually are in small-satellite programmes. Thus, a different approach is needed.

The general approach to minimizing the risks associated with the use of COTS technologies is to rely more on *sound design practices*, experience and *system-level testing* rather than on (expensive) programmes of testing individual devices. However, this requires the design engineers to have a good knowledge of the (space) environment to which their systems will be subjected—and to use appropriate design techniques and margins as necessary. Some of the measures employed may be summarized as follows:

- Minimize the variety of devices/materials.
- Avoid toxic, volatile or potentially explosive substances.
- Keep the interfaces simple.
- Minimize moving parts—use of body cells, use of passive thermal control.

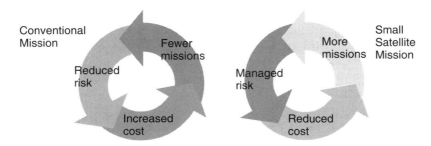

Figure 18.3 Consequences of conventional mission *versus* small-satellite mission design philosophy

- Use previously flown designs and components in essential systems.
- Be realistic with safety margins.
- Ensure systems are capable of independent operation—avoid chains.
- Use carefully selected high volume components where possible.
- Avoid software in the loop.
- Knowledgeable use of volume production and modern components.
- Use a layered, failure-resilient system architecture.
- Ensure a thorough burn-in prior to flight.

Solving the technical challenges associated with the design, construction, test and operation of a microsatellite is only half of the story. In parallel with the technical considerations of the mission, *effective project management* is crucial to the realization of a successful low-cost, sophisticated small-satellite project.

Affordable small satellites require a very different approach to management as well as to technology if cost, performance and delivery targets are to be met. Several attempts by traditional aerospace organizations to produce such satellites have failed because of the rigidity of management structure and 'mind-set'. Small teams (typically 25 persons), working in close proximity with good communications, and well-informed and responsive management, are essential.

These characteristics are best found in small companies or research teams rather than large aerospace organizations, who may find it difficult to adopt or modify procedures necessary to produce affordable small satellites using staff and structures that are designed for conventional aerospace projects. The main ingredients for a successful small-satellite project can be summarized as follows:

- Highly innovative technical staff.
- Small, motivated teams.
- Personal responsibility for work rigour and quality.
- Good team communications and close physical proximity.
- Well-defined mission objectives and constraints.
- Technically competent project management.
- Appropriate level of documentation.
- 'Best-practise' procedures.
- Short timescale (which prevents escalation of objectives).

A crucial aspect for a successful programme is the appropriate use of documentation and 'best practice' processes and procedures carefully selected from industry (both space industry and others). This ensures that the documentation is actually useful and read, and the processes and procedures are streamlined and sensible. This may appear to be a startling glimpse of the obvious, but there are many instances of conventional satellite projects coming unstuck owing to critical information lost in a deluge of unread documentation and out-of-date processes and procedures, followed blindly or used to provide ill-founded confidence.

18.3 SMALL-SATELLITE SYSTEM DESIGN

The spacecraft should be designed such that, where possible, essential platform subsystems are fully redundant. However, this does not mean simply duplicating systems, but rather to build a *layered architecture* in which each successive layer of redundancy relies on different systems comprising increasingly well-proven technologies. The upper-layer systems use state-of-the-art high-performance device types—often without *flight heritage*—which give a very high degree of functionality. The lower-layer systems, on the other hand, use device types that have been flown and tested in previous spacecraft and that are able to carry out most of the same functions—albeit with a possible loss of performance. In this way, problems caused by an inherent system design fault or by the systematic failure of a particular device type will not be duplicated in the redundant layers. If the newer devices prove successful, they become effectively *flight-qualified* for use in future spacecraft, allowing the capabilities of the satellites to be enhanced generation by generation in an evolutionary approach. Table 18.2 shows an example of this type of evolution with respect to the main on-board processors used on some of Surrey's 'UoSAT' microsatellites.

It is the *on-board data handling (OBDH)* system (see also Chapter 13) that is the key to the sophisticated capability of the microsatellite. For example, at the heart of the OBDH system of a current generation UoSAT microsatellite is an 80C386 *on-board computer (OBC)*, which runs a real-time multitasking operating system performing various tasks ranging from payload operations, basic *housekeeping* functions, to sophisticated attitude control and orbit maintenance. There is also (usually) a secondary OBC to share computing-intensive tasks and act as a complete back-up if necessary. In addition, all major subsystems and payloads have their own in-built microcontrollers that provide local intelligence, low-level data processing and that also support *telemetry* and *telecommand* (TTC) functions. In this way, the architecture of the microsatellite OBDH system provides similar functionality to that of a modern conventional spacecraft using, for example, the ESA standard OBDH bus—albeit using COTS data bus standards [in this case the

Table 18.2 Small-satellite computer processors

Microsatellite	Main on-board computer processors (subsystem microcontrollers and payload processors not included)			
	1802	Z80	80C186	80386EX
UoSAT-2 (1984)	Primary			
UoSAT-3 (1990)	Primary (back-up)	Secondary (experimental)	Primary (experimental)	
UoSAT-5 (1991)		Secondary	Primary	
KITSAT-1 & S80/T (1992)		Secondary	Primary	
PoSAT-1 & HealthSat-2 (1993)			Primary	
Cerise (1995)			Primary	
FASat-Bravo & Thai Phutt (1998)			Secondary	Primary (experimental)

controller area network (CAN) bus] with bespoke 'cut-down' software interfaces that avoid the large overheads associated with the *Consultative Committee for Space Data Standards* (CCSDS) TTC formats. However, unlike typical practice with conventional spacecraft, a primary feature of the microsatellite OBDH philosophy is that *all* the primary software on board the microsatellite is loaded *after* launch and can be upgraded and reloaded at will by the control ground station thereafter.

Normally, the satellite is operated *via* the primary computer and the real-time multitasking operating system. All telecommand instructions are formulated into a 'diary' at the ground station and then transferred to the satellite OBC for execution either immediately or, more usually, at some future time. The OBC can also issue its own telecommands autonomously in response to sensor inputs. Telemetry from on-board platform systems and payloads is gathered and monitored by the OBC and is transmitted to the ground immediately and/or is stored in on-board memory while the satellite is out of range of the control station. The OBCs also operate the attitude control systems according to control algorithms that take input from the various attitude sensors and then act accordingly. Thus it is this OBDH environment that allows such a tiny microsatellite to operate in a highly complex, flexible and sophisticated manner, enabling fully automatic and autonomous control of the satellite's systems and payloads.

Surrounding the OBDH system are attitude determination and control systems, power generation and conditioning systems, communications systems, as illustrated in Figure 18.4—all of which support the mission payloads housed in a mechanical structure.

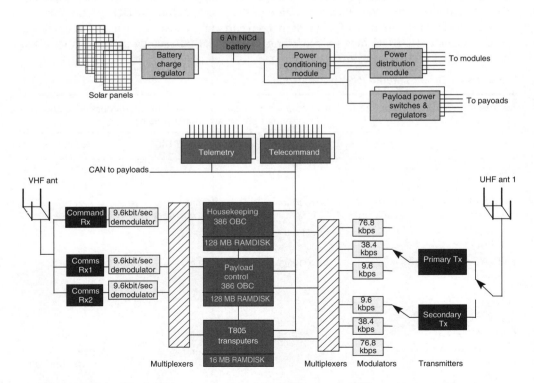

Figure 18.4 Block diagram of typical UoSAT microsatellite system architecture

18.4   COTS COMPONENTS IN THE SPACE ENVIRONMENT

The space environment (see also Chapter 2) can be particularly harmful to COTS devices. Once in orbit, the devices will experience high-vacuum conditions, potentially extreme hot or extreme cold temperature conditions, and relatively high levels of ionizing radiation.

Vacuum effects

Many COTS parts contain plastic materials, which may *out-gas* under vacuum (see Section 2.4.1 of Chapter 2). These either cannot be flown or require further encapsulation before they can do so. The plastics used to encapsulate integrated circuits are generally stable under vacuum, and to date, no problems due to use of plastic-encapsulated parts have been encountered in Surrey's spacecraft—despite common usage. However, prior to flight, care must be taken to store these parts under controlled temperature and humidity conditions in order, in particular, to prevent the ingress of water.

Plastic encapsulation is thought to increase the risk of *electrostatic discharge* (ESD) damage (see also Section 16.4.1 of Chapter 16)—particularly in high orbits such as the geostationary Earth orbit (GEO). However, this effect is insignificant in small spacecraft in low Earth orbit (LEO).

Thermal control issues

Many COTS parts are only rated to operate at temperatures between 0 and +70 °C. Thus, particular care must be taken with the spacecraft's thermal design. Virtually all microsatellites make use of passive thermal control techniques (see also Chapter 11) to maintain interior temperatures at moderate levels (10–30 °C), with interior temperature variations kept to a few degrees Celsius per orbit in order to reduce thermal stress. Where available, COTS parts with extended temperature ranges, for example, *industrial spec* (−20 to +85 °C) or *Mil-spec* (−55 to +125 °C) are preferred. Exterior systems necessarily experience greater thermal cycling during an orbit, with variations of the order of 50 to 100 °C not being unusual.

Prior to flight, extensive thermal-cycle *burn-in* testing is carried out at module level, and mandatory thermal-vacuum testing is performed on the spacecraft as a whole in order to screen the COTS parts for reliability under simulated space conditions.

Effects of ionizing radiation

COTS devices may be particularly susceptible to the deleterious effects of the ionizing radiation environment encountered in space [4]. Thus, particular attention must be paid to the design of COTS-based systems in order to cope with the resultant *total dose effects* (TDEs) and *single-event effects* (SEEs) (see also Section 2.4.1 of Chapter 2 and Section 13.7 of Chapter 13). The trapped radiation belts (*Van Allen belts*) are a very serious threat to satellites, both in terms of radiation dose and SEEs. There is a single proton belt, comprising high-energy protons, which affects LEO satellites in a region known as the *South Atlantic Anomaly* (SAA) (see Figure 18.5). The electron belts are divided into two, the inner electron belt being more or less coincident with the proton belt and the outer electron belt being at very high altitude. LEO satellites pass through

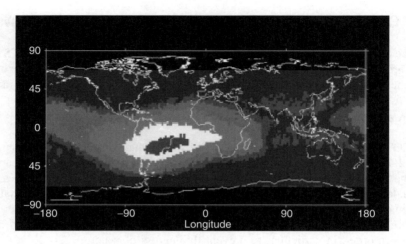

Figure 18.5 South Atlantic anomaly proton flux at 1330-km altitude measured by KITSAT-1 cosmic ray experiment

the inner belt electrons in the SAA, and high-inclination LEO satellites will also pass through the outer belt electrons near the poles. GEO satellites orbit within the outer belt, and elliptical orbits such as *Geostationary Transfer Orbit* (GTO), and to some extent *Molniya* orbits (see Section 5.7.2 of Chapter 5), cross both inner and outer belts. While trapped electrons are not a problem as far as SEEs are concerned, they are (potentially) a major source of radiation dose.

In addition to the trapped particles, there are also *galactic cosmic rays* (GCRs) originating in deep space. These comprise protons and heavy ions with very high energies (typically a few GeV). Satellites in high Earth orbits are essentially totally exposed to GCRs, but satellites in LEO receive some shielding effect from the Earth's magnetic field—except at high latitudes. There is also a shielding effect from the solid Earth and its atmosphere. In any case, the flux of GCRs is relatively low, and so they do not constitute a serious source of radiation dose. However, the small heavy-ion content is very effective at causing SEEs.

Solar-flare particles are similar to GCRs, comprising mainly protons with a few percent heavy ions. Their energies tend to be slightly lower than GCRs. The occurrence of flares is correlated with the 11-year solar (Sun-spot) cycle (see also Chapter 2). Major flares occur around the time of *solar maximum* and can produce intense particle fluxes at Earth for a day or so, posing a very serious threat to satellites. GEO and other high-altitude satellites are particularly at risk, and one major flare may give the equivalent of a year's radiation dose. Polar, and other high-inclination LEO satellites, will also encounter these particles at high latitudes. As with GCRs, solar-flare particles are an effective source of SEEs.

TDEs are produced by the accumulated charge and/or displacement damage caused by particles' energy deposition within device structures. For example, changes in threshold voltage and increases in leakage current occur as a result of hole-trapping within the field and gate oxides of metal oxide semiconductor (MOS) devices exposed to ionizing radiation.

The total dose tolerance of COTS devices varies widely. Some parts fail at less than 5 krad(Si) total dose, while others may survive as much as 100 krad(Si). A figure of

5 krad(Si) can be taken to be a *reasonable* design limit for untested COTS parts. However, where possible, total dose test data on parts should be obtained, either from published sources or from the results of ground testing using, for example, a Cobalt-60 γ-ray source.

Where a component is likely to receive more than its failure dose within the planned mission lifetime, the use of *spot shielding* by high-density metals (e.g. copper, tungsten or tantalum) should be considered, or if this is still insufficient, the part should be replaced altogether with a *rad-hard* version. Even so, total dose damage will accumulate, and thus design margins must be built into the spacecraft's systems to cope with the expected changes—particularly in terms of voltage level shifts and increased current consumption.

SEEs occur as a result of the charge deposited along the track of ionizing particle passing through a device structure. These effects may be temporary in nature or may lead to permanent damage.

SEEs can be reproduced in proton or ion-beam tests, but generally these tests are expensive to perform and require specialist high-energy particle accelerator facilities, which are not widely available. Useful test data may be found in published literature [e.g. that resulting from the annual Nuclear and Space Radiation Effects Conference (NSREC) published in the *IEEE Transactions on Nuclear Science*] or in on-line radiation effect databases. However, such data is no guarantee that an individual component is safe to fly as there are often considerable batch-to-batch variations in radiation tolerance for COTS devices.

In any case, it should be assumed that SEEs *will* occur and that the spacecraft systems must be designed with this in mind. SEEs include *single-event upset* (SEU) and *single-event latch-up* (SEL). SEUs are unexpected but impermanent changes in a device's state. They can be corrected by *error detection and correction* (EDAC) or majority voting circuits. To prevent the accumulation of SEUs in semiconductor memories, the memory should be *washed* (i.e. the contents read, corrected and rewritten) on a regular basis. Figure 18.6 illustrates the occurrence of SEUs, as detected and corrected during such a wash cycle. The correlation with the trapped protons of the SAA (see Figure 18.5) is obvious.

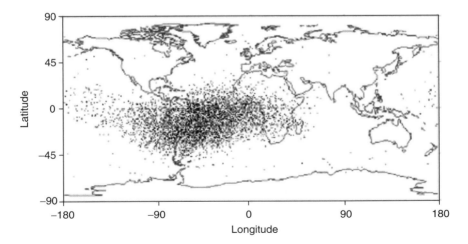

Figure 18.6 Single-event upsets in the S80/T on-board computer program memory at 1330-km altitude

Special attention should be paid to whether or not the part is SEL-sensitive as this is a destructive effect, which is hard to counter. SELs are usually permanent failures unless the power can be switched off rapidly, whereupon the device may recover. SEL-susceptible parts should be avoided if at all possible. In particular, parts that latch-up readily under proton bombardment should be rejected as unsuitable for use in space.

Mechanical environment

Once in orbit, spacecraft experience a *micro-gravity* environment, which imposes relatively little mechanical stress on a spacecraft structure—thermal effects excepted. However, the act of launching a satellite does impart quite severe loads on the satellite's structure—not only from the acceleration of the launch vehicle but also from the associated vibration and acoustic loads (see also Chapter 8). During the ascent phase, the launch vehicle will undergo stage separation, achieved through the firing of pyrotechnic devices, which may impart substantial shock loads on the spacecraft. Thus, as with any spacecraft, microsatellites have to be designed to be mechanically robust. This does not just apply to the structure of the spacecraft. The electrical components also have to be robustly mounted—and particular attention needs to be paid to the use of COTS electronic devices, where there can easily be insufficient strength in the soldered connections to mechanically hold the device under the imparted loads. Devices should not be mounted too high off the printed circuit boards (PCBs), and 'heavy' devices (e.g. large capacitors, crystals etc.) should have additional support in terms of straps to hold them in place. Plastic (vacuum-rated) *conformal coatings* and foams can also play a useful role in providing extra mechanical support. Vacuum-rated RTV silicone rubber is also useful in supporting wired connections to PCBs.

One factor that has to be considered very carefully is that of *mechanical resonance*. Small satellites often fall in a mass-stiffness range that leads to their having resonant frequencies of the order of a few tens of hertz (e.g. 30–45 Hz). This is just the frequency range in which launch vehicles tend to produce large vibrational energies, and so satellites may experience significant amplification (or *Q-factor*) of the imparted loads. The mechanical structure should therefore not be too stiff and should include mechanisms to damp down vibrations and to dissipate energy.

Any new microsatellite structure must undergo *qualification vibration testing* and *shock testing*, which is representative of the intended launch vehicle or vehicles. Once built, the flight-model satellite must also undergo a further set of vibration and shock tests to *acceptance level* to satisfy the launch agency that the satellite is sufficiently robust.

18.5 MICROSATELLITE PLATFORMS

The first modern microsatellites such as *UoSAT-1*, shown in Figure 18.7, used a conventional mechanical structure with electronic module boxes clustered around a central thrust column. Solar panels were body-mounted to stringers supported from the end (Z-facet) honeycomb panels. However, the need to accommodate a variety of payload customers within a standard launcher envelope, coupled with increased demands on packing density, electromagnetic compatibility, economy of manufacture and ease of integration, led to the development at Surrey during 1986 of a novel, modular design of a

Figure 18.7 UoSAT-1 (1981) showing interior and exterior
with three out of four body-mounted solar panels attached

multi-mission microsatellite platform, where the electronics modules actually formed the
primary structure of the spacecraft. The same scheme was adopted by US AMSAT and
other University groups for their 1990-launched microsatellites, and it has subsequently
been used by many countries developing their own small spacecraft. Indeed, the modular microsatellite platform has been the mainstay of the commercial/technology-transfer
microsatellite programmes undertaken by Surrey's commercial arm: SSTL. As of 2002,
this platform has been used successfully on 17 different SSTL missions, each with different payload requirements, and allowing the spacecraft to proceed from order to orbit
in typically 10 to 12 months.

The SSTL modular microsatellite, shown in Figure 18.8, has no 'skeleton' but rather
a series of identical outline machined module boxes, stacked one on top of the other, to
form a body onto which solar panels and instruments may be mounted. The modules are
held together by tie-rods that pass through the whole stack and allow some dissipation
of vibrational energy. While this does limit the maximum potential height of the stack,
in practice this is not a problem in the context of microsatellite mass constraints. Larger
spacecraft can use multiple stacks (see Section 18.6).

At the lower end of the microsatellite mass scale, the $\sim$10 kg AMSAT microsatellites
typically have five stacked module trays, forming a cubic structure with body-mounted
solar panels on all six facets. These panels typically provide between 6 and 16 W of electrical power. Communications are provided by standard amateur radio uplinks/downlinks
(e.g. 2-m, 70-cm, 23-cm and/or 13-cm wavelengths) using the AX.25 packet protocol.
This protocol was developed in the mid-1980s by AMSAT groups (including Surrey)
as a wireless-compatible extension of the standard X.25 wide-area network computer
communication protocol. The 'packetized' nature of the transmissions—both to and from

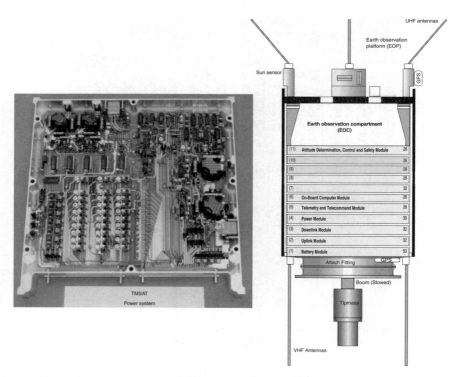

Figure 18.8 Example module box, with schematic showing the construction of a typical SSTL modular microsatellite

the ground—give the ability to support several 'virtual' links to the spacecraft operating over a single RF carrier. Thus telemetry, telecommand, software uploads and payload data can all be supported over a single duplex RF link.

Such small, low-cost microsatellites have limited attitude control—usually confined to passive alignment to the local magnetic field achieved *via* permanent magnets. Sometimes the antennas are black-and-white painted so as to impart a slow spin to aid thermal control. The use of more or less omnidirectional antennae means that the satellites can carry out their primary role as amateur radio communication transponders without the need for accurate pointing control or knowledge. However, high-gain communications transponders and Earth-observation payloads do require an Earth-pointing facet or platform. Thus, a typical 50-kg SSTL microsatellite is maintained to within 1° of nadir by employing a combination of gravity-gradient stabilization using a pyro-released 6-m boom (see also Section 9.4.3 of Chapter 9) and closed-loop active damping using electromagnets operated by the on-board computer. More recent spacecraft have used momentum wheels instead of gravity-gradient booms to provide even more accurate attitude control (typically ∼0.2°)—however, this does introduce moving parts, which are inevitably less reliable. Thus, a gravity-gradient boom is usually retained, ready to be deployed should the wheels ever fail. Attitude determination is provided by the Sun, geomagnetic field sensors (flux-gate magnetometers) and star field cameras, while the orbital position is determined autonomously to within ±15 m by on-board *global positioning system* (GPS) receivers. Electrical power is typically generated by four body-mounted GaAs solar array panels,

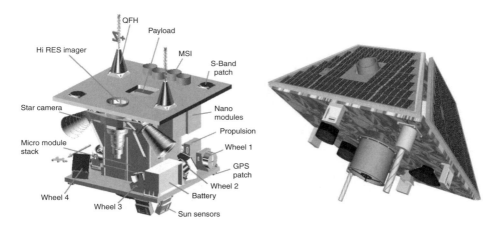

Figure 18.9 High capability 'enhanced' microsatellite

each generating ~35 W, and is stored in a 7 A-h NiCd rechargeable battery. Communications are supported by very high frequency (VHF), ultra high frequency (UHF), L-band and/or S-band uplinks/downlinks, operating on amateur, commercial or military frequency bands depending on the mission context. These satellites also use fully error-protected AX.25 packet link protocols operating at data rates between 9.6 and 156 kbps.

At the higher end of the microsatellite mass scale, the latest generation of 100- to 150-kg SSTL 'enhanced microsatellites' are based on the proven flight heritage of the 50-kg microsatellites, but are designed to support missions requiring larger or more demanding payloads—still at relatively low cost. The $600 \times 600 \times 500$ mm bus structure accommodates larger solar panels and provides flexible internal and external payload accommodation. Nevertheless, this bus fits comfortably on low-cost secondary payload carriers, such as the new Ariane 5 *Ariane structure for auxiliary payloads (ASAP)*.

Four tilted body-mounted GaAs solar panels provide 38-W average power. A full 3-axis attitude control system allows payloads to be pointed at any terrestrial or celestial target with an accuracy of better than $0.2°$. A network of OBCs and embedded controllers automates all telemetry and telecommand, communications, payload control and data management functions. S-band downlinks at 8 Mbit/s provide high-speed communications for remote-sensing payloads, using international standard CCSDS ground/space link techniques. Figure 18.9 shows a schematic of such a microsatellite platform.

18.6 MINISATELLITE PLATFORMS

There has been a steady stream of primarily scientific and military satellites in the few 100-kg 'minisatellite' class. Since the early 1990s, a number of such satellites have been built for just a few tens of millions of dollars—low by conventional space standards. However, in response to growing payload demands for power, volume and mass—but within very small-scale financial budgets—there has been increasing interest in flying minisatellites constructed on the same principles as the microsatellites previously described.

SSTL's minisatellite platform, shown in Figure 18.10, has been designed according to such cost-effective principles—resulting in a basic platform cost of around £6 million.

Figure 18.10   SSTL's 312-kg UoSAT-12 minisatellite, launched in 1999 (Surrey Satellite Technology Ltd)

This is compatible with a range of affordable launch options—on Ariane, CIS, Chinese and US (Pegasus) launchers to meet a variety of mission objectives and capable of
operating in many different orbits. Its primary features are as follows:

- Up to 400-kg total mass.
- 150-kg payload capacity.
- 1.2-m diameter, 1-m height.
- 3-axis, 0.1°, attitude control.
- GPS autonomous orbit and attitude determination.
- 1-Mbps L/S-band communications links;
- On-board propulsion for orbit manoeuvres.
- Cold-gas thrusters for attitude control.
- 300-W orbit average power, 1-kW peak power.

UoSAT-12, launched in April 1999, provided the first demonstration of the capability of this minisatellite platform. In an age of decommissioning of nuclear weapons, it is interesting to note that UoSAT-12 was launched successfully by a converted SS18 Inter-Continental Ballistic Missile (ICBM) (Dnepr). Three-axis control is provided by a combination of magnetorquers, momentum wheels and cold-gas N_2 thrusters, while an experimental electric N_2O *resisto-jet* thruster provides orbit trimming and maintenance demonstrations for future network constellations. Much of the technology proved on this mission has been incorporated in the current generation of enhanced microsatellites—demonstrating the important interplay between technology demonstration missions and the development of operational spacecraft.

18.7 NANOSATELLITE PLATFORMS

At the opposite end of the mass scale, the continuing trend in the miniaturization of technology has raised interest in the development of really tiny spacecraft of just a few kilograms or less. Much of this interest has resided in the United States, particularly at the Aerospace Corporation and amongst the military—the Air Force Office of Scientific Research (AFOSR) and the Defence Advanced Research Projects Agency (DARPA), who have stimulated a programme of nanosatellite and picosatellite research in US Universities—most notably Stanford. Similar interest resides in European Universities including Surrey. Early examples of such satellites include the 3-kg SPUTNIK-40—a 40th anniversary commemorative one-third scale functional replica of the original Sputnik, which was deployed from the MIR space station in November 1997. SPUTNIK-40 was built by French students from l'Aeroclub of France (responsible for the radio transmitter) and staff from the Russian Aeronautical Federation (responsible for the structure). This was followed in 1998 by the Technical University of Berlin's 8-kg TUBSAT-N and 3-kg TUBSAT-N1, launched on a Shtil-1 converted missile from a Russian nuclear-powered submarine. The follow-up SPUTNIK-41 was also launched in 1998. However, it is the new millennium that has ushered in the new era of sophisticated nanosatellites and picosatellites, starting with a group of six ~250- to 500-g picosatellites launched from Stanford's OPAL microsatellite in January 2000. This launch also saw the flight of Arizona State University's 5-kg ASUSAT-1, and the US military JAWSAT nanosatellites, which enjoyed mixed success.

In June 2000, the UK launched its first nanosatellite—the 6.5-kg SNAP-1, designed and built at Surrey in just nine months. SNAP-1 is a highly integrated and sophisticated spacecraft carrying advanced microminiature GPS navigation, complementary metal oxide

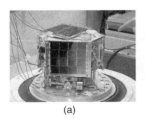 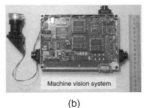 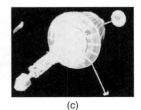

(a) (b) (c)

Figure 18.11 (a) The SNAP-1 nanosatellite, (b) the SNAP-1 MVS payload and (c) the Russian Nadezhda search-and-rescue satellite imaged by SNAP-1 in orbit (Surrey Satellite Technology Ltd)

(a)

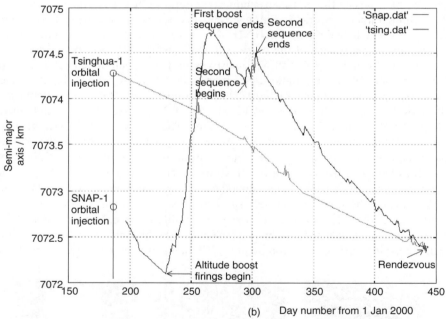

(b)   Day number from 1 Jan 2000

Figure 18.12   (a) SNAP-1 butane cold gas thruster and (b) SNAP-1 orbital manoeuvres: SNAP-1 semi-major axis compared with target spacecraft

semiconductor (CMOS) video camera technology, on-board computing, propulsion and attitude control technologies. Figure 18.11(a) shows the spacecraft undergoing mechanical acceptance testing. To get a sense of scale, the circular plate it stands on is approximately the size of a dinner plate.

SNAP-1's primary payload is a *machine vision system* (MVS) that enables it to inspect other spacecraft in orbit. The MVS consists of three ultra-miniature wide-angle CMOS video cameras and one narrow-angle CMOS video camera—each weighing just a few grams, together with sophisticated image-processing electronics supported by a 220-MHz StrongARM RISC processor (see Figure 18.11b). The MVS has been used to image the Russian Nadezhda satellite (see Figure 18.11c), which carried SNAP into orbit, and in particular, was used to monitor the deployment of another SSTL microsatellite into space. Subsequently, the MVS has been used to provide medium (500-m) resolution near-IR images of the Earth from SNAP-1's 650-km altitude, near-polar orbit.

SNAP-1's miniature cold-gas propulsion system, which uses butane as a propellant, has been used to manoeuvre SNAP-1 with respect to the Chinese Tsinghua-1 microsatellite. Figure 18.12(a) shows the miniature thrusters and fuel reservoir, developed by SSTL and Polyflex Ltd and Figure 18.12(b) shows a graph of the time history of the altitude of SNAP-1 and Tsinghua-1, as recorded by their respective on-board GPS navigation systems. Tsinghua-1, which has no propulsion, falls naturally under the action of atmospheric drag. SNAP-1 falls rapidly at first (a feature of nanosatellites due to their poor *ballistic ratio*), but climbs rapidly once autonomous orbital manoeuvres are initiated. Intense solar activity meant that SNAP-1 fell more rapidly than planned, and so a second sequence of firings was initiated before SNAP was allowed to fall onto Tsinghua-1's orbit.

Future applications for the nanosatellite include the remote inspection of satellites (e.g. the international space station), monitoring of deployments systems in orbit and carrying small space science instruments requiring measurements with spatial diversity.

18.8 AFFORDABLE LAUNCHES FOR SMALL SATELLITES

A sustained, commercial, low-cost small-satellite programme must also be matched by correspondingly inexpensive and regular access to orbit through formal launch service contracts—as it makes little sense to construct sophisticated yet inexpensive microsatellites if the launch costs remain prohibitively high. The majority of microsatellites have been launched as 'piggy-back' or *secondary payloads* accompanying a larger primary payload into orbit—where the primary payload pays for the majority of the launch cost. Such launch services are available on a diversity of expendable boosters, as shown in Figure 18.13 (see also Chapter 7).

| Delta | Ariane | Tsyklon | Zenit | SS18/Dnepr | Cosmos | Athena |

Figure 18.13 Example launch vehicles used for small-satellite missions (Surrey Satellite Technology Ltd)

Figure 18.14 Microsatellites mounted as auxiliary payloads on the ASAP ring

Early microsatellites (particularly the amateur radio satellites) were launched virtually for free on a 'favour' basis by the USA and former USSR [now Confederation of Independent States (CIS)], but these launch opportunities were infrequent and unpredictable. The breakthrough came in 1988 when Arianespace developed the Ariane Structure for Auxiliary Payloads (ASAP) ring specifically to provide, for the first time, regular and affordable launch opportunities for 50-kg microsatellites into both LEO and GTO on a commercial basis. Figure 18.14 shows two UoSAT and four US AMSAT microsatellites mounted on the first ASAP, which flew with the French SPOT-2 spacecraft as primary payload. To date, some 18 microsatellites have been launched via the ASAP but, while it has been key in providing microsatellite launches worldwide, Ariane alone cannot now provide the number of launch opportunities into LEO needed to meet the burgeoning growth of small satellites. Alternatively, inexpensive launch options from the CIS (on Tsyklon, Zenit

Figure 18.15   Microsatellites mounted directly on the primary payload spacecraft

Figure 18.16   Silo launch of a minisatellite using a converted SS-18 ICBM

and Cosmos) are now being used increasingly for microsatellites/minisatellites, with the microsatellites often directly attached to the primary payload (see Figure 18.15).

Within the last few years, the large stockpiles of ICBMs in the CIS have become available for use as small launchers through the de-militarization programme (e.g. SS-18/Dnepr; SS-19/Rockot; SS-25/START). Indeed, SSTL cooperated with ISC Kosmotras (Moscow) to convert the SS18 ICBM into the first Dnepr small-satellite launcher for the successful launch of the UoSAT-12 minisatellite from a silo at Baikonur in April 1999, as illustrated in Figure 18.16.

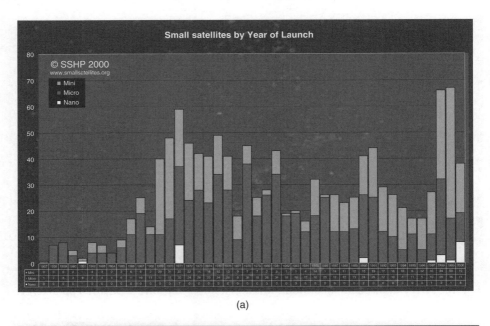

(a)

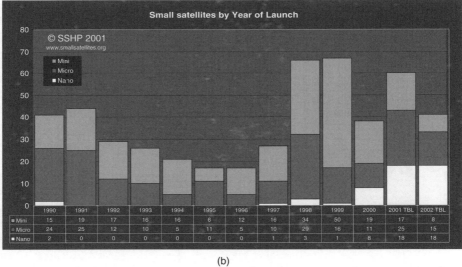

(b)

Figure 18.17 (a) Total number of small satellites launched and (b) small-satellite launch numbers for the past decade (Reproduced by permission of Small Satellite Home Page)

The total number of small satellites launched is shown in Figures 18.17 (a) and (b). These figures include conventional small satellites as well as low-cost small satellites. It is clear from the plots that there has been a sustained interest in small satellites since the late 1960s. Most of these are conventional small military or scientific satellites, with a few amateur radio satellites. The surge in early 1990s reflects the coming of age of 'microsatellite' technology—and in particular, reflects the increasing involvement of US groups.

It is also no coincidence that 1990 marked the introduction of the ASAP ring on the Ariane-4 launch vehicle, which provided the first regular access to space for microsatellites. The even larger surge in the late 1990s reflects the increasing commercialization of small-satellite technology with orbital constellations coming on stream.

It is also interesting to note the rapid rise in nanosatellite activity as we enter the new millennium. There is perhaps some evidence that nanosatellites are taking over some of the roles previously carried out by microsatellites, while microsatellites are growing to acquire some of the capabilities only previously available on minisatellites, such as propulsion and orbit control.

18.9   IN-ORBIT OPERATIONS

The last element of the low-cost small-satellite equation is *in-orbit operations* (see also Chapter 14). Compact and low-cost mission control ground stations have been developed to operate the microsatellites once in orbit. An example is shown in Figure 18.18. These ground stations are based on PCs and are highly automated—interacting autonomously with the microsatellite in orbit—to reduce manpower requirements and to increase reliability. For example, the SSTL Mission Control Centre at Surrey currently operates 11 microsatellites in LEO with just a single operator. Such facilities rely on the spacecraft themselves to have a high level of *autonomy*, as, in practice, the satellites only spend a small fraction of their time in range of any single ground station (typically 1 h per day for a LEO polar orbiter over a mid-latitude ground station). Thus, the investment in ground station and satellite software should not be underestimated. Indeed, the total amount of software on board a typical microsatellite is of the order of several hundreds

Figure 18.18 Mission control ground station at Surrey Space Centre (Reproduced by permission of Surrey Satellite Technology Ltd)

of kilobytes—perhaps an order of magnitude greater than that in a typical conventional spacecraft. Similarly, the ground station tracking and data handling software comprises many thousands of lines of code.

18.10   SMALL-SATELLITE APPLICATIONS

18.10.1   Small satellites for communications

Satellite communications have become synonymous with large geostationary satellites for transparent real-time wideband services. Satellites in LEOs are closer to the user and the consequent reduction in transmission loss and delay time are attractive, holding out the promise of less expensive ground terminals and regional frequency reuse. Nevertheless, the communications characteristics associated with a LEO constellation pose quite different and demanding problems, such as varying communications path and links, high Doppler shifts and handover from satellite to satellite. The use of early internet-like communication techniques was pioneered on microsatellites by the US military: GLOMR, (1985), MACSAT (1990) and at Surrey: UoSAT-2 (1984), UoSAT-3 (1990), to provide worldwide non-real-time digital data store-and-forward e-mail connectivity—especially to remote regions where the existing telecommunications infrastructure is inadequate or non-existent [5]. Two SSTL microsatellites (HealthSat-1 and 2) were procured by Satel-Life (USA) to provide routine e-mail communications for medical teams and aid workers in the Third World. As another example, PoSAT-1 was used to provide military e-mail communications for Portugal during the Bosnia crisis. Figure 18.19 shows a polar scientist using portable hand-held equipment to communicate home via an SSTL microsatellite.

The minisatellite platform provides a means of carrying real-time communications transponders into geostationary orbit.

Figure 18.19   Using hand-held equipment to send e-mails from remote areas via microsatellites

18.10.2   Small satellites for space science

Microsatellites and minisatellites can offer a very quick turnaround and relatively inexpensive means of exploring well-focused, small-scale science objectives (e.g. detecting X-rays, monitoring the space radiation environment, updating the international geomagnetic reference field etc.)—examples being the UK's ARIEL series and the 74-kg AMPTE-UKS (1984), the 286-kg Swedish VIKING (1986) and ASTRID series and the 62-kg Danish OERSTED (1999). They can also provide a means of *proof-of-concept* prior to the development of large-scale instrumentation in a fully complementary manner to expensive, long-gestation, large-scale space science missions. This not only yields scientific data quickly but also provides opportunities for young scientists and engineers to gain 'real-life' experience of satellite and payload engineering—an invaluable experience for future large-scale missions. For example, a doctoral student can initiate a programme of research, propose and build an instrument and retrieve orbital data for analysis and presentation in a thesis within a normal period of postgraduate study. Five Surrey microsatellites have carried payloads to monitor the near-Earth radiation environment (see Figure 18.5). Ground-based numerical models have been validated with flight data, and simultaneous measurements have been made of the radiation environment and its induced effects upon on-board systems. A collaborative microsatellite mission with Chile, FASat-Bravo (1998), carried UV-imaging cameras and UV-radiometers as part of the *Ozone Layer Monitoring Experiment (OLME)*, which provided unique data on the ozone concentrations and structure in the Earth's polar regions, as shown in Figure 18.20.

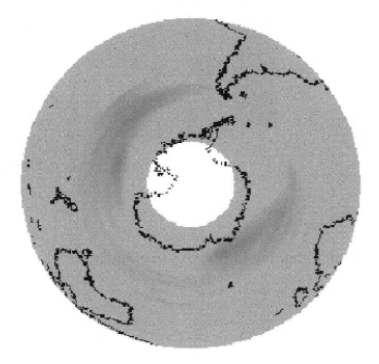

Figure 18.20 Total ozone concentration measured by the FASat-Bravo Ozone Layer Monitoring Experiment

18.10.3 Small satellites for technology verification

Microsatellites also provide an attractive, low-cost and rapid means of demonstrating, verifying and evaluating new technologies or services in a realistic orbital environment and within acceptable risks, prior to a commitment to a full-scale, expensive mission. The UK's Defence Evaluation and Research Agency's (now known as QinetiQ) STRV-1 (1994) series of microsatellites are prime examples—designed to test new technologies in the harsh environment of GTO. For example, one of their experiments tested the performance of new solar cells. Satellites depend upon the performance of solar cell arrays for the production of primary power to support on-board housekeeping systems and payloads throughout their 7 to 15 years operational lifetime in orbit. Knowledge of the long-term behaviour of different types of cells in the radiation environment experienced in orbit is, therefore, essential. Unfortunately, ground-based, short-term radiation susceptibility testing does not necessarily yield accurate data on the eventual in-orbit performance of the different cells and hence there is a real need for evaluation in an extended realistic orbital environment. UoSAT-5 (1991) carried a precursor *Solar Cell Technology Experiment (SCTE)* designed to evaluate the performance of a range of 27 samples of GaAs, Si and InP solar cells in LEO from a variety of manufacturers, as shown in Figure 18.21.

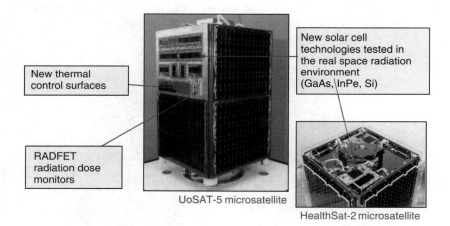

Figure 18.21 New technology demonstration experiments on UoSAT-5

18.10.4 Small satellites for Earth observation

Microsatellites have really brought about a revolution in Earth observation. Conventional Earth-observation and remote-sensing satellite missions are extremely costly—$500 million is not unusual. Thus, there are relatively few such missions and the resulting data, while providing impressive spatial and spectral resolution, yield poor temporal resolution (revisit) of ground targets owing to the small number of these spacecraft actually in orbit. A new opportunity for remote sensing using inexpensive small satellites has come with the availability of (1) high-density two-dimensional array, semiconductor charge-coupled device (CCD), optical detectors (as used in consumer video and digital still cameras) and (2) low-power consumption yet computationally powerful microprocessors. In fact, UoSAT-1 (1981) and UoSAT-2 (1984) both carried experimental first-generation 2D-CCD Earth-imaging cameras. These paved the way for the first operational cameras on board UoSAT-5 (1991), which was able to image the oil well fires in Kuwait resulting from the 1991 Gulf war—the only privately owned Earth-imaging satellite able to do so. The Tsinghua-1 microsatellite launched in June 2000 provides remarkable 35-m resolution images in 4 spectral bands (compatible with LANDSAT) with the capability of ±15° (±200 km) off-nadir imaging coverage upon demand—all at a total mission cost of £3.5 million, launched into orbit. Minisatellites and enhanced microsatellites have been able to achieve even better results. For example, the 103-kg Korean KITSAT-3 satellite, launched in 1999, produces 17-m ground-resolution images in red, green and near-IR bands, while the 312-kg UoSAT-12 minisatellite, also launched in 1999, is able to resolve down to 10 m with its panchromatic Earth-imaging payload—constructed entirely from COTS optical components. Examples of small-satellite Earth-imaging capability are shown in Figure 18.22.

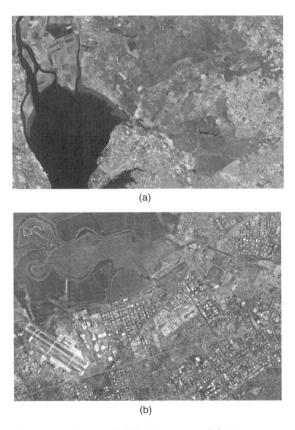

(a)

(b)

Figure 18.22 UoSAT-12 imagery (a) 32-m reso-lution four-band multi-spectral image of Lisbon, Portugal and (b) 10-m resolution panchromatic image of San Jose, California

18.10.5   Small satellites for military applications

Much of the use of small-satellite technology over the years has been driven by the requirements of military agencies—particularly in the US and the former USSR. While these spacecraft have tended to be conventional—albeit small—there has been increasing interest in taking a more cost-effective approach.

The demands of military-style satellite procurement and the cost-effective approach to microsatellite engineering might, at first sight, appear incompatible. However, while retaining the essential characteristics of low-cost and rapid response, a military version of the SSTL microsatellite platform with deployable solar panels has been developed to support various military payloads. The main difference between the 'commercial' and 'military' versions of the platform is in the specification and procurement of components and, particularly, in the amount of paperwork that traces hardware and procedures. An optimum trade-off between the constraints of a military programme and economy has been sought, which results in an increase factor for cost and timescale of approximately 1.5 when compared with the 'commercial' microsatellite procurement process.

Practical examples of where small satellites and groups of small satellites have been employed in military scenarios include the following:

- US TRANSIT programme, in which small satellites were employed within a military navigation system.
- US MACSAT and Portuguese PoSAT missions, in which satellites were used by small divisions for digital store-and-forward communications.
- US MICROSAT constellation of seven satellites for battlefield communications.
- French military CLEMENTINE and CERISE spacecraft, which have been employed in electronic intelligence gathering.
- TechSAT-21 mission, which will employ small satellites for radar imaging and geolocation.

The first use of the SSTL military microsatellite platform was on the CERISE mission designed and built for the French Defence Ministry and launched into a 700-km low Earth orbit by Ariane in July 1995 (see Figure 18.23). After a year of perfect operations, CERISE made history as the first operational satellite to be (knowingly) struck by a piece of space debris (a rocket fragment), which severed its stabilization boom. However, owing to the flexibility of the microsatellite systems, SSTL engineers were able to re-stabilize CERISE by uploading new attitude control algorithms, returning it to operations. A second microsatellite for the French Defence Ministry (Clementine) was launched into LEO in 1999 and a third microsatellite (called PICOsat) was launched successfully on 30 September 2001 for the United States Air Force (USAF), carrying advanced technology payloads for the US Department of Defense.

The use of microsatellites to provide agile, 2.5-m *ground-sample distance* (GSD) high-resolution imaging for military applications is to be demonstrated in 2004 on the TOPSat microsatellite being built jointly by SSTL, Rutherford Appleton Laboratory (RAL) and QinetiQ. Figure 18.24 shows a schematic of the spacecraft that gives an idea of the relatively large optical bench needed to obtain such high-resolution images.

Figure 18.23   The CERISE spacecraft, built for the French Defence Ministry (Surrey Satellite Technology Ltd)

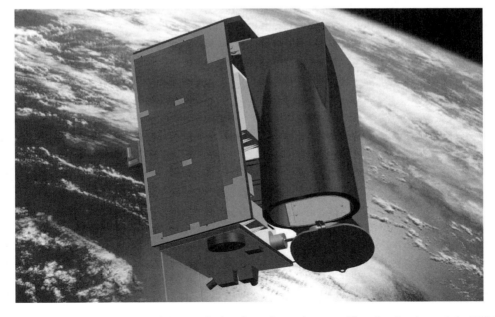

Figure 18.24   TOPSat — a high-resolution imaging microsatellite due for launch in 2004

18.11 PICOSATELLITES AND RECENT ADVANCES IN MINIATURIZATION

The drive towards small hand-held computers and mobile communications equipment has stimulated even further miniaturization—specifically in low-power microprocessors, efficient transmitters, low-mass and high-capacity battery technology and low-power GPS receivers.

This has been followed by specific developments of mobile multimedia systems such as digital cameras, spurring low-cost, low-power, high-density data storage as well as high-resolution imaging sensors. Recent trends have also tended towards full integration of electronics onto a single chip (*System-On-a-Chip*), which is particularly applicable to picosatellite technology (*Spacecraft-On-A-Chip*). Other technological advances in these areas can be summarized as follows.

Micro-Electro-Mechanical Systems (MEMS) use integrated circuit manufacturing techniques to develop miniature mechanical assemblies together with electronics and have found early mass market uses in air-bag deployment sensors and ink-jet printer heads.

Custom devices can be manufactured to order, and in conjunction with application-specific integrated circuit (ASIC) technologies will soon enable entire spacecraft avionics systems, including sensors and actuators, to be assembled within 'sugar cube' volumes. Already, inertial guidance platforms of the order just a few millimetres across are available, as are some micro-propulsion systems. Much of this work is taking place in US research facilities such as Sandia National Laboratories. Wafer-scale integration of electronic and MEMS technology expands the opportunities and is the basis of a highly advanced and integrated nanosatellite/picosatellite proposed by the Aerospace Corporation in the US.

Some technologies are driven by space commercialization, including advances in ground segment systems and highly efficient, radiation-tolerant multi-junction solar cells.

Over the next decade, the outcome of these technologies should be the availability of tiny, sophisticated and intelligent satellites that can be mass-produced cheaply. Such satellites could (potentially) be launched by the thousand, and could have sufficient on-board and distributed intelligence to be self-organizing—essentially synthesizing functions through mass action. Quite what the applications of such satellites will be remains to be seen.

18.12 CONCLUSION

The modern concept of the microsatellite appeared in 1981 with the application of advanced microprocessors and other COTS technologies (see Figure 18.25) to the kind of low-cost satellite platforms developed by amateur radio enthusiasts during the 1970s. The 1990s saw the rapid commercialization of these technologies, and now low-cost microsatellites are playing an important role in enabling emerging space nations to carry out increasingly sophisticated space missions including communications, Earth observation, space science and technology demonstration. It would be wrong to directly relate low cost to the small size of the satellites. Cost-effectiveness comes through the application of a particular design philosophy—not only to the satellites but to all aspects of the mission. It is this design philosophy that really defines the concept of a 'small' satellite mission.

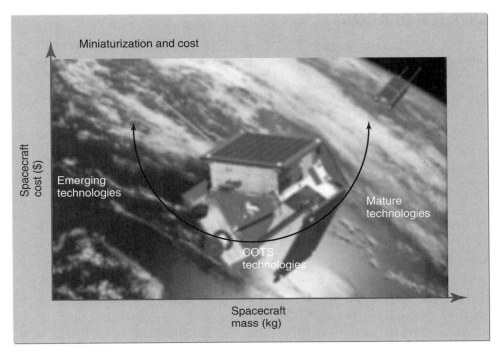

Figure 18.25 The use of COTS technologies to optimize spacecraft mass/cost

The contrast between the small-satellite approach and that of conventional satellites has meant that the *smaller, faster, cheaper* ideal has seen most success in small companies and research groups, rather than in large aerospace organizations.

As we enter the new Millennium, rapid progress in the miniaturization of COTS electronics and the advent of MEMS technology has ushered in a new era of even smaller nanosatellites and picosatellites, which are opening up new mission scenarios—often involving the use of clusters of satellites to synthesize a particular function. The next decade is likely to see viable satellites mass-produced at the 'chip' level.

REFERENCES

[1] Davidoff, M. (1990) *The Satellite Experimenter's Handbook*, American Radio Relay League, Newington, CT, USA. Chapter 2, pp 2–1 to 2–15, Chapter 3, pp 3–1 to 3–9, Chapter 4, pp 4–1 to 4–23.

[2] Sweeting, M. N. (1982) UoSAT—an investigation into cost-effective spacecraft engineering, *J. Inst. Electr. Radio Eng.*, **52**(8, 9), 363–378.

[3] Sweeting, M. N. (1987) The University of Surrey UoSAT-2 Spacecraft Mission, *J. Inst. Electr. Radio Eng.*, **57**(5), (Supplement) S99–S115.

[4] Underwood, C. I. (1996) Single Event Effects in Commercial Memory Devices in the Space Radiation Environment, PhD Thesis, University of Surrey, Guildford, Surrey, UK.

[5] Ward, J. W. (1991) Microsatellite for global electronic mail networks, *Electron. Commun. Eng. J.*, **3**(6), 267–272, December 1991.

FURTHER READING

Wertz, J. R. and Larson, W. J. (1996) *Reducing Space Mission Cost*, ISBN 0-881883-05-1, Micro-cosm Press, Torrance, Calif. and Kluwer Academic Publishers, Dordrecht, The Netherlands.

Boden, D. G. and Larson, W. J. (1996) *Cost-Effective Space Mission Operations*, ISBN 0-07-006379-6, McGraw-Hill, New York.

Small Satellites Home Page, www.ee.surrey.ac.uk/SSC/SSHP.

19 SPACECRAFT SYSTEM ENGINEERING

Adrian R. L. Tatnall[1], John B. Farrow[2] and C. Richard Francis[3]

[1] Aeronautics and Astronautics, School of Engineering Sciences, University of Southampton
[2] International Space University, Strasbourg
[3] European Space Research and Technology Centre (ESTEC), European Space Agency

19.1 INTRODUCTION

The first chapter of this book gave an introduction to the system view of the spacecraft, and the relationship between the subsystems and the overall system was explained. In the following chapters the subsystems have been discussed and the interfaces between them have been considered. All these aspects are drawn together in this final chapter, in which the system engineering process required to ensure that the overall spacecraft meets the mission objectives is explained, and some specific spacecraft examples considered. There are many other examples that could have been chosen but this small sample provides a good demonstration of the breadth and complexity of mission drivers that arise in the search for optimum solutions to such challenging mission objectives. The similarity in the design process used is emphasized by considering the mission objectives, mission requirements and design drivers for each case and showing how the design has evolved as a result of many system trade-offs.

The *mission requirements* are the top-level requirements on the spacecraft. They are quantitative in nature and must be specified by the customer or user and they are an assessment of the performance required to meet the mission objectives. For the space-craft system design these requirements are translated into engineering parameters. This translation can be complex, depending on the particular application.

For a communication spacecraft such as one of the Intelsat series described in Section 19.3.2, the translation is relatively straightforward since the user coverage and data requirements can readily be used to define the satellite parameters. The communications link budget described in Chapter 12 is at the heart of this process. On a satellite such as EnviSat, however, described

Spacecraft Systems Engineering (Third Edition). Edited by P. W. Fortescue, J. P. W. Stark and G. G. Swinerd
© 2003 John Wiley & Sons Ltd

in Section 19.3.4, the translation of the user requirements on geophysical parameters, such as atmospheric pressure and temperature, into instrument specifications is complex and involves many system algorithms. In this case the process involves assumptions about other related parameters such as the level of processing required and the data available from other sources. During the design process these assumptions may not prove to be acceptable and the engineering requirements will have to be modified. It may also be necessary to modify the requirements as a result of a clearer understanding of the impact they have on the spacecraft design.

This iterative process is essential to ensure that the most relevant and realistic requirements are used for the spacecraft design. There are plenty of examples in which the engineering requirements have become 'tablets of stone' at the start of the design and the overall system has suffered because of unwillingness to question them as the design has evolved. It is always necessary to define how much quality is needed or how much 'science' is enough in order to hold down mission costs and avoid unnecessarily restrictive requirements. In other examples, technological constraints, such as the inability to space-qualify critical parts or processes, may dictate a revision of requirements.

There is another issue that can determine the design approach taken. The procurement of commercial communication satellites is targeted towards meeting market needs perceived to be arising through increased capacity requirements. Since a commercial organization is generally driven by the need to satisfy customers' needs, in competition with other suppliers, timeliness is the essence. This commercialism is demonstrated by Intelsat (described in Section 19.3.2, but also see Chapter 12), which has recently become a private company, in order to compete with other (private) communications carriers.

As a result of these commercial considerations, the principal mission requirements for communications satellites are specified at the outset. Engineering solutions must be proposed to solve these specific requirements, although clearly some level of discussion takes place between contractors and customers to define them precisely. These programmes therefore do not go through the iterative process associated with missions such as ERS-1, which are principally experimental in nature and used to establish the user market. Instead, their design iterations are more directed towards having a major impact on the timescale of the programme and its cost. Even in this market, however, there needs to be an appreciation by the user community of the implications of their requirements on the overall design. It is this appreciation that has led to the pressure in the US and Europe to reduce the scale (in terms of both size and schedule) of the spacecraft proposed and designed for science and Earth-observation applications over the past 10 years. Further discussion of the issues associated with the design of a large multi-purpose spacecraft can be found in Section 19.3.4 on Envisat. On the other hand, communications spacecraft, driven by different commercial constraints, have continued to grow in size.

19.2   SYSTEM ENGINEERING

19.2.1   Programme phases

The overall process of system engineering may be conveniently understood by considering the way in which space programmes are broken down into individual phases.

The total time from the initial conception of a complex mission to its launch and operation can extend over a period of 10 years, and sometimes even longer when delays are introduced because of funding limitations. The Rosetta mission, for example, was approved in 1993. It is to be launched in 2003 and will rendezvous with the Wirtanen comet on November 29, 2011. Many missions fall by the wayside through failure to gain approval for funding. As a measure of the competition that exists, in October 1999, ESA released a Call for mission proposals for the second and third flexi-missions (F2 and F3) of the Horizons 2000 programme. Of the 49 proposals received, the ESA advisory bodies selected 6 for an assessment study phase. Of these, two—the Next Generation Space Telescope and the Solar Orbiter—were selected.

Several factors influence the duration of the planning cycle for the new mission selection. Firstly, political and financial constraints can impede the smooth flow of a programme (shown in both the Meteosat and Envisat programmes below—see Sections 19.3.3 and 19.3.4). Secondly, launch opportunities can be limited, particularly for interplanetary missions. The Giotto spacecraft, for example, presented a once in a lifetime opportunity to encounter Comet Halley, and this imposed critical deadlines for development and launch. In some cases there will be the need to demonstrate or develop new technology before the feasibility of the mission can be adequately assured. This provides the most impact upon the 'engineering' timescale of a programme.

The spacecraft programme is traditionally divided into several distinct phases as outlined below. It is worth noting that this format is not adopted for small satellites and it may be significantly shortened in the case of commercial programmes, particularly if existing buses are being used.

Phase A (feasibility)

This phase, lasting typically 8 to 12 months, is termed the feasibility or outline design phase. Its objectives include

- selection of an optimum (and cost-effective) system concept from the range of options under consideration;
- demonstration of the feasibility of the project by design and analysis;
- definition of a technical solution to the extent necessary to generate and substantiate realistic performance, schedule, planning and cost data for all subsequent phases.

This phase is sometimes split further into a 'pre-phase A', wherein specific aspects of the system are considered in outline. In the selection of the next ESA Earth Explorer Core Mission, for example, five missions were selected for an assessment prior to further selection for a Phase A study. Sometimes different contractors will perform parallel Phase A studies, so that many concepts can be evaluated. The cost of performing this phase is relatively low, but the cost implications arising from it are large. The ultimate cost of the mission will depend on decisions made during the early phases of the design.

In a spacecraft Phase A study the customer normally provides preliminary information on mission, launcher and payload requirements as well as a target performance specification. System engineering plays a central role in the selection of a preferred concept in these early phases as indicated in the flow diagram of Figure 19.1 (by their very nature some of the design tasks are iterative rather than sequential). The system engineer provides an overview of the entire system in order to advise the programme manager and the customer on the overall technical feasibility.

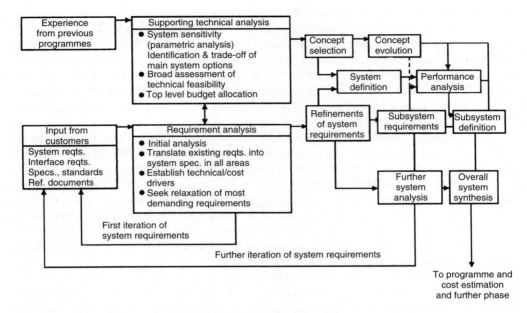

Figure 19.1 Phase A system engineering flow diagram

Once the programme has gained approval or a contract is awarded to a 'prime contractor', the subsequent phases summarized below will occur.

Phase B (detailed definition)

This phase will typically last 12 to 18 months. Significant extension of the phase may occur in large agency (ESA, NASA etc.) programmes if financial and political considerations slow down the process of full mission commitment. This phase is associated with detailed design and definition. Specific activities include

- definition of the system and subsystem designs in sufficient detail to allow the main design and development Phase C/D to proceed with the minimum of problems;
- production of subsystem requirements and design specifications, subsystem and equipment design and development plans, programme schedules and a full proposal for the subsequent C/D phases;
- initiation of advanced C/D activities such as ordering of long-lead items or detailed design of critical parts.

The reviews in this definition phase are the system requirements review, the system design review and the non-advocate or independent review.

Phase C/D (development, manufacture, integration and test)

This is the longest phase and may extend for three to five years. It encompasses development, manufacture, integration and test. The specific activities include

- completion of all design and analyses;
- preparation of manufacturing drawings and special procedures;
- completion of all development and qualification testing;
- manufacture of the flight hardware and acceptance testing.

The *preliminary design review (PDR), critical design review (CDR), test readiness review (TRR)* and *flight readiness review (FRR)* all take place in this phase.

Phase E (the mission operations and data analysis)

While this phase is sometimes referred to as the launch campaign and includes delivery of the spacecraft to the launch site and support of the launch campaign, it is more typically associated with the support of the in-orbit operations from launch through the nominal mission life. This phase of the mission accounts for a large percentage of the overall mission cost and it is particularly important that an account is taken of this part of the mission in the system optimization (see Chapter 14 and Section 19.2.2).

19.2.2 System engineering techniques

The techniques used in system engineering are broadly those defined in Table 19.1.

At the start of the programme the requirements imposed by the customer or end-user are likely to be defined at a fairly high level, possibly covering only the overall mission objectives and the main payload interfaces. Figure 19.2 shows how it is then necessary to expand these top-level requirements into specifications covering the entire range of system and subsystem engineering parameters. It also shows the importance of establishing, in parallel, budget data.

Table 19.2 is intended to act as a checklist of the full range of parameters that are likely to be specified in later, more detailed phases of a programme.

In the early conceptual design phases of a new mission, choices may have to be made between system-level options for the class of orbit, the launcher type, the propulsion system for orbit adjustment if required, the attitude control concept, the data retrieval mode and the spacecraft configuration.

The case studies described later in this chapter illustrate many of these factors. Here we briefly consider a single example—the choice of orbit for an astronomy mission. This highlights some of the key points that must be taken into account in concept selection and optimization.

Table 19.1 System engineering techniques

• Requirements identification/analysis	• Concept selection
• System specification	• Budget allocation
• Options identification	• Performance analysis
• Mission assessments	• System optimization
• Trade-offs	• Interface specification
• Feasibility assessment	• System definition
• Cost comparison	• Cost estimation

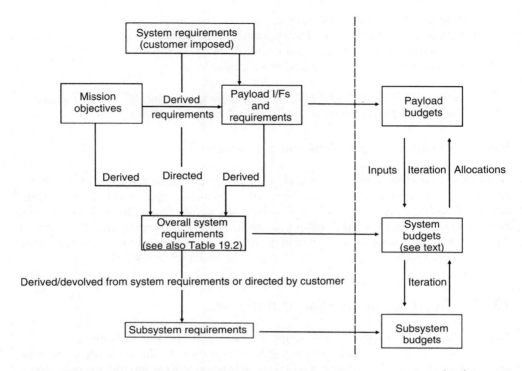

Figure 19.2 Expansion/translation of top-level requirements to system and subsystem level

Figure 19.3 is a tree diagram showing the possible orbits about the Earth, and beyond, which could be adopted for an astronomy mission. With the exception of the lunar surface, all of the listed orbits have actually been selected as indicated by the mission names in the diagram.

Key factors influencing the selection of a preferred orbit are listed in Table 19.3. The main point to emerge from it is that the choice of orbit for this class of mission is by no means clear-cut. Much depends on the existing infrastructure in the form of launcher availability, data retrieval capabilities or existing spacecraft bus designs. For example, NASA's major observatories—*Hubble Space Telescope (HST)* and *Gamma Ray Observatory (GRO)*—can be well satisfied in a circular low Earth orbit (LEO) provided that the Space Transportation System (STS)/Shuttle is available for launch/servicing and the Tracking and Data Relay Satellite System (TDRSS) can be used for data retrieval. However, ESA's astronomy missions, such as the *Integral* and the *X-ray Multi-Mirror Mission (XMM-Newton)*, have selected highly elliptical orbits to guarantee long periods of uninterrupted observation away from trapped radiation in the Earth's proton and electron belts. Other missions such as *GAIA, Darwin* and *First/Planck* have selected orbits around a point about 1.5 million km from Earth in the direction away from the Sun known as the L_2 Lagrangian point (see Chapter 4). In this orbit, advantage can be taken of the fact that the benign thermal and radiation environments are ideal for long-distance observations.

Table 19.2   Checklist of system requirements

Mission requirements (Chapter 5)
Launch windows
Orbit (transfer; operating–nominal
   and back-up)
Operations
   -Launch & early orbit phase
   -Operational phase
   -End-of-life
Lifetime
(Retrieval/repair/re-supply)*
Autonomy
Reliability/availability
Ground segment
	Physical requirements (Chapters 3, 8, 9, 10)
	Axes definition
	Configuration constraints
	Dimensions
	Mass
	Mass properties
	Internal torques
	Disturbances
	Power/energy

Environmental requirements (Chapters 2
   & 16)
Ground activities
Launch & ascent conditions
Transfer & operating orbit environment
(Re-entry, descent)*
Structural/thermal inputs, loads, ranges
Environmental protection
Cleanliness/contamination
EMC
DC magnetic fields
Radiation
Spacecraft charge
Atomic oxygen*
Autonomy
	Performance requirements (Chapters 5, 8,
	   9, 12, 13)
	Orbit maintenance
	Ranging accuracy
	Timing accuracy
	Pointing accuracy
	Measurement accuracy
	Stability
	Pointing range
	Slew rate
	Data rate
	Data storage capacity
	On-board processing
	Link budget margins
	Telemetry/telecommands
	Strength/stiffness
	Thermal control
	Reliability

PA requirements (Chapters 2 & 17) Reliability Availability Maintainability Safety test philosophy Parts, materials, processes Cleanliness Storage, handling, transport Configuration management Software	*AIV programme requirements* (Chapter 17) Schedule Model philosophy Test philosophy GSE requirements Facilities usage *Cost constraints*

*For some missions only.

Note: AIV: assembly, integration and verification; EMC: electro-magnetic compatibility; PA: product assurance.

Reuse of existing designs of spacecraft equipment offers very significant savings compared to new developments. For example, the Service Module (SVM) of Integral is a rebuild of the one developed for XMM-Newton. The cost of the satellite bus, the launcher and the ground segment operations over the mission life feature very significantly in the total programme cost and can have a dominant effect on early trade-offs. Clearly the ensuing spacecraft system and subsystem design for LEO, HEO or other options can differ greatly as a result of the very different environmental constraints and orbit conditions.

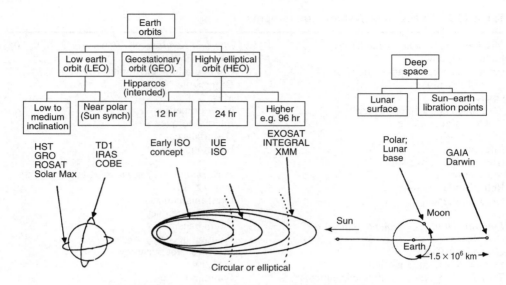

Figure 19.3 Orbit options diagram for astronomy missions

There are also a number of domain-specific tools that are used for particular aspects of the spacecraft design. A list of the tools used within ESA programmes is shown in Table 19.4.

The implementation of these tools in the context of system engineering is summarized in the next section.

19.2.3 Developments in system engineering

While the goal of system engineering has not changed over the past decade, the means of implementing these goals have been aided considerably by the developments in computer technology. An example of this improvement is given by the current emphasis on *concurrent engineering*. Many organizations such as NASA/JPL and ESA/ESTEC have demonstrated the benefit of this approach in reducing proposal times and enabling the evaluation of a large number of missions. Concurrent engineering is the application of classical system engineering in an integrated computer environment. It enables multiparameter optimization of performance, schedule, operations and cost. In order to find the true system optimization, the interactions of the design, fabrication, test, assembly, launch and operation elements must all be considered concurrently with the interactions of the mission, the spacecraft and the operations elements. Specialized facilities known as design centres have enabled concurrent engineering, using specific processes and a common suite of tools, to reduce the time and the cost required for engineering design. This in turn has allowed many more design iterations to be done quickly. Many of these are orientated to the users, so they can make informed decisions about the impact of their requirements on the design.

Despite the progress that has been made, and the increasing number of tools and models available, there are still occasions when there are failures in spacecraft design that can be

Table 19.3 Main factors in orbit selection for astronomy missions

Factor	General preference	Comments
Scientific objectives/performance		
Viewing efficiency (area occulted by earth on celestial sphere)	HEO/GEO	Very poor in LEO because of large angle occulted by the Earth
Uninterrupted observing time (on a chosen source)	HEO/GEO	Generally poor in LEO because of short orbit period
Observatory mode of operation	GEO, HEO or LEO plus DRS	See *data retrieval* below
Radiation (background noise)	LEO	HEO acceptable above 40 000 km (GEO marginal)
Contamination	HEO/GEO	Generally no problem for LEO above a few hundred kilometers
Cryogen boil-off (for cooled infra-red astronomy missions)	HEO/GEO	Only applicable to missions such as IRAS and ISO with stored cryogens
Launch and operations		
Launcher availability and capability	LEO	Generally much higher injection mass than HEO/GEO
Scope for sharing launcher	HEO/GEO	Low inclination HEO only via GTO (e.g. with communication satellite)
Data retrieval		
Ground station coverage	GEO or HEO	If there is no Data Relay Satellite and a need for real-time data
Complexity of on-board data handling and communications	HEO	LEO acceptable if DRS facility available or if real-time link is not imperative
Spacecraft design		
Eclipse duration and frequency	GEO or HEO	Depends on specific orbit parameters and power demand; generally worst for LEO
Thermal control	GEO or HEO	Thermal control generally simpler in HEO than LEO
Orbit raising and maintenance	LEO	Depends on specific orbit parameters and power demand; generally worst for HEO
Solar array degradation	LEO or GEO	Avoids passage through radiation belts
AOCS disturbances	LEO or GEO	Depends on minimum orbit altitude and pointing requirements

Note: AOCS: attitude and orbit control system; DRS: data relay satellite; GTO: geostationary transfer orbit; HEO: highly elliptical orbit; IRAS: infra-red astronomical satellite.

ascribed to problems with the system engineering. The formalization of the procedures and the improvement in tools and models does not detract from the need for systems engineers who have a genuine understanding of the whole system. The failure of the *Mars Climate Orbiter* was attributed to an inadequate consideration of the entire mission as a total system and the lack of robustness of the system engineering function. One of

Table 19.4 Domain-specific software tools used in space-craft design

Domain	Tool
Structural design, configuration and accommodation	CATIA
Thermal	ESATAN and ESARAD
AOCS	Matrix X
Mission analysis	IMAT
Mission simulation and visualization	EUROSIM
Programmatics	MS-Project

the recommendations of the *Ariane 501* failure review board was that close engineering cooperation, with clear-cut authority and responsibility, was needed to achieve system coherence, with simple and clear interfaces between partners.

It was the growing concern about the development time required and the corresponding escalating costs that led to the *faster, better, cheaper (FBC)* approach adopted by NASA (see Chapter 17). This approach is based on a set of principles that includes streamlined management, flexibility and technological capability that can be matched to the science objectives and requirements for a given mission. It has, however, also been equated to compromising mission outcomes in order to reduce cost, and the loss of the *Wide-Field Infra-red Explorer, Mars Climate Orbiter* and *Mars Polar Lander* has called into question some elements of the FBC philosophy.

19.2.4 Design drivers

The payload is, in most cases, the single most significant driver of the satellite design. The purpose of the satellite bus is to provide the support required for the payload to ensure that it can operate in the required orbit and environment. Power, heating and cooling, structure and communication are all provided to ensure that the payload can operate satisfactorily and relay its data to ground. The propulsion, the Attitude and Orbit Control System (AOCS) subsystems and the mission analysis provide the means of getting the payload into the right position to make its measurements. In the case of *Giotto*, the spacecraft that intercepted Halley's comet, for example, the retrograde orbit of Halley around the Sun meant that any interception had to occur at extremely high relative velocity (70 km/s) and the spacecraft had to have some means of protection against particles moving at this speed. The design of the bumper shield therefore became a key design driver. It also became necessary to spin the spacecraft to provide gyroscopic rigidity to ensure that it remained in the right orientation for the payload to view the nucleus and transmit data to the Earth. *Rosetta*, the next ESA cometary mission, on the other hand, is due to intercept the comet Wirtanen and move along with it as it comes closer to the Sun. This comet has a prograde orbit. It is therefore possible to have a three-axis-stabilized configuration since the environmental disturbances are expected to be much less. In both cases the design of the spacecraft stems from the requirements and constraints imposed by the payload.

19.2.5 Trade-offs

It is common to make use of trade-off tables to 'score' the alternative options in early concept studies. Major evaluation criteria for such trade-offs include

- cost, which is generally a dominant factor;
- satisfaction of performance requirements (e.g. image quality in an Earth-observation mission);
- physical characteristics, notably mass, size and power, which, in turn, impact on cost and feasibility;
- availability of suitable hardware technology and timescales for any predevelopment;
- compatibility with launcher, ground segment and other system elements and the complexity of interfaces;
- flexibility to encompass alternative mission options;
- reliability and availability.

If some of these criteria are considered more important than others, then a weighted trade-off can be performed. Regardless of whether a trade-off is weighted or not, it should only be used as a guide to allow the focus of resources and effort on elements of a design that are found to be critical. It is impossible to guarantee that a trade-off is entirely objective and that the evaluation criteria are exhaustive and independent. Cost, for example, is influenced by all the criteria above and its use as an independent parameter is highly questionable.

Only some of these factors lend themselves to quantitative evaluation, and hence direct comparison to select a preferred solution. Other factors in the list are fairly subjective and rely more on broad engineering judgment than on the results of parametric analyses and optimization methods. Still others may be quantitative in nature but, in the early stages of a programme, the estimation of the respective values may be too coarse to allow a single solution to be selected with confidence from the range of options. Finally, it should be noted that, as in the example of orbit selection for an astronomy mission, there may not necessarily be a single preferred choice and two or more alternative approaches may be equally viable.

19.2.6 Budgets

An important system engineering tool is that concerned with system budgeting. At spacecraft level the main system budgets to be compiled include

- mass and mass properties (e.g. centre-of-mass, inertias),
- ΔV and associated propellant quantity,
- power and energy,
- pointing and stabilization,
- alignment/boresight error (e.g. of a payload sight line with respect to nominal spacecraft axes),
- telemetry and telecommand budgets,
- data rate and storage capacity (if appropriate),

- communication link budgets,
- reliability.

Almost invariably mass and power are at a premium in space programmes, but just how critically they are constrained depends on the type of mission. The design drivers—such as payload power demand and data rate for EnviSat or pointing and stabilization for an astronomical telescope—must be established from the outset. Devolving the key system requirements and budgets to subsystem level, and iterating these as necessary, is intimately linked to other steps in the system engineering processes in the early programme stages. These steps traditionally include the identification of alternative concepts, the trade-offs of these options against the main evaluation criteria and the shortlisting of preferred solutions. Parametric analyses at system and subsystem level are necessary to establish system sensitivities and technical drivers.

Note that the system requirements listed in Table 19.2, and the associated budgets, are applicable to almost all types of spacecraft except where noted. If one were to consider a specific mission, it would be necessary to extend the list of mission and payload requirements still further. For example, a remote-sensing payload is generally characterized by parameters including spectral and spatial resolution, dynamic range, signal-to-noise ratio, calibration accuracy and the area and repeat frequency of coverage. Only a few of these parameters are themselves subject to compilation of budgets but they all play a part, in a very interactive manner, in establishing the optimum system design.

The system engineering tasks must be applied throughout the overall programme and not just in the early phases. In particular, the following aspects must be addressed:

- compliance of all physical and performance parameters with the system requirements;
- devolution of the system-level specification and budgets to subsystem level;
- resolution of conflicts at detailed design level to establish the most cost-effective compromise solution satisfying overall mass, risk, cost and other targets;
- definition, monitoring and control of interfaces between all system elements and between subsystems;
- detailing of operational procedures for the mission;
- management and maintenance of budgets for mass, power, fuel and other key resources;
- analysis of 'system-level' factors such as reliability, EMC and chemical cleanliness.

19.3 CASE STUDIES

19.3.1 Introduction

Three spacecraft—*Intelsat, Meteosat and EnviSat*—have been chosen to serve as case studies. These provide specific examples of generic classes of mission. They also illustrate the procurement differences between a space agency, such as ESA, and spacecraft organizations, such as INTELSAT or EUMETSAT. In each case the background for the

Table 19.5   Summary characteristics of selected case studies

Satellite	Orbit	Application	Stabilization	First launch
Intelsat	GEO	Communications	Spin/dual spin/three-axis	1965
Meteosat	GEO	Weather	Spin	1977
EnviSat	LEO	Earth observation	Three-axis	2002

particular satellite is given, before considering specific design drivers. The characteristics of the three satellites are shown in Table 19.5.

19.3.2   The Intelsat spacecraft

Background

The International Telecommunications Satellite Organisation, Intelsat, is an international communications company offering Internet, broadcast, telephony and corporate network solutions around the globe through its current fleet of 20 satellites. In July 2001, Intelsat changed from a quasi-governmental satellite organization governed by 144 countries to a private company with 200 shareholders, most of whom are large telecommunications companies such as Lockheed Martin Global Telecommunications, BT, Deutsche Telekom and France Telecom. This change in status was considered necessary for Intelsat to remain competitive in an extremely volatile market. Satellite operators are expecting that satellite technology will retain its niche status for remote communication and multicasting/broadcasting. Its competitiveness for mainstream multimedia applications will depend on its ability to decrease user costs and provide a rapid deployment capability.

The evolution of the Intelsat series of spacecraft has demonstrated the growing requirement for satellite communication and the advances in satellite technology since *Intelsat 1* (Early Bird) was launched in 1965. Table 19.6 gives the characteristics of the Intelsat series (see also Table 12.1). There was initially a steady increase in the size and capability of spacecraft, but after Intelsat VI, while there was a small increase in capability, the size of the spacecraft decreased. This reflected the trend that has taken place across other types of spacecraft since the 1990s. The latest versions of the Intelsat spacecraft Intelsat IX and Intelsat X, however, are larger than ever and demonstrate the different pressures on commercial satellites.

The first two in the series, *Intelsat I and II*, were simple spin-stabilized spacecraft with toroidal antennas and a simple hydrogen peroxide gas-jet system for attitude control.

They provided the evidence that a geosynchronous spacecraft could provide a high-capacity link between North America and Europe. At this stage the idea of using geosynchronous satellites rather than LEO satellites to provide regular communications was not universally accepted, and there are a number of comparative studies of different systems. The *Intelsat III* launch, in 1968, marked the end of the debate and the start of the global network of geosynchronous satellites. It is interesting that the same debate started again in 1993 with a number of proposals to provide a worldwide mobile communications system. While the LEO constellations, such as Iridium, were expected to corner this market, rapid development in worldwide telecommunications and overambitious plans have meant that

Table 19.6 Intelsat series characteristics

	Intelsat I	Intelsat II	Intelsat III	Intelsat IV	Intelsat V	Intelsat VI	Intelsat VII	Intelsat VIII	Intelsat IX	Intelsat X
Launch year	1965	1967	1968	1971	1980	1989	1992	1998	2001	2003
Contractor	Hughes	Hughes	TRW	Hughes	Ford Aerospace	Hughes	Ford Aerospace	Lockheed Martin	Loral/Space systems	Astrium
Launcher	Thor Delta	Thor Delta	Thor Delta	Atlas Centaur	Ariane or Atlas Centaur	Ariane 4, Titan or shuttle	Ariane or General Dynamics	Thor Delta	Thor Delta	Ariane, Proton, Sea Launch
Approximate launch mass (kg)	68	160	295	1415	1930	3675	3650	3245	4725	5000
Power BOL (W)	40	83	160	600	1800	2600	3968	6400	8600	8000
Lifetime (years)	1.5	3	5	7	7	13	15	15	15	13
Transponders	2C	2C	2C	12C	21C/ 4 Ku	38C/ 10Ku	26C/ 10Ku	38C/ 6Ku	44C/ 12 Ku	36C/20 Ku

the GEO/LEO debate still continues. The new series of Intelsat III spacecraft provided increased payload capacity but the major improvement concerned the introduction of the mechanically despun antenna. This allowed the use of a high gain antenna, pointing continuously towards the Earth. There were also changes to the attitude control system, and hydrazine was used as the propellant rather than hydrogen peroxide.

At this stage there was evidence that the market could support a considerable increase in capacity and that the next Intelsat satellite, *Intelsat IV*, would have to be considerably larger than the previous ones. It also became clear that a long-term operational programme was required. As a result, negotiations started with potential contractors in 1966 and a contract was awarded to Hughes Aircraft Company. The long interval between the start of the negotiations and the contract award reflected the uncertainties in the design, particularly with regard to the stabilization concept to be used. A major trade-off was between gravity-gradient stabilization and spin stabilization, and a considerable amount of analysis had to be done before it became clear that the proposed spacecraft, with its large despun platform, could use spin stabilization, without going into a flat spin (see Section 3.4.3 of Chapter 3). There had been spin-stabilized platforms with despun antennas before, but the size and scope of the antennas proposed for Intelsat IV were new and heralded the increasingly complex antenna systems that were to be used in the 1970s. The Intelsat IV also had sufficient power to enable it to be the first spacecraft to be frequency-limited rather than power-limited, although spot beams allowed the same frequency to be used in two separate beams, thus increasing the power available.

A contract was awarded to Ford Aerospace and Communications Corporation to construct seven *Intelsat V* satellites in 1976. This was later than first intended because in 1972 a decision had to be made whether to procure a modified version of Intelsat IV—Intelsat IVa—instead of a new Intelsat V because of the shorter delivery time and the reduced technical risk. This policy of selecting the low-risk option in preference to a solution with greater long-term potential but more immediate risk is often made in spacecraft design. There is an understandable tendency, particularly in commercial programmes, to prefer to minimize risk and mitigate against failure by relying on older, proven technologies and occasionally by over-engineering the spacecraft, particularly if the consequences of failure are so catastrophic. On the other hand this conservative use of new technology has proved frustrating to the innovative engineer and makes radical changes in spacecraft design difficult to achieve. Intelsat V was a three-axis-stabilized spacecraft that was first launched in 1980. It was designed to be compatible with Ariane or Atlas Centaur launchers.

Mission objectives and design drivers of Intelsat VI

In 1980, to cope with the increasing demand for telecommunications capacity, Intelsat issued a request for a proposal for a new series of spacecraft. The mission objectives were similar to the objectives of the previous spacecraft series but the total capacity required was increased to the equivalent of 120 000 simultaneous telephone channels and at least 3 television channels. This represented a threefold increase on the capacity of Intelsat V. An increase in capacity requires an increase in the number of transponders and consequently an increase in power requirements. There is also a requirement on the zones of coverage and a sixfold frequency reuse. For example, the size of the antennas at C-band is determined by the need to maintain a minimum interbeam spacing of $2.15°$

and an isolation of 27 dB. These requirements provide the prime mission requirements and enable the power necessary to meet the C/N requirements of the ground stations to be calculated from the link budget (see Section 12.2.10 of Chapter 12).

A contract for five satellites was awarded initially with the option for another six. Owing to the growing non-US involvement in the Intelsat programme, both the two competing consortia, led by Hughes and Ford, had non-US subcontractors. This development is similar to constraints that ESA imposes on contractors to involve certain countries, in order to ensure that they receive a return on their investment, although it does not necessarily lead to the optimum technical solution.

The following were key influences on the design:

- compatibility with Space Shuttle, Ariane 4 and Titan launcher
- power of 2.6 kW, 2.2 kW End of lifetime (EOL)
- 1×3.2 m C-band antenna
- 1×2.0 m C-band antenna
- 2×1.0 m K-band antennas
- lifetime of 13 years.

This was the first of the Intelsat series that was required to be compatible with a Space Shuttle launch. In order to achieve its operational orbit from LEO, where it would be placed by Shuttle, the satellite had to be capable of being deployed with an additional perigee stage. This stage establishes the geostationary transfer orbit, which is then circularized using an Apogee Boost Motor (ABM). On Ariane 4 the satellite is placed directly into a geostationary transfer orbit and so it is only then required to circularize the orbit using an ABM (see Sections 7.4 and 7.5 of Chapter 7 for details and Sections 5.2 and 5.6 of Chapter 5 for propellant budgeting).

Intelsat VI spacecraft design

The two consortia bidding for the Intelsat VI contract proposed very different solutions based on their particular expertise. Ford proposed a three-axis-stabilized solution like their Intelsat V design, although considerably larger, to accommodate the larger antennas and the higher power required. Hughes proposed a spinning spacecraft with a despun platform like the Intelsat IV solution. In April 1982, Hughes was awarded a $700 million contract to build five satellites with an option for six more that was never taken up.

Intelsat VI consists of a spun section and a despun section interconnected by a bearing and power transfer assembly (see Section 15.3.1 of Chapter 15). The solar array and the propulsion system are mounted on the spun section and the antennas and all the communications subsystem are mounted on the despun section. A mass budget is shown in Table 19.7 for Intelsat VI alongside that of Intelsat V. The mass of Intelsat VI is greater than that of Intelsat V in all areas except attitude control. This reflects the additional size and complexity of the spacecraft required to accommodate antennas, which are considerably larger than those on Intelsat V.

The relatively low mass of the attitude control subsystem is due to the relative simplicity of attitude control required for a dual-spin spacecraft. But this configuration contributes to the large increase in the mass of the power subsystem because only a fraction of the solar cells are illuminated at any time (see Section 10.3 of Chapter 10). In order to meet

Table 19.7 Intelsat V and VI mass budgets

Subsystem	Intelsat VI (kg)	Intelsat V (kg)
Antenna	309	59
Repeater	326	175
TT&C	80	28
Attitude control	70	73
Propulsion*	120	96
Electric power	330	142
Thermal control	52	26
Structures	280	157
Wire harness	99	40
Balance mass	23	15
Spacecraft dry	*1689*	*750*
Margin	90	24
Residual propellant	28	—
End of life	*1807*	*835*
Station keeping	420	173
BOL mass	*2227*	*1008*
AMF propellant	1439	861
Reorientation	10	—
Separation mass	*3676*	*1869*

*Includes apogee motor inert mass.
Note: BOL: beginning of life; TT&C: tracking, telemetry and command.

the power and the launch vehicle accommodation requirements, Hughes had to adopt a novel approach for solar array design. In a spinning spacecraft the solar array is mounted on the drum of the spacecraft, whose diameter is limited by that of the shroud of the launch vehicle, which is about 3.8 m in the case of Ariane 4. This means that the drum must be about 6 m high since a power of 2.2 kW EOL requires about 22 m² of array to be perpendicular to the Sun. In order to reduce this size in the launch vehicle, a 'dropped skirt' solar array was designed. This is shown in Figure 19.4.

The inner solar panel is 2.2 m in length and is fixed to the body, while the outer panel of 3.8 m length is deployable but is placed over the inner panel during launch. This reduces the overall height of the spacecraft, including the folded antennas, to 5.3 m in the launch configuration, and this is able to fit in the Ariane launch vehicle. In addition to the novel solar array, Intelsat VI also has two 44 A-h Ni–H_2 batteries that automatically supply power to maintain services during eclipse. These batteries were preferred to the more traditional Ni–Cd batteries because of the higher specific performance due to the higher depth of discharge permitted and the longer life (see Section 10.4 of Chapter 10).

The propulsion subsystem is used for apogee boost, orbit maintenance and attitude control as discussed in Chapter 6. As the satellite has a long lifetime requirement, optimization of the propellant usage is of particular importance, and the additional complexity of an integrated bipropellant propulsion system is justified. This system consists of nitrogen tetroxide (N_2O_4) and mono-methylhydrazine (MMH) as the propellant, four 22-N radial thrusters to provide E-W station-keeping and two 490-N thrusters for apogee boost and reorientation.

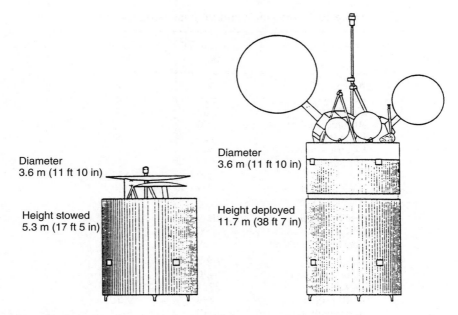

Figure 19.4 Intelsat VI solar array design (Reproduced by permission of Hughes Aircraft Company)

The first of the Intelsat VI series was launched on 27 October 1989 on an Ariane 4 from Kourou and was finally placed at 63° East longitude over the Indian Ocean. On 14 March 1990, the second in the series was launched from Cape Canaveral aboard a Titan III. With this launch vehicle the spacecraft had to be supplemented with a perigee motor that could not, however, be fired because the launch vehicle's upper stage failed to separate from the spacecraft and its perigee motor. The only option was for the Intelsat controllers to command the spacecraft to separate from the upper stage and the perigee motor, thereby allowing the satellite to be placed in a LEO. While this orbit did not permit its operation, it did allow astronauts on the first flight of Endeavour to perform a spectacularly successful rescue mission in May 1992. They attached a new perigee motor that took the satellite into the correct orbit. This rescue cost Hughes $96 million, although this has since been considered by some as an underestimate of the true cost of the mission.

Intelsat VII

The main requirement of the next generation of Intelsat spacecraft was to provide the Pacific region with increased power and coverage for smaller ground stations. There was also scope for providing increased flexibility through independently steerable Ku-band spot beams. In terms of complexity, however, Intelsat VII was expected to be between Intelsat V and VI and reversed the trend of increasingly complex spacecraft.

This relative conservatism on the part of the operators was driven in part by increasing competition and the need to avoid unnecessary risks.

Despite this, the demanding power/mass ratios made it very difficult to use a dual-spin spacecraft, and all the designs proposed were three-axis-stabilized. In June 1988, Ford

Aerospace was selected, principally on the basis of a superior technical and price proposal. The bus proposed was based on the earlier Intelsat V, although the accommodation of eight antennas including a 2.44 m, 4 GHz antenna and the need to avoid blockage placed additional constraints on the configuration.

The trend towards more powerful spacecraft has continued with the next generation of Intelsat spacecraft, Intelsats VIIA and VIII. The Intelsat VII and VIIA series were nearly identical except for an increase in the number of Ku-band transponders in the VIIA series. The three-axis-stabilized spacecraft was based on the Space Systems/Loral 1300 bus and used a hydrazine propulsion system. Two large solar panels with one-axis articulation provided 3900-W EOL. The payload consisted of 26 C-Band and 10 Ku-Band transponders with the capacity to provide 18 000 telephone calls and 3 colour TV broadcasts simultaneously or up to 90 000 telephone circuits using digital circuit multiplication equipment (DCME). The additional power provided by a fourth solar array panel enabled the EOL power of the Intelsat VIIA to be increased to 4800 W. Nine Intelsat VII/VIIAs were actually built but one was destroyed in a failed launch attempt.

Intelsat VIII

The first of the Intelsat VIII series, Intelsat 801, was successfully launched from Kourou, French Guiana in February 1998. The Intelsat VIII/VIII-A spacecraft series, shown in Figure 19.5, provided improved power and coverage at C-band enhancing Intelsat's ability to meet customer growth requirements. All of the Intelsat VIII satellites were built by Lockheed Martin on the basis of the Series 7000 bus. A series of 6 satellites was delivered to Intelsat within a 15-month time period. Of the six satellites, four were launched on Ariane 4 and two on Atlas. Different launch vehicles not only provide different accommodation but also place the spacecraft into slightly different transfer orbits. In order to compensate for differences in the GTO of the two launch vehicles, the ΔV provided had to vary from 1450 to 1700 m/s and the burn plan had to vary from 3 to 4 burns.

Figure 19.5 Intelsat VIII (Reproduced by permission of Intelsat)

Other features of the Intelsat VIII satellites include

- two independently steerable Ku-band spot beams that can be pointed anywhere on the surface of the earth that is visible from the spacecraft;
- interconnected operation between C- and Ku-bands;
- expanded satellite news gathering (SNG) service provided by the capability to connect spot beams to global beams;
- transportable SNG stations for voice/data communications;
- a total of 44 transponders, 38 at C-band and 6 at Ku-band, for optimized coverage of landmass areas for Intelsat's voice/data and video services.

Intelsat IX

Space Systems/Loral was originally awarded a contract on March 20 1997 for $600 million to build two satellites with options for additional satellites. Each of the Intelsat IX series satellites carries 44 C-band and 12 Ku-band operating transponders, and its solar arrays generate more than 8.6 kW of power (beginning of life). The launch mass is 4723 kg and the dry mass 1972 kg. The Intelsat IX series of spacecraft carries a greater percentage of high-power amplifiers and generates more solar array power than their predecessors with a small increase in dry mass. They replace the Intelsat VI satellites and provide enhanced global voice, video and data transmission services. The Ku-band spots provide direct-to-home TV services to particular geographical areas, while the increased C-band power enables broadcasters to use smaller, lower-cost antennae.

The Intelsat IX has a generic feed array, which is a suitable configuration for its orbital location. It includes advanced features such as selectable split uplink for SNG and Ku-Band overdrive control. It will provide Internet, video and public-switched telephone service for the Americas, Europe, the Middle East and Africa.

The Intelsat 901 satellite was launched on June 9, 2001.

Intelsat X

After an international competitive procurement process, the Intelsat Board of Governors awarded Astrium the contract to build the largest and most powerful satellite ever procured by Intelsat. This spacecraft will be located in the West Atlantic Ocean Region, at 310° East longitude, to provide a variety of communications services. This spacecraft, Intelsat 10-01, will have a launch mass of 5000 kg. Thirty-six C-band transponders will be dedicated to cover the North, Central and South American continent, as well as Western Europe. It will also have 20 Ku-band transponders for services in Latin America. The power of its payload will be in the order of 8 kW. The satellite will be launched in 2003 and will have a 13-year orbital design life. It is compatible with the launch vehicles Ariane, Proton and Sea Launch, the latter being an international company with American, Russian, Ukrainian and Norwegian partners.

The Intelsat 10-01 spacecraft will use the new high-powered version of Astrium's Eurostar series, the Eurostar 3000 (see Section 8.8.1 of Chapter 8), and will use Hall electric propulsion thrusters for station-keeping (see Section 6.4 of Chapter 6).

Conclusion

The growth and changes that have taken place in the Intelsat satellites since the first satellite was launched parallel the developments that have taken place in telecommunications. At the outset the accent was on innovation and the market was small. As the market has developed, the trend has been towards the application of the technology in the most efficient way to reduce costs and increase the market share.

19.3.3 Meteosat

Background

Strictly speaking, the name Meteosat refers only to the operational series of satellites in Europe's geostationary meteorological programme, launched between the late 1970s and the 1980s. In this section we broaden the discussion to include later designs—the closely similar *Meteosat Operational Programme (MOP)* series, which followed in the late 1980s, and the more advanced *Meteosat Second Generation (MSG)* series from 2002 onwards. It is also interesting to compare European designs with the US *Geostationary Orbit Environmental Series (GOES)*. Table 19.8 summarizes the European and US systems. The European spacecraft are illustrated in Figure 19.6 and the US spacecraft in Figure 19.7. In the late 1980s, the possibility of changing to a three-axis-stabilized solution was considered and a number of solutions were proposed. It was subsequently decided to implement a less radical change in design and continue with the spinning spacecraft concept. Since 1994, the GOES spacecraft have been three-axis-stabilized.

There are several close similarities between the Meteosat series described here and the INTELSAT series covered in Section 19.3.2. In terms of operational status, both programmes have evolved over more than two decades and now form an essential part of an

Table 19.8 European and US geostationary meteorological satellites

Spacecraft (operating life in brackets)	GEO location	Stabilization concept
European series		
Meteosat/MOP series (late 1970s to 2002)	0°	Simple spinner with electrically despun antenna (EDA) (Figure 19.6 left)
Meteosat second generation (MSG) (from early 2000s onwards)	0°	Various concepts assessed; enlarged spinner selected (Figure 19.6 right)
US series		
Early missions (late 1960s and 1970s)	75° W and 135° W	Simple spinner (similar to Meteosat)
GOES series (up to 1990s)	75° W and 135° W	Dual spin (with mechanically despun communications antennas)
Current GOES (from mid-1990s)	75° W and 135° W	Three-axis-stabilized (Figure 19.7)

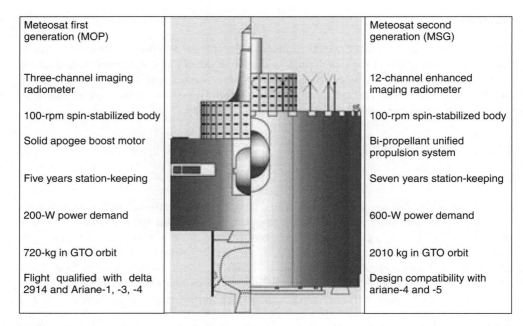

Meteosat first generation (MOP)	Meteosat second generation (MSG)
Three-channel imaging radiometer	12-channel enhanced imaging radiometer
100-rpm spin-stabilized body	100-rpm spin-stabilized body
Solid apogee boost motor	Bi-propellant unified propulsion system
Five years station-keeping	Seven years station-keeping
200-W power demand	600-W power demand
720-kg in GTO orbit	2010 kg in GTO orbit
Flight qualified with delta 2914 and Ariane-1, -3, -4	Design compatibility with ariane-4 and -5

Figure 19.6   A comparison of Meteosat first and second generation (Copyright EUMET-SAT 2002, reproduced with permission)

operational service to a worldwide community of users. The Meteosat programme, like that of INTELSAT, is funded by contributions from many nations to a central agency. *EUMET-SAT*, founded in 1985, is responsible for the exploitation and operation of meteorological satellites. The original development costs were borne by ESA, who contributed to the costs for the development of the second-generation spacecraft. Both series require attainment of a geostationary orbit and operation there over a long mission life. This requirement strongly influences the mission design for launch, boost phase and operational orbit maintenance and drives several design features of the satellite bus, notably propulsion, thermal control and communication subsystems. The full range of platform-stabilization options have been considered in design assessments and in the chosen concepts. These options range from simple spin through dual spin to three-axis stabilization.

While recognizing the obvious difference between meteorological and communication satellite payloads, there are also large areas of common ground, and in both cases the trend towards greater performance has been accompanied by increases in the demand for mass, power, data rate, pointing capabilities and other platform 'resources'.

Mission objectives and design drivers

Global and regional weather forecasting, and longer-term climatological studies, require frequent, regular, repeated inputs from a range of sensors on satellites in both geostationary and low-Earth, high-inclination orbits, as illustrated in Figure 19.8. The characteristics of LEO missions are described in Chapter 5 and will be touched upon again in Section 19.3.4 as part of the description of the EnviSat programme. Here we concentrate on the GEO missions.

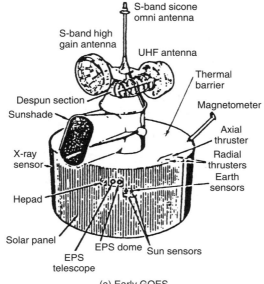

(a) Early GOES

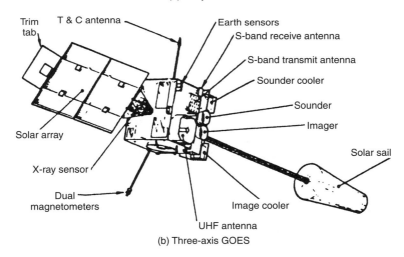

(b) Three-axis GOES

Figure 19.7 US GEO meteorological spacecraft (Reproduced by permission of NOAA [National Oceanographic and Atmospheric Administration])

The primary functions of the Meteosat satellite system and of the second-generation MSG systems shown in Figure 19.9 are:

1. to provide imaging in visible and infra-red channels over that part of the Earth's hemisphere which is accessible from a fixed geostationary orbit location;
2. to disseminate images and meteorological data to users;
3. to collect information from remote *'Data Collection Platforms' (DCPs)* and send this down with image data.

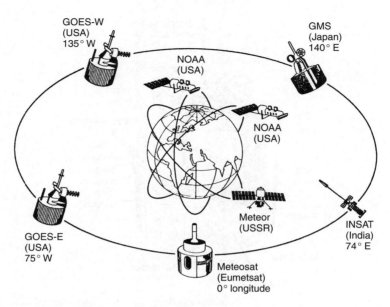

Figure 19.8 GEO and LEO meteorological spacecraft

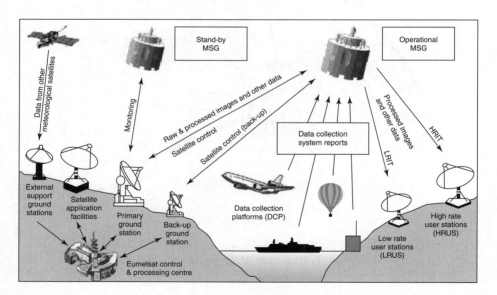

Figure 19.9 Meteosat space and ground segment (Copyright EUMETSAT 2002, reproduced with permission)

The associated ground segment includes primary and secondary data user stations, a network of DCPs (on buoys and in remote areas) and all the necessary facilities for processing, distributing, archiving and retrieving meteorological data products.

To the general public, the most familiar of these products are the visible channel images featured on television weather forecasts. These are built up line-by-line in the manner

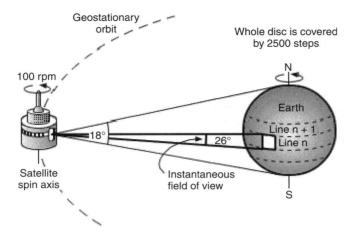

Figure 19.10 Meteosat spin-scan concept (Copyright EUMET-SAT 2002, reproduced with permission)

shown in Figure 19.10. The instantaneous fields of view (IFOV) of each channel of the Meteosat imaging radiometer are defined by an array of small detector elements in the focal plane of an optical telescope, each corresponding to a spatial resolution at the Earth's surface of only a few kilometres. These IFOVs are swept across the Earth in a longitudinal direction as a result of the satellite spin (at 100 rpm about an axis parallel to the Earth's spin axis). After each revolution of the spacecraft, the IFOVs are mechanically stepped through their own width in an N-S direction so that successive scan lines are contiguous (i.e. they lie parallel and adjacent to one another).

A few thousand such steps build up the entire Earth image over a period of half an hour and the cycle is then repeated throughout the life of the spacecraft. On the current series of Meteosat spacecraft, the N-S scan steps are achieved by gimballing the entire telescope, which is accommodated at the centre of the drumlike spacecraft, viewing through a cut-out in the cylindrical solar array. This spin-scan concept has the advantage of simplicity, compared to the more sophisticated two-axis scanning mechanism, which would be required on a three-axis-stabilized platform. However, it is inefficient in useful observing time since the Earth subtends an angle at GEO altitude corresponding to only 5% of each 360° revolution. Thus, for 95% of the time, the sensor is viewing space rather than the Earth. This shows the intimate association between payload and bus design.

The broadband visible channel (0.5–0.9 μm) on Meteosat measures solar radiation reflected from the Earth's surface and its atmosphere. The output signal is processed to form the familiar images in which the ocean appears dark and the cloud tops are white. As well as being an indicator of rain, the cloud images provide information on wind from their movement during the period between one image and the next.

The thermal infra-red channel operates in an atmospheric 'window' at wavelengths between 10.5 and 12.5 μm. Hence it measures the amount of radiation emitted from the Earth's surface or cloud tops. With suitable calibration, the processed detector outputs provide information on the effective temperatures over the Earth's disc. Another infra-red channel operates in a spectral band between 5.7 and 7.1 μm corresponding to water vapour absorption in the atmosphere. This channel measures radiation absorbed and re-emitted by water vapour in the middle and the upper tropospheric layers of the atmosphere. Signal strength in this channel provides an indication of humidity.

To give sufficient sensitivity (signal-to-noise ratio) to allow the effective temperatures of adjacent scene elements to be distinguished it is necessary:

- to collect the incoming radiation in a telescope of at least 40-cm aperture,
- to divide the radiation between separate spectral channels and focus it on the detectors with high transmission efficiency,
- to cool the infra-red detectors below 100 K to minimize 'thermally' generated noise.

These requirements enable design drivers to be identified and the impact of each of these can be traced through to the final design solution.

Meteosat satellite design

The resulting Meteosat instrument design is shown in Figure 19.11 alongside an exploded view of the satellite. Cooling of the infra-red detectors is achieved by passive means, on the basis of radiative exchange with cold space from a conical surface with a large view factor centred on the North-South spin axis. One end of the satellite is dominated by an *electrically despun antenna* assembly that is required to provide sufficient gain (see Section 13.3.3 of Chapter 13) to send the raw data stream to ground. The communications payload acts as a repeater for the distribution directly to users over a wide coverage area, and also for processed data sent back up from the central ground station at different frequencies.

The other main feature to note in Figure 19.11 is the propulsion system, which comprises a solid ABM plus a monopropellant hydrazine system to provide all subsequent orbit manoeuvres and attitude control. The ABM is jettisoned after burn-out to allow the radiation cooler to view cold space over almost an entire hemisphere, the axis of which is normal to the equator. As with any GEO mission achieved by boosting from a GTO orbit, the propulsion system mass is a high proportion (40–50%) of the total wet mass at launch.

Meteosat second generation (MSG)

The mission objectives for the MSG programme are

- to provide continuity with Meteosat/MOP for imaging, data distribution, DCPs and so on;
- to extend the capabilities in imaging and to provide some capability for atmospheric 'sounding', that is, derivation of vertical profile information of temperature and water vapour by inclusion of more infra-red channels;
- to ensure a high level of availability of operational data over more than 10 years, commencing in 2002;
- to provide the means to accommodate an additional small 'scientific' instrument for Earth radiation budget assessments.

To achieve these objectives, the imaging payload has to be enhanced to accommodate additional channels with higher spatial and spectral resolution and a shorter frame time.

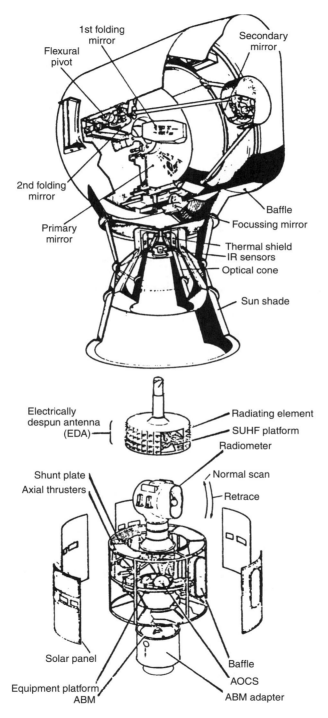

Figure 19.11 Meteosat imaging radiometer and spacecraft

These lead to a significant increase in the meteorological payload mass and power demand, and, because of the much higher data rate, a corresponding increase in the complexity of the communications payload.

To limit the programme costs to an acceptable level, the series of MSG spacecraft (like the Meteosat/MOP satellites) must be compatible with the overall mass and size constraints of a shared Ariane launch into GTO. In this way the total launch costs over the entire programme are reduced by many tens of millions of euros compared to dedicated launchers.

Pre-Phase A and Phase A studies of MSG payload, spacecraft and ground segment continued over several years, during which the following major systems options were assessed.

Payload

- Various levels of improvement in performance parameters, and consequently in the instrument design.
- Various groupings of instruments (imagers, sounders—separate or combined— visible and infra-red only or including microwave spectral coverage).
- Infra-red detector cooling either passively (radiative exchange) or actively (closed cycle refrigerator).
- Various levels of on-board processing of payload data.
- Alternative designs for the associated communications payload, including options on carrier frequency and antenna design.

Spacecraft

- Attitude control options including simple spin, dual spin, three-axis stabilization and a slow oscillation mode in an East-West direction to avoid the need for instrument scanning in that plane.
- Alternative propulsion systems, which in turn have a significant impact on the choice of overall spacecraft configuration.
- Various levels of physical and functional modularity between the bus and the payload elements.
- Various levels of heritage (e.g. from communication satellite programmes) through adaptation of existing designs of platforms or entire subsystems.
- Various levels of bus and payload reliability/redundancy and hence overall spacecraft design life, leading to different numbers of spacecraft to meet the availability target for the entire programme duration.

Ground segment

- Various levels of enhancement beyond Meteosat/MOP
- Centralized versus de-centralized concepts for the ground segment architecture

The decisions about the options selected were based on technical and financial considerations as described in Section 19.2.2, although the solutions adopted show the complexity of the issues that arise in practice. Following the trend set by communication satellites, the decision was taken fairly early to adopt a more efficient, unified bipropellant system in place of a solid ABM plus hydrazine RCS.

The choice of the stabilization method was not so straightforward as it was so intimately linked to the composition, design, complexity and efficiency of the instrument scanning concepts. In the USA the GOES spacecraft now adopt the three-axis-stabilized solution to allow 100% viewing efficiency of Earth, rather than only 5% from a spinner in GEO. The chosen configuration (shown in Figure 19.7) results from the requirement to minimize blockage of the field of view of the passive infra-red cooler in a north or south direction, which is incompatible with the usual symmetric solar array arrangement deployed along that axis.

A single-sided solar array (i.e. one wing only) is only acceptable if means are provided to counteract the solar radiation pressure torque due to the resulting lack of symmetry (see Section 9.4.5 of Chapter 9). Since the fuel mass over the mission life would be excessive, a 'solar sail' is deployed to minimize the offset of the centre of pressure from the centre-of-mass, while presenting an acceptably small blockage of the cooler's view to cold space. The failure of a 'solar sail' to deploy on one of the first Indian GEO applications spacecraft (INSAT) caused the early loss of a mission. Active cooling of infra-red detectors, with lower demands on radiator area and view factor, would allow the adoption of a conventional, symmetric array geometry. Stirling cycle coolers of suitable performance are now operating in space and are likely to feature increasingly in future infra-red missions.

Partly as a result of problems experienced in the USA in the early 1990s in develop-ing the three-axis GOES spacecraft (with very significant schedule and cost overruns compared to the original targets), Europe selected the simpler spin-stabilization con-cept for MSG. The current configuration of MSG spacecraft is shown in Figure 19.6 alongside the earlier Meteosat design. Because the same design drivers exist, there is clearly considerable similarity in outward appearance between MSG and Meteosat. This is most noticeable for the solar array, with its cut-out for the radiometer field of view, and the communications payload. The apogee boost manoeuvre on MSG will use a pair of 400 N engines. These use the same bipropellant supply as the smaller thrusters (10 N) sized for orbit maintenance and attitude control. The two 400 N engines are mounted symmetrically away from the central axis, since that area must be left free for the radia-tive cooler.

The total instrument mass to be carried on MSG is rather more than twice that on Meteosat. Total spacecraft power demand is roughly trebled. The required increase in solar array area is achieved by increasing both the diameter and the height of the main cylindrical section (by about 50% each), taking advantage of the greater volume offered on a shared launch of Ariane 4 compared to Ariane 1 to 3 vehicles used for the Meteosat/MOP series.

Initially a series of three MSG satellites was being built by European industry to ensure continuity of observations over a period of at least twelve years. The launch of MSG-1 took place in August 2002, some two years later than planned at the start of the main development phase (C/D). In fact the flight model programme of MSG-1 was completed in 2000 and the spacecraft was put 'in storage'. Unusually, the cause of the launch delay was not the space hardware development but the ground segment, in which upgrades to handle the much-increased volume and types of data have required more time to implement than anticipated. This illustrates the central importance of managing interfaces and considering all system elements from the start of the programme and throughout its duration. Fortunately the existing Meteosat MOP series was able to provide good service during the delay.

19.3.4   EnviSat

Introduction

On March 1 2002 **EnviSat**, the largest spacecraft ever built in Europe, was launched successfully on Ariane 5 from Kourou in French Guiana. Given the huge scale of the project, both in terms of cost and schedule, the success of this mission was critical to the European Earth-observation community, and indeed to the company, Arianespace, responsible for the launch. As such, the successful launch was a considerable relief to all involved. As the mission has progressed, calibration of the instruments has taken place (at the time of writing), and it is expected that a long-term, reliable service to commercial and scientific users will begin in January 2003. When Artemis, the geostationary satellite used for relaying data from EnviSat, is available from February 2003, this will complete the operational infrastructure of the mission.

EnviSat is a very large satellite. No further Earth-observation satellites of this size are planned in Europe. The emphasis now is on much smaller, thematic satellites. There is little doubt that it has absorbed so much of the resources available to European Earth observation that its failure would have had a major impact on further European development in this area. How did this situation arise?

The EnviSat story started in the mid-1980s at a time when the ERS-1 programme of the European Space Agency (ESA) was in development.

In January 1984, President Reagan of the United States announced the start of a project to develop an *International Space Station (ISS)*[1]. During 1984, the member states of ESA considered if and how they would participate in this programme, and at the Ministerial Council Meeting of Rome, in January 1985, agreed to establish the *Columbus Programme*. The nature of this European participation in the Space Station was rather fluid during these early years, but it was recognized that the manned environment of a large space station would degrade experiments in microgravity, while the low inclination orbit would be unsuitable for Earth-observation purposes. Consequently the Phase B1, which started in July 1985, included an attached pressurized module (APM) as well as a free-flying 'polar platform'. The Inter-Governmental Agreement of November 1988 specified that the European governments, through ESA and the Columbus Programme, would provide the *Attached Pressurized Module* for material and life experiments, a *Man-Tended Free Flyer (MTTF)* for microgravity experiments and a *Polar Platform (PPF)* dedicated to the study of the Earth and its environment.

The PPF was foreseen as a large and versatile autonomous platform, which could be instrumented with Earth-observation equipment. It would periodically make major orbit manoeuvres in order to rendezvous with the STS for maintenance. Such activities would include replenishment of fuel as well as refurbishment or even replacement of instruments.

From the point of view of the Columbus programme, it offered enhanced scientific justification. For the Earth-observation community, it offered the chance to 'piggyback'

---

[1] It took several years to evolve this name—in 1988, the President chose the name *Freedom* from a short list of 3 identified by a naming committee from 600 proposals. In 1993, when Russia joined the programme, it was renamed *Alpha*.

on another programme and maximize the resources available for instrument development and exploitation. The idea, at this stage, was much larger than EnviSat is today. Large platforms were in fashion at the time. NASA were on tour promoting the *Earth Observing System (EOS)*, which was envisaged to consist of a series of very large, multipurpose Earth-observation platforms. EOS was descoped several times and is now a pair of EnviSat-class satellites and a series of smaller missions.

Users were canvassed for ideas to equip the platform. In November 1988, the European scientific community was called to ESA Headquarters, in Paris, for a User Consultation Meeting. Following this meeting the Agency succeeded in obtaining approval for the development of ERS-2, the follow-on to ERS-1 in order to bridge the gap until the PPF would be ready for launch. Two series of PPF missions were defined, the first, *EPOP-M1*, to be launched in 1997, would focus on meteorology, ocean and climate as well as having a suite of atmospheric chemistry sensors. An identical satellite, EPOP-M2, would be launched in 2002, followed by others at five-year intervals. The second series, *EPOP-N1, -N2*, and so on, would focus on land resources, with a first launch in 1999.

The payload proposed to the second User Consultation Meeting in May 1991, for the first mission, had three parts. The *Operational Meteorological Instruments* were improved versions of instruments already flying on the NOAA polar-orbiting satellites. NOAA and EUMETSAT would provide them in order to ensure continuity of observations in the 'morning' orbit after 1997, at which time, responsibility for the polar-orbit meteorology service would be shared between these organizations. The *ESA Developed Instruments* represented the core scientific mission and would be funded by ESA's Earth-Observation Directorate. Finally the *Announcement of Opportunity Instruments* were to be provided by external organizations and had been selected following a competitive process.

The Programme was endorsed and ESA continued work on the first mission, now called *POEM-1*, an acronym for the first *Polar-Orbiting Earth-observation Mission*. This caused some confusion amongst many who had been involved with earlier polar-orbiting Earth-observation missions, such as ERS-1, which was launched two months after the meeting, in July 1991.

The Granada meeting in 1992 was a landmark. The ambitious PPF, and its series of payloads, was reduced to two smaller satellites, *EnviSat* and *MetOp*, which shared the POEM-1 payload. Furthermore, the PPF became a separate programme managed by the Earth-Observation Programme Board instead of the Columbus Programme Board. Nevertheless, the contractual and financial basis of the PPF and EnviSat remained distinct, with the EnviSat payload still being regarded as the first payload for the PPF design. However, the MetOp Programme, which started several years after EnviSat, did not in fact reuse the PPF design. MetOp, initially a series of three satellites, is a joint venture with EUMETSAT and subject to many constraints. Amongst these is a strong requirement to maintain compatibility with several launch vehicles and this has effectively ruled out reuse of the PPF, which relies on the large Ariane 5 launcher. In fact, MetOp is now expected to be launched by Soyuz.

The EnviSat mission

Despite the descoping with respect to the POEM-1 concept, EnviSat remains a large satellite with a wide-ranging set of Earth-observation instruments. It provides continuity of the measurements from ERS-1 and ERS-2, which, between them, have provided continuous measurements from the Synthetic Aperture Radar (SAR), the scatterometer, the radar altimeter and the Along-track Scanning Radar (ATSR) for more than a decade. ERS-2, launched in 1995, introduced a further instrument for the measurement of atmospheric chemistry, particularly ozone, and this measurement series will also be continued by EnviSat (and MetOp).

EnviSat supports four fundamental types of measurements:

- high-resolution radar imaging of all global surfaces (land, ocean, ice) using SAR techniques;
- medium-resolution optical/infra-red imaging over open oceans, coastal zones and land;
- measurements of atmospheric composition, and particularly trace gases;
- high-precision measurements of the elevation of ocean, ice and land surfaces.

An artist's impression of EnviSat in orbit is shown in Figure 19.12. Individually the instruments on EnviSat are large and capable. Together they represent a comprehensive and synergistic package that will be able to provide the measurements needed to support Earth science, environmental monitoring, operational Earth observation and commercial applications. A ground segment that is able to deliver data products to users in near-real time if necessary, and that supports extensive user services for archived data, complements the payload. Seven hundred scientific projects, involving thousands of scientists worldwide, have already been selected following the first call for proposals for data exploitation.

Figure 19.12 Artists impression of EnviSat in flight (Reproduced by permission of European Space Agency)

As a large European project, EnviSat has also been identified as a key contributor to the recent European Commission initiative called *Global Monitoring of our Environment and Security (GMES)*. Within this, EnviSat will provide monitoring of pollution and disasters.

The orbit of EnviSat is, like ERS-1 and ERS-2, Sun-synchronous (see Section 5.4 of Chapter 5). The mission of ERS-1 was characterized by several changes in the orbital repeat period, in response to the demands of users. However, the majority of its lifetime was spent in the so-called multi-disciplinary orbit, with a 35-day repeat cycle (see Section 5.4 of Chapter 5). This has proved to be a very good compromise and has been used throughout the life of ERS-2, and will be adopted for EnviSat as well. The ground tracks will be identical to the Earth Resources Satellite (ERS) orbits, and phased such that EnviSat will precede ERS-2 along the tracks by half an hour, to improve cross-calibration of measurements between the two satellites.

The EnviSat system is composed of several parts:

- The space segment, which comprises the PPF (the bus based on the SPOT series—see the section on the Polar Platform) and the payload instruments.
- The launch vehicle is Ariane 5. EnviSat fully exploits the large mass and size capabilities of this launcher, including the large 2536-mm diameter launch vehicle adaptor.
- The ground segment that is divided into the *Flight Operations Segment*, which performs command and control, and the *Payload Data Segment*, which receives, processes and distributes the instrument measurement data.

The establishment of the compatibility between Ariane 5 and EnviSat, particularly with respect to mechanical shocks generated by the fairing release and the separation of the upper stage from the main stage below has been a lengthy process.

The EnviSat payload

The composition of the EnviSat payload was initially defined as a subset of the EPOP-M1 candidate payload identified above, but various evolutions during the programme definition led to a slightly modified payload. The final composition is as follows:

ESA-developed instruments:

ASAR	advanced synthetic aperture radar
MERIS	medium-resolution imaging spectrometer
MIPAS	Michelson interferometer for passive atmospheric sounding
GOMOS	global ozone monitoring by the occultation of stars
RA-2	radar altimeter-2
MWR	microwave radiometer
LRR	laser retroreflector

Announcement of opportunity instruments:

SCIAMACHY	scanning imaging absorption spectrometer for atmospheric cartography
AATSR	advanced along-track scanning radiometer
DORIS	Doppler orbitography and radio positioning integrated by satellite

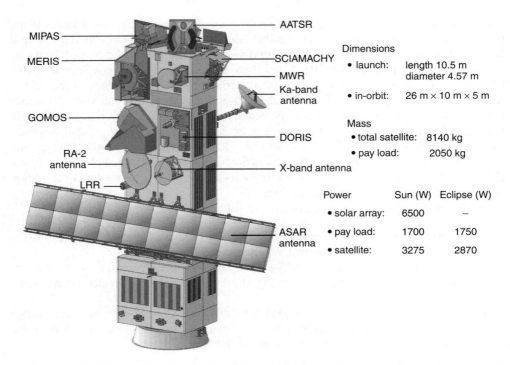

Figure 19.13   EnviSat (Reproduced by permission of European Space Agency)

These are shown in Figure 19.13.

Rather than follow this listing, broken down according to instrument supplier, the payload instruments will be described according to the grouping of fundamental measurement types identified earlier.

Radar imaging (ASAR)

The Radar Imaging mission provides continuity with the SAR on ERS-1 and ERS-2. This type of sensor provides high-resolution images of terrestrial surfaces in all types of weather, day and night. The antenna beam is fan-shaped, to illuminate a swath in the cross-track direction with relatively little illumination along-track.

The ERS radars proved to be extremely stable so that colour composite images made from images of the same scene on different dates (*multi-temporal images*) could be generated, facilitating change-monitoring to extract surface topography. They proved spectacularly successful in monitoring very small changes in scenes, for example, before and after earthquakes.

The Advanced SAR (ASAR) on EnviSat offers a number of enhancements compared to the ERS SAR. These are

- a wide swath (400-km), low-resolution mode (150-m), using the ScanSAR technique;
- steerable elevation angle for the normal resolution (30-m) mode (100-km swath, selectable over a range of 485-km width);

- alternating polarization, allowing scenes to be imaged simultaneously in horizontal and vertical polarization.

Optical imaging (MERIS, AATSR)

The Medium Resolution Imaging Spectrometer (MERIS) had been a high priority for post-ERS satellites. MERIS addresses the needs of three application domains: primarily, oceanographic and secondly, atmospheric and land observations. MERIS, complemented by the ASAR, RA-2 and AATSR provides a synergistic mission for bio/geophysical characterization of the oceans and coastal zones.

The MERIS concept exploits the characteristics of two-dimensional Charge Coupled Device (CCD) arrays. The optical system features a camera looking down at the Earth, with a narrow slit in the optical path. The orientation is such that this slit lies across the satellite track, so that a narrow strip of the Earth's surface is imaged. As the satellite moves forward, this strip moves, recording a swath. This type of geometry is called *push-broom*, and it acts in much the same way as a broom sweeps a swath of floor. The light from this slit is directed to a diffraction grating that splits it into a spectrum. This is focused onto the two-dimensional CCD such that the cross-track pixels lie along a row. Each wavelength is imaged on a different row, so that a column represents the spectrum of a single pixel on the ground.

There are five identical cameras, each employing these principles. The five cameras image adjacent (slightly overlapping) swaths, so that the overall instrument swath-width is 1150 km and have a full spatial resolution of 300 m at the nadir.

An important requirement for this instrument, common for the EnviSat payload, is the need for calibration of its measurements. During calibration, the Earth-view port is closed and the Sun-view port is opened to provide, in the case of radiometric calibration, a spectrally and spatially uniform radiance source, and in the case of spectrometric calibration, a radiance source with a known spectral signature.

Each camera images an across-track strip of the Earth's surface onto the entrance slit of an imaging optical grating spectrometer. This entrance slit is imaged through the spectrometer onto a two-dimensional CCD array, thus providing spatial and spectral information simultaneously.

The advanced along-track scanning radiometer

The AATSR is based on heritage from the very similar instruments that were flown on ERS-1 and ERS-2. These earlier instruments were combined with a microwave radiometer for the determination of water vapour at the on-board data-handling level. For EnviSat the microwave radiometer is a separate instrument, described later. The ERS-1 instrument, the ATSR, was a purely infra-red instrument, while for ERS-2, the ATSR-2 had three visible channels added.

The primary objective of the AATSR is to measure the global sea-surface temperature for climate-research purposes. Its absolute accuracy is better than 0.5 K when averaged over areas of 50 km × 50 km, assuming that 20% of pixels within the area are cloud-free. For the cloud-free pixels, of 1 km × 1 km, the relative accuracy is about 0.1 K.

To achieve these objectives, the AATSR was designed as an imaging radiometer with four co-registered channels with wavelengths of 1.6, 3.7, 11 and 12 μm, defined by beam splitters and multi-layer interference filters. The IFOV at the nadir on the Earth's surface is a 1 km × 1 km square, which is imaged onto the detectors via an f/2.3 parabolic mirror. These detectors, fixed onto a focal-plane assembly, are cooled to 80 K by a Stirling cycle cooler in order to reduce their background noise to an acceptable level.

The IFOV is scanned over a 500 km swath by a rotating plane mirror in such a way that it gives two Earth views; at nadir, and 57° forward. In order to calibrate the optical and electrical signal chain, two black bodies (one hot and one cold) within the AATSR are also scanned during the rotation.

Atmospheric chemistry (GOMOS, MIPAS, SCIAMACHY)

Ozone plays a central role in the chemistry of the atmospheric. It is largely responsible for stratospheric heating through absorption of harmful UV radiation and it largely determines the oxidation capacity of the troposphere and is an important greenhouse gas. The discovery of the ozone hole over Antarctica (and later over the northern hemisphere) also drew attention to the global ozone budget.

The EnviSat payload includes three instruments that measure ozone and other trace gases using a range of techniques. It is intended that measurements from these sensors will enable the necessary measurements to be made with the improved accuracy needed to significantly contribute to our understanding of atmospheric chemistry.

GOMOS

The key attribute of the GOMOS instrument is the accuracy with which it can measure ozone concentration. As its name suggests, it makes its measurement by observing the variations in stellar spectra as the stars are occulted by the Earth's atmosphere.

The instrument line-of-sight will be successively oriented towards pre-selected stars and maintained by autonomous tracking, while the star sets behind the Earth's atmosphere. During the star occultation, the ultraviolet, visible and near-infra-red spectra of the star will be continuously recorded. During the occultation, the stellar spectrum becomes more and more attenuated by absorption of the various gases in the atmosphere, each of which is characterized by a known, well-defined spectral signature. The attenuated spectra may be compared with the unattenuated stellar spectrum measured at the start of the occultation, enabling the absorption spectra to be derived very accurately.

GOMOS measures the atmospheric constituents by analysis of the spectral bands between 250 and 675 nm, 756 and 773 nm and 926 and 952 nm. In addition, two photometers operate in two spectral channels—between 470 and 520 nm and 650 and 700 nm.

The primary GOMOS mission objectives are

- measurement of profiles of ozone, NO_2, NO_3, OClO, temperature and water vapour;
- day and night side measurement capability;
- global coverage with typically over 600 profile measurements per day;
- altitude measurement capability between the tropopause and 100 km;
- altitude resolution of better than 1.7 km.

MIPAS

The MIPAS instrument is a Fourier transform spectrometer for the measurement at high resolution of gaseous emission spectra at the Earth's limb. It operates in the near-to-mid infra-red (from 4.15 to 14.6 μm) where many of the atmospheric trace gases playing a major role in atmospheric chemistry have important emission features. Because of its high spectral resolution and low noise, the detected features can be spectroscopically identified and used as input to suitable algorithms to extract atmospheric concentration profiles of a number of target species.

MIPAS is designed for limb observations to give maximum sensitivity and good vertical resolution. The IFOV, realized with an anamorphotic telescope, is only 3 km high to achieve a good vertical resolution, but 30 km wide to collect sufficient radiance.

As a result of the limb-viewing geometry, the distance between instrument and tangent point is about 3300 km. Thus, in order to measure at a predetermined limb height, a line-of-sight pointing knowledge with respect to nadir of better than 0.01° (1-sigma) is required. This is a key driver for the satellite design and led to the introduction of an attitude control system based on star trackers (see later).

The objectives of MIPAS are

- measurement of profiles of ozone, water vapour, methane, N_2O and HNO_3;
- day and night side measurement capability;
- global coverage with typically 75 elevation scans per orbit;
- altitude measurement capability between 5 and 200 km;
- altitude resolution of 3 km.

SCIAMACHY

SCIAMACHY measures the spectrum of sunlight and moonlight that is transmitted, reflected or scattered by the Earth's atmosphere using a combined limb, nadir and occultation strategy. It is designed for the ultraviolet, visible and near-infra-red wavelength domains (240–2380 nm), covering this range with a resolution of 0.24 to 1.5 nm.

SCIAMACHY is a passive remote-sensing instrument, designed to measure atmospheric constituents and parameters of importance in the stratosphere and troposphere. It comprises four subsystems:

- a scan-mirror system that determines the instrument's observational mode;
- a spectrometer that breaks down the incoming signal into its spectral components;
- a cooling system that maintains the spectrometer and its detectors at selected temperatures to optimize the signal-to-noise ratio in the measured spectra;
- an electrical subsystem that controls and operates SCIAMACHY and interfaces to the EnviSat platform.

The high resolution and the wide wavelength range makes it possible to detect many different trace gases even when their concentration is very low. The large wavelength range is also ideally suited for the detection of clouds and aerosols.

Altimetry (RA-2, MWR, DORIS, LRR)

In contrast to the ASAR, the Radar Altimeter-2 deliberately reused as much of the ERS altimeter technology as possible.

The Radar Altimeter (RA-2)

The Radar Altimeter is a nadir-pointing pulse radar designed to make precise measurements of the return echoes. Over ocean surfaces the echo characteristics of interest are

- time delay with respect to the transmitted pulse, which provides the measure of altitude;
- slope of the echo leading edge, which is related to the height distribution of reflecting facets and thus to the ocean wave height;
- the power level of the echoed signal, which depends on small-scale surface roughness and thus on wind speed.

The effective pulse width is 3 ns, which is equivalent to about 45 cm in two-way range. The radar is said to be 'pulse-width-limited' because not all of the target is illuminated simultaneously by the short pulse, and the received power is controlled by the illumination. Over ocean surfaces, the distribution of the heights of reflecting facets is Gaussian or near Gaussian, and the echo waveform has a characteristic shape that can be described analytically. It is a function of the standard deviation of the distribution, which is closely related to the ocean wave height. Different echo waveforms occur over ice surfaces. Over sea ice, there is generally a strong specular component, while the rough topography of continental ice sheets at the margins leads to complex return waveforms. In central ice sheet areas, the height distribution becomes more regular and echoes similar to ocean returns are observed.

The constraints of available peak transmit power and required pulse width again determined that a pulse-compression technique be used to spread the required energy over time, allowing reduced peak power.

Microwave radiometer (MWR)

The main objective of the MWR is to measure the atmospheric integrated water content (vapour and liquid) in order to compute the most problematic part of the tropospheric path delay in the Radar Altimeter's signals. The MWR has two channels, operating at 23.8 and 36.5 GHz, each with a bandwidth of 400 MHz. The instrument is nadir-viewing, using an offset antenna. Calibration measurements are made by a sky horn pointing to cold space and internal hot loads.

Doppler Orbitography and Radio-positioning Integrated by Satellite (DORIS)

DORIS is capable of determining satellite orbital position with an accuracy better than 0.05 m. It relies on accurate measurement of the Doppler shifting of radiofrequency signals transmitted from ground beacons and received on-board the satellite. Measurements at 2.03625 GHz are used for precise Doppler measurements, and measurements at 401.25 MHz for ionospheric correction of the propogation delay. DORIS is used in conjunction with the Radar Altimeter to determine spatial and temporal ocean-surface topography changes and variations in ice coverage.

The laser retro-reflector (LRR)

The LRR is a passive device used as a target by ground-based laser-ranging stations. They measure the round trip time of laser pulses reflected from this array of corner cubes, mounted on a tower (necessary to ensure the field of view) on the Earth-facing side of the payload module.

The polar platform

Although British Aerospace was the prime contractor for the PPF, the platform to be used for EnviSat, the avionics and architecture of the SVM were based on the SPOT-4 satellite, being developed at that time by Matra under contract to CNES and closely based on the earlier satellites in the SPOT series. This had some clear advantages of commonality with SPOT as well as ERS, which was based on the earlier SPOT bus. Eventually, a clear family tree of SPOT SVMs became evident with evolution from the Mk I to the Mk II used for EnviSat, and eventually the Mk III for MetOp. This tree is shown in Figure 19.14.

The PPF, however, required a substantially larger infrastructure for payload support than that offered by SPOT-4:

- 10 instruments with different needs for fields of view, physical accommodation and so on;
- mounting surface of $43\,m^2$ externally and $10\,m^2$ internally;

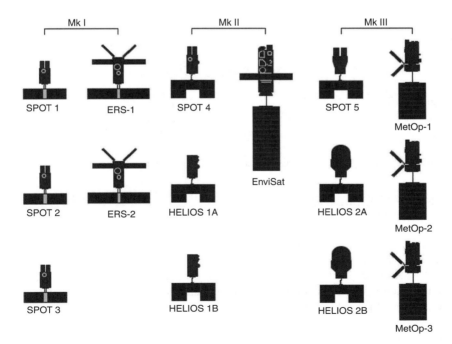

Figure 19.14 The family of satellites based on the SPOT bus (Reproduced by permission of European Space Agency)

- total instruments mass of 2200 kg;
- total instruments average power 1.9 kW over one orbit, 3-kW peak;
- attitude pointing better than 0.1°;
- attitude measurement better than 0.04°;
- handling of 10 low/medium rate (up to 32 Mbps) and one high rate (100 Mbps) data streams.

The **Service Module (SVM)**, with a new structure and solar array, but with avionics largely recurrent from SPOT-4, accommodates most of the satellite support subsystems:

- power generation, storage and distribution;
- Attitude and Orbit Control System (AOCS);
- communication on S-Band for TT&C;
- support structure and launcher interface.

The *Payload Module (PLM)* [which is further subdivided into the *Payload Equipment Bay (PEB)* and the *Payload Carrier* (PLC)] accommodates the payload instruments and provides dedicated payload support subsystems:

- instrument control and data handling;
- communication on X- and Ka-Band;
- power distribution;
- support structure.

Configuration and structure

The major drivers for the overall satellite configuration are

- reuse of the SPOT Mk II service module;
- provision of adequate mounting area on the Earth-facing surface;
- accommodation of a cluster of instruments requiring a view to deep space for passive radiative coolers;
- physical constraints of the Ariane 5 fairing and interface.

In flight, the major spacecraft longitudinal axis (the X_s axis) is normal to the orbit plane, the $-Y_s$ axis is closely aligned to the velocity vector and the $-Z_s$ axis is Earth-pointing. This configuration concept provides a large, modular, Earth-facing mounting surface for payload instruments and an anti-Sun face for radiative coolers, free of occultation by satellite subsystem equipments.

The SVM structure is built around a central cone, which attaches to the large, 2435 mm diameter, launch vehicle adaptor at the lower end and tapers to the 1.2 m diameter tubular interface to the PLM. This 1.2 m central tube continues through the length of the PLM. The remaining structural elements of the SVM are made from conventional aluminium skin honeycomb panels. This cone and tube, made from Carbon fibre reinforced plastic (CFRP) skin honeycomb, are empty except for the four fuel tanks supported at the interface between the modules.

The PLM itself is divided into four similar sections of 1.6 m in length (this was reduced from five when POEM-1 was split into EnviSat and MetOp). These sections are constructed with CFRP skin/aluminium honeycomb panels that are removable for the installation of internal units. To assist in the frequent connection and disconnection of these panels to and from the main harness, there are large connector brackets attached to the edges. The use of CFRP skins (an electrical insulator) prevented the use of the structure for grounding purposes, and so substantial aluminium grounding rails run throughout the structure.

Many of the instruments are externally mounted onto the outer panels and they form part of the primary structure for mechanical loads. In the extreme case, the ASAR antenna, with a mass of almost 800 kg, required very careful design and placement of the inserts in order to avoid unacceptable loads during the launch.

The thermal control approach complements the overall modular design of the satellite, the payload, platform and battery compartment being thermally insulated from one another as far as practicable, allowing separate analysis and testing. The individual modules are also insulated from the external environment by multi-layer insulation blankets, except for the radiators. These radiator areas have been optimized for the extreme hot and cold operating conditions that are encountered in nominal attitude (Earth-pointing) and during the safe mode (Sun-pointing) in which the payload would be inert. Active heater systems provide autonomous thermal control, predominantly by on-board software in nominal modes and by thermostats otherwise.

Electrical power subsystem

This comprises two Power Distribution Units (PDUs), one for the PEB and the other for the Payload and the Heater Switching Unit (HSU), to control all PEB internal temperatures.

The power demands of EnviSat significantly exceed the capabilities of SPOT-4 and a substantial electrical power subsystem had to be developed. The very large solar array, based on BSFR silicon cells, was developed by Fokker Space. It has 14 panels (the multimission design can have up to 16), measures 14 m × 5 m and provides 6.5 kW of power at end-of-life. A shorter version, with eight panels, will be used on MetOp.

Attitude and orbit control subsystem

The attitude control system uses a star tracker–based reference system.

EnviSat is a three-axis-stabilized, Earth-pointing satellite. Its yaw axis is parallel to the local vertical defined by an ellipsoid that takes the Earth's oblate shape into account. The pitch axis oscillates slightly during each orbit to keep it oriented normal to the composite ground velocity vector, taking account of the Earth's rotation to provide doppler compensation. This, nominal, mode of the attitude control system is called the Yaw-Steering Mode (YSM), and it is at the top of a tree of modes whose different properties enable the spacecraft to efficiently transfer from any state (e.g. separation from the launch vehicle) into operation. The residual attitude errors in YSM are less than $0.06°$ on each axis.

There are several attitude sensors on board. The prime AOCS operational mode, known as stellar yaw steering, uses two of the three star trackers for attitude measurement and derives the satellite rates from two of the four available gyroscopes. The other sensors are used as back-up. There is, in addition, the capability of operating the spacecraft in a YSM similar to that used on ERS-1 and 2. In this case, gyroscopes, Earth sensors and Sun sensors are used for attitude determination. Finally, there are two wide-field Sun sensors for use in the initial stages of attitude acquisition, and in safe mode, when the satellite would be Sun-pointing rather than Earth-pointing.

The primary means of attitude control is the set of five reaction wheels, each with a capability of 40 Nms and nominally at rest. They can be spun in either direction, exchanging angular momentum with the satellite in the process. Angular momentum is continuously dumped from the wheels. The on-board computer contains a simple model of the Earth's magnetic field and is able to control the current in a pair of orthogonal magneto-torquers (see Section 9.4.2 of Chapter 9), which generate torques by interacting with the Earth's geomagnetic field. Using a servo loop and the built-in field model, the computer continuously adjusts the magneto-torquers to control the maximum spin rate of the reaction wheels.

EnviSat also has 16 hydrazine thrusters, aligned about the spacecraft's three primary axes. They are used in different combinations to maintain and modify the satellite's orbit and to adjust its attitude during certain non-nominal operations. This is done by using pairs of thrusters to provide in-plane thrust when maintaining or changing the orbital height or by turning the spacecraft in yaw to obtain out-of-plane thrust when maintaining or modifying the orbital inclination.

Data-handling subsystem

There are two *On-Board Data Handling (OBDH)* systems (see also Chapter 13), one for the SVM and one for the PLM. These are identified in Figure 19.15, which illustrates the overall data-handling system.

The SVM system is the master with a digital serial OBDH bus connecting to all of the SVM equipment. The PLM system is also slaved to this bus, with its own OBDH bus servicing all the PLM equipment.

The SVM data-handling subsystem performs the processing and storage of housekeeping data, the management of telecommand and housekeeping telemetry, data-bus control, satellite alarm management and clock generation for all OBDH peripherals, including the PLM. It is based on a *Central Communication Unit (CCU)*, which handles the exchange of information between the ground and the SVM equipment via the OBDH data bus. The CCU also serves as the main computer for the AOCS, running dedicated control algorithms and monitoring functions in the CCU software.

The TT&C link to the ground relies on a bidirectional S-band link with an uplink rate of 2000 bps and a downlink rate of 4096 bps. It uses two S-band antennas with cardioid radiation patterns and opposite circular polarizations. Omnidirectionality of the link allows communication in any non-nominal attitude. There are two S-band transponders, connected via a passive hybrid to the antennas.

The PLM data-handling subsystem is centred on the *Payload Module Computer (PMC)*, which performs the command and control functions for the payload instruments. Its tasks include scheduling of the commanded mission, and monitoring of the PLM systems (PEB avionics and payload instruments). The PMC gathers and formats housekeeping telemetry from the PLM and its payload, which is merged into overall satellite housekeeping telemetry.

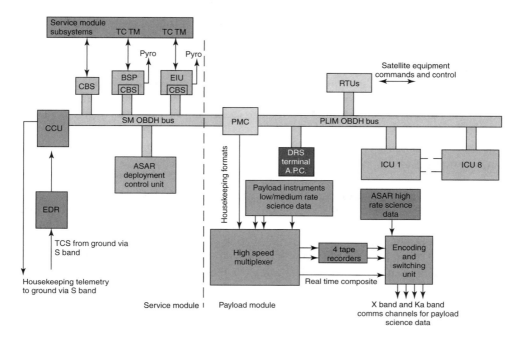

Figure 19.15 Block diagram of EnviSat's on-board data-handling system (Reproduced by permission of European Space Agency)

Instrument Control Units (ICU) are responsible for the instrument command and control functions, including the management of instrument internal data buses and communication with secondary scientific processors. Each payload instrument, and the DRS terminal, is equipped with an ICU. This set of interdependent computers fulfils a critical requirement. EnviSat is required to have substantial autonomy, and this is achieved by providing intelligent payload elements controlled by a capable central computer. A basic concept in this philosophy is the macrocommand, a coded instruction expanded and acted upon by the ICU. In this way the ICU relieves the PMC of many detailed tasks related to internal instrument configuration and operations. The PMC manages two types of macrocommands; immediate macrocommands, for immediate execution, and time-tagged commands. Mission-planning commands for autonomous operations are stored as a queue of time-tagged commands.

The instruments generate data in the form of source packets: a logical division of telemetry data from the instrument point of view. These are coded following the Consultative Committee for Space Data Systems (CCSDS) recommendations. These are acquired by the High-Speed Multiplexer (HSM) that formats and encodes them into Virtual Channel Data Units (VCDU). The VCDUs are multiplexed together to produce two streams of Channel Access Data Units (CADU), one for recording by one of the two solid-state recorders (SSRs) (or by the tape recorder) and the other for real-time downlink.

Payload data must be delivered not later than 3 h after data generation. To meet these constraints, the PLM carries two SSRs, each providing a dynamic RAM capacity of 70 Gbit. These two SSRs allow independent recording (4.6 Mbps) and dumping (50 Mbps) of the EnviSat low–bit rate composite, of the ASAR high–bit rate data (100 Mbps record and dump) and of the MERIS full-resolution data (22 Mbps record and 50 Mbps dump). Initially, the PLM had only four conventional tape recorders, each of 30 Gbit capacity; however, during the satellite development it was agreed to replace three of these by the two SSRs while retaining a single tape recorder as an additional spare capability to record and to dump the low–bit rate composite.

This on-board data recording capacity enables the satellite to maintain a global mission despite the limited visibility from the primary ground station at Kiruna, supplemented by the Artemis visibility.

Communications subsystem

The EnviSat platform's telemetry needs are served by the S-band TT&C subsystem described above. Because of the very high bit rates involved, the science data cannot use this link and the payload therefore includes its own data-handling system, which has two distinct communication subsystems, at X-band and Ka-band.

The Kiruna X-band ground station provides only about 10 min of satellite visibility per orbit. The add-on capability of data relay via the Artemis satellite provides an additional 20 min of visibility per orbit.

The X-band communication subsystem, used for data transmission direct to the ground station, has three frequency channels operable independently and, if need be, simultaneously. The frequencies used are 8.1, 8.2 and 8.3 GHz, each with an occupied bandwidth of about 70 MHz for the 100-Mbps data rate. Together they occupy virtually all of the available bandwidth (8.050–8.400 GHz) allowed for space-ground data transmission in this part of the spectrum. Fortunately they are only transmitting when EnviSat is in direct line-of-sight to Kiruna. The modulation used is Quadrature Phase-Shift Keying (QPSK); after amplification and band-pass filtering, the modulated signals are added through a waveguide output multiplexer.

The fact that the X-band transmission is required to have a minimum power-level fluctuation during the satellite pass led to the design of a shaped-beam antenna able to compensate for losses at low satellite elevation angles, when the distance to the ground station is long, and the attenuation due to the atmosphere's water content is high. To achieve this, the antenna reflector is shaped so that its radiation pattern compensates for the inverse-square-law variation in received power with distance as the satellite passes across the sky at the ground station.

The Ka-band assembly also has three channels. The assembly uses a 2 m deployable mast equipped with an azimuth and elevation pointing mechanism, which steers a 90 cm diameter Cassegrain antenna towards the Artemis satellite. Pointing commands are generated by an Antenna Pointing Controller, which is accommodated together with the RF part of the assembly within the PEB. The subsystem carrier frequencies are 26.85, 27.10 and 27.35 GHz. The system operates open loop.

19.4 CONCLUSION

In this chapter it has only been possible to illustrate the design process of a few spacecraft, but this does not reflect the extraordinary range of spacecraft that have been designed and manufactured over the near-half-century of the Space Age. In terms of mass alone, spacecraft can range from a few to many thousands of kilograms and there are similar ranges over orders of magnitude in volume, power and data rate. Many spacecraft are required to operate in LEO, at a few hundred kilometres altitude, whilst others, interplanetary spacecraft, must operate over distances of many millions of kilometres from Earth. Given this diversity, it is therefore inevitable that there should be considerable variety of the system engineering approaches employed in design. This has been particularly evident in the design approaches used for small spacecraft, and the addition of Chapter 18 in this book recognises this trend, and illustrates these methodologies.

In most spacecraft, however, the fundamental technologies and issues explained in this book have remained the same, although there has recently been an increasing emphasis on the use of Micro Electro Mechanical Systems (MEMS) or Micro Systems Technology (MST) in current and future satellite systems. Over the next decade, the trend towards the increasing use of small, capable spacecraft is expected to continue, although predicting trends is fraught with hazards. The recent decision of Teledesic to stop all satellite manufacture until the telecommunications market shows signs of up-turn, indicates how rapidly trends can change. Only a few years ago it was anticipated that Teledesic would develop an 840-satellite constellation and be able to capitalise on the production-line benefits of building large numbers of spacecraft. At the heart of all these issues is cost. Satellite costs are relatively easy to estimate-nearly all satellite designs are based on the utilisation of existing systems—however, it is much more difficult to estimate the return on the investment, and the size of budgets which sponsoring organizations require to support programmes. Innovative approaches to design are often driven by the desire to continue to provide a full range of new scientific missions despite falling budgets.

It is not the fact that manned exploration of the moon last took place over thirty years ago that is extraordinary, as at the time the political will and financial resources available were unprecedented. The extraordinary fact is the extent to which spacecraft designers have adapted since to the changing financial, political and commercial climates, to maintain a thriving industry that has a major role to play in our future development in a broad range of scientific and commercial applications.

Index